The Interior of the Earth: its structure, constitution and evolution

In his hand are the deep places of the earth:
the strength of the hills is his also

Ps. 95. 4

The Interior of the Earth: its structure, constitution and evolution

Second edition

Martin H. P. Bott

M.A., Ph.D., F.R.S.
Professor of Geophysics, University of Durham

 Edward Arnold

© Martin H P Bott, 1982

First published 1971
Reprinted 1974
Second Edition 1982

by Edward Arnold (Publishers) Limited
41 Bedford Square
London WC1 3DQ

British Library Cataloguing in Publication Data
Bott, Martin H. P.
   The interior of the earth.—2nd ed.
   1. Physical geology
   I. Title
   551.   QE28.2

ISBN 0–7131–2842 9

Photo Typeset by Macmillan India Ltd., Bangalore.
Printed and bound by Thomson Litho, East Kilbride, Scotland

Preface

This book has been written to bring up to date *The Interior of the Earth*, published in 1971 following the plate tectonic revolution. Plate tectonic theory has now been widely accepted as a unifying theory for the origin of the Earth's major surface features, and the broad viewpoint taken in 1971 has been vindicated. The last ten years, however, has been a period of continuing rapid advance in earth sciences. The structure of the Earth's interior has become much better defined. A new understanding of the physical processes within the Earth which permit the escape of heat from the deep interior, drive the geomagnetic dynamo and cause the plate motions, is emerging. Consequently, much of the original text has had to be re-written to produce this new book.

The same general plan as that of *The Interior of the Earth* is followed. An introductory chapter reviews the Earth as a planet, describing its broad structure, composition, origin and rotation. The physical and chemical structure of the crust, mantle and core are described in the following five chapters. Chapter 5 on continental margins has been newly introduced; and the discussion of continental drift, sea-floor spreading and plate tectonics is incorporated in Chapters 2 and 3 on the crust rather than forming a separate chapter later in the book. The last three chapters deal with processes in the lithosphere and underlying parts of the mantle. The new understanding of the interior supply of heat and its mechanisms of escape from the deep interior is described in Chapter 7. This is followed in Chapter 8 by a discussion of rheology, where the discoveries of modern physics have had an important bearing on the acceptance of the new mobilistic view of the Earth. The final chapter speculates on the mechanism of global tectonics, which is seen as a bye-product of the escape of heat from the interior. Ways in which this may cause plate motions and continental splitting are discussed. An extensive list of references at the end of the book has been subdivided according to chapter to assist those wishing to follow up specific topics in greater detail.

The book has been written for undergraduate and graduate students of geology and of geophysics, and for earth scientists requiring a general account of the discoveries of solid earth geophysics. In order that it can be comprehensible to all who are interested in the broad structure of the Earth, the mathematical treatment has been omitted although a few essential formulae are included.

In any rapidly advancing branch of science, it is certain that further important findings and theories will have been made before the time of publication. Every effort has been made to make the book broadly up to date at the time of going to press, but in a compilation of this breadth it is impossible to give full coverage to the increasingly voluminous literature. Selection has been necessary, and this has been done in an attempt to give a balanced picture of the present state of knowledge.

M. H. P. Bott

Durham, 1982.

Acknowledgements

I am most grateful to my colleagues in the University of Durham for specialist advice on a number of topics during preparation of the book. My wife, and my son, Andrew, have assisted with the proofs and the indexing, and Dr C. H. Emeleus, Dr G. A. L. Johnson, Dr J. G. Holland, Dr R. E. Long, Prof. H. N. Pollack and Dr G. K. Westbrook critically read sections of the manuscript. I am indebted to my publishers for their courtesy and helpful advice during the production of the book. I would like to thank all of these and many others who have given me practical assistance, advice and stimulating discussion over many years on the subjects covered in this book.

I am also very grateful to the authors and publishers who have given me permission to use material in the Figures. Some of these have been carried over with little or no modification from the *Interior of the Earth* of 1971 and others are new to this book. The source of each of the figures is acknowledged in the captions.

M. H. P. B.

Contents

1 The broad structure and origin of the Earth

1.1 Introduction

Rapid advance of science during the second half of the twentieth century has enabled man to make a successful start on the exploration of planetary space. The deep interior of the Earth, however, remains as inaccessible as ever. This is the realm of solid-Earth geophysics, which still mainly depends on observations made at or near the Earth's surface. Despite this limitation, a major revolution in knowledge of the Earth's interior has taken place over the last twenty years. This has led to a new understanding of the processes which occur within the Earth that produce surface conditions outstandingly different from those of the other inner planets and the Moon. How has this come about? It is mainly the result of the introduction of new experimental and theoretical techniques into geophysics, together with opportunities to make observations on a much wider scale than before.

At the outset, here is a summary of some of these recent advances in geophysics. During the 1950s, palaeomagnetic studies gave strong support to the hypothesis of continental drift, which had previously been a subject of inconclusive debate. During the 1960s, oceanic geophysical investigations led to the theories of sea-floor spreading and plate tectonics, thereby providing an integrated explanation of continental drift and the origin of the Earth's major surface features. During the 1970s, the thermal history of the Earth has become much better understood in terms of convection in the mantle and conduction through the lithosphere. The mantle convection associated with the escape of heat gives rise to global tectonic activity but the mechanism of coupling between the convection currents and the lithosphere is still controversial. Seismology has been revitalized by new techniques and a vast improvement of the world-wide network of stations, greatly increasing knowledge of the Earth's internal structure. No longer can the Earth be treated as a rigid body possessing radial symmetry, for large lateral variations have been recognized in the upper mantle as well as in the crust. Great progress has been made in understanding the generation of the main geomagnetic field and its reversals by dynamo action within the outer core, although the geomagnetic dynamo still cannot be properly modelled. The age of the Earth has been convincingly determined and its origin has become less speculative. One of the most encouraging features of the present state of knowledge is that the old controversies in geology and geophysics are being resolved as an overall pattern of processes in the crust and mantle is beginning to emerge.

Before turning to the study of the structure and processes of the Earth's interior, this introductory chapter examines some of the features of the Earth as a planet, notably its shape, layering, composition, age and origin.

1.2 The shape of the Earth

It was Isaac Newton who first showed that the Earth, because of its rotation, ought to be an ellipsoid of revolution slightly flattened at the poles (i.e. an oblate spheroid). As a result of early observations, the French geodesist Cassini inclined to the opposite view that the equator is flattened and the poles bulge like an egg. To settle this controversy, French geodetic expeditions were sent to Peru and Lapland in the 1730s to measure the radius of curvature of the Earth's surface near the equator and near a pole. They did this by measuring the length of an arc along the Earth's surface in a north-south direction by triangulation, and by measuring the difference between the directions of the vertical at each end of the arc by astronomical surveying. This enabled the radius of curvature to be calculated near pole and equator. The result conclusively proved that the Earth is flattened at the poles as predicted by Newton.

The theoretical argument for flattening of the poles is as follows. The surface of a rotating mass of fluid of uniform density acted on by its own gravitation and centrifugal force alone is an ellipsoid of revolution. An increase of density towards the centre such as is known to occur within the Earth would only cause a departure from a true ellipsoid of 3 m maximum deviation. Thus the Earth's shape ought to be very nearly an ellipsoid of revolution flattened at the poles provided that it is in hydrostatic equilibrium.

Although the Earth is not completely in hydrostatic equilibrium, the hydrostatic hypothesis is sufficiently close to reality to make it reasonable to treat the Earth's surface as an ellipsoid of revolution. Consequently it is convenient to divide the theory of the shape of the Earth into two parts: (*i*) determining the shape and dimension of the ellipsoid which gives the best fit to the sea-level surface (this ellipsoid is called the spheroid); and (*ii*) determining deviations of the sea-level surface (which is called the geoid) from the spheroid. The shape of the spheroid is defined by its flattening according to the following equation

$$f = (a - c)/a$$

where f = degree of flattening, a = mean equatorial radius and c = polar radius.

The old method of determining the Earth's flattening was to measure the radius of curvature by geodetic surveying as described above. A much better approach nowadays is to deduce the flattening from measurements of the variation of the Earth's gravity field or potential with latitude. Prior to the 1960s, variation of sea-level gravity with latitude was used, but much improved accuracy is now obtained by using the response of the orbits of artificial satellites to this gravity variation.

The gravity potential of a hypothetical radially symmetrical Earth at an external point distance r from the centre would be GM/r where G is the gravitational constant and M is the mass of the Earth; an artificial satellite would then maintain an accurately elliptical orbit. However, the Earth deviates from a perfect sphere because of the equatorial bulge and other lateral density inhomogeneity so that the gravity potential averaged along lines of latitude must be written

$$U = (GM/r)\left[1 - \sum_{n=2}^{\infty} J_n P_n(\cos\theta)\right]$$

where θ = angular distance from the north pole (approximately the co-latitude) and J_n are constants. $P_n(\cos\theta)$ is a polynomial of degree n in $\cos\theta$ known as the *Legendre polynomial*. The flattening can be calculated to adequate accuracy from the coefficient J_2 using the relationship

$$f = \tfrac{3}{2}J_2 + \tfrac{1}{2}m + \tfrac{15}{28}mJ_2 + \tfrac{9}{8}J_2^2 + \tfrac{3}{56}m^2$$

where $m = \omega^2 a^3 (1 - f)/GM$ and ω = angular velocity of rotation of the Earth. m is approximately the ratio of centripetal acceleration to gravity at sea-level on the equator.

The equatorial bulge causes the point at which an inclined orbit of a satellite crosses the equator to migrate at constant rate (Fig. 1.1a) linearly dependent on J_2 and to a smaller extent on higher even harmonic coefficients J_4, J_6 etc. By using several satellites of differing orbital inclination, J_2 can be accurately determined and thus f can be found.

The value of f calculated by Newton for an Earth of uniform density was 1/230. The eighteenth-century French geodetic expeditions obtained observational values ranging from 1/310 to 1/178. Before the satellite observations, a typical estimate of f obtained by JEFFREYS (1959) based on surface gravity measurements was 1/297·3, with an accuracy of 1 part in 300. Observations on satellite orbits yield an estimate of 1/298·257 with an accuracy of about 1 part in 200 000 (MORITZ,

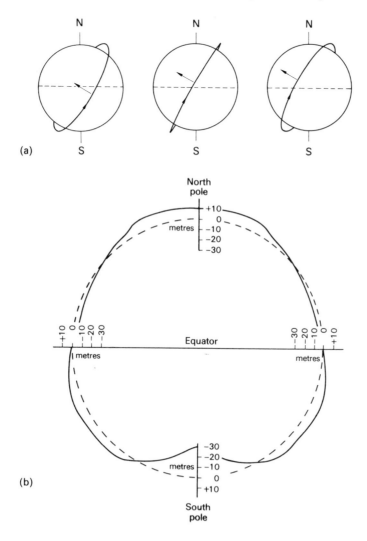

Fig. 1.1   (a) The Earth's equatorial bulge makes the orbit of an eastward moving artificial satellite regress towards the west. The rate of regression enables the flattening of the Earth to be calculated. Redrawn from KING-HELE (1967), *Scientific American*, **217**, No. 4, p. 70. (b) Height of the geoid (solid line) relative to a spheroid of flattening 1/298·25 (broken line) assuming the Earth to be axially symmetrical about the polar axis. Note exaggeration. Redrawn from KING-HELE (1969), *Royal Aircraft Establishment Technical Memorandum Space 130*, Fig. 4.

1976). The Earth's mean equatorial radius is 6378·140 km and the polar radius is 6356·755 km.

Turning to the deviation of the geoid from the spheroid, this used to be estimated from variation of gravity over the Earth's surface using a theorem of Stokes (JEFFREYS, 1976, Chapter 4). Such an approach is still the best for relatively local variations, say over distances less than about 2000 km. However, the longer wavelength variations of the geoid are most accurately determined from the orbital parameters of suitable artificial satellites. An even more recent approach which is starting to yield accurate geoid maps of oceanic regions is the use of radar altimetry from satellites to determine directly the elevation of the sea surface (MARSH and others, 1976). The average deviation of the geoid from the spheroid round an arbitrary line of longitude is shown in Fig. 1.1(b). This displays the widely publicized pear-shape. In fact the pear-shaped deviation of the geoid from the spheroid is less than 20 m whereas the equatorial bulge is over 20 km. A recent global map of the geoid based on satellite data supplemented by surface gravity measurements is shown in Fig. 1.2.

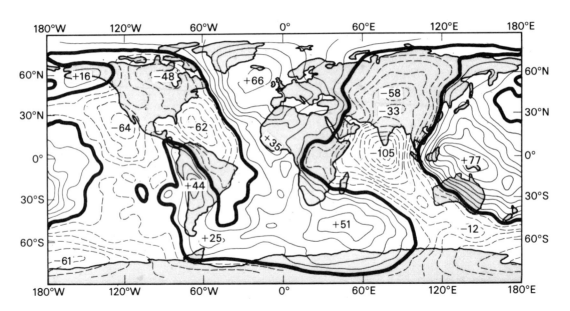

Fig. 1.2 Height of the geoid referenced to an ellipsoid of flattening 1/298·255, contour interval 10 m. Redrawn from WAGNER and others (1977), *J. geophys. Res.*, **82**, 911.

The study of the shape of the Earth is one branch of geophysics which has greatly benefited from the introduction of a new technique, in this case the technique being the use of artificial satellites. The increased accuracy has important repercussions on our knowledge of the state of stress within the Earth's interior.

1.3 The Earth's mass and moments of inertia

The mass of the Earth can be estimated from the value of gravity at the surface, after correction for the small contribution to gravity caused by rotation. The method in simplified form is as follows. The Earth is treated as a radially symmetrical sphere, not rotating, when it can be shown that the gravitational attraction at an external point is given by GM/r^2, where r is the distance from the centre, M is the mass and G is the gravitational constant. The gravitational attraction (gravity) is accurately known and so is the radius. G has been determined by experiment to an accuracy of about 0·06 %. The product GM has recently been most accurately determined from space probes and lunar laser data, with an accuracy of better than 1 part in 10^6. Thus M can be obtained to an accuracy of about 0·06 %, the limitation being the experimental accuracy of G. For this reason, the determination of G is sometimes spoken of as 'weighing the Earth'. From the mass and volume the mean density can also be obtained. The results are

$$\text{mass} = 5\cdot974 \times 10^{24}\,\text{kg}, \text{ mean density} = 5515\,\text{kg m}^{-3}.$$

If the Earth is assumed to be symmetrical about the polar axis, then there are two principal moments of inertia, defined as A about an equatorial axis and C about the polar axis. These can be estimated from astronomical observations and the observed flattening through two steps as follows (O'KEEFE, 1965):

(i) The Earth's axis does not remain in a fixed direction in space, but it precesses about a fixed direction describing a cone of angle 23°27'. The period of precession is 25 735 y which is accurately known from astronomical observations. From the theory of the motion of a symmetrical body, the period of precession enables the precession constant $H = (C - A)/C$ to be obtained, yielding a value of 0·003275 for the Earth (COOK, 1967).

(ii) The quantity $(C - A)$ can be estimated from a theoretical relation connecting the principal moments of inertia of a body and the second harmonic of the external gravity potential which is known as MacCullagh's formula (RAMSEY, 1940, p. 87). In turn, the second harmonic of gravity potential can be expressed in terms of the flattening, rotation, mass and dimension of the Earth. The relation is (COOK, 1967)

$$(C - A)/Ma^2 = \tfrac{1}{3}(f - \tfrac{1}{2}m) = 0\cdot001\,082\,65.$$

Combination of the two above equations enables A and C to be estimated. The flattening f is the least accurately known term in the equations and it therefore defines the accuracy of the estimates of the principal moments of inertia. The improvement in the estimate of the flattening stemming from observations of satellite orbits also improves the accuracy of estimates of the moments of inertia. Thus C/Ma^2 is found to be 0·3308. If the Earth were a uniform sphere it would be 0·4.

Both the mean density and the moment of inertia show that there must be a strong increase in density with depth within the Earth. Knowledge of the mean density and moment of inertia is important in studying the internal density distribution within the Earth, because any acceptable model of density must satisfy these observations. The moments of inertia are also important because they are needed to calculate the shape the Earth would have if it were in perfect hydrostatic equilibrium. The theoretical flattening can then be calculated from an equation arising from the theory of the internal gravity field of the Earth (JEFFREYS, 1963). Jeffreys estimated that the Earth would have a flattening of $1/(299\cdot67 \pm 0\cdot05)$ if it were in perfect hydrostatic equilibrium and KHAN (1969) using a modified method obtained 1/299·75.

1.4 Internal layering of the Earth

The branch of geophysics dealing with the origin and propagation of elastic waves within the Earth is called seismology. It is seismology which has revealed the internal layering of the Earth. It has therefore provided the framework for most other types of investigation of the interior, and is of fundamental importance to geophysics. Instrumental seismology was begun near the beginning of the present century and a few years later the three main subdivisions of the Earth had been discovered. A second stage of rapid advance in seismology was stimulated by the need to detect underground nuclear explosions and by the Upper Mantle Project. This has vastly increased our knowledge of the Earth's interior since about 1960, by revealing much greater detail in the layering and by showing up lateral variations. It is, however, with the early discoveries of seismology that this section is concerned.

When an earthquake or an explosion occurs within the Earth, part of the energy released takes the form of elastic waves which are transmitted through rocks with a definite velocity depending on density and elastic moduli. There are two main types, body waves and surface waves (Fig. 1.3). Body waves conform to the laws of geometrical optics, being reflected and refracted at interfaces where the velocity changes. The two types of body waves correspond to transmission of (i) compressions and rarefactions (P waves), and (ii) shear displacement (S waves). The velocities are given by

$$V_P = \sqrt{\left(\frac{k + \frac{4}{3}\mu}{\rho}\right)} \text{ and } V_S = \sqrt{\frac{\mu}{\rho}}$$

where V_P = velocity of P waves, μ = rigidity modulus,
$\quad V_S$ = velocity of S waves, ρ = density.
$\quad k$ = bulk modulus,

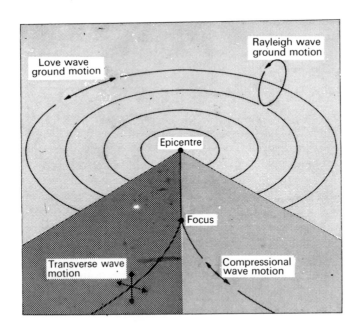

Fig. 1.3 Diagram illustrating the focus and epicentre of an earthquake, and showing the fundamental types of seismic wave originating from the earthquake. Redrawn from DAVIES (1968), *Science Journal*, Nov. 1968, p. 79.

It follows that P waves always travel faster than S waves, and that S waves cannot be propagated through liquid. In general, denser rocks have higher body wave velocities since the elastic moduli increase with increasing density more rapidly than the density itself does.

Surface waves are restricted to the vicinity of a free surface, or in exceptional circumstances, an internal interface. The two main types are Rayleigh waves, with the particle motion confined to the vertical plane containing the direction of propagation, and Love waves, with motion in a horizontal direction perpendicular to the direction of propagation. These are described in more detail in Chapter 4.

Elastic waves are detected by seismographs, which respond to ground displacement or velocity depending on design. Short-period instruments (about 1s period) are used to detect body waves, and long-period instruments (15s or longer period) are used for surface waves. A normal seismograph station has three short-period and three long-period seismographs, to detect the three components of ground motion.

The elastic waves emanating from an earthquake are usually taken to originate from a single point within the Earth, although more realistically the source is often extended over a distance of 10–100 km. This point is called the *focus* (or *hypocentre*). The point on the surface vertically above the focus is the *epicentre*. It will be shown later (Chapter 8) that most earthquake foci are above a depth of 100 km but that deep events occur down to about 700 km depth. The location of the focus and epicentre of a given earthquake is determined from the arrival times of seismic waves at a selection of seismological observatories – the more the better. Recent improvements to the world-wide network of stations, supplemented by the use of computers, have considerably increased the accuracy of locating earthquake foci.

Before about 1950, most of the major discoveries of seismology came from studies of the time of travel of body waves from earthquakes. The Earth was assumed to possess radial symmetry and thus many different earthquakes could be used to build up a table of travel-times of *P* and *S* waves over the range of possible angular distances from the epicentre (0° to 180°).

The existence of a central core within the Earth was deduced by OLDHAM (1906) as the result of his observation that *P* waves recorded near the angular distance of 180° from the earthquake epicentre arrive much later than expected. This was attributed to delay introduced by passage through a low velocity core. The discovery was later substantiated by the following more detailed evidence (Fig. 1.4). Up to an angular distance of about 103° from the epicentre, *P* and *S* waves are observed and

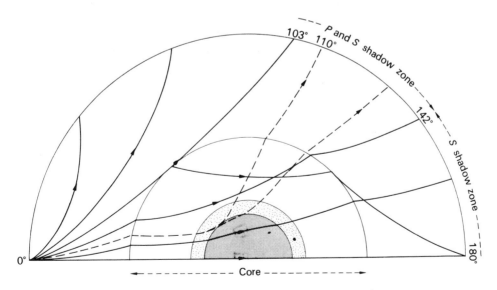

Fig. 1.4 Selected ray paths for *P* waves passing through the Earth. The *P* and *S* shadow zones are shown, and the dashed ray paths represent the weak *P* arrivals within the shadow zone which provide the main evidence for an inner high *P* velocity division of the core. Redrawn from GUTENBERG (1959), *Physics of the Earth's Interior*, p. 104, Academic Press.

indicate a progressive increase in velocity with depth through the major shell of the Earth which is known as the *mantle*. However, between 103° and 142°, *P* and *S* waves are both substantially absent, causing a 'shadow zone'. From 142° to 180° delayed *P* arrivals occur but *S* is absent. This shows that about half way to the centre of the Earth there is a major discontinuity beneath which the *P* velocity drops abruptly and *S* is absent, suggesting a fluid *core* in contrast to the solid, overlying *mantle*. The depth of the discontinuity between can be obtained by a variety of methods, including use of reflections from it. In 1914, Gutenberg obtained a depth of 2900 km for it (although recent revisions reduce this estimate by 14 km). The corresponding value for the mean radius of the core is 3470 km. The core-mantle boundary is known as the *Gutenberg discontinuity*.

A second major discontinuity at shallow depth was discovered by MOHOROVIČIĆ (1909) through study of seismograms of the Yugoslavia earthquake of October 8, 1909, out to distances of a few hundred kilometres. He observed two *P* and two *S* pulses, and he interpreted these as direct and refracted *P* and *S* arrivals caused by a low velocity *crust*, about 50 km thick, overlying a higher velocity substratum now called the *mantle*. The intervening boundary is called the *Mohorovičić discontinuity* (or just the *Moho*). Seismic refraction studies using artificial explosions have greatly amplified and extended this discovery (p. 33). Early studies on the dispersion of surface waves suggested that the crust is thinner beneath oceans than continents, and this also has been substantiated by later refraction surveys.

The three major subdivisions of the 'solid' Earth were known by 1910. Between the world wars, the main effort of seismology was devoted to obtaining and refining the velocity-depth distribution throughout the Earth for *P* and *S* waves. This depended on building up detailed knowledge of the travel-times of the waves covering all angular distances from the epicentre. If radial symmetry is assumed for the Earth, then, provided that velocity increases continuously with depth, the velocity-depth curves can be computed from the *P* and *S* travel-time curves by a mathematical process described in section 4.2. This method was used by JEFFREYS (1939a) to construct the velocity-depth curves for *P* and *S* through the mantle. Jeffreys obtained the velocity distribution in the core by trial and error.

The velocity-depth distribution of JEFFREYS (1939a) is compared with a more recent distribution of HART and others (1977) in Fig. 1.5. Above 700 km depth, the more recent distributions show the presence of a low velocity zone between about 80 and 300 km depth, first postulated by GUTENBERG (1953). Both distributions show a steep increase in velocity with depth between about 370 and 720 km but the recent distribution shows that this consists of a series of steps rather than being regular. This is known as the *mantle transition zone*, and it is overlain by the *upper mantle* and underlain by the *lower mantle*. Below 800 km depth, the two distributions shown in Fig. 1.5 only differ in detail except near the steep rise in velocity with depth in the core at about 5100 km depth. This was discovered by LEHMANN (1936) from observations of weak *P* arrivals in the shadow zone between 110° and 143° angular distance from the epicentre. These arrivals could not be properly explained by diffraction at the core-mantle boundary but require a strong increase of *P* velocity within the core. This used to be regarded as a transition zone but is now known to be a sharp boundary with transitional regions on either side of it. It separates the low velocity *outer core* from the high velocity *inner core* which is now known to transmit slow *S* waves. The Earth's layering is summarized in Table 1.1 and Fig. 1.6.

The boundaries between crust, mantle and core represent relatively abrupt changes in elastic properties and density, and they are normally regarded as marking major compositional changes. Another type of boundary of great importance occurs at a depth of about 100 km within the upper mantle and marks a change in rheological properties rather than in composition. This is the boundary between the relatively strong uppermost shell of the Earth known as the *lithosphere* and

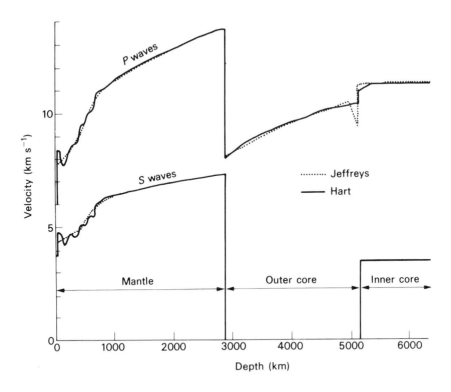

Fig. 1.5 The broad velocity-depth distribution of Jeffreys compared with a more recent determination of HART and others (1977).

Table 1.1 The layers within the Earth, based on the seismic velocity-depth distribution (Fig. 1.5). The letters used to identify the concentric shells follow the nomenclature of BULLEN (1963) with omission of F (core transition zone) which is now recognized as a sharp boundary.

	Region	Depth range (km)	
CRUST	A	0–20 (variable thickness)	
Mohorovičić discontinuity			
	B	20–370	upper mantle
MANTLE	C	370–720	transition zone
	D	720–2886	lower mantle
Gutenberg discontinuity			
CORE	E	2886–5156	outer core
	G	5156–6371	inner core

the weaker underlying region called the *asthenosphere*. The lithosphere comprises the crust and the topmost mantle, and the asthenosphere extends down to about 300 km beneath which the rheological properties are less well known; these are average depths as there are lateral variations. The initial recognition that there must be a weak layer at depth where lateral flow can occur stems from observed vertical movements and their explanation in terms of the theory of isostasy (p. 50). The concepts of lithosphere and asthenosphere are now of key importance in plate tectonics.

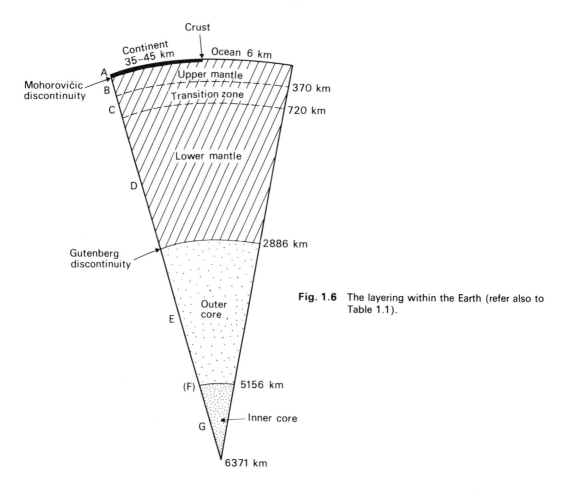

Fig. 1.6 The layering within the Earth (refer also to Table 1.1).

The gradual weakening of rocks with depth is primarily attributed to raised temperatures. The lithosphere is thus a relatively cool zone and the asthenosphere represents a region where temperatures most closely approach the melting point. The intervening boundary cannot be accurately defined as can the crust-mantle boundary and it is certainly likely to be gradational. It has often been assumed that the boundary can be identified with the top of the seismic low velocity zone in the upper mantle. Despite the difficulties over the definition and identification of this boundary, the concepts of lithosphere and asthenosphere have proved to be very useful. They are discussed in greater detail later in the book.

1.5 Radiometric dating methods, meteorites, and the age of the Earth

In the late nineteenth century it was widely accepted that the Earth was about 20–80 My old. This was Lord Kelvin's estimate, and was based on the premise that the Earth's outward heat flow represents the cooling of an initially hot body. However, the discovery of radioactivity at the turn of the century put a new complexion on the problem of the age of the Earth in two ways. Firstly, it provided a method for dating rocks which soon showed that many rocks are older than 80 My. Secondly, the energy given off during radioactive decay of uranium, thorium and potassium provides sufficient heat to explain the outward heat flow without need for the cooling hypothesis of Kelvin. The age of the Earth can now best be estimated from the isotopic composition of lead in meteorites and in terrestrial occurrences of known age.

Radiometric dating methods

The rate of decay of a radioactive isotope is proportional to the number of atoms present. Expressing this algebraically, we obtain

$$\frac{dN}{dt} = -\lambda N$$

where N = number of atoms of parent isotope, t = time and λ = decay constant. The time taken for half the original number of atoms to decay is called the 'half-life' or $t_{\frac{1}{2}}$. It can be shown that $t_{\frac{1}{2}} = (\ln 2)/\lambda$ where the logarithm is to the base e. The common radioactive isotopes with sufficiently long half-lives to be useful in dating rocks are shown in Table 1.2. Rubidium and potassium decay to their respective daughter isotopes by a single step, but uranium and thorium involve more complicated decay series. If, at time of crystallization, a mineral contains N atoms of the parent isotope, then at some later time t there will remain N_r atoms of the parent isotope and N_s atoms of the daughter isotope will have been produced. These are related by the equation

$$t = \frac{t_{\frac{1}{2}}}{0.693} \ln (1 + N_s/N_r)$$

Table 1.2 Radioactive isotopes commonly used for dating rocks. The uranium and thorium isotopes decay to the lead isotopes by a series of stages, but the half-lives are controlled by the first stage of decay because the later stages occupy negligible time by comparison. The table is mainly based on the recommendations of the subcommision on geochronology (STEIGER and JÄGER, 1977).

Parent isotope		Daughter isotope	Decay constant ($\times 10^{-3}$ My^{-1})	Half-life (My)	Remarks
^{238}U		^{206}Pb	0.155125	4468	decay series
^{235}U		^{207}Pb	0.98485	704	decay series
^{232}Th		^{208}Pb	0.049475	14010	decay series
^{87}Rb		^{87}Sr	0.0142	48800	beta decay
^{147}Sm		^{143}Nd	0.00654	106000	alpha decay
^{40}K	(89.05%)	^{40}Ca	0.4962 }	1265	{ beta decay
^{40}K	(10.95%)	^{40}A	0.0581 }		{ capture of orbital electron

where 0.693 = ln 2. By measuring N_s and N_r and using the experimentally determined value of $t_{\frac{1}{2}}$, the age t is determined. It may, of course, be necessary to correct for initial presence of the daughter isotope.

The rubidium-strontium method provides a good example of the ways in which dating methods can be used. The four naturally occurring isotopes of strontium have mass numbers of 84 (0.55%), 86 ($\sim 10\%$), 87 ($\sim 7\%$) and 88 (82.6%). The proportion of ^{87}Sr increases with time relative to the three stable isotopes as a result of beta decay of ^{87}Rb. If a set of rocks or minerals form contemporaneously from a common source, such as a cooling magma or part of the condensing solar nebula in which the isotopic composition is homogeneous and uniform, then the initial ratio

$^{87}Sr/^{86}Sr$ will be the same for all samples, but because of fractionation processes the Rb/Sr ratio would vary from sample to sample. As a first approximation the age could be determined by choosing a sample which contained a negligible quantity of strontium on formation. However, a more satisfactory procedure uses several samples of differing Rb/Sr ratio with a plot of their $^{87}Sr/^{86}Sr$ versus $^{87}Rb/^{86}Sr$ ratios as determined by mass spectrography. Assuming that each sample has acted as a closed system to Rb and Sr since formation, then the points should lie on a straight line called an *internal isochron* which steepens with time. The gradient of the isochron is $(e^{\lambda t} - 1)$ at time t after formation, and the measured value of this gradient thus yields the age of formation of the samples from an isotopically homogeneous source. The observed isochron also yields the *initial ratio* of $^{87}Sr/^{86}Sr$ in the source at the time of formation. Initial ratios have been used, not always without controversy, to determine whether a magma originates by melting in the crust or mantle, as preferential concentration of rubidium in the crust by a factor of about ten leads to higher initial ratios for crustal than mantle sources.

Another method of using rubidium-strontium dating, particularly applicable to lunar samples, is to use a bulk homogenized sample such as the lunar soil to calculate its age of formation from a source of specified initial strontium ratio such as the solar nebula. This method assumes that the bulk sample has remained a closed system since then, although internal redistribution of Rb and Sr may have occurred. This yields a so-called *model age*, which is usually an upper estimate of the date when the rock was last an open system, and can be used to date the Moon subject to the assumptions made. The model age is generally greater than the ages of individual constituents such as lava fragments.

The uranium-lead methods were once widely used in dating rocks and are still employed in isotopic studies of various types, in particular for estimating the age of the Earth (see below). However, the potassium and rubidium methods now predominate for dating of rocks as they can be much more widely applied. The recently developed method based on alpha decay of samarium-147 to neodymium-144 is closely comparable to the rubidium-strontium method in methodology, but has the advantage that the two rare earth elements exhibit closely similar geochemical behaviour, in contrast to the differences between rubidium and strontium.

These methods have been used to construct a time-scale for the geological column (Table 1.3). The base of the Cambrian, for instance, has been shown to be about 570 My ago (HARLAND, SMITH and WILCOCK, 1964). The methods have also helped greatly in unravelling the stratigraphy of Precambrian rocks. The oldest Precambrian rocks dated yield ages of about 3800 My, showing that the Precambrian lasted more than six times as long as the period from the base of the Cambrian to the present.

Meteorites

Meteorites are small bodies in orbit round the Sun which sometimes fall onto the Earth. It is thought that they were formed by the relatively recent break-up of small planetary bodies such as the asteroids. They provide the best samples we have of the material which forms the inner planets Mercury, Venus, Earth, Mars and the asteroids. The study of meteorites yields important evidence on the age, origin and composition of the Earth.

There are two main groups of meteorites. (*i*) The *irons*, or siderites, are mainly composed of an iron-nickel alloy (90% iron). They amount to about 10% of observed falls. (*ii*) The *stones*, or aerolites, are mainly composed of silicates and are subdivided into:

(a) chondrites, with small rounded grains known as chondrules;
(b) achondrites, which lack chondrules.

Table 1.3 The geological time-scale. Ages Cambrian to recent taken from HARLAND, SMITH and WILCOCK (1964).

Eras	Periods	Beginning of periods (My)
QUATERNARY	Recent	0·01
	Pleistocene	1·5–2·0
TERTIARY	Pliocene	7
	Miocene	26
	Oligocene	38
	Eocene	54
	Palaeocene	65
MESOZOIC	Cretaceous	136
	Jurassic	190–195
	Triassic	225
PALAEOZOIC	Permian	280
	Carboniferous	345
	Devonian	395
	Silurian	430–440
	Ordovician	500
	Cambrian	570
PRECAMBRIAN	oldest known rock	3800
	origin of Earth	4600
	origin of meteorites	4600
	solar nebula condenses*	4600
	origin of galaxy*	13 000–20 000
	origin of universe*	13 000–20 000

* See section 1.6

90 % of the stones are chondrites. They have a fairly uniform composition resembling an ultrabasic rock. The average mineral composition is

46 % olivine: $(Mg, Fe)_2 SiO_4$,

25 % pyroxene: e.g. $(Mg, Fe)SiO_3$ or $(Ca, Mg, Fe, Al)_2 (Al, Si)_2 O_6$,

11 % plagioclase: $NaAlSi_3 O_8 – CaAl_2 Si_2 O_8$

and   12 % nickel-iron phase.

Achondrites are poorer in olivine and nickel-iron phase and they have a bulk composition closer to basalt. Intermediate between the irons and the stones are the siderolites. The glassy, silica-rich tektites form a further small group.

As will be discussed more generally in the following section, the inner planets and the meteorite parent bodies are believed to have formed by accretion of solid grains which condensed out of an initially hot gaseous planetary nebula as it cooled from about 2000 to 400 K. The parent bodies of most of the chondrites and some of the irons did not undergo subsequent differentiation, although some have suffered heating and metamorphism. The meteorites thus give us the only direct method of measuring the age of condensation of planetary material out of the nebula. They yield remarkably closely-grouped ages of about 4600 My, spreading between 4500 and 4700 My. The scatter is partly due to errors of observation and factors such as metamorphism, but it possibly conceals a small but real variation in age and initial isotopic composition.

Meteorites yield the initial isotopic ratios of strontium and lead at the time of condensation of the planetary nebula. Chondrules from the Allende meteorite containing negligible rubidium have yielded the lowest observed $^{87}Sr/^{86}Sr$ ratio of 0·698 77. Iron meteorites with negligible uranium and thorium provide us with an estimate of the isotopic composition of primeval lead which condensed from the planetary nebula, and this is used to estimate the Earth's age as described below.

The age of the Earth

The Earth must be older than 3800 My, the age of the oldest known rocks. It must be younger than the time needed for all the ^{207}Pb in the Earth to form by decay of ^{235}U, which according to geochemical estimates of terrestrial abundances is about 5500 My. A more accurate model age of the Earth can be determined using the relative abundances of the lead isotopes.

Common lead consists of four isotopes with mass numbers of 204, 206, 207 and 208. The abundance of ^{204}Pb has remained constant but the other three isotopes have progressively grown in abundance by decay of uranium and thorium. The products of the decay series $^{238}U \rightarrow {}^{206}Pb$ and $^{235}U \rightarrow {}^{207}Pb$ are jointly used to estimate a model age for the Earth. The two uranium isotopes are probably inseparable in nature and the present ratio of $^{238}U/^{235}U$ is 137·8. Because ^{235}U decays more than six times faster than ^{238}U, the ratio has progressively increased with time. Its value at any specified time in the past can be computed accurately from the present value using the decay constants of the two series.

Following earlier rough estimates by Holmes, the modern approach (PATTERSON, 1956) assumes that the initial lead isotope ratios for the Earth are identical to those determined on meteorites. The method also makes the fundamental assumption that there has been no chemical fractionation of uranium and lead between the time of formation of the Earth and the date of withdrawal of the lead from its source deep within the Earth to form a lead deposit of known age. The abundances of the lead isotopes within such a source region at time t after the Earth's formation can then be related to their initial values by the equations

and
$$\left[{}^{206}Pb\right]_t - \left[{}^{206}Pb\right]_i = \left[{}^{238}U\right]_t (e^{\lambda_1 t} - 1)$$
$$\left[{}^{207}Pb\right]_t - \left[{}^{207}Pb\right]_i = \left[{}^{235}U\right]_t (e^{\lambda_2 t} - 1)$$

where λ_1 and λ_2 are the respective decay constants. Knowing the date at which the lead was withdrawn from the Earth's interior, the relative abundance of the two uranium isotopes can be computed and eliminated from the two equations. The initial and observed abundances of the lead isotopes are in practice expressed as ratios to the abundance of ^{204}Pb and the resulting equation then enables the unknown t to be determined.

PATTERSON (1956) used the lead isotope abundances in oceanic sediments to show that the Earth's age is probably the same as that of the meteorites which he estimated to be 4550 ± 70 My. More recent values using improved techniques slightly increase the estimated age of the Earth to 4580 My. The model age of the Moon using the rubidium-strontium method is about 4600 My (WASSERBURG and others, 1977).

It should be emphasized that the model ages for the Earth and Moon are at best only estimates of the age when the condensed material forming these bodies had the specified initial ratios appropriate to meteorites. The values are also in doubt because the interiors probably did not remain strictly closed systems. However, the consistency between the initial ratios and ages determined for meteorites, the Earth and the Moon strongly supports the concept that they all formed approximately contemporaneously from the planetary nebula about 4600 My ago.

1.6 The origin of the Earth

The universe and our galaxy

To put the origin of the Earth in its proper setting, a short description of its place in the history of the universe will be given. The mean distance of the Earth from the Sun is 149.6×10^6 km and the average velocity of the Earth in its orbit is 29.8 km s^{-1}. Some numerical facts about the Sun and its planets are given in Table 1.4.

Table 1.4   Table of planetary constants based on tables in the *British Astronomical Association Handbook*, 1970.

	Mean distance from the Sun* (× 10⁶ km)	Eccentricity of orbit*	Sidereal period* (days)	Inclination to ecliptic*	Equatorial radius (km)	Mass (Earth = 1)	Density (kg m⁻³)	Siderial period of axial rotation (days)
Sun					696 000	332 958	1409	25·380†
Moon	0·384 400 (from Earth)	0·0549	27·321 661	5° 08′ 43·4″	1738	0·0123	3342	27·322
Mercury	57·91	0·2056	87·969	7 00 15·0	2420	0·054	5410	59
Venus	108·21	0·0068	224·701	3 23 39·6	6150	0·8150	4990	244·3‡
Earth	149·60	0·0167	365·256	0 00 00·0	6378	1·0000	5517	0·997
Mars	227·94	0·0934	686·980	1 50 59·5	3395	0·107	3940	1·026
Jupiter	778·34	0·0485	4332·59	1 18 17·3	71 400	317·89	1330	0·410†
Saturn	1427·0¶	0·0556	10 759·20	2 29 22·1	59 650	95·14	706	0·426
Uranus	2869·6	0·0472	30 685·0	0 46 23·2	23 550	14·52	1700	0·451‡
Neptune	4496·7	0·0086	60 190·0	1 46 22·2	22 400	17·46	2260	0·625
Pluto	5900	0·25	91 000	17 08 24	2950	0·10	5500 (?)	6·39

*   Some of the orbital elements are affecting by long period variations, which are most noticeable for the outermost planets. Except for Pluto, these are given for Epoch January 0·5, 1969 E.T.

†   At equator (period varies with latitude).

‡   Retrograde.

The Sun, with its planetary system, is situated in the outer part of a lens-shaped disc of stars and interstellar gas and dust which is our galaxy, visible to us as the Milky Way. The Sun is about 27 000 light years from the centre of the galaxy (one light year is about 10^{13} km) and it is rotating about the centre with a velocity of about 230 km s^{-1}. It takes 220 My to complete one orbit. Our galaxy forms only a minute part of the whole universe. There are a multitude of other galaxies which have been observed up to 4500 million light years away. Light now reaching us from the most distant galaxies started on its way at about the time when the solar system was being formed!

The spectral lines from distant galaxies are shifted towards the red end of the spectrum. This is usually interpreted as a Doppler shift of wavelength caused by movement of the sources away from us with velocities proportional to distance, as measured by apparent brightness. Taken at face value, this suggests that the universe is expanding. Extrapolating back in time, expansion is variously estimated to have started from a concentrated nucleus between 10 000 and 20 000 My ago, with the main uncertainty depending on distance measurement. For instance, HANES (1979) used the brightnesses of globular clusters in the galaxies to estimate their distances, and assuming a uniform rate of expansion his results yield an age of 12 500 My for the universe. On the other hand, there is evidence (see below) that our galaxy is over 13 000 My old, suggesting that the universe must be at least as old as this.

Support for the hypothesis that the universe was spontaneously created in a 'big bang' somewhat over 13 000 My ago comes from three other types of observations as follows: (*i*) radio-astronomers have detected a background microwave radiation of the type predicted by Gamow for a hot big-bang; (*ii*) the distribution of distant radio sources indicates that these may have been much closer together in the past as the theory predicts; and (*iii*) after the 'big bang' the universe would initially consist of fundamental particles which would combine on cooling within about one hour to produce an initial mixture of about 25 % helium and 75 % hydrogen, as appears to be borne out by observations of the present composition.

The cosmic abundances of uranium isotopes, rhenium and their decay products as estimated by chemical analyses of meteorites can be used to estimate the age of the galaxy at the time of formation of the solar system (CLAYTON, 1964). A recent study based on the abundances of rhenium and osmium isotopes in iron meteorites yields a present age for our galaxy of between 13 000 and 22 000 My (LUCK and others, 1980). Thus the galaxy was in existence long before the solar system began to form. One idea is that it formed by gravitational collapse of a turbulent body of hydrogen and helium gas, possibly soon after the initial 'big bang'.

Astronomers think that stars, similar to the Sun, are forming at the present time in the galaxy (SPITZER, 1963). They form in clusters from clouds of interstellar gas and dust which have become sufficiently dense to be gravitationally unstable. The interstellar material is partly hydrogen and helium dating back to the formation of the galaxy, and partly helium and heavier elements synthesized by nuclear reactions in the interior of stars and in supernovae and other explosions as described below. Gravitational collapse of the cloud first forms a cluster of protostars, and later the protostars themselves collapse to become young stars. As collapse occurs, gravitational energy released heats up the interior of the star and causes it to radiate and become luminous. Eventually the interior temperatures become high enough for nuclear fusion reactions to commence and the star ceases to contract as it joins the 'main sequence'. The Sun is a rather typical main sequence star which underwent its contraction from interstellar material about 4600 My ago. When the available nuclear fuel within a star becomes exhausted, if it is very small it may die a cold death, but more commonly sudden collapse initiates a supernova explosion which results in scattering of the material of which it is formed into interstellar space. In this way, the chemical composition of a galaxy progressively becomes enriched in elements heavier than hydrogen as a result of the nucleosynthetic processes described in the following section. This explains why there are many stars in the galaxy having a more primitive, hydrogen-rich composition than the Sun.

Chemical evolution of the galaxy

The hydrogen and most of the helium in the interstellar cloud which condensed to form the solar system probably date back to the products of the initial 'big bang'. The heavier elements were subsequently synthesized from hydrogen and helium within the galaxy prior to the Sun's formation. Two main types of process have taken place, described in more detail by TRURAN (1973). *Thermonuclear fusion reactions* within stellar interiors can produce elements up to about ^{62}Ni in a series of reactions requiring progressively higher temperatures: (*i*) hydrogen burning produces helium starting at about 10^7 K; (*ii*) helium burning produces carbon and oxygen above about 2×10^7 K; (*iii*) carbon and oxygen burn above about 10^9 K to produce predominantly silicon; (*iv*) reactions involving silicon and other nucleons can occur at even higher temperatures to produce most nuclear species up to iron and nickel. Most of the carbon and oxygen were probably formed by slow helium burning in the interior of stars which have subsequently exploded, and most of the nitrogen is probably a byproduct of catalytic hydrogen burning. The nuclei from ^{20}Ne to ^{62}Ni were probably mainly formed by explosive carbon, oxygen and silicon burning in supernovae.

Most of the nuclei heavier than iron have been produced by *neutron capture* followed by beta decay. Slow neutron capture (*s*-process) may occur within evolving stars but fast capture (*r*-process) takes place only at the high temperatures involved in catastrophic events such as supernovae explosions. A few elements are apparently not formed by either fusion reactions or neutron capture. The few heavy elements by-passed by the neutron capture processes may be formed by proton capture (*p*-process) at exceptionally high temperature. The three light elements lithium, beryllium and boron are unstable at the high temperatures of all the fusion reactions and have probably been formed mainly by cosmic ray bombardment of the interstellar material.

Much of the nucleosynthesis within the galaxy has probably occurred during supernovae and other explosive events. Such explosions also have the effect of dispersing the products of nucleosynthesis into the interstellar space. The evolution of several generations of massive stars within the galaxy has probably been an important factor in the synthesis of the elements, larger stars having shorter lifespans and higher internal temperatures. Thus the Sun, with a predicted lifespan in the main sequence of about 8200 My, is still in middle age and in the hydrogen burning stage, but a star only three times more massive lasts only about 230 My before its catastrophic termination.

The anomalous isotopic composition of xenon and magnesium in some meteorites may provide an estimate of the time interval between the last nearby supernova explosion and the condensation of the solar nebula. ^{129}Xe is the decay product of ^{129}I which has a half-life of 17 My; ^{244}Pu produces ^{136}Xe by fission. Both parent isotopes are produced by neutron capture in an exploding star. According to SCHRAMM and WASSERBURG (1970) the abundances of the daughter xenon in some meteorites yields an estimate of 150–200 My for this time interval.

A much shorter time interval between the last nucleosynthetic event and condensation of meteoritic material has been inferred from the occurrence of excess ^{26}Mg correlating with the Al/Mg ratio in an inclusion in the Allende meteorite (LEE and others, 1976; LATTIMER and others, 1977). The excess ^{26}Mg is assumed to have formed by decay of ^{26}Al which has a half-life of 0.74 My. The $^{26}Al/^{27}Al$ ratio is estimated to have been 5×10^{-5} when the inclusion condensed. While some ^{26}Al may have been formed by irradiation in the solar nebula, this ratio requires a more prolific source for most of it, such as explosive carbon burning in a supernova. The theoretical ratio of the aluminium isotopes produced by carbon burning of about 10^{-3} suggests that the supernova exploded between 2 and 3·7 My before condensation of the meteorite inclusion (TRURAN and CAMERON, 1978). Anomalous proportions of some stable isotopes, including those of oxygen and neon, have also been observed in some meteorites. The anomalous isotopes are preserved in meteorites because the parent bodies have mostly not been homogenized later like the Earth and the Moon. These observations suggest that a supernova explosion occurred nearby a few million years before the solar nebula condensed, unevenly injecting fresh gas and dust of anomalous isotopic composition into the cloud of interstellar gas and dust which had accumulated over the lifespan of the galaxy. We may speculate that this supernova triggered the initial condensation of interstellar gas and dust which formed the solar system.

Estimates of the abundances of elements in the solar system are obtained mainly from meteorites for the non-volatile elements and from the solar spectrum for the volatile elements (Fig. 1.7). Details of the observed distribution agree at many points with the theoretical predictions of nuclear physics. The result of the galactic nucleosynthesis was that about 98 % of the solar nebula consisted of hydrogen and helium mainly inherited from the pre-galactic period, about 1.4 % consisted of the nebular ice-forming elements carbon, nitrogen and oxygen, and only about 0.25 % consisted of the most abundant rock-forming elements silicon, iron and magnesium.

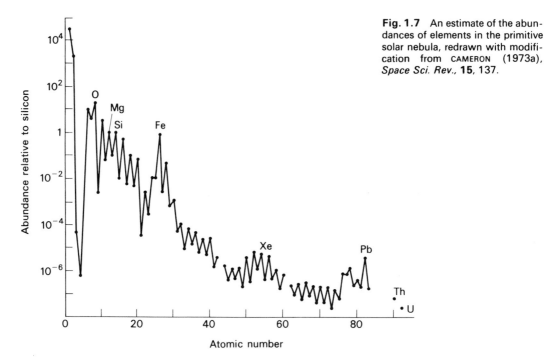

Fig. 1.7 An estimate of the abundances of elements in the primitive solar nebula, redrawn with modification from CAMERON (1973a), *Space Sci. Rev.*, **15**, 137.

Formation of the Earth and planets

According to modern opinion, the planets and their satellites formed at the same time as the Sun by condensation of the solar nebula of gas and dust. This is a modified version of the old condensation hypothesis which goes back to Descartes and Kant but which was out of favour during the first half of the present century. The rival hypothesis in favour between 1900 and 1950 attributed the formation of the planets to a catastrophic event such as a nearby supernova explosion or the close approach of another star to a pre-existing Sun; it was supposed that this event caused a filament of the Sun to be drawn out and to condense to form planets. The catastrophic hypotheses are now confronted by some insuperable difficulties such as that pointed out by SPITZER (1939) that a filament large and hot enough to form the planets would dissipate itself into space in a matter of an hour, long before it could cool enough to begin condensation. Hence the nebular hypothesis has again come into favour.

Let us look at some of the facts about the solar system which need explaining. Foremost is the regular pattern of rotation and the distribution of angular momentum. All planets except Pluto occupy nearly circular orbits which lie close to a single plane and they move round the Sun in the direction that the Sun itself rotates. Most satellites orbit in the same direction as their primary planet rotates and in the equatorial plane, although there are exceptions which may be attributed to orbital capture or tidal friction. The planets are spaced at approximately regular intervals from the Sun as expressed by the modified Titius-Bode law. The mean distance of the nth planet from the Sun, counting the asteroids together as a single planet, is given by $r_n = r_0 m^n$, where $m = 1.89$. The physical significance of the law is not yet understood.

The distribution of angular momentum in the solar system has created difficulty for all theories. 98 % of the angular momentum is possessed by the planets and the Sun itself rotates slowly with a period of 24·65 days. On the other hand, most of the mass of the system is contained in the Sun. The problem is to explain how the angular momentum has been transferred from the central body to the outer parts of the system. It was this difficulty which led to the abandonment of the nebular hypothesis of Kant early this century, because the angular momentum of a contracting disc of rotating gas and dust would remain firmly anchored to the main mass condensing to form the Sun itself. The catastrophic theories were put forward to meet this difficulty, but in fact they failed to do it satisfactorily. With the revival of the nebular hypothesis, it is now recognized that the outward transfer of angular momentum can be explained by turbulence or other induced currents within the nebula.

Until UREY (1952) pointed out its importance, chemical evidence bearing on the origin of the solar system was largely ignored. As pointed out above, three groups of elements form the major constituents of the solar system: (i) the gaseous elements H and He, (ii) the ice-forming elements C, N and O which are found as solid NH_3, CH_4 and H_2O in the cold outer parts, and (iii) the rock-forming elements Mg, Fe and Si. The rock-forming elements combined with oxygen are the major constituents of the four inner planets, the Moon and the asteroids. Jupiter and Saturn, although much larger than the inner planets, have much lower densities (Table 1.4) showing that they are mainly H and He around a core which is probably rocky. Uranus and Neptune, having intermediate densities, predominantly consist of the icy elements in the form of solid ammonia, methane and water. Variations in the densities of the terrestrial planets, after correction for compression, indicates a high iron content for Mercury and deficient iron for the Moon and Mars. The Sun itself is mainly formed of H and He.

According to modern versions of the nebular hypothesis, the planets formed by condensation from primitive material having the same bulk composition as the Sun. Consequently the planetary part of the nebula must have lost large quantities of material, predominantly hydrogen, into space at a late stage of development. From the compositions of the planets it can be estimated that the planetary nebula must have contained a minimum mass of primitive material amounting to 3% of the mass of the Sun. Until quite recently, such a minimum planetary nebula surrounding a proto-Sun was typically assumed. However, CAMERON (1973b, 1975) has now suggested that the Sun and planets formed together from a much more massive nebula of about twice the Sun's mass. CAMERON and PINE (1973) have carried out numerical simulations of the development of such a nebula. Such a massive solar nebula might evolve to form the solar system by the following sequence of events.

(1) The starting point is thought to be a contracting cloud of interstellar gas and dust which rotated through incorporation of one or more turbulent eddies. As the cloud contracted, its angular velocity increased and it became somewhat flattened into a rotating lens-shaped disc, pinched-in near the spin axis. Contraction ceased when gravity was balanced by the combined effects of centrifugal force and the gas pressure gradient. The primitive nebula would be dense enough to trap the heat produced by the collapse, resulting in a temperature of about 3000 K in the central region decreasing to very low values near the surface and edge of the rotating disc. The interstellar dust evaporated to gaseous state in the hot inner regions but remained solid in the cool outer parts.

(2) The gas pressure gradient in the nebula helped to balance out the gravitational force near the central plane but its effect was negligible near the surface of the nebula. This effect caused the gas at given distance from the spin axis to rotate more slowly near the central plane than at the outer surface. In certain parts of the nebula, thermal convection or turbulence mixed the gas between the central plane and outer surface. Because of the pressure gradient effect, this interchange of gas

caused some of it to rotate too fast and some of it too slow for rotational equilibrium. The result was that gas with excess angular momentum migrated outwards from the spin axis, and gas with deficient angular momentum migrated inwards towards the spin axis. This caused an outward migration of the angular momentum of the nebula and inward migration of slowly rotating material which concentrated near the spin axis to form the proto-Sun. According to CAMERON (1975), both the cooling of the nebula by convection and the outward transfer of angular momentum would be expected to occur on a time-scale of a few thousand years. He suggested that the lifespan of the primitive nebula was extended to about 100 000 years (needed for condensation)· by continuing accretion from space.

(3) As the nebula cooled, solid material condensed out of it as small grains. Silicate grains would condense in the inner parts and icy material in the outer parts. The history of condensation of rocky silicate material out of the hot gaseous nebula as it cooled from about 2000 K to 400 K has been evaluated using thermodynamic data and observations on the mineralogy of meteorites. The cooling silicate grains apparently remained in chemical equilibrium with the gases of the nebula until accretion occurred at about 400 K to form the parent meteorite bodies. Based on accounts by GROSSMAN and LARIMER (1974) and WASSON (1972), the following stages appear to have taken place:
(i) Condensation of oxides and silicates of the refractory elements Ca, Al and Ti occurred between 1800 K and 1500 K with preferential removal to cooler parts of the nebula;
(ii) Fe, Mg and Si condensed between 1500 K and 1300 K to form metallic iron grains with nickel and magnesium silicates of pyroxene and olivine structure, and the alkali metals condensed around 1200 K to be taken into pre-existing calcium feldspar;
(iii) A major fractionation of silicate and metal grains took place between 1000 K and 700 K, possibly triggered by the onset of magnetism;
(iv) Below 750 K metallic iron started to be taken into silicate minerals, and soon after some of it reacted with gaseous H_2S to form the FeS phase troilite;
(v) Chondrules formed by flash melting of the silicate dust and solidification of the molten drops, with loss of volatiles and some reduction of troilite to metal, possibly resulting from lightning or impact melting;
(vi) Hydration of silicates occurred by reaction with water below about 350 K, and accretion of the grains started at about 400 K.

(4) The solid grains could not be supported by the gaseous pressure gradient in the nebula and thus they fell by gravity towards the central plane. The gas resistance had the important effect of restricting the relative velocities of nearby grains, making it possible for them to stick together on collision. Magnetism may have assisted the agglomeration of metallic iron grains. Thus solid bodies of progressively increasing size rapidly grew near the central plane of the nebula, at first forming planetesimals and then larger sized bodies such as the meteorite parent bodies. During the later stages, gravitational attraction would assist this accretion process. Eventually, a limited number of rotating planetary sized bodies would form by accumulation of bodies of slightly differing orbits. Within the cold outer fringes of the nebula, mainly methane, ammonia and ice rapidly accumulated to form Uranus and Neptune. In the inner parts where rocky material alone had condensed from the hot nebula, the inner terrestrial planets were formed. In the intermediate region, rocky and icy material accumulated to form the cores of Jupiter and Saturn, which were massive enough to retain large quantities of hydrogen.

(5) The final stage of nebular evolution involved the rapid collapse of the proto-Sun as it developed into a normal main sequence star. During this stage, an intense solar wind cleaned-up

the planetary region by dispersing the surplus unaccreted gas and dust into interstellar space. This is known as the T-Tauri stage. Stars of this type, surrounded by a cloud of nebular material much of which is being dispersed into space, can now be observed in the sky.

Thus the main process of forming the Sun and its planetary system probably took much less than one million years and was complete about 4600 My ago. The pattern of the system has not much changed since then, although there may have been some capture of satellites by planets, and some slowing down of planetary rotation by tidal friction especially affecting Mercury, Venus and Earth.

1.7 Chemical composition of the Earth

The mantle and core together form about 99 % of the Earth's volume. They are not accessible to chemical analysis. The rocks of the crust do not represent the mean composition of the Earth for they have much too low a density, even allowing for increase of density on compression. One useful guide to the overall composition of the Earth is to draw analogy with meteorite compositions. Thus the presence of metallic and silicate phases in meteorites may suggest that the Earth has an iron core containing some nickel and a mantle similar in composition to the silicates of the chondrites. The recent recognition that both the Earth and the meteorite parent bodies formed by accretion of condensed material from the inner part of the planetary nebula strengthens this argument. More detailed discussion of the composition of the crust, mantle and core is given in later chapters.

Meteorites now provide an indirect method of estimating the bulk composition of the Earth. The chondritic meteorites are believed to give a good approximation to the composition of the rocky fraction of the planetary nebula. A volatile-rich group known as the C1 type of carbonaceous chondrites yields the best estimate of the abundances of the non-volatile elements in the condensing planetary nebula, as borne out by close agreement with abundances determined from the present solar spectrum. However, there were pronounced regional variations in the composition of the condensed material. These were partly caused by primary variations as evidenced by the variability of oxygen and magnesium isotopic compositions in meteorites, but

Table 1.5 Estimated percentage abundances of prominent elements in the Earth and Moon, adapted from ANDERS (1977).

Element	Earth	Moon
Iron	35·9	9·0
Oxygen	28·5	41·4
Silicon	14·3	18·6
Magnesium	13·2	17·4
Nickel	2·0	0·5
Calcium	1·9	6·4
Sulphur	1·8	0·4
Aluminium	1·8	5·8
Cromium	0·5	0·1
Sodium	0·16	0·09
Titanium	0·10	0·34
Potassium	0·017	0·010
Thorium	65 parts/10^9	210 parts/10^9
Uranium	18 parts/10^9	65 parts/10^9

mainly result from fractional separation of some of the basic ingredients during condensation. Thus the composition of the Earth cannot be identical to that of chondritic meteorites or of the primitive rocky fraction of the nebula.

ANDERS (1977) has overcome this difficulty in estimating the bulk composition of the Earth by speculatively postulating that the composition of both chondrites and individual inner planets can be expressed in terms of the following seven basic ingredients which may be present in varying proportions: (i) early refractory condensate, (ii) unmelted metal, (iii) metal remelted during chondrule formation, (iv) the sulphide troilite, (v) unmelted silicates, (vi) silicates remelted during chondrule formation, and (vii) a volatile-rich late-formed component. Anders used constraints based on rock chemistry and internal structure of the Earth and Moon to estimate the relative proportions of the basic ingredients in these bodies, from which their overall chemical composition could be computed. The results of the analysis (Table 1.5) must of course be treated with some caution, but are probably the best estimates we have at present. It is notable that the Earth shows enrichment in the refractory component and iron, and there is depletion of volatiles relative to the primitive abundances.

1.8 The rotation of the Earth

The Earth's rotation about its spin axis has been inherited from the relative motions of the fragments which accreted to form it 4600 My ago. The rotation has subsequently been affected by long term (secular) changes and by short term periodic and irregular variations. These are observable as (i) changes in the direction of the spin-axis in space known as precession and nutation, (ii) movement of the instantaneous pole of rotation known as polar motion, and (iii) changes in the length of day resulting from variation of the rate of rotation. Such changes may variously result from application of external couples to the Earth, internal transfer of angular momentum between solid and fluid divisions, internal redistribution of mass, and possibly from long term change of the gravitational constant. The Earth's rotation, past and present, is of interest to geophysicists because of its relevance to the internal structure of the Earth and past history of the Earth-Moon system. The three main types of anomalous rotational behaviour of the Earth, reviewed in detail by ROCHESTER (1973), are summarized as follows:

(i) *Precession and nutation*: the Earth's spin axis makes an angle of 23·5° with the normal to the plane of the ecliptic. Consequently the Sun and Moon exert gravitational couples on the Earth's equatorial bulge which cause the spinning Earth to precess like a top with a period of 25 700 years, so that the spin axis describes a cone in space. The period of the precession is used in determining the Earth's moments of inertia (p. 5). Irregularities in the orbital motions of the Earth and Moon cause small periodic variations of the couples, giving rise to small ripples which are superimposed on the precession. These are known as nutation, the most prominent one being of 18·6 year period.

(ii) *Polar motion*: the Earth's axis of instantaneous rotation is affected by various wobbles and by long term secular motion. Polar motion has been observed since the beginning of this century by recording variation of latitude with time at a number of special observatories. New techniques such as tracking of artificial satellites and lunar laser ranging will eventually replace the older methods. The average displacement of the poles from their mean position is about 3 m and the corresponding angular disturbance to the instantaneous axis of rotation is about 0·1″ (Fig. 1.8). Part of the observed wobble is a forced oscillation with a period of about a year which is probably caused by atmospheric phenomena. Most of the oscillation, however, is a free oscillation of 435 day period which is called the *Chandler wobble*. This appears to be a randomly excited damped oscillation with a decay time of the order of 40 year. The wobble is probably excited by the

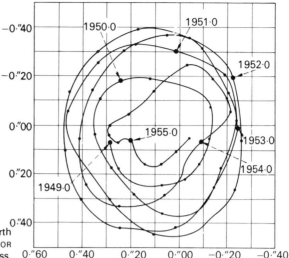

Fig. 1.8 Polar motion: spirals described by the North Pole between 1949 and 1955. Redrawn from MELCHIOR (1957), *Physics Chem. Earth*, **2**, 222, Pergamon Press.

redistribution of mass associated with large earthquakes (MANSINHA and SMYLIE, 1968) and possibly with motion of air and water masses. The period of the Chandler wobble provides evidence for the fluid nature of the core (p. 237). An outstanding problem is to understand the mechanism of the strong damping (p. 169). Turning to the secular motion of the pole, an observed drift of about 0·2″ over the last 70 years towards Newfoundland can be attributed to sea-level fluctuations. On a geological time scale, the apparent migration of the pole relative to the surface is known as *polar wandering* (p. 338).

(*iii*) *Changes in the length of day* are caused by variations in the rate of rotation of the Earth's surface. Prior to about 1955, the length of day was measured by ephemeris time based on the motions of Sun, Moon and planets over several centuries which provided a standard independent of the Earth's motion. Since about 1955, atomic time has been used as the standard with much improved accuracy and resolution. Several of the observed periodic changes in length of day are attributable to known meteorological, tidal and other hydrological phenomena. The most rapid irregular changes in l.o.d. can be attributed to meteorological effects, but irregular changes taking place over about ten years are probably caused by transfer of momentum between core and mantle by a mechanism which still remains controversial (p. 264). The secular variation of the rate of the Earth's rotation can on the short term be attributed to changes of sea-level, etc. but on a long term the most widely accepted but still controversial mechanism is tidal friction which is discussed in the following paragraphs.

Past history of the Earth-Moon system

The Earth and its satellite Moon form a system which is to some extent unique in the solar system. This is because the ratio of the mass of the Moon to that of the Earth is 1/81·3 which is exceptionaly high for a satellite. Over 80% of the angular momentum of the system is tied up in the orbital motion of the Moon. In all the other satellite systems most of the total angular momentum is possessed by the rotating planet itself.

When we trace the Moon's orbital history back into the geological past, we find that the Moon has apparently been progressively receding from the Earth. There has been a complementary increase in the length of day as the Earth's spin rate has slowed down. Both phenomena occur as a result of tidal interaction between the Earth and Moon. The rotational history of the Earth-Moon system has been inferred (*i*) over the past 250 years by direct observations recently improved by the introduction of atomic clocks, (*ii*) over the past 3000 years using ancient records of eclipses, and (*iii*) back to the Lower Palaeozoic and earlier using fossil 'clocks', particularly corals.

Tidal interaction between the Earth and Moon causes a progressive loss of rotational energy from the system. The Moon has already been brought to a 'standstill' so that the same side always faces the Earth. However, the total angular momentum of the system must be conserved. The loss of angular momentum as the Earth slows down is balanced by an equal increase in the angular momentum of the Moon's orbital motion, which means that the Moon progressively recedes from the Earth. The rotational energy lost by the Earth is partly used to increase the orbital energy of the Moon but is mainly dissipated as heat by tidal friction in the shallow seas and to a small extent by tides in the solid Earth.

Tidal friction works by exerting a couple on the Earth, which slows down the rate of rotation. The slight bulge on the Earth caused by the tides also exerts a couple on the Moon which increases its velocity in orbit, thereby causing it to recede from the Earth, which increases its period of revolution (Fig. 1.9).

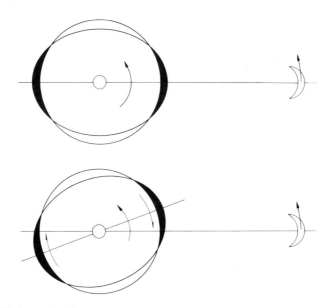

Fig. 1.9 How tidal friction works. The upper diagram shows the tidal bulge on the Earth which would be produced by the Moon if there is no tidal friction. This is produced because the Moon's gravitational attraction on the Earth only exactly balances the centrifugal force at the centre. The lower diagram shows how tidal friction delays the high tide. This produces a couple on the Earth, slowing down the axial rotation, and a force on the Moon, speeding up its velocity in orbit and causing it to recede. Redrawn from MUNK and MACDONALD (1960), *The rotation of the Earth*, p. 199, Cambridge University Press.

Variation of the apparent angular velocity of the Moon as viewed from the rotating Earth is caused *both* by variation in the Moon's period of orbital rotation *and* by variation in the Earth's rate of rotation. This is because Universal Time, itself dependent on the Earth's rate of rotation, has been used as a standard. These two contributions to the Moon's apparent motion were first separated by SPENCER JONES (1939) who used past observations of the celestial positions of the Sun,

Mercury and Venus over the last 200 years to provide a time standard unaffected by the Earth's rotation. More accurate separation of the two contributions can be carried out over the past 25 years using atomic clocks as the independent time standard, and estimates can be carried back to 1375 B.C. using ancient eclipses, albeit with lesser accuracy. MULLER and STEPHENSON (1975) have reviewed the data, contributing their own analysis of the ancient eclipses. They conclude that within error the secular deceleration of the Moon has been constant over the last 3000 years and they estimate it to be 40.75 ± 2.5 seconds of arc per century^2 corresponding to a retreat of the Moon from the Earth of about 6 cm per year. Their estimate for the secular increase in length of day is 2.7 ± 0.3 millisecond per century.

The rate of slowing down of the Earth's rotation caused by lunar tidal friction can be calculated accurately from the Moon's secular deceleration by the principle of conservation of momentum, without reference to the mechanism of tidal friction. Tidal interaction between the Sun and Earth contributes a further increase in length of day. The exact value is not known because the Sun's response cannot be measured, but it is estimated to be between 20 % and 30 % of the lunar effect. Taking the above value of the Moon's secular deceleration, it can be estimated that tidal friction is increasing the length of day by 4·1 milliseconds per century. Comparing with the observed value of 2·7, there must be a non-tidal decrease in the length of day of about 1·4 milliseconds per century. This could be explained by a slight speeding up of the rate of rotation caused by a decrease in the Earth's moment of inertia. The melting of the Pleistocene icecaps could contribute to this decrease in moment of inertia. Another possibility is transfer of angular momentum between core and mantle by electromagnetic coupling associated with an 8000-year period change in the geomagnetic dipole moment revealed by archaeomagnetic studies (YUKUTAKE, 1972).

The story of the Earth's rotation and the Moon's orbit can be carried much further back into the past by using fossil 'clocks'. Some fossil rugose corals of Palaeozoic age show a characteristic banding on the outer skin (epitheca). WELLS (1963) was able to recognize tentatively both a daily and annual banding in corals of Middle Devonian age (Fig. 1.10). He calculated that there were 400 ± 7 days in the Middle Devonian year, which was about 375 My ago. This gives an average increase in the length of day between then and now 2.4×10^{-3} seconds per century, which is significantly less than the estimate for historical time. Later SCRUTTON (1964) recognised apparent monthly banding in Middle Devonian corals and he was able to suggest that the Devonian year was divided into 13 lunar months of 30·5 days each. More recently the method of fossil clocks has been applied to other periods in the past, including the use of stromatolites in the Precambrian (PANNELLA, 1975).

These estimates of the length of day and lunar month 375 My ago are particularly interesting because they enable the slowing down of the Earth by tidal friction to be separated from changes in

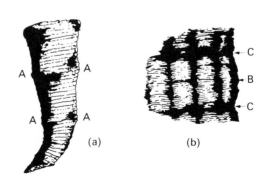

Fig. 1.10 Diagrammatical representation of a fossil coral 'clock'; (a) shows a coral with three complete annual growth cycles each divided into thirteen monthly bands; (b) shows an enlargement of a single monthly band subdivided into thirty daily growth ridges. Reproduced from SCRUTTON (1967), *International Dictionary of Geophysics*, Vol 1, p. 1, Pergamon Press.

the Earth's rotation rate caused by other processes (RUNCORN, 1964). The length of the lunar month makes it possible to estimate the angular momentum of the Moon's orbital motion. Because angular momentum is conserved, the slowing down of the Earth's rotation caused by lunar tidal friction between then and now can be estimated. Runcorn suggested that any residual effect may be the result of a change in the moment of inertia of the Earth caused by growth of the dense core. In fact he found that nearly all the change in length of day can be attributed to tidal interaction and that the moment of inertia in the Devonian does not differ significantly from the present value.

The above results suggest that the lunar month was about 4% shorter in the Devonian than now. Kepler's third law of planetary motion shows that the Moon's semi-major axis must have been about 6% less than now. Tidal friction probably depends on the sixth power of the distance between the interacting bodies. Consequently it becomes more effective the further back in geological time one goes, provided that the energy dissipation occurred under similar conditions. Assuming that tidal friction was equally effective before and after the Lower Palaeozoic, it can be calculated that the Moon must have been very close to the Earth at a date somewhere between 1000 and 2000 My ago. Based on numerical analysis along these lines, GERSTENKORN (1955) suggested a major event in the Earth-Moon system some 1400–1600 My ago. MUNK (1968) graphically described it as follows: 'A heavy hot atmosphere over a darkened Earth. Giant tides on a 5h day, with steaming tidal bores following the Moon on a 7h polar orbit. Mr Gerstenkorn may not appreciate having his name attached to this era.' In fact there is no record of such an event either in the geological evidence from the Precambrian going back 3800 My or from the history of the Moon going back 4400 My. Such an event probably did not happen, indicating that tidal friction prior to the Devonian was much less effective than subsequently resulting from a different style of shelf-sea palaeogeography. In view of the discrepancy between the relatively large historical value and the significantly smaller average slowing down from Devonian to present, such an interpretation is plausible. Nevertheless, the Moon probably was much closer to the Earth early in the history of both bodies, and the Earth probably rotated with a period closer to 5h than to 24h when it was first formed.

1.9 The Moon

Structure and development

The Moon is our nearest planetary neighbour and speculations about its origin have excited interest for a long time. This interest has been intensified by the outstanding success of exploration of the lunar surface starting with the Apollo 11 mission of July 1969. Results stemming from the American and Russian programmes have revolutionized our knowledge of the structure and history of development of the Moon although its origin still remains controversial. The new knowledge has relevance to our understanding of the origin and early history of the Earth. Recent appraisals of the results of lunar exploration are given by TAYLOR (1975) and by MASSEY and others (1977).

The lunar surface is blanketed to a few metres depth by a 'soil' of fine-grained rock and mineral fragments known as the regolith. This has been broken down, reworked and added-to by meteorite impacts. The lunar craters are now known to be almost all of impact origin, and the mare basins are interpreted as having been excavated by huge impacts prior to their infilling by basaltic rocks of variable composition which are chemically distinct from terrestrial basalts. The highland regions appear to be formed of coarser-grained igneous rocks which are chiefly the plagioclase-rich anorthosites but also include the gabbroic varieties of norite and troctolite. These highland rocks represent the primitive lunar crust. Breccias resulting from impact metamorphism are common, indicating the highly bombarded nature of the rocks underlying the lunar soil.

The internal structure of the Moon has been investigated by highly sensitive seismograph stations left operating at four of the Apollo sites (LAMMLEIN and others, 1974; LAMMLEIN, 1977). The recordings, which are telemetered back to Earth, show that the Moon is exceptionally quiet seismically. The observed events include small moonquakes with foci clustering around 900 km depth, a few larger shallow moonquakes, meteorite impacts, and artificial events such as the impact of discarded lunar modules. The records (Fig. 1.11) differ markedly from terrestrial

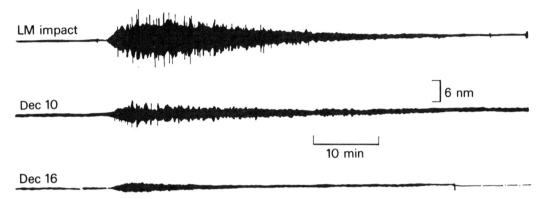

Fig. 1.11 Seismic signals received on the long-period vertical component seismometer from the Lunar Module impact on November 20 and from natural events on December 10 and 16, 1969. Redrawn from LATHAM and others (1970), *Science, N.Y.*, **167**, 456.

seismograms in that they show a relatively slow build-up of seismic energy over a period of about 5–10 minutes followed by a slow decay, the whole event continuing for about an hour. The character of these records may be explained by the trapping of the seismic energy in a low velocity near-surface layer where the elastic waves are strongly scattered but very weakly attenuated. This layer can be modelled by a P velocity distribution starting at about 0.3 km s^{-1} just below the soil, increasing to 4 km s^{-1} at about 5 km depth, and grading into a crustal velocity of about 6 km s^{-1} at 25 km depth. The low velocity layer has been attributed to the impact brecciation of the Moon's surface that occurred early in its history.

Seismic recordings of the man-made impacts have revealed a major discontinuity under the Oceanus Procellarum at about 60 km depth where the P velocity increases from 6·3–7·0 km s^{-1} above, to about 8·0 km s^{-1} below, values which are closely similar to those across the Earth's crust-mantle boundary. Here, the Moon's crust is almost twice as thick as the Earth's average continental crust, and the higher elevation of the far side suggests on isostatic grounds that the crust there may be 100 km thick or thereabouts. Beneath 60 km depth, the Moon's mantle extends almost to the centre (Fig. 1.12). It can be subdivided into three zones (LAMMLEIN, 1977): an upper mantle about 250 km thick with P and S velocities of about 8·1 and 4·7 km s^{-1} respectively, a middle mantle between 300 and 1000 km depths with a reduced shear-wave velocity, and a lower mantle below about 1000 km depth where shear waves are highly attenuated. The lunar lithosphere has been identified with crust and mantle down to 1000 km depth, with an asthenosphere below. An innermost zone of 170–360 km radius with much reduced P velocity may possibly represent a small molten core, chiefly of iron sulphide. The absence of a sizeable core

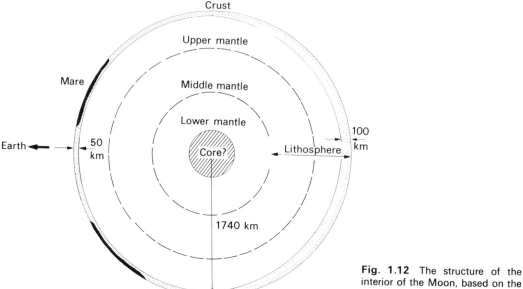

Fig. 1.12 The structure of the interior of the Moon, based on the model of LAMMLEIN (1977).

accounts for the observed mean density of 3340 kg m^{-3} and the uniform density-depth distribution implied by the moment of inertia of $0.395\ Mr^2$.

The geochemical and geophysical results of lunar exploration suggest that the Moon was formed by homogeneous accretion from the planetary nebula and that it subsequently differentiated into a feldspathic crust and a mantle formed mainly of the ferromagnesian minerals olivine and pyroxene. The mare basalts formed later by partial fusion of the Moon's upper mantle. Relative to the cosmic abundance pattern of primitive carbonaceous chondrites, the Moon is strongly deficient in iron and related siderophile and chalcophile elements, enriched in the refractory large-ion lithophile elements, and depleted in volatiles (Table 1.5). Relative to the Earth, it is iron deficient but could be regarded as having a compositional and genetic relationship to the Earth's mantle, despite deficiency in volatiles.

Radiometric dating has revealed the following approximate chronology of lunar events:

(*i*) The moon formed from the planetary nebula about 4600 My ago.

(*ii*) Differentiation into crust and mantle occurred about 4500 My ago and was virtually complete 200 My after the Moon's formation.

(*iii*) Widespread bombardment, producing cratering, brecciation and metamorphism at and near the surface was at a maximum between about 4000 and 3800 My ago, producing the major basins such as Imbrium possibly after a previous relatively quiet period.

(*iv*) Mare basalts were extruded between about 3900 and 3000 My ago.

(*v*) The Moon has apparently been volcanically and tectonically quiescent since about 3000 My ago in strong contrast to the Earth where the early history has been obscured by such activity.

Origin of the Moon

The Earth and its satellite Moon form a system which is to some extent unique in the solar system. This is because the ratio of the mass of the Moon to that of the Earth is 1/81·3 which is exceptionally high. Over 80 % of the angular momentum of the system is tied up in the orbital motion of the Moon, although the proportion was somewhat less in the past because of tidal friction. In all the other satellite systems of the solar system, most of the angular momentum is possessed by the rotating planet itself. It is generally recognized that the Moon formed by accretion from the planetary nebula and the main problem is to understand how the Earth and Moon came to be associated with each other. The hypotheses are briefly reviewed here. A fuller account has been given by KAULA and HARRIS (1975). The three main hypotheses are:

(i) *The Moon formed by fission from the Earth*, early in its history. Darwin's original version of the hypothesis attributed the Moon's separation to the resonant build-up of a very large solar tide when the Earth rotated at the same period as its free vibration. The fatal objection, pointed out by Jeffreys, is that friction would prevent the separation of the tidal protuberance. Most modern versions of the hypothesis are based on the geochemical similarity of the Moon and the Earth's mantle, allowing for loss of volatiles from the Moon. The most widely discussed version, suggested by RINGWOOD (1970) and named the *precipitation hypothesis*, postulates that silicate material was evaporated from the proto-Earth because of the high surface temperature produced during accretion. This subsequently precipitated into a ring of 'sediment', orbiting the Earth and then coagulating to form the Moon. This hypothesis is probably the most acceptable one for chemists but some dynamical difficulty is encountered because of the excessively high initial angular momentum which needs to be postulated for the Earth. In common with the binary system hypothesis, the uniquely large size of the Moon relative to the Earth and the obliquity of the Moon's orbit to the Earth's equatorial plane remain unexplained.

(ii) *The Moon and Earth formed together as a binary system* within the planetary nebula, the Moon being formed by coagulation of planetesimals trapped in orbit round the proto-Earth as it accreted (KAULA and HARRIS, 1975). The problem of differing bulk compositions of the Earth and Moon may be met by postulating selective capture of smaller silicate, rather than larger iron planetessimals. Other problems concern the size of the Moon and the obliquity of its orbit. This hypothesis is complex and has so far defied accurate mathematical modelling. KAULA and HARRIS (1975) favoured it perhaps mainly because they find the other theories less plausible.

(iii) *The Moon was captured by the Earth*. Such an event is most likely to have happened shortly after formation as we have no evidence of the violent disruptions which would have left their record had they occurred later. The Moon would probably have been captured initially into an oblique retrograde orbit which would at first shrink by tidal friction. On approaching within a few radii of the Earth, the Moon would be deflected towards a polar orbit rather than coming closer and breaking up into fragments. It would finally move into a prograde orbit with tidal friction then causing it to recede progressively towards the present-day location. The capture hypothesis raises chemical problems because of the deviation of the Moon's composition from that inferred for the primitive planetary nebula, but the anomalous size and obliquity of orbit are readily explained. However, to avoid impact or break-up, the Moon would need to approach the Earth from a distance with a relative velocity of less than about $2 \, \text{km} \, \text{s}^{-1}$. The unlikelihood of such an event and the chemical problems are the main obstacles to wide acceptance of this otherwise attractive hypothesis.

These three hypotheses were formulated in primitive form prior to the manned and unmanned exploration of the lunar surface and interior. Important geochemical constraints arising from the lunar sampling have led to their modification. All three hypothesis in their modified and varied forms are still held as there has been no decisive evidence in favour of one in particular.

2 The continental crust

2.1 Introduction

A century or more ago, it was believed that the Earth consisted of a thin rigid crust overlying a hot fluid substratum which provided magma to feed volcanoes. This concept was being questioned before 1900. It was finally abandoned when early seismological research showed that the Earth is normally solid down to 2900 km depth. It is now known that magma is not drawn from a permanently fluid region within the Earth but forms by local fusion of the normally solid rocks of the upper mantle and crust.

The crust is nowadays almost always defined as the region above the Mohorovičić discontinuity (or Moho). This discontinuity has been found to be almost universally present beneath continents and oceans. It was defined by STEINHART (1967) as the level where the compressional wave velocity first increases rapidly or discontinuously to a value between 7·6 and 8·6 kms.$^{-1}$ In absence of a recognizable steep gradient, it is taken as the level where the P velocity first exceeds 7·6 km s^{-1}. GIESE (1976a) added further precision by suggesting the use of the level where the velocity-depth gradient is steepest, as recognizable using the deepest penetration of the large amplitude lower crustal arrivals known as the P_mP group of waves (p. 35). This allows practical identification of the boundary and extends the definition to exceptional regions outside the scope of Steinhart's range of values.

Defined in this way, the crust forms less than 1% of the Earth by volume and less than 0·5% by mass. But it is the only major subdivision as yet directly accessible to man and its importance is out of all proportion to its size. The topmost part of the continental crust is the most fully investigated part of the Earth and study of it has provided most of the evidence we have of the past history of the Earth. In contrast, surprisingly little is known of the structure of the lower part of the continental crust.

Two important facts about the structure of the uppermost part of the Earth were known before the Moho had been discovered. *Firstly*, it had been recognized that the mean density of the Earth is substantially greater than that of rocks at or near the surface, suggesting the existence of a low density layer near the surface. *Secondly*, measurements of the local variations in the vertical direction near mountain ranges led to the discovery of the theory of isostasy during the eighteenth and nineteenth centuries. This shows that there are large lateral variations in density within the upper layers of the Earth and that the near surface layer of relatively strong and brittle rocks now called the *lithosphere* must be underlain by a weaker substratum which deforms by flow, now called the *asthenosphere*. Occasionally the term 'crust' has been used for the lithosphere, but its lower boundary does not in general coincide with the Moho and is gradational rather than sharp.

Geological structure of the uppermost crust

Over a century of geological investigation has given us a detailed knowledge of the surface rocks forming the continents. Typically a variable thickness of partly consolidated sedimentary rocks overlies a strongly folded and metamorphosed basement, or alternatively the basement itself crops out at the surface. Most of the sedimentary rocks have been formed by erosion of pre-existing sedimentary rocks. Over large areas, such as parts of the Precambrian shields, unmetamorphosed sediments are absent. At the other extreme, local accumulations of sediment may exceed 10 km in geosynclines and deep basins. According to POLDERVAART (1955), the average thickness of sediments in regions of young fold belts is about 5 km and in continental shield areas it is 0·5 km. Lava flows and minor igneous intrusions commonly occur in sedimentary sequences, but penetration by large igneous intrusions is relatively rare.

The underlying rocks of the basement are metamorphosed sedimentary and igneous rocks which are locally penetrated by large igneous intrusions, especially granites and granodiorites.

It is convenient to subdivide the continental regions into structural provinces based on the history of deformation over the last hundred million years or more. The main subdivisions are:

(i) *Stable regions*, sometimes called *cratons*, which show little evidence of vertical or horizontal movement apart from broad warping and a few minor faults. Included in this category are the *Precambrian shields* which are gently arched regions of large areal extent where Precambrian rocks are found at the surface, and *platforms* where the basement rocks are overlain by a thin cover of flat-lying sediments.

(ii) *Semi-mobile regions*, which are characterized by relatively strong differential vertical movement including the formation of sedimentary basins. Great Britain has been a semi-mobile region since the end of the Palaeozoic mountain-building movements.

(iii) *Mobile belts*, or young mountain ranges, which have been strongly deformed with indication of powerful vertical and horizontal movement. The two main mobile belts are the circum-Pacific belt of mountain chains and island arcs which forms a ring round the Pacific Ocean and the Alpine-Himalayan mountain belt. Metamorphism and emplacement of large granite batholiths occur in mobile belts. Most parts of the continental crust have been mobile belts at some time during the Precambrian or later.

2.2 Earthquake seismology and the discovery of the crust

Earthquake seismology laid the foundation for the modern study of the crust. As a method of investigating the thickness and internal structure of the crust, it has now been largely superseded by refraction and reflection studies using artificial explosions.

The starting point was Mohorovičić's discovery of the discontinuity at the base of the continental crust, now called the Moho. He recognized two P and two S pulses on seismograph records of the Croatia earthquake of October 8, 1909, at observatories within a few hundred kilometres of the epicentre. Near the epicentre, the slower travelling P and S pulses which are now usually called P_g and S_g were prominent and arrived first. They progressively died out. Beyond 200 km P_g and S_g were overtaken by P_n and S_n pulses travelling with a higher apparent velocity, which could be identified as the normal P and S phases observed up to an angular distance of 103° from the epicentre. Mohorovičić interpreted P_g and S_g as the waves transmitted direct from the focus to the seismograph station through an upper low velocity layer which is the continental crust. The faster P_n and S_n arrivals were interpreted as so-called head waves which had travelled for most of their path in a higher velocity underlying medium, which is the mantle. The principle, applied to a surface source, is shown in Fig. 2.1. MOHOROVIČIĆ (1909) obtained the following results:

$$P_g = 5·6 \, km \, s^{-1} \quad P_n = 7·9 \, km \, s^{-1}, \text{ and crustal thickness} = 54 \, km.$$

The next important step was made by CONRAD (1925) who recognized two new body wave phases

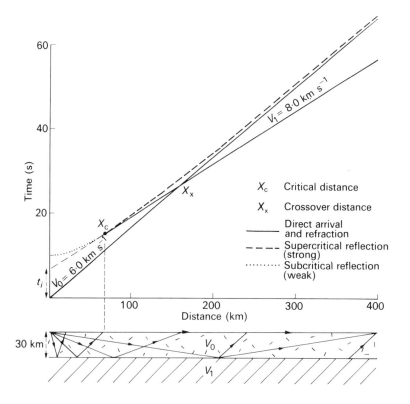

Fig. 2.1   Time-distance graph for direct, refracted and reflected arrivals for an explosion source at the surface of a uniform horizontal layer overlying a higher velocity half-space. The velocities are equal to the reciprocal gradients of the respective first-arrival segments, and the depth of the interface is given by the following relationships

$$d = \frac{t_i}{2} \frac{V_0 V_1}{\sqrt{(V_1{}^2 - V_0{}^2)}}, \quad d = \frac{X_x}{2} \sqrt{\left(\frac{V_1 - V_0}{V_1 + V_0}\right)} \quad \text{or} \quad d = \frac{X_c}{2} \frac{\sqrt{(V_1{}^2 - V_0{}^2)}}{V_0}.$$

P^* and S^* with intermediate velocities of 6·29 and 3·57 km s^{-1} respectively by studying records of the Tauern earthquake of November 28, 1923. On this evidence, he subdivided the crust into an upper layer which transmits P_g and S_g overlying an intermediate layer giving rise to the refracted arrivals P^* and S^*. The intervening interface is called the *Conrad discontinuity*.

From a quite different standpoint, Daly had suggested earlier that beneath the continents there is an upper silicon-aluminium-rich layer (the SIAL) providing the source of granite magma overlying a silicon-magnesium-rich layer (the SIMA) which is the source of basalt magma. Earthquake seismologists took over Daly's model because they found that the seismic velocities in granite and basalt were closely similar to the observed velocities of the upper and intermediate layers respectively. They called these two layers the 'granitic layer' and the 'basaltic layer'. It will be shown later in the chapter that the upper crust is not granitic and that the lower crust is probably not basaltic, nor is the twofold subdivision found everywhere. It is now preferable to use the non-committal terms 'upper crustal layer' and 'lower crustal layer' wherever a twofold subdivision of the continental crust is made.

Until the advent of crustal explosion seismology about 1950, earthquake body wave studies were widely used to investigate the structure and thickness of the continental crust. The dispersion of earthquake surface waves was used to investigate broad structure across wide tracts of country (PRESS and EWING, 1955) and latterly regional variations in crustal structure (EWING and PRESS, 1959). Although explosion seismology has been found to be a much more effective method for probing crustal structure, earthquake studies have not been entirely superseded. They do still hold one advantage, which is that earthquakes occur at different depths in the crust and below it, and thereby can provide complementary information on lower crustal structure unattainable by surface sources.

2.3 Explosion seismology and the structure of the crust

The refraction method

Nearly all our knowledge of the structure of the continental crust has been obtained since about 1950 by explosion seismology. The advantages of using artificial explosions rather than earthquakes are that the time and position of the shot are accurately known. Experiments can be planned in relation to geological structure and regions without earthquakes can be investigated without difficulty.

Most applications of explosion seismology to investigation of crustal layering and thickness make use of the refraction method of seismic prospecting (e.g. DOBRIN, 1976). The basic method is to determine the travel-time of P waves between shot points and seismic recorders at varying distance apart up to several times the depth of penetration required, usually along straight line profiles. Figure 2.1 shows the simplest possible situation, with a horizontal layer of velocity V_0 overlying a higher velocity substratum (V_1). From the shot point to the crossover point X_x the direct wave arrives first. Beyond the critical distance X_c the refracted head wave from the underlying layer reaches the surface and it becomes the first arrival beyond X_x. A wave reflected from the interface also reaches the surface; this is a relatively weak arrival below the critical distance, but its amplitude increases strongly near X_c. Beyond X_c the supercritical reflection (as it is called) may be the largest amplitude arrival on the record. In this simple horizontal two-layer case, the velocities are given by the reciprocal gradients of the first arrivals on the time-distance graph and the depth to the interface can be calculated from t_i (the intercept time), X_x or X_c.

The dip of a plane interface can only be determined by shooting in both directions. The velocities and the dip and depth to the interface are computed from the two travel-time graphs as shown in Fig. 2.2. The same procedure can be extended to the multi-layer case provided the velocity increases downwards at each interface.

In practice, seismic refraction lines need to be about 200–300 km long to determine continental crustal structure. It is nowadays standard practice to reverse the lines. The simple method of interpreting is to plot a reversed time-distance graph and to fit straight-line segments to the first arrivals by least squares. The inverse gradients of the segments and their intercepts on the time axis are measured. Interpretation proceeds on the assumption that the underlying structure consists of layers of uniform velocity separated by plane interfaces (which may dip).

Well established methods of interpretation of refraction surveys can take into account the deviations of the segments of the time-distance graph from straight lines. For instance, the method of HAGEDOORN (1959) enables the shape of a refractor to be determined along a reversed profile. A generalization of this approach to surveys where the shots and recorders are not along a line is known as the 'time-term' method and it has proved useful in some crustal structure investigations (WILLMORE and BANCROFT, 1960). In theory, the time-term method enables the subsurface shape of one or more refractors to be mapped provided the overlying velocity distribution is known and one shot point and recording point are common.

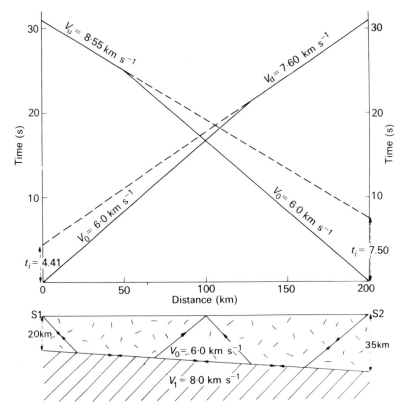

Fig. 2.2 Time-distance graph for reversed profiles along the surface of a dipping layer (velocity $= V_0$) overlying a substratum (velocity $= V_1$). V_u and V_d are the apparent velocities of the refracted head wave shooting up-dip and down-dip respectively; these apparent velocities are equal to the reciprocal gradients of the corresponding segments on the graph. The dip of the interface is

$$\theta = \tfrac{1}{2}\left\{ \sin^{-1}\left(\frac{V_0}{V_d}\right) - \sin^{-1}\left(\frac{V_0}{V_u}\right)\right\}.$$

The true velocity of the underlying layer is

$$V_1 = 2V_u V_d \cos\theta / (V_u + V_d).$$

The depth of the interface at S_1 (or S_2) is

$$d = \frac{t_i}{2}\frac{V_0 V_1 \sec\theta}{\sqrt{(V_1{}^2 - V_0{}^2)}}.$$

Commonly in crustal refraction surveys only the direct upper crustal phase P_g and the upper mantle phase P_n are observed as first arrivals. Using the above methods only, the results from such a survey would be interpreted in terms of a single-layered crust of uniform velocity (Figs 2.1 and 2.2), or as a simple two-layered crust if P^* could also be recognized. Such an interpretation would not be unique because widely different velocity-depth distributions could give rise to exactly the same first arrival segments. It could only be regarded as a rough approximation to the true crustal structure. Layers which are thin or represent small increases of velocity would remain undetected and low velocity layers would be missed. Velocity may in reality vary with depth within layers and the discontinuities inferred between layers may be steep velocity gradients rather than sharp boundaries. Further complications may arise from lateral variation of velocity and structure.

In modern crustal refraction studies, some of the above difficulties are partially overcome by use of the large amplitude second arrivals which are usually observed on the records. These may originate by supercritical reflection from a discontinuity or more commonly by refraction at a steep downward increase in velocity with depth. Such arrivals can be used to distinguish discontinuities from velocity gradients and to yield information on the velocity-depth distribution. They may also reveal the presence of low velocity layers and place limits on the width and velocity distribution within such layers. Those large amplitude arrivals arising by reflection or refraction at the Moho will be referred to as $P_m P$ and those arising at the Conrad discontinuity as $P_1 P$.

At a discontinuity marked by an increase in velocity, the amplitude of the reflected ray is relatively small if the angle of incidence is less than the critical angle. In contrast, for angles of incidence above the critical angle, large amplitude reflections occur because of total reflection. According to ray theory, the maximum amplitude occurs at the critical distance X_c, at which point the time-distance segment of the refracted arrival is tangent to the reflected arrival. However, ČERVENÝ (1966) has shown that ray theory is an oversimplification and that the curved wavefront causes the maximum amplitude to occur somewhat beyond this point.

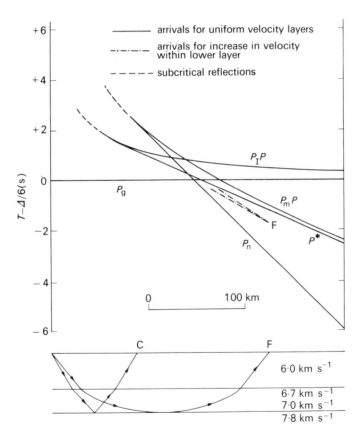

Fig. 2.3 Reduced time-distance graph showing refractions and reflections for a two-layered crust with (**a**) uniform velocity layers, and (**b**) increase in velocity with depth in the lower crustal layer (giving rise to the segments meeting at F). The increase in velocity with depth changes the curve considerably from the uniform case, $P_m P$ arriving earlier but not being observed beyond F.

Reduced time-distance graphs like this, where (travel-time minus distance × velocity) is plotted against distance, are commonly used in presenting crustal refraction results because they allow the time-scale to be expanded without letting the graph become unmanageable. Redrawn from EATON (1963), *J. geophys. Res.*, **68**, 5795.

Time-distance curves for the reflected and refracted arrivals for a single-layered and two-layered crust are shown in Figs 2.1 and 2.3 respectively. The amplitude-distance curves for the head waves P_g, $P*$ and P_n, and for the reflected arrivals P_IP and P_mP (using ray theory) for a two-layered crust overlain by a thin veneer of sediments are shown in Fig. 2.4; these show that the reflected arrivals beyond the critical distance have larger amplitudes than the head waves by one to two orders of magnitude.

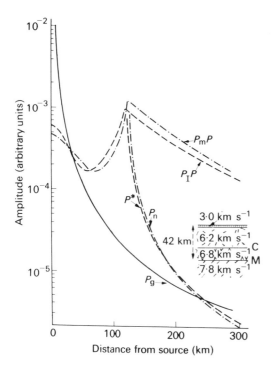

Fig. 2.4 Theoretical amplitudes of refracted and reflected rays for the model of crustal structure as shown. Redrawn from BERRY and WEST (1966), *The Earth beneath the continents*, p. 474, American Geophysical Union.

A similar pattern of large amplitude arrivals is produced by rays which reach their deepest level of penetration in a steep downward increasing velocity gradient. In Fig. 2.5(a), the rays which do not penetrate deep enough to reach the steep gradient form the normal segment OB which is convex upwards. The retrograde segment BC, which is concave upwards, marks the emergence of the large amplitude arrivals which have reached their deepest level in the steepening upper part of the gradient. The critical point C marks the emergence of the ray which has penetrated to the level of the steepest gradient. Beyond C, a normal segment of lower amplitude arrivals is formed by emergence of rays which have bottomed in the lower decreasing part of the steep gradient. Assuming that the layering is horizontal, the apparent velocity at the point of emergence of a ray is the true velocity at its depth of maximum penetration.

Large amplitude second arrivals such as P_mP are used in several ways in modern crustal studies. *Firstly*, they can be used to distinguish between a sharp discontinuity and a steep velocity gradient. Each of these gives rise to a distinct relationship between apparent velocity and distance along the segment BC (Fig. 2.5a) which can be recognized using simple criteria (GIESE, 1976c), and to a characteristic pattern of amplitudes. *Secondly*, the apparent velocity at the critical distance can be used to estimate the velocity at the Moho as defined by GIESE (1976a), although caution needs to be

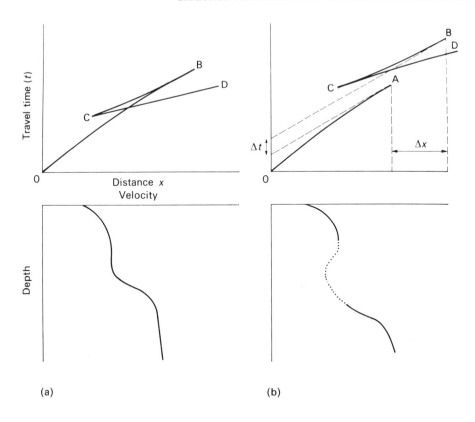

Fig. 2.5 Time-distance curves for continuous variation of velocity with depth, showing (**a**) triplication with retrograde branch BC caused by steep velocity-depth gradient, and (**b**) the effect of a low velocity zone causing a gap in arrivals between A and B, offsets Δx and Δt between A and B, and retrograde branch BC from the steep velocity-depth gradient below the low velocity zone. Adapted from GIESE (1976b), *Explosion seismology in Central Europe*, p. 132, Springer-Verlag.

taken to use the phase rather than the group velocity. *Thirdly*, wide angle reflections can be used to estimate the depth of the discontinuity and the mean velocity down to it using the t^2/x^2 method (p. 91), and rough estimates of these quantities can similarly be obtained for steep gradients as described by GIESE (1976c). *Fourthly*, if the time-distance curve OBCD is uninterrupted, then the Herglotz-Wiechert method of inversion (p. 140) can be used to determine the velocity-depth distribution down to the gradient and through it; if the curve is interrupted because of observational gaps or low velocity layers, then the distribution can be inferred over the depth range where the observed arrivals bottom. *Fifthly*, such arrivals from a steep gradient or discontinuity below a low velocity layer can be used to identify its presence. In Fig. 2.5(b), the effect of the low velocity layer is to cause the arrivals which do not reach down to it to die out at A. No rays can reach their deepest level within the low velocity layer, and the shallowest ray to emerge from below it reaches the surface at B with the same apparent velocity as the one at A. The presence of a low velocity zone is indicated by delay of the tangent at B relative to the parallel one at A, and the zone is particularly pronounced if the distance gap between A and B is short or even reversed. By measuring the apparent velocity at A or B and the time and distance gaps between A and B, the maximum possible thickness of the inversion zone and its mean velocity can be calculated although the velocity distribution within it cannot be uniquely determined (see GIESE, 1976c).

S waves always occur as second arrivals and because of the difficulty in picking accurate arrival times their use in crustal refraction studies has been relatively rare. The more widespread recent use of three component seismometer stations and processing methods which enable separation of the ground motion characteristic of P or S is now leading to increasing use of S arrivals. Such arrivals are important because if the ratio r of P to S velocity is known then the Poisson's ratio σ can be computed using the formula

$$\sigma = (\tfrac{1}{2}r^2 - 1)/(r^2 - 1)$$

Poisson's ratio is a useful diagnostic property for distinguishing between certain rock types.

The methods so far described make only qualitative use of amplitudes to identify phases such as $P_m P$ and to determine the critical distance. However, the relative amplitudes of the different phases and their variation with distance can provide a sensitive method of studying the finer details of crustal structure provided that the observations can be compared with theoretical predictions. A recently developed method of carrying out this comparison is to construct synthetic seismograms for the body wave arrivals (FUCHS and MÜLLER, 1971; BRAILE and SMITH, 1975). Computational methods based on elastic wave theory make it possible to predict the seismic waveform above a horizontally layered half-space at varying distance from a source of specified form. The procedure of interpretation using this method is one of trial and error. A layered velocity model for the crust and topmost mantle is set up, the observed stacked records are compared with the synthetic seismograms covering the same distances from source to receiver, and the model is progressively adjusted until acceptable agreement is obtained. It should be emphasized that this method does not overcome all ambiguity and that its use can be vitiated by lateral variations of structure or by differing seismometer characteristics. Nevertheless, it has opened up the opportunity of studying finer details of crustal layering including the nature of the Moho. An example of a synthetic seismogram, computed for a Basin and Range type structure with a shallow low velocity layer in the upper crust, is shown in Fig. 2.6.

In concluding this section, an example of stacked records obtained at a single seismic station from a line of shots is shown in Fig. 2.7. This shot line formed part of an extensive sea-to-land

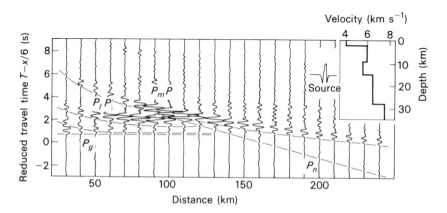

Fig. 2.6 Stacked record sections of synthetic seismograms computed for the generalized Basin and Range (western U.S.A.) crustal model with velocity-depth distribution shown in the inset. The source wavelet is also shown and critical points for $P_l P$ and $P_m P$ are indicated. Redrawn from BRAILE and SMITH (1975), *Geophys. J. R. astr. Soc.,* **40**, 170.

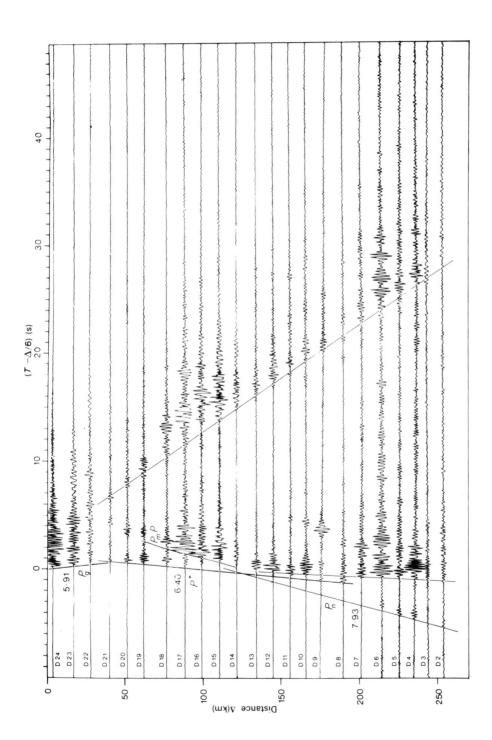

Fig. 2.7 Stacked records of shots fired across the North Scottish shelf along line A (Fig. 2.11a) observed at a seismic recording station at Cape Wrath at the southern end of the line. From SMITH and BOTT (1975). *Geophys. J. R. astr. Soc.,* **40,** 193.

crustal seismic investigation of the structure between Scotland and Iceland carried out in 1972 and called the North Atlantic Seismic Project (NASP). The station was situated at Cape Wrath at the northwestern tip of mainland Scotland and the 200 kg shots were fired at sea along a line across the shelf extending to the Shetland Islands (Fig. 2.11). Over twenty land stations observed the shots so that the time-term method was useful in interpretation. The first arrivals at distances up to about 40 km from Cape Wrath are the upper crustal phase P_g having an estimated true velocity of $6·10 \pm 0·15$ km s^{-1}. This travels in the metamorphic basement rocks of Caledonian and earlier age which occurs beneath a variable thickness of later sediments. The first arrival between about 40 and 120 km range is interpreted as a local P^* phase with an estimated velocity of $6·48 \pm 0·06$ km s^{-1}. This intra-crustal refractor which is widely observed to occur beneath North Scotland is tentatively identified with Scourian type granulites of the Archean Lewisian formation. The first arrival beyond 120 km is the Moho head wave P_n which has an estimated velocity of $7·99 \pm 0·02$ km s^{-1}. The large amplitude late arrival between 80 and 110 km range is the wide-angle Moho reflection $P_m P$ and that beyond 200 km is probably from a lower crustal interface or velocity gradient. Using results obtained at this and other land stations, SMITH and BOTT (1975) interpreted the structure beneath this shot line in terms of a crust 26 ± 2 km thick with a $6·5$ km s^{-1} refractor varying in depth between 2 and 16 km.

The normal incidence reflection method in crustal seismology

Another seismic method of investigating crustal structure is to look for reflections of waves which initially travel downwards in a nearly vertical direction. Such vertical-incidence reflections can only occur where the discontinuity is sharp in relation to the wavelength; in theory, this should make it possible to determine whether the discontinuities within the crust and at its base are sharp or gradational. Vertical-incidence reflections are less easy to recognize than the wide-angle reflections discussed in the last section because of their relatively small amplitude in relation to the background noise produced by other phases. Their detection therefore requires sophisticated field and interpretational techniques, preferably with the digital recording of the data. Such techniques, however, are available in the seismic reflection prospecting method used in particular in exploration for petroleum.

One of the more convincing early applications of the seismic reflection method to crustal seismology was that of DIX (1965), who obtained a cluster of reflections spread out over about 0·8 s from the expected depth of the Moho, in the Mohave desert region of California. This suggested that the Moho here may be spread over about 2 km as a series of discontinuities. Subsequently, substantial contributions have been made by several national groups, particularly by German crustal seismologists who have been using the reflection method since about 1960 and the more recently initiated investigations in the U.S.A. by the Consortium for Continental Reflection Profiling (COCORP).

The earlier German work (e.g. LIEBSCHER, 1964; DOHR and FUCHS, 1967) relied on statistical examination of late reflections obtained in commercial prospecting. Reflections were read from a large number of records in a region and the number of reflecting horizons in a given depth interval were plotted as a frequency histogram. Convincing peaks in the numbers of reflections were found to occur for some, but not all, of the regions studied (Fig. 2.8). The reflections in Fig. 2.8 occurring between 10 and 11 s two-way travel time are interpreted as the Moho and those at shallower depth as the Conrad and Förtsch discontinuities. More recent German work based on digital recording and processing has greatly improved the resolution of the deep reflectors. In general, this work has indicated that the transition from crust to mantle is marked by a zone of reflectors indicating interlayering at the Moho. The German results also display prominent

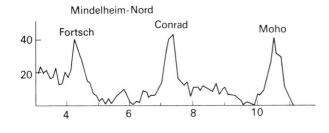

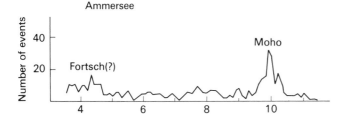

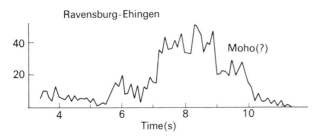

Fig. 2.8 Frequency polygons showing the number of events observed as a function of two-way travel-time in normal-incidence seismic reflection surveys in south Germany. The peaks are interpreted as major discontinuities within the crust (Förtsch and Conrad) and at the base (Moho). Based on LIEBSCHER (1964) and redrawn from JAMES and STEINHART (1966), *The Earth beneath the continents*, p. 314, American Geophysical Union.

occurrence of horizontal reflectors in the lower crust (e.g. GLOCKE and MEISSNER, 1976), which may arise by ductile flow in the lower crust having the effect of streaking out a horizontal layering. Similar results have been obtained by MAIR and LYONS (1976) in the Cordilleran region of Canada. Figure 2.9(a) shows a sample of the results they obtained using the Vibroseis technique (see below), in which both lower crustal horizontal layering and reflections inferred to come from the Moho at about 11 second two-way travel time are well displayed.

Since 1975, the COCORP programme has applied the normal incidence seismic reflection method to a variety of crustal problems in U.S.A. Progress has recently been reviewed by SCHILT and others (1979). This programme makes use of the Vibroseis technique, in which 8–32 Hz mechanical vibrators are used in place of explosives as the seismic source. The results so far obtained emphasize the heterogeneity of the continental crust on a scale of a few kilometres, as revealed by the lateral variability of crustal reflectors and the variable occurrence of diffractions. The Moho is only recognizable in a few locations, where it may be represented either by simple reflectors or more commonly by a zone of complex reflectors indicating interlayering. In contrast to the German results, major intracrustal discontinuities other than fracture planes are conspicuously absent and horizontal layering of the lower crust is not prominent in most of the regions studied.

The most spectacular results from the COCORP programme are the successful tracing of major fault planes within the crust. A good example is the thrust fault which forms the southern boundary of the Wind River uplift in Wyoming (SMITHSON and others, 1979), as shown in Fig. 2.9(b). The reflection profiles show that the thrust plane extends downwards at a dip of about 35° to a depth of at least 25 km, that is nearly to the base of the crust. Gravity evidence, however, suggests that the Moho itself is not displaced by the faulting. The implications are that the Wind River uplift was formed by crustal shortening in response to horizontal compression, and that almost the whole crust here has responded to the stress by brittle fracture. The Moho in this region is not clearly recognizable on the reflection records. Even more striking results have subsequently come from a reflection traverse across the southern Appalachians, where the presence of a sub-horizontal thrust plane of at least 225 km offset and overlain by 6 to 15 km of crystalline rocks has been detected (COOK and others, 1979). Across the Rio Grande rift of New Mexico, COCORP profiles show that the graben boundary faults do not flatten at depth and the presence of a sheet-like magma chamber within the crust has been confirmed. It is to be expected that further important findings will in the future come from COCORP and other crustal reflection programmes.

Structure and depth of the continental Moho

Explosion seismology has been used extensively to investigate the crustal structure beneath U.S.A., Russia, Central Europe and Japan since about 1960, and quite large programmes have been carried out in several other countries. Useful compilations of results include STEINHART and SMITH (1966), HART (1969), MUELLER (1973), GIESE, PRODEHL and STEIN (1976) and HEACOCK (1977). The Deep Seismic Sounding method (DSS) developed by the Russians is described by KOSMINSKAYA and others (1969). This section and those following summarize some of the more prominent conclusions stemming from explosion seismology.

The base of the continental crust can generally be recognized by a substantial velocity discontinuity or steep gradient at the Moho, where compressional velocity increases downwards from typical crustal values of $7 \cdot 5 \ \mathrm{km \, s^{-1}}$ or less to mantle values of about $8 \cdot 0 \ \mathrm{km \, s^{-1}}$. According to the definition of the Moho given by GIESE (1976a), the velocity at the boundary is recognizable using the critical point of the $P_m P$ phase, whether it is a sharp discontinuity or a steep gradient. The depth to the Moho can then be estimated subject to a possible error which may arise from uncertainties in the crustal velocity-depth profile.

The P_n phase penetrates deeper than $P_m P$ and travels in the topmost mantle with a velocity typically ranging between $7 \cdot 8$ and $8 \cdot 3 \ \mathrm{km \, s^{-1}}$. P_n occurs as the first arrival beyond the crossover distance of about 120–200 km distance from the shot depending on crustal thickness. It dies out beyond about 250–350 km where deeper penetrating phases of slightly higher apparent velocity take over as the first arrivals. BAMFORD (1977) has used extensive explosion recordings from western Germany to show that P_n varies with direction by about 7 % beneath this region, being about $8 \cdot 3 \ \mathrm{km \, s^{-1}}$ on average in the direction $20°$ E of north and about $7 \cdot 75 \ \mathrm{km \, s^{-1}}$ in the perpendicular direction. On the other hand, no significant velocity anisotropy appears to be present beneath North Britain (BAMFORD and others, 1978). Where such velocity anisotropy occurs, it can best be explained by preferred alignment of olivine crystals in the topmost mantle (p. 185).

The fall-off of amplitude of P_n with distance is generally less than predicted for a simple head-wave refracted at a sharp boundary between uniform layers. It is better explained by a slight penetration of the rays into a weak positive gradient just below the Moho, or alternatively by supercritical reflection at a small discontinuity a few kilometres below it. The study of the $P_m P$ phase and near-vertical reflections indicates that the Moho in many regions is gradational, possibly partly resulting from interlayering of crust and mantle near their contact. The gradation typically occurs over a depth range of about 2 km. However, beneath some anomalous regions such as the axial region of the Alps and the Sierra Nevada of western U.S.A. where the $P_m P$ phase is not well developed, the gradational transition may range over 10 km (GIESE and PRODEHL, 1976; PRODEHL, 1976).

The continental Moho occurs between depths of about 20 and 80 km. Away from young mountain ranges, the average thickness of the continental crust is about 35–40 km although an exact estimate is difficult to obtain. In Central Europe north of the Alps and the British region the average thickness is about 30 km, but in the United States and most Precambrian shield region it is about 40 km thick. Roots of exceptionally thick crust underlie young folded mountain ranges, reaching 80 km beneath the Himalayas, 70 km beneath the Andes and 55 km beneath the Alps. In contrast, the young plateau uplifted regions of East Africa and western U.S.A are not underlain by anomalously thick crust.

A sufficiently large number of refraction surveys have been done to show that there are regional variations in the depth to the Moho which correlate quite well with the boundaries between structural regions. An important example is the variation in crustal thickness across the western part of U.S.A. shown in section in Fig. 2.10 (PAKISER, 1963). This shows that local thickening of the crust occurs beneath the Sierra Nevada mountains and that each of the main geological provinces is associated with a characteristic crustal thickness as follows:

Coast Range, California	25 km	Colorado province	40 km
Central Valley, California	20 km	Great Plains	45–50 km
Basin and Range province	25–30 km		

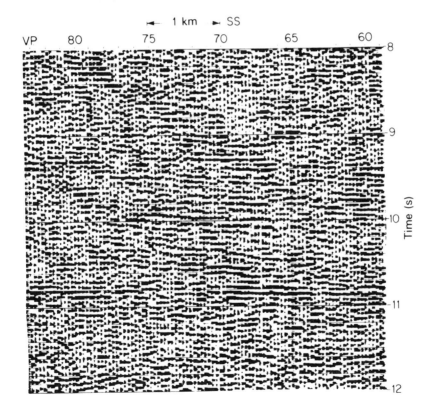

Fig. 2.9 Examples of results obtained in crustal seismic reflection surveys:
(a) Expanded view of a section of Vibroseis data obtained in the Cordilleran region of Canada (at location D_1, Fig. 2.11a), showing horizontal reflections from the lower part of the crust and prominent reflections at about 11 seconds two-way travel-time coming from about the depth of the Moho as observed in refraction survey in the region. After MAIR and LYONS (1976), *Geophysics*, **41**, 1289.

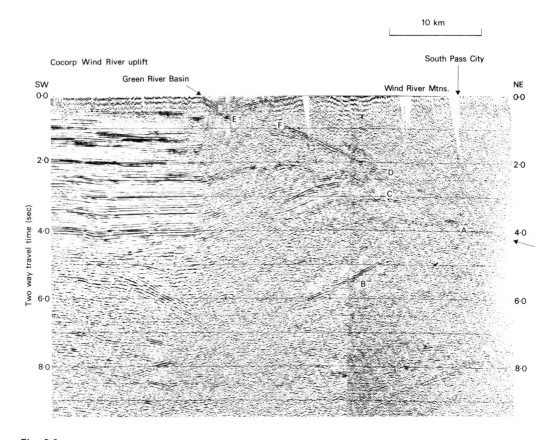

Fig. 2.9
    (b) Close-up view of seismic reflection results across the Wind River thrust (Wyoming, Fig. 2.11 (a), D₂). The thrust plane is marked by arrows. A–deepest faulted sedimentary rocks; B–an anomalous event; C–folded sedimentary rocks; D–thrust plane reflections; E–flattening of thrust near surface. After SMITHSON and others (1979), *J. geophys. Res.*, **84**, 5964.

A second example of local variation of crustal thickness which correlates with geological provinces is found across Scotland. Beneath the foreland to the north-west of the ancient Caledonian mountain range the crust is about 26 km thick, but it thickens to about 35 km beneath the southern part of the Scottish Highlands and the Midland Valley of Scotland as shown in Fig. 2.11 (SMITH and BOTT, 1975; BAMFORD and others, 1978), suggesting the presence of a small relict root to the Caledonian range. Further south, beneath the Hercynian granite terrain of south-west England, the crust is estimated to be about 27 km thick (BOTT and others, 1970).

Seismological structure of the upper crust

Seismic refraction surveys typically yield a velocity of 5·9–6·2 km s⁻¹ for the direct wave P_g travelling in the crystalline basement rocks which form the upper part of the crust. The P_g arrivals may be delayed by up to one or two seconds by thick sediments of low velocity overlying the basement. When examined in detail, the P_g phase may appear to be complex; for instance, GIESE (1976d) described the initial P_g phase in Central Europe as dying off at about 100 km range, beyond which a series of shorter segments of higher apparent velocity of up to 6·4 km s⁻¹ occur out to the crossover distance, each delayed relative to the previous one.

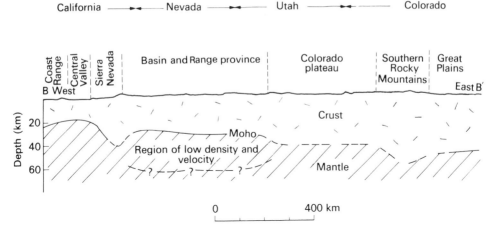

Fig. 2.10 Variations in crustal thickness from San Francisco, California (B) to Lamar, Colorado (B') based on crustal refraction surveys (BB' is shown in Fig. 2.11a). Redrawn from PAKISER (1963), *J. geophys. Res.*, **68**, 5751.

The P_g velocity determined by explosion seismology is significantly higher than the value of $5.6 \, \text{km s}^{-1}$ yielded by the old earthquake studies of crustal structure. This discrepancy between earthquake and explosion studies has been explained in the following ways:

(i) Lack of the exact time and focus of an earthquake may lead to serious errors in the estimate of P_g. For instance, WOOD and RICHTER (1931, 1933) using quarry blasts in California found P_g to be $5.9–6.2 \, \text{km s}^{-1}$ when they knew the time of the blast, but estimated it to be $5.5 \, \text{km s}^{-1}$ when they did not know it.

(ii) Earthquake studies of crustal structure have been concentrated in the active seismic belts where crustal structure may not be typical.

(iii) Earthquake waves may travel in a low velocity channel within the continental crust while the direct wave in explosion studies travels in the higher velocity layer above (GUTENBERG, 1954).

Most seismologists used to favour (i) or possibly (ii) as the explanation of the discrepancy. Gutenberg, however, brought forward further arguments for (iii). He called attention to the horizontally polarized shear waves of 4 second period and $3.5 \, \text{km s}^{-1}$ velocity which are commonly observed in trains of earthquake waves which travel entirely along continental paths, and are known as L_g. These are channel waves propagated in the crust. It is thought that they propagate by total internal reflection at the upper and lower boundaries of the channel. Gutenberg considered that L_g is propagated in a low velocity channel within the middle crust. The more generally favoured interpretation is that the boundaries of the channel are the Earth's free surface above and the Moho below.

MUELLER and LANDISMAN (1966) have revived the idea of a low velocity layer within the upper crust. They based this on their interpretation of a strong amplitude supercritically reflected phase which they name P_c, which occurs in refraction experiments in Germany and elsewhere starting at a distance of about 50–60 km from the shot (this being the emergence of the critical reflection). On the seismograms, P_c follows P_g by about one second. They argued that a discontinuity at a depth of 10 km marking a downward increase in velocity can explain both the reflections at about 4 seconds

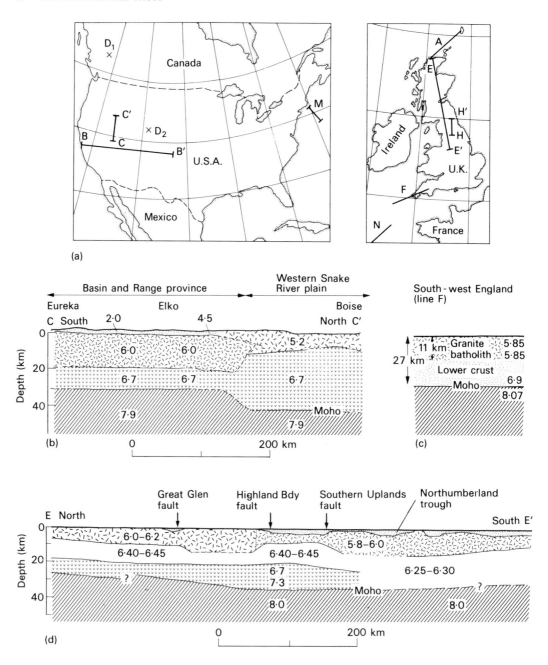

Fig. 2.11 Some different types of crustal "layering". (**a**) Key maps to show locations of profiles illustrated in Figures 2.7, 2.9, 2.10, 2.11, 2.18, 5.2 and 5.3. (**b**) Crustal structure along CC' across the boundary between the Basin and Range province and the western Snake River plain, illustrating a lower crustal layer which changes thickness across a structural boundary. Redrawn from HILL and PAKISER (1966), *The Earth beneath the continents*, p. 410, American Geophysical Union (**c**) A model of crustal structure beneath the granite batholith of south-west England, along the land part of line F, after BOTT and others (1970). (**d**) Crustal structure of northern Britain along line EE', adapted from BAMFORD and others (1978), *Geophys. J. R. astr. Soc.*, **54**, 58.

observed by Liebscher (Fig. 2.8) and the supercritically reflected phase P_c. In order to reconcile both sets of observations, and also the almost constant time difference between P_g and P_c, they considered that the discontinuity must mark the base of a low velocity channel a few kilometres thick. The model of crustal structure incorporating this idea is shown in Fig. 2.12. In it the velocity decreases at a depth of about 6 km and then it abruptly increases again at the Förtsch discontinuity at 10 km depth to a value at least 0.2 km s^{-1} higher than above the channel.

Because of the geological complexity of the upper crust, some caution needs to be exercised in recognition of crustal velocity inversions. However, there is convincing evidence that an upper crustal low velocity zone resembling that of Fig. 2.12 does apparently occur widely beneath the Hercynian terrain of south Germany (GIESE, 1976e). A strongly developed velocity inversion is evidently present beneath the axial region of the Alps between 10 and 30 km depths, with minimum P velocity of between 5.0 and 5.5 km s^{-1} (GIESE and PRODEHL, 1976). In western U.S.A., crustal velocity inversions have been detected beneath the Basin and Range province, the Cascade Range and the Rocky Mountains, but are absent or undetected beneath the Coast Range, the Sierra Nevada and the Colorado Plateau (PRODEHL, 1976). Low velocity zones in the crust appear to be generally absent beneath the Precambrian shield and platform regions and have not been detected beneath Britain. In summary, a low velocity zone in the upper crust is present in some but not all continental regions, apparently being best developed in young orogenic or volcanic regions.

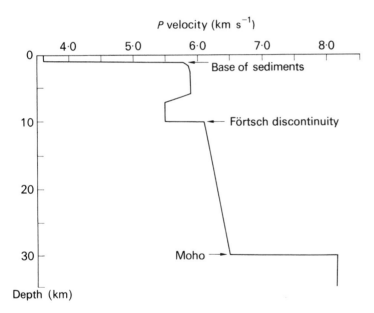

Fig. 2.12 Model of P velocity distribution within the crust, showing the postulated low velocity layer in the upper crust. Redrawn from MUELLER and LANDISMAN (1966), *Geophys. J. R. astr. Soc.*, **10**, 530.

Structure of the lower crust

Present-day ideas on the lower crust and its layering are much less definite than those of thirty years ago. Following Conrad's recognition of P^* and S^*, these phases became widely recognized in near-earthquake studies and they were attributed to the refracted head waves from the Conrad discontinuity which was interpreted as the boundary between the upper and lower crust. As late as 1950–60, earthquake seismologists generally believed that the Conrad discontinuity is universally present in the continental crust (GUTENBERG, 1959; BYERLY, 1956) although a minority of them such as JEFFREYS (1959) thought that the evidence was far from clear-cut.

Explosion seismologists have been much more sceptical about the widespread existence of the Conrad discontinuity. At the extreme, TATEL and TUVE (1955) failed to find any evidence from explosion seismology in widely separated regions of U.S.A. for layering within the crust. However, they did show that an increase in velocity with depth was required to reconcile the travel-times of refracted and critically reflected arrivals from the Moho. Later work has confirmed the general increase in velocity with depth through the continental crust, but it has also shown that layering does occur in many regions. Sub-Conrad discontinuities are found also in some regions.

With present techniques, there are two methods of convincingly recognizing the Conrad discontinuity where it exists. *Firstly*, it may give rise to a refracted first arrival where the lower crustal layer is thick and the velocity contrast is relatively large. *Secondly*, even if there are no first arrivals, the large amplitude phase P_1P may show up the existence of a discontinuity or steep velocity gradient in the middle of the crust. The criteria for recognizing P_1P are (*i*) large amplitude, and (*ii*) the travel-time curve for it becomes parallel to that of P_g at large distances and touches that of P^* at the critical distance. Where there is confusion between P_mP and P_1P phases, synthetic seismograms are of great assistance in convincingly recognizing and studying the phase.

The variability of lower crustal structure is well illustrated by a selection of examples from North America. In eastern Colorado and eastern New Mexico there is evidence for a lower crustal layer about 20 km thick which in some places gives rise to first arrivals. Further west in U.S.A., a distinct lower crustal layer is found beneath the northern part of the Basin and Range province, the Snake River basalt plateau (Fig. 2.11) and parts of the Rocky Mountains, but elsewhere such a layer is not well defined (PRODEHL, 1970). The lower crustal layer is relatively thin beneath the northern Basin and Range province where it was originally recognized from a well developed P_1P phase (HILL and PAKISER, 1966), the velocity increasing to 6·8–7·0 km s^{-1} over a 3–5 km wide transition at about 18–22 km depth. Beneath the Snake River plateau the velocity increases abruptly from about 6·4 to 6·7–6·9 km s^{-1} at about 11–17 km depth giving rise to a first arrival P^* phase. Beneath the Canadian Shield, the Conrad discontinuity occurs on average at about 20 km depth where the velocity rapidly increases to 6·85 km s^{-1} (HALL and HAJNAL, 1973). Beneath the highly anomalous Lake Superior region, 5–10 km of Keeweenawan sediments and volcanic rocks overlie a lower crustal layer which gives first arrivals of 6·8 km s^{-1}; here there is an unusually great variation in the interpreted depths of the Moho ranging from 25 to 60 km.

Turning to Great Britain (Fig. 2.11), a distinct lower crustal layer appears to be better developed beneath and to the north of the Caledonian belt than to the south of it. SMITH and BOTT (1975) detected a 6·5 km s^{-1} layer which locally gives first arrivals at depths of 2 to 16 km beneath the north Scottish shelf mostly underlain by the Caledonian foreland; this belt reaches most close to the surface beneath a belt of positive gravity anomalies west of the Shetland Isles and was interpreted in terms of Archean granulites. BAMFORD and others (1978) have shown that this layer extends south along the LISPB profile shown in Fig. 2.11 beneath most of the Caledonian belt. They also find some evidence for a 6·7 km s^{-1} layer at about 20 km depth beneath the Caledonian belt, with the velocity increasing downwards to about 7·3 km s^{-1} above the Moho. South of the Caledonian belt, the lower part of the upper crust along the LISPB line attains a velocity of 6·3 km s^{-1}, beneath which velocity increases with depth without recognizable discontinuities. Beneath Cardigan Bay, BLUNDELL and PARKS (1969) found evidence for a 7·3 km s^{-1} layer at about 24 km depth. Beneath the granite belt of south-west England, starting at about 10 km depth the velocity increases from about 5·9 to 6·9 km s^{-1} without evidence for any layering of the crust beneath the granite (BOTT and others, 1970).

The Conrad discontinuity is only weakly developed in south Germany and is apparently not present in recent interpretations of the crustal structure beneath the Alps. In contrast, high velocity lower crustal layers are a well-developed feature of the Baltic shield.

To summarize the results of explosion seismology, a series of crustal velocity-depth profiles is shown in Fig. 2.13. These show the great variability in upper and lower crustal structure and in crustal thickness. This variability extends to the mean velocity of the crust, which may vary between extremes of about $6.0\,\mathrm{km\,s^{-1}}$ characteristic of young orogenic regions and about $6.7\,\mathrm{km\,s^{-1}}$ appropriate to Precambrian shield and platform regions.

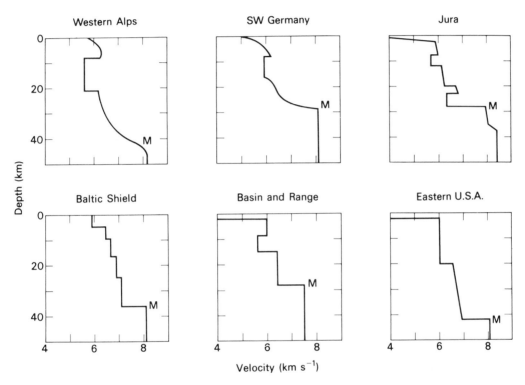

Fig. 2.13   Some postulated crustal velocity-depth distributions. Western Alps, SW Germany and Baltic Shield taken from GIESE and others (1976), Jura taken from MUELLER (1977), and Basin and Range and eastern U.S.A. taken from BRAILLE and SMITH (1975).

2.4 Gravity anomalies and crustal structure

The discovery of isostasy

The original observations which led to the discovery of the principle of isostasy were made between 1735 and 1745 during the measurement of an arc of meridian in Peru by the French geodetic expedition under Bouguer's leadership. They recognized that the Andes would cause a horizontal attraction on the plumbline which would result in local variations in the vertical direction. On investigation, they found that the observed deflection of the vertical was much smaller than the value computed theoretically from the known topography of the Andes. Bouguer originally noted this discrepancy and a few years later Boscovitch postulated attenuation of matter beneath the mountains to explain it. Next century, similar results were found near the Himalayan mountain chain, and it is now known to be a fairly general phenomenon associated with the Earth's major surface features.

For both Andes and Himalaya, the underlying mass deficiency needed to explain the observed deflection of the vertical is approximately equal to the surface load represented by the mountain ranges. The term 'isostasy' was introduced by Dutton in 1889 to explain this phenomenon.

To elaborate a little, the principle of isostasy states that beneath the 'depth of compensation', pressures within the Earth are hydrostatic. This means that the weight of the overlying columns of unit cross-section must all be equal at and below the depth of compensation, allowance being made for a small correction for the Earth's curvature. If there is an excess load on the Earth's surface such as a mountain range or an ocean ridge or an icecap, then if isostatic equilibrium has been reached there must be an equivalent compensating mass deficiency beneath the surface feature but above the depth of compensation; and vice versa for deficient loads such as oceans.

Isostasy is merely the application of Archimedes' principle to the uppermost layers of the Earth. The existence of isostatic movements and of other types of vertical movement affecting the crust shows that lateral flow must be able to occur in the relatively weak region below the depth of compensation, which is commonly called the *asthenosphere*. In contrast, the overlying relatively strong *lithosphere* must reach isostatic equilibrium either by elastic bending or by a combination of fracture and flow.

The two main hypotheses of isostasy were both put forward in 1855. Each of these attempts to explain the shape of the underlying mass distribution which compensates the surface topography. These hypotheses (Fig. 2.14) are as follows:

(i) *Pratt's hypothesis* (1855) assumes that the density within the shell of the Earth above the depth of compensation varies laterally depending on the elevation of the overlying topography. The condition of isostasy requires that

$$\rho(h + D) = \text{constant},$$

where D = depth of compensation, h = height of topography and ρ = the underlying density. A small correction needs to be applied if the Earth's curvature is taken into account. According to Pratt's hypothesis, mountain ranges are underlain by anomalously low density rocks which extend downwards to the depth of compensation; oceans are underlain by relatively high density rocks. The American geodesist Hayford developed this hypothesis in the early part of this century; he arbitrarily took the depth of compensation to be 113·7 km.

(ii) *Airy's hypothesis* (1855) assumes that the uppermost shell of the Earth is a low density 'crust' overlying a higher density substratum. The 'crust' and substratum are each assumed to have uniform density. The relatively rigid 'crust' or lithosphere is assumed to float on the fluid substratum (i.e. the asthenosphere). In the original form of the hypothesis, the base of the low density crust is identical with the boundary between rigid lithosphere and weak asthenosphere, although the more realistic situation where the boundaries are distinct can be incorporated without difficulty. Compensation occurs as a result of variation in thickness of the low density crust; mountain ranges are considered to be underlain by a thicker crust than normal (a *root*) and oceans by a thinner crust than normal (an *antiroot*). The condition for isostasy, neglecting the Earth's curvature, is

$$r = h\rho_c/(\rho_s - \rho_c)$$

where r = depth of root, h = height of topography, ρ_c = density of 'crust' and ρ_s = density of substratum. There are several variations on Airy's hypothesis. For instance, Vening Meinesz put forward the idea of regional compensation, in which the root has a wider lateral extent than the surface feature it is compensating.

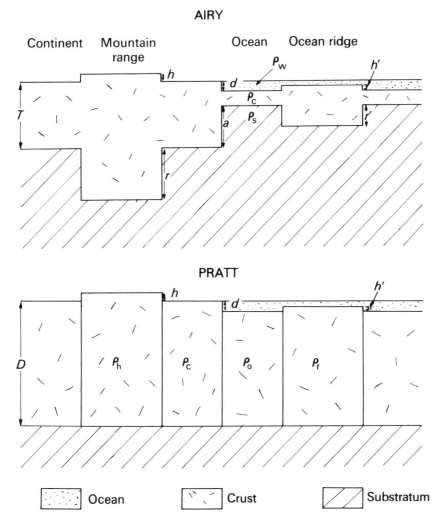

Fig. 2.14 Isostatic compensation according to the Pratt and Airy hypotheses,
where ρ_c = density of crust, ρ_h = density of crust beneath mountain of height h (Pratt),
   ρ_w = density of sea-water, ρ_o = density of crust beneath ocean of depth d (Pratt),
   ρ_s = density of substratum,   and ρ_r = density of crust beneath ocean ridge of height h' (Pratt).

Testing isostasy by gravity measurements

The old method of testing isostasy was to compare the observed deflections of the vertical with the theoretically computed values according to a specified hypothesis of isostasy. During the present century, gravity measurements have been used instead because they can be made much more rapidly and they give the same basic information on the subsurface mass distribution.

There are two main problems. The first is to test to what extent isostatic equilibrium occurs, irrespective of the hypothesis. The second is to attempt to distinguish between the different hypotheses and their versions. Gravity measurements are an effective method of tackling the first problem but have only very limited success in dealing with the second.

Before gravity observations can be interpreted, they need to be corrected for latitude and elevation. The three main types of anomaly used for interpretation are as follows:

$$\text{Bouguer anomaly} = g_{obs} - g_\phi + \text{FAC} - \text{BC} + \text{TC},$$
$$\text{Free air anomaly} = g_{obs} - g_\phi + \text{FAC},$$
$$\text{Isostatic anomaly} = \text{Bouguer anomaly} - \text{computed anomaly of root},$$

where g_{obs} = observed value of gravity at a point on the Earth's surface;

g_ϕ = theoretical gravity on the spheroid at latitude ϕ of the point, as given by the International Gravity Formula;

FAC = the free air correction, allowing for the variation in gravity with height above the spheroid;

BC = the Bouguer correction, which is the attraction of the rock between sea-level and the height of the point, treating it as a slab of uniform thickness;

and TC = the correction for deviations of the topography from a flat plateau.

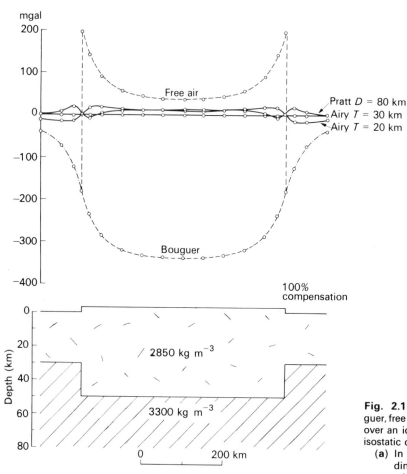

Fig. 2.15 The computed Bouguer, free air and isostatic anomalies over an ideal mountain range with isostatic compensation as follows:
(a) In exact equilibrium according to the Airy hypothesis, with crustal thickness $T = 30$ km;

The Bouguer anomaly shows up the gravitational effect of the lateral variations in density below sea-level. It is the most convenient basis for interpreting local and regional gravity anomalies of the continents and shelf seas in terms of subsurface mass distributions. The free air anomaly is useful for interpreting gravity anomalies of the oceans and across continental margins, and also provides a useful rough test of isostasy.

Figure 2.15 shows how gravity anomalies are used to test isostasy. Figure 2.15(a) shows all three types of anomaly over a mountain range in perfect isostatic equilibrium according to the Airy hypothesis, with a crustal thickness of 30 km. In Fig. 2.15(b) the surface topography is only 75% compensated by the root, and in Fig. 2.15(c) there is no compensation at all.

The most thorough method of testing isostasy over a given surface feature, such as a mountain range, is to compare the observed Bouguer anomaly with the predicted gravitational effect of the compensating mass deficiency according to both Airy and Pratt hypotheses, allowing for different depths of compensation. This is equivalent to computing the isostatic anomaly. Thus in Fig. 2.15(a) the gravity effect of the predicted root for $T = 30$ km is exactly equal to the observed Bouguer

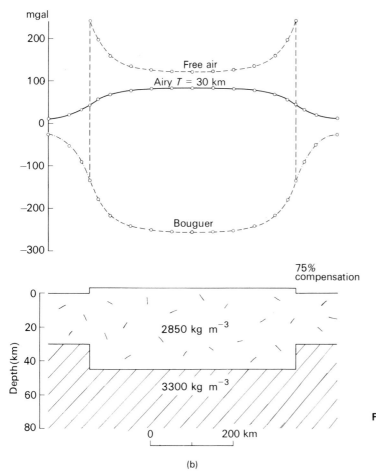

(b)

Fig. 2.15 cont.

(b) 75% compensation according to the Airy hypothesis, the depth extent of the root being 25% less than in (a);

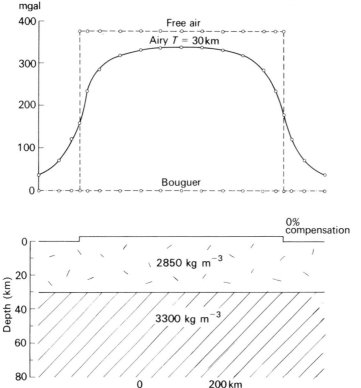

Fig. 2.15 cont.
(c) No isostatic compensation at all.

anomaly, and the corresponding isostatic anomaly is zero everywhere. Ideally this approach should enable us to determine (*i*) whether or not the mountain range is in isostatic equilibrium, and (*ii*) which hypothesis gives the best agreement with the gravity observations. In practice, the isostatic anomalies provide an accurate test of the extent to which isostatic equilibrium occurs; but they do not clearly distinguish between the different hypotheses and depths of compensation. This is because the gravity anomalies lack sensitivity to the exact geometry of the compensating mass deficiency and because of the ubiquitous presence of disturbing anomalies of shallow origin which are typically of larger amplitude than the differences we would be looking for.

A rough, but effective, method of testing isostasy over a topographic feature which is wide in comparison with the depth of compensation is to use the free air anomaly. This would be approximately zero near the centre of the feature if it is in equilibrium. This is because the gravitational effect of a wide root is approximately equal to the Bouguer correction, and thus the free air anomaly will be approximately zero. This method breaks down near the margin of the feature and also in regions of rugged topography, but is reasonably effective provided that the feature is about ten or more times wider than the depth of compensation.

Isostasy and crustal structure

Gravity observations have shown that most of the Earth's major surface features are approximately in isostatic equilibrium, but they cannot unambiguously reveal the form the compensation takes. On the other hand, seismic refraction studies give the crustal structure and thickness but cannot give information on isostatic equilibrium. Put together, these two sets of geophysical observations enable us to infer the form of the compensating masses, thereby giving much more information than the two methods treated separately could do. Because of this, the intense seismic investigations of crustal structure since about 1950 enable a new assessment of the hypotheses of isostasy to be made.

Looked at in the most general way, isostatic compensation occurring near the Earth's surface may occur in one or more of three basic ways. These are: (*i*) by lateral variation in the mean density of the crust; (*ii*) by variation in the thickness of the low density crust; and (*iii*) by lateral variation of density within the upper mantle. (*i*) and (*iii*) could be regarded as modified forms of the Pratt hypothesis, and (*ii*) is the classical form of the Airy hypothesis.

The oceans and continents are in broad isostatic equilibrium with each other. Seismic crustal studies show that this occurs principally according to variations in crustal thickness, although there may also be some lateral variation in the mean density of the crust and upper mantle. Similarly, young fold mountains have been shown to have thickened crust beneath, approximately as predicted by the Airy hypothesis. In contrast, the ocean ridges and some continental uplifted areas such as East Africa and western U.S.A. are compensated by low density rocks within the upper mantle.

The concentrated crustal structure investigations in the western U.S.A. (STEINHART and MEYER, 1961; PAKISER, 1963) are particularly significant in showing the different forms that isostatic compensation can take. They show that crustal thickness is not necessarily related to topographical elevation as the Airy hypothesis predicts. For example the Great Plains are elevated 1 km above sea-level and are underlain by a crust of 45–50 km thickness; but the Basin and Range province with an average elevation of about 2 km has a crust only 25–30 km thick, quite contrary to the predictions of the Airy hypothesis. Pakiser concluded that changes in crustal thickness across the boundaries of the major geological provinces of western U.S.A. bear little direct relation to the change in altitude unless the velocity of P_n remains constant across the boundary. The whole region is in approximate isostatic equilibrium, the compensation between major provinces occurring at least partly in the upper mantle. On the other hand, within the Basin and Range province P_n remains approximately constant and changes in topographical elevation do appear to be related to changes in crustal thickness. Similarly the Sierra Nevada and Rocky Mountains appear to be underlain by local thickened crust suggesting Airy type compensation.

No single universal hypothesis of isostasy can explain all the Earth's major surface features. The form of compensation beneath any given feature needs to be investigated by combined gravity and seismic studies rather than to be assumed. It is quite invalid, for instance, to assume the Airy hypothesis and then use gravity studies alone to determine variations in crustal thickness as has sometimes been done in the past. Knowledge of the form of the compensation is especially important in discussing the cause of vertical movements at the Earth's surface, and the fact that compensation can occur in the mantle as well as the crust adds a further dimension to our understanding of the origin of the Earth's surface features.

Gravity anomalies and the density of the crust

Quite large gravity anomalies of local areal extent are a characteristic feature of the continental crust. Negative anomalies are caused by thick accumulations of low density sediments in basins, and both positive and negative anomalies occur over large igneous intrusions in the basement. The negative anomalies which occur over granite and granodiorite intrusions are important for crustal structure in that they show that the mean density of upper crustal basement rocks is considerably

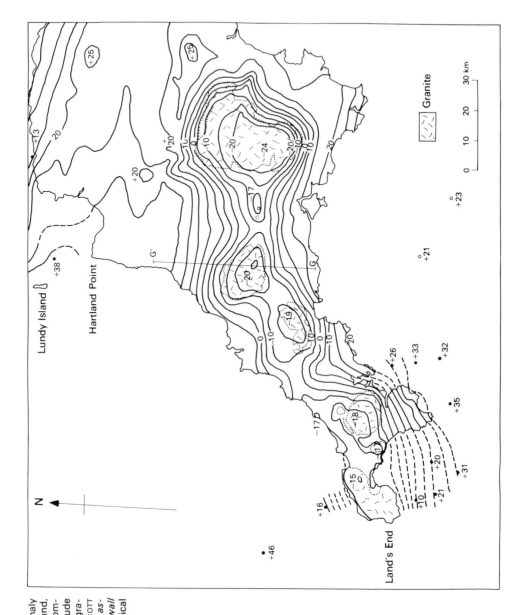

Fig. 2.16 The Bouguer anomaly map of south-western England, showing the belt of negative anomalies of about − 50 mgal amplitude associated with the Armorican granites. Redrawn from BOTT and SCOTT (1964), *Present views of some aspects of the geology of Cornwall and Devon*, p. 28, Royal Geological Society of Cornwall.

higher than that of granite, thereby throwing doubt on the old idea of a granitic layer.

The decrease in Bouguer anomaly over a typical post-tectonic granite batholith is between 15 and 60 mgal (1 mgal = 10^{-5} m s^{-1}). As an example, the Bouguer anomaly map over the granites of south-west England is shown in Fig. 2.16. To explain the steep marginal gradients of this − 50 mgal anomaly, the granite itself must be at least 100 kg m^{-3} lower in density than the intruded basement rocks—in fact the contrast is probably about − 160 kg m^{-3}. An interpretation of the profile across the Bodmin Moor granite is shown in Fig. 2.17, which shows that the granite batholith has outward sloping contacts and must extend to a depth of at least 10 km. The granite itself is known to have a density of about 2600 kg m^{-3}, which means that the basement rocks have an average density of about 2750 kg m^{-3} or thereabouts, and that this density must extend down to 10 km depth which is a third to half of the crustal thickness in this region (p. 44).

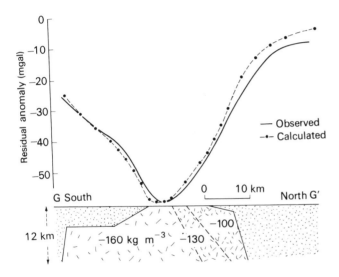

Fig. 2.17 Interpretation of the gravity anomaly profile across the Bodmin Moor granite (south-western England) in terms of sub-surface shape of the low density granite batholith, along line GG′ (Fig. 2.16). Redrawn from BOTT and SCOTT (1964), *Present views of some aspects of the geology of Cornwall and Devon*, p. 31, Royal Geological Society of Cornwall.

The density differential of 100–160 kg m^{-3} between granite and basement rocks is confirmed by surface measurements of rock density. For instance, WOOLLARD (1966) found the average density of 1158 samples of basement rock from North America to be 2742 kg m^{-3}. This is substantially higher than the density of 2670 kg m^{-3} which used to be assigned to the 'granitic layer'. Clearly, the term 'granitic layer' is a misnomer for the basement rocks of the upper crust, which have a density intermediate between that of acid and basic igneous rocks (BOTT, 1961).

As an example, the structure of the upper third of the crust beneath the northern Pennines, England, is shown in Fig. 2.18. This has been deduced from gravity anomalies, and shows two of the common structural features, namely sedimentary basins and granites.

The knowledge of crustal thickness determined by seismic surveys makes it possible to estimate the density difference between crust and topmost mantle. For a long time it was traditionally assumed by isostasists that the contrast is 600 kg m^{-3} (crust = 2670, topmost mantle = 3270). However, WOOLLARD (e.g. 1966) has made a careful study of the variations in crustal thickness of regions in isostatic equilibrium, and he finds that on average a topographical elevation of 1 km corresponds to a root at the base of the crust of 7·5 km. This indicates a density differential of 390 kg m^{-3} or thereabouts between crust and mantle. Woollard also found that the density differential varies from region to region, partly depending on the density of the underlying mantle

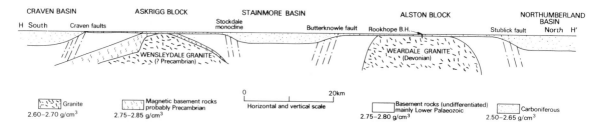

Fig. 2.18 The structure of the uppermost third of the continental crust beneath the northern Pennines, England, along line HH' (Fig. 2.11a) as interpreted from gravity and magnetic anomalies. Redrawn from BOTT (1967a), *Proc. Yorks. geol. Soc.*, **36**, 165.

and partly on the density of the crust. Allowing for an increase in density with depth, Woollard estimated the mean density of the continental crust to be in the range 2870–3300 kg m^{-3}.

2.5 Special regions of the continental crust

Mountain ranges

There are two present-day belts of young fold mountains, namely the circum-Pacific belt and the Alpine–Himalayan belt. They show geological evidence for strong horizontal and vertical movements affecting the rocks involved during the Tertiary. Other belts of the continental crust have been active as young fold mountains during the geological past, but these have been partly or completely obliterated as surface features by erosion and repeated uplift. In some respects, the crustal structure beneath the modern mountain ranges differs from the normal continental crust, as has been revealed by gravity and seismic observations briefly summarized here.

Gravity anomalies show that the present-day mountain belts are in general in approximate isostatic equilibrium. To give an example, Fig. 2.19 shows the gravity anomaly profile across the eastern Alps. Although there are quite large local anomalies disturbing the profile, it is clear that the

Fig. 2.19 The Bouguer, free air and Airy isostatic anomalies across the eastern Alps. The central negative anomaly is caused by a shallow source in the upper crust, probably a granite, and its presence shows how local anomalies make it difficult to use gravity to distinguish between isostatic hypotheses. The topography is too rugged for sensible use of the free air anomaly to test isostasy. Comparison of the isostatic and Bouguer anomalies suggests that the mass deficiency beneath is slightly larger than needed to compensate the mountains (i.e. they are overcompensated), but that equilibrium is at least 90% attained. Redrawn from HOLOPAINEN (1947), *Publs isostatic Inst. int. Ass. Geod.*, No. 16.

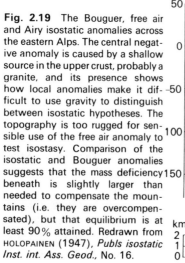
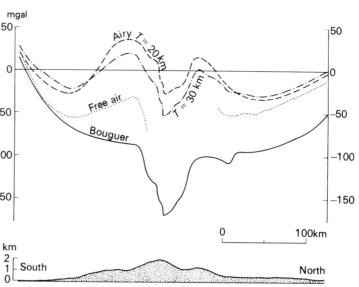

Airy isostatic anomalies depicted are of much smaller amplitude than the negative Bouguer anomaly. This shows that isostatic equilibrium has been reached to an extent of 90–95 % at least. The Airy anomaly for a crustal thickness of 20 km gives the best fit despite later seismic work showing that 30 km would be the better estimate. The reason for this discrepancy is that the sharp negative anomaly near the crest of the Alps is caused by a mass deficiency within the crust, probably a granite (BOTT, 1954), and if this is allowed for it is found that the Airy T = 30 km gives a better fit; this serves as a warning against placing too much confidence in use of isostatic anomalies to estimate crustal thickness. Pratt type isostatic anomalies would equally well account for the observed profile.

The gravity anomalies of the western Alps provide a less clear-cut example (Figs 2.20 and 2.22). These anomalies show that the main Alpine mountains are in isostatic equilibrium, but that the belt of strong positive anomalies to the east (known as the Ivrea zone) and the strong negative anomaly over the Po basin further east represent large deviations from equilibrium.

It has been widely assumed for many years that mountain ranges are compensated by roots of thickened low density crust beneath, according to the Airy hypothesis. But before about 1950 there was scarcely any evidence to confirm this. The isostatic equilibrium could be explained equally well by the Pratt hypothesis. Opinion had been influenced by the surface geological evidence of strong overfolding and thrusting suggesting crustal shortening by compression, but the geological evidence can also be interpreted in terms of vertical movement and gravity tectonics. Seismic evidence on crustal thickness is needed to distinguish between the isostatic hypotheses.

Seismic refraction crustal structure investigations have now been done in some mountain ranges. These generally confirm the presence of roots of thickened crust beneath. The Alps have been studied in much greater detail than other major mountain ranges and investigations up to about 1974 are summarized by GIESE and PRODEHL (1976). The crust is about 30 km thick beneath the northern and western foreland and beneath the Italian plains to the south-east of the Alps. It thickens from both sides to about 55 to 60 km beneath the axis of the Alps (Figs 2.21 and 2.22). The upper crust beneath the Alps is characterized by a strongly developed velocity inversion where P velocities decrease to a minimum of about 5·0 to 5·5 km s^{-1} between 10 and 30 km depths beneath the axial region; this is probably caused by partial fusion or a near approach to it. The Conrad discontinuity is not observed but velocity increases with depth through the lower crust and the Moho is transitional over a depth range of about 10 km in contrast to a fairly sharp Moho beneath the foreland. The splitting of the P_mP group of arrivals into branches indicates the possible presence of narrow but intense velocity inversions in the lower crust just above the Moho. The thickening of the crust beneath the Alps is about what would be expected according to the Airy hypothesis. There is therefore no need for a large mass deficiency in the upper mantle beneath of the sort which occurs beneath continental plateau uplift regions and ocean ridges (Chapter 3), but a slightly anomalous upper mantle cannot be ruled out.

The crust beneath the western Alps is further complicated by presence of an anomalous zone of 7·2 to 7·4 km s^{-1} rocks which reach within a few kilometres of the surface at the Ivrea zone. These high velocity rocks are the source of the high positive gravity anomaly of the Ivrea zone. They are underlain by a strong velocity inversion and dip eastwards connecting to the lower crust and upper mantle beneath the Po plain (Fig. 2.20). The anomalous Ivrea rocks do in fact crop out at the surface, giving us a possible exposure of the crust-mantle boundary.

The crust beneath the central Andes of Peru and Bolivia reaches about 70 km thickness, as determined by explosion seismology (OCOLA and MEYER, 1972) and surface wave dispersion (JAMES, 1971). On the basis of the refraction records, low velocity zones were postulated within the 6·1–6·2 km s^{-1} upper crust and the 6·8–6·9 km s^{-1} lower crust. Russian deep seismic sounding has revealed

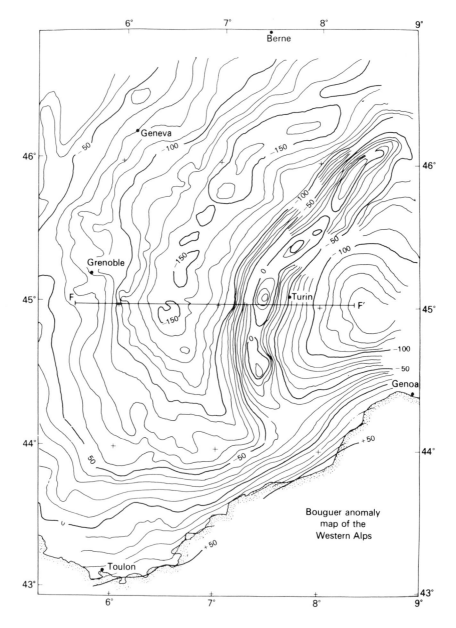

Fig. 2.20 Bouguer anomaly map of the western Alps, with contours at 10 mgal interval. Redrawn from CORON (1963), *Seismologie*, ser. XXII, **2**, 33.

a crust about 65 km thick beneath the Pamirs and 55 km thick beneath the Caucasus (KOSMINSKAYA and others, 1969). A thick crust beneath the Himalayas is also indicated by surface wave dispersion and other earthquake studies. The early refraction investigations in western U.S.A. (PAKISER, 1963)

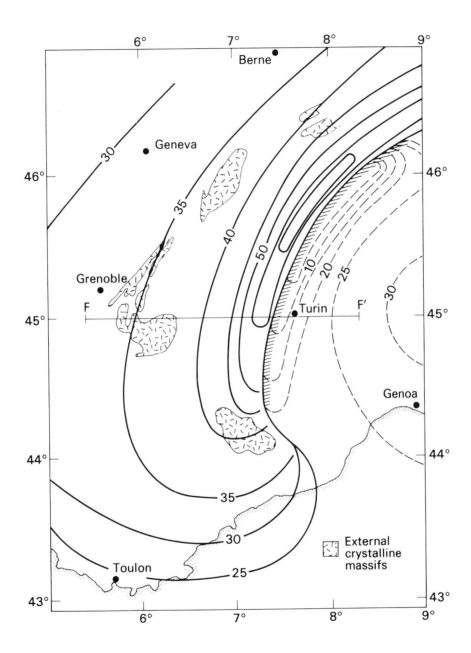

Fig. 2.21 Contour map of the Mohorovičić discontinuity beneath the western Alps, as marked by the steepest velocity gradient within the range 7·5 to 8·2 km s⁻¹. The dashed contours towards the south-east mark the Ivrea surface, the actual Moho being at greater depth here. This map covers an identical area to Fig. 2.20. Adapted from GIESE and PRODEHL (1976), *Explosion seismology in central Europe*, p. 375, Springer-Verlag.

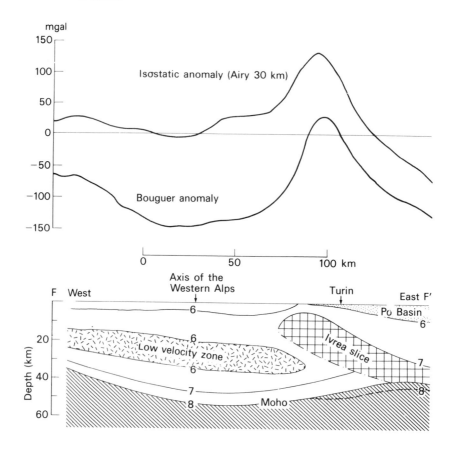

Fig. 2.22 Crustal structure, Bouguer and isostatic anomalies along FF' (Figs. 2.20 and 2.21) across the western Alps and the Ivrea zone. Adapted from GIESE and PRODEHL (1976) and CORON (1963).

show thickened crust beneath the Sierra Nevada and beneath the Rocky Mountains (Fig. 2.10). Thus where seismic surveys have been done, the existence of roots of thickened crust has been established beneath most young mountain ranges. However, much more seismic work on the deep structure of mountain ranges is needed.

The conventional interpretation of the root beneath a mountain range is that it is caused by crustal thickening. The old hypothesis that crustal shortening resulted from a contracting Earth is no longer tenable, because the amount of shortening which could be produced in this way would be quite inadequate. It is now apparent, within the plate tectonic framework (p. 130), that the crustal thickening characteristic of young continental mountain ranges can originate in two different ways. In Alpine–Himalayan mountain ranges it occurs by crustal shortening as a result of continent–continent collision. In Andean type mountain ranges it probably occurs without significant crustal shortening as a result of addition of basaltic and andesitic igneous material to the crust, this having been produced from the underlying mantle (p. 221). Turning to the ancient Appalachian mountain range, the COCORP seismic reflection results demonstrating the presence of a major sub-horizontal thrust plane underlying the fold belt at about 10 km depth (COOK and others, 1979, see also p. 42) clearly indicates that major crustal shortening occurred during the orogeny.

Rift valley systems and associated uparching of the crust

The East African system of interrelated rift valleys stretches from Zimbabwe to the Gulf of Aden with eastern and western branches encircling Lake Victoria. It extends further northwards beyond the Red Sea into the Dead Sea rift system, attaining a total length of over 6000 km. Another linked series of rift depressions extends north-eastwards and also westwards from Lake Baikal in south-central Siberia and is over 2000 km in length. These two rift systems, which have been active during the Tertiary, form major linear tectonic features of the continental crust. The East African rift system also appears to form a continental extension of the ocean ridge system. Other less extensive rift systems of Tertiary to Recent age include the Rhine graben and the Rio Grande rift belt of Colorado and New Mexico. The Basin and Range province of western U.S.A., consisting of sub-parallel north-south aligned graben and horsts occurring within a region of plateau uplift, is probably of similar origin. The Oslo graben of Permian age is a well-known example of an ancient rift system.

Continental rift systems typically occur in regions which have undergone broad uparching of the crust which is unrelated to folding. For instance, the East African rift region stands about 2 km above sea level and the Basin and Range province is at similar elevation. In addition to the broad uparching, narrow rim uplifts may border the rift depressions on either side as a result of the fault movements. Rift regions are normally affected by basaltic volcanism which may be of alkaline character; for instance, the Oslo graben is noted for the intense alkaline volcanism that occurred at the time of its formation. The main problems of rift systems and their associated features concern the nature and origin of the faulting, the cause of the uparching, and the relationship between faulting, uparching and volcanism.

It has now been shown beyond reasonable doubt that the faults bounding the rift depressions are of normal type. This has been demonstrated by geological observations of the faults, by gravity profiling across them (e.g. GIRDLER, 1964), and most recently by the COCORP seismic reflection profiling across the Rio Grande rift (BROWN and others, 1980). Furthermore, the normal faults bounding the Rio Grande rift do not flatten at depth and are therefore not of listric type. The demonstration of normal faulting indicates that rift systems form in response to horizontal deviatoric tension in the crust with maximum tension occurring perpendicular to the fault lines (p. 303). This is confirmed by earthquake mechanism studies.

Gravity surveys show that the uparched regions are in isostatic equilibrium but that negative isostatic anomalies occur locally over the sediment-filled rift troughs. For instance, BULLARD (1936) found that the East African plateau is in approximate isostatic equilibrium, but that some individual rift valleys show negative anomalies reaching down to about -80 mgal. Similarly, there is a large local negative isostatic anomaly across the Lake Baikal rift depression. The negative anomalies are substantially caused by the low density sedimentary infill, which may reach up to 5–6 km in thickness.

More recent seismological and gravity investigations have now added greatly to knowledge of the deep structure beneath the Gregory rift of Kenya (Fig. 2.23). Near-earthquake arrivals recorded at the Kaptagat seismological array station, situated about 10 km west of the rift zone at $0.5°$ N, show that the rift system is located within shield-type continental crust about 43 km thick, which is subdivisible into upper and lower crustal layers with estimated P velocities of 5·9 and 6·5 km s^{-1} respectively (MAGUIRE and LONG, 1976). This crust is underlain by a normal sub-Moho velocity of about 8·0 km s^{-1} which reaches to the western margin of the rift zone. A strongly contrasting type of structure underlies the Gregory rift itself. A gravity survey indicates the presence of a high density igneous body about 20 km wide penetrating the upper crust along the rift zone (SEARLE,

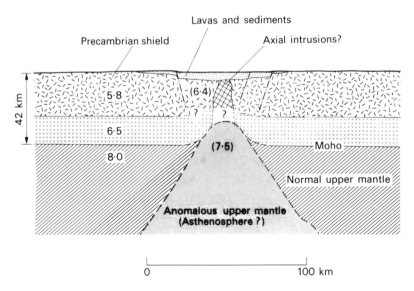

Fig. 2.23   A schematic cross section of the structure of the crust and topmost mantle beneath the Gregory rift, Kenya, East Africa, showing the rift valley, its axial intrusion, crustal attenuation and anomalous upper mantle (thinning of the lithosphere?). Seismic velocities are shown in $km s^{-1}$.

1970), in addition to low density sediments. A refraction line reveals $7 \cdot 5\ km\ s^{-1}$ material at about 19 km depth beneath a $6 \cdot 4\ km\ s^{-1}$ crust (GRIFFITHS and others, 1971). The $7 \cdot 5\ km\ s^{-1}$ layer has been interpreted as the top of an anomalously low velocity upper mantle directly underlying the thin crust of the rift zone. This anomalous upper mantle has been shown from the arrival pattern of teleseismic P waves at Kaptagat to widen downwards into a broadly ellipsoidal structure at 150 km depth which appears to underlie the Kenya domal uparching (LONG and BACKHOUSE, 1976). This anomalous upper mantle has been interpreted as an upward penetration of the hot and low density asthenosphere into the lithosphere (GIRDLER and others, 1969).

As the Kenya dome is in approximate isostatic equilibrium and the crust outside the rift zone has apparently not been modified or thickened, the Tertiary uparching must be supported by low density rocks within the underlying upper mantle, such as also occur beneath the plateau uplift region of western U.S.A. This low density region is probably broadly correlatable with the low velocity upper mantle, both features resulting from raised temperatures which cause thinning of the lithosphere.

The crustal and upper mantle structure observed beneath other present-day rift systems is similar to that beneath the Gregory rift except that the crustal thinning and igneous activity are generally less intense. The Baikal rift system lies in the north-western part of an extensive uplifted region underlain by a low velocity mantle, but the crust is only slightly thinner than beneath the adjacent Siberian platform (PUZYREV and others, 1978). The crust beneath the Rhine graben is similar in character but slightly thinner than that beneath the adjacent uparched region and the Moho appears to be more gradational (PRODEHL and others, 1976); the existence of a $7 \cdot 6\ km\ s^{-1}$ cushion near the base of the crust has now been discounted. Gravity surveys indicate that a thinned crust and anomalously dense lower crust probably underlie the ancient Oslo graben of Permian age (RAMBERG, 1976).

Turning to the origin of the rift depressions, the main problem has been to explain the mechanism of subsidence in view of the large local deviation from isostatic equilibrium. The depressions appear to be held down despite the buoyancy indicated by the negative isostatic anomalies. On the older compression hypothesis which has now been abandoned, it was assumed that horizontal compression would hold down the rift block bounded by reverse faults against the isostatic buoyancy. An explanation in terms of the tension hypothesis has been developed by Vening Meinesz (HEISKANEN and VENING MEINESZ, 1958). Referring to Fig. 2.24, the stages in development are as follows:

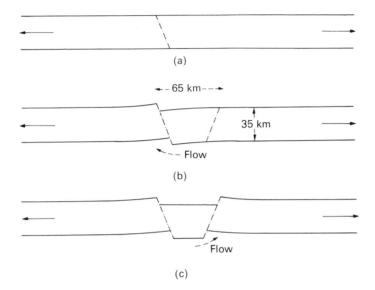

Fig. 2.24 Vening Meinesz' hypothesis for the formation of a rift valley by tension and normal faulting. Adapted from HEISKANEN and VENING MEINESZ (1958), *The Earth and its gravity field*, p. 390, McGraw-Hill.

(*i*)   Under conditions of crustal tension, a normal fault is produced approximately perpendicular to the maximum tension (Fig. 2.24(a));

(*ii*) As a result of the movement on the fault, the crust is warped and the bending sets up further stresses. Maximum tension is produced at the surface on the downthrown side where the curvature is greatest;

(*iii*) A second normal fault develops at this position (Fig. 2.24(b)), which according to the theory of bending of thin sheets would occur about 65 km from the first fault if the elastic crust is 35 km thick. If the two normal faults dip towards each other, a rift valley is formed;

(*iv*) Once the two fault planes have been established, persistent tension will cause repeated subsidence of the rift wedge. At the same time, rim uplifts will be formed by upbending of the adjacent elastic crust.

The above mechanism does not violate the principle of isostasy, because a wedge of crust narrowing downwards floats in a fluid at a lower level than a rectangular block would do. Vening Meinesz showed that a subsidence of several kilometres can occur by this mechanism, provided that the subsidence is aided by a load of infilling sediments.

The recent observations on the deep structure beneath rift systems show that the original Vening Meinesz model of rifting by subsidence of a crustal wedge is an oversimplification, in that the crust is characteristically thinned beneath rift zones rather than protruding into the mantle at the bottom of the subsiding wedge. Nevertheless, the wedge subsidence hypothesis is still a valid explanation of graben formation as it can readily be applied to the brittle upper part of the crust rather than to the

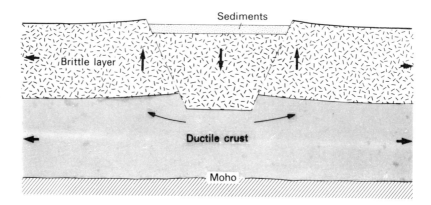

Fig. 2.25 Graben formation by subsidence of a downward narrowing wedge of the brittle upper crust, with complementary elastic upbending of the adjacent regions. Adapted from BOTT (1981), *Tectonophysics*, **73**, 3.

crust as a whole (Fig. 2.25). This can apply because the lower part of the crust in rift regions is probably hot enough to deform by ductile flow rather than fracture, as suggested by ARTEMJEV and ARTYUSHKOV (1971). The energy transactions associated with the subsidence of an upper crustal downward-narrowing wedge have been studied by BOTT (1976), who showed that the theoretical amount of subsidence possible is greater for narrower graben than for wider ones. The subsidence is increased by a factor of two to three as a result of sediment loading, depending on mean sediment density. The amount of subsidence possible also increases almost linearly with the applied deviatoric tensile stress, with stresses of the order of 200 MPa being required to cause subsidence of 5 km in a sediment loaded graben about 40 km wide. It is important to emphasize that substantial deviatoric tensile stresses are needed to cause graben subsidence of the observed amplitude.

The uparching of the continental crust in regions of doming or plateau uplift is the isostatic response to the development of a region of low density rocks beneath. Such uplift occurs without folding or crustal thickening, and it affects regions which were previously at relatively low elevation. The best explanation is that temperature is substantially raised in the underlying upper mantle and that the uplift occurs as a result of thermal expansion. The occurrence of high temperatures in the upper mantle beneath such regions is indicated by the associated volcanism which requires ongoing partial fusion. Suppose that the average temperature down to 100 km depth is raised by 500 K and that lateral constraints restrict the thermal expansion to the vertical direction. Taking the coefficient of thermal expansion to be $3 \times 10^{-5} \, K^{-1}$, the resulting isostatic uplift of the surface would be 1·5 km and this may be further increased if exothermic phase transitions occur in the hot upper mantle. Thus broad uparching of the observed type can readily be explained by thermal expansion.

The development of a hot mantle region below the continental lithosphere appears to be a necessarily preliminary stage to continental uparching (Fig. 2.26). Such a hot spot in the upper mantle is probably caused by convective upwelling from the deeper part of the mantle, during which partial fusion is likely to occur as pressure is reduced in the rising material. If the continental lithosphere above such a hot spot is heated and thinned just by thermal conduction from below, then the time scale of uplift becomes unrealistically long, of the order of 50 My. However, the lithosphere can be net veined by rising magma, and blocks can thus break loose, subsiding to be replaced by hot asthenospheric material upwelling from below. In this way, the continental

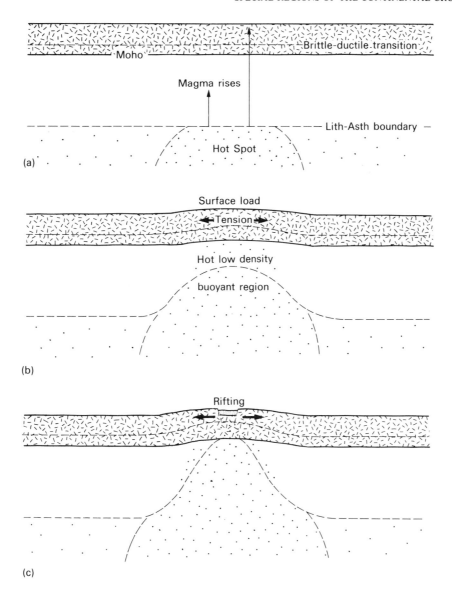

Fig. 2.26 Stages in the development of an uparched and rifted structure:
  (a) Hot spot forms below the continental lithosphere by upwelling from the deeper parts of the mantle;
  (b) The continental lithosphere becomes heated and thinned, with consequent isostatic uplift and development of tensile stress system in the upper crust;
  (c) Graben formation starts when the tensile stresses become sufficiently large.
From BOTT (1981), *Tectonophysics*, **73**, 7.

lithosphere may become thinned on a relatively short time scale, and uplift can occur shortly after the onset of volcanism.

The common association of rifting and uparching suggests that there may be a genetic association between them. One possibility is that doming occurs first and that rifting follows as a consequence of stresses developed by the uplifted structure. Alternatively, cracking of the lithosphere may occur first, causing upwelling of mantle material beneath with consequent raising of temperatures and isostatic uplift (e.g. OXBURGH and TURCOTTE, 1974). Geological evidence from, for instance, the Rhinegraben (ILLIES, 1977), the Baikal region (KISELEV and others, 1978) and Ethiopia (DAVIDSON and REX, 1980) shows that the earliest volcanism precedes doming and that doming and plateau uplift precede rifting. This suggests that the rifting may be a secondary consequence of the doming (Fig. 2.26).

What, then, is the origin of the tensile stress which causes the rifting? If it is caused by drag of underlying convection currents, or by membrane stresses (p. 330) as suggested by OXBURGH and TURCOTTE (1974), then the observed time sequence of uparching followed by rifting would be difficult to understand. The time sequence is better explained if the uparching itself causes the tension. The most obvious candidate is the bending stress in the elastic part of the lithosphere caused by the uparching, but it can be shown that the associated strain is much too small to account for the graben subsidence. There is, however, another type of stress system which must be associated with uplifted continental regions which are in isostatic equilibrium. This is caused by the surface loading of the uplifted topography and by the upthrust of the low density compensating region in the upper mantle beneath. BOTT and KUSZNIR (1979) showed that if the region below the uppermost 10–20 km of the crust deforms by visco-elastic flow, then tensile stresses of the order of 200 MPa* can develop in the uppermost elastic part of the crust by this mechanism (Fig. 8.21). Such stresses are adequate to account for the observed rifting and graben formation in regions such as East Africa and the Basin and Range province. A hypothesis for the origin of rift systems based on these ideas is shown in Fig. 2.26.

2.6 Interpreting the continental crust

Principles of interpretation

The observed pattern of elastic wave velocities, incomplete as it is, provides the main basis for interpreting the chemistry and mineralogy of the crust beneath depths penetrable by boreholes. The link between the observed velocity structure and its interpretation depends on experimental measurements of the physical properties of rock types over the appropriate range of temperature and pressure. At best, we can only hope to obtain a broad picture of the variation of composition with depth.

Both pressure and temperature increase with depth through the crust. The confining pressure is mainly caused by the weight of the overburden, which at depth d is $\rho g d$ where ρ is the mean density of the overlying rocks and g is gravity. In practice the confining pressure at a point may be increased or decreased slightly by tectonic overpressure. Assuming the crustal density to be 2900 kg m^{-3}, the pressure increases by 30 MPa km^{-1}* and at typical Moho depth of 35 km it is 1000 MPa. Anticipating the discussion in Chapter 7, the average temperature gradient in basement rocks is 25 K km^{-1}, but this falls off with depth to about half its surface value at the Moho because of radioactive heat sources within the crust. The average temperature at the Moho is probably about 700 to 800 K, but significantly higher temperatures occur beneath hot regions such as the Basin and Range province of western U.S.A. and significantly lower temperatures occur at the Moho beneath shield regions. A slightly above average continental geothermal gradient is shown in Fig. 2.27.

It is convenient to relate the composition of rocks at different depths in the crust (and upper mantle) to igneous rocks and their high pressure forms. The average compositions of the main groups of igneous rocks and of eclogite, which is the high pressure form of gabbro and basalt, are

---

* 1 Megapascal (MPa) = 10^6 N m^{-2} = 10 bar = 10^7 dyne cm^{-2} = 9·869 atmosphere.

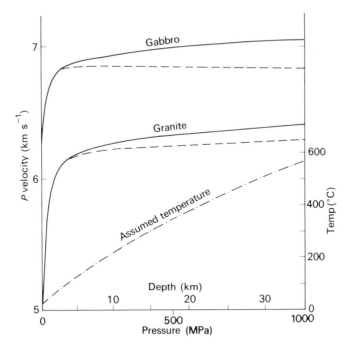

Fig. 2.27 The compressional wave velocity (P) in granite and gabbro as a function of (i) pressure alone (solid line), and (ii) pressure and temperature (dashed line), assuming geothermal gradient as shown. The graphs are based on the experimental results of Birch and his co-workers. Redrawn from PRESS in CLARK (1966), *Handbook of physical constants*, revised edition, p. 208, Geological Society of America.

Table 2.1 Average composition of igneous rock groups and eclogites. The average compositions of the igneous rock groups were taken from NOCKOLDS (1954), and of eclogites from LAPADU-HARGUES (1953). The number of samples in each group is shown at the head of the column.

	Alkali granite (48)	Granodiorite (137)	Intermediate igneous rock[†] (635)	Gabbro (160)	Peridotite (23)	Eclogite (water free) (34)
SiO_2	73·9	66·9	54·6	48·4	43·5	49·0
TiO_2	0·2	0·6	1·5	1·3	0·8	—
Al_2O_3	13·8	15·7	16·4	16·8	4·0	14·5
Fe_2O_3	0·8	1·3	3·3	2·6	2·5	3·8
FeO	1·1	2·6	5·2	7·9	9·8	9·1
MnO	0·1	0·1	0·2	0·2	0·2	—
MgO	0·3	1·6	3·8	8·1	34·0	8·9
CaO	0·7	3·6	6·5	11·1	3·5	11·5
Na_2O	3·5	3·8	4·2	2·3	0·6	2·5
K_2O	5·1	3·1	3·2	0·6	0·3	0·7
H_2O	0·5	0·7	0·7	0·6	0·8	—
P_2O_5	0·1	0·2	0·4	0·2	0·1	—

† excluding nepheline types.

shown in Table 2.1. The densities of coarse-grained igneous rocks are shown in Tables 2.2 and 2.3. It should be pointed out that the problem of adequate random sampling introduces errors into the quoted average compositions and densities. In particular, later work has shown that Daly's values for the densities of granite and granodiorite are probably too high.

The *P* velocities in a wide range of rocks up to pressures of 1000 MPa have been determined by BIRCH (1960, 1961a) and his co-workers (Table 2.3 and Fig. 2.27). The velocity increases strongly with pressure up to about 50–100 MPa as the pores collapse. However, the effect of increasing temperature is to cause the velocity to decrease. The computed variation of velocity of granite and gabbro with depth, as determined by BIRCH (1958) using a realistic geothermal gradient, is shown in Fig. 2.27. Beneath about 5 km depth, the effects of temperature and pressure tend to cancel each other out. Thus the *P* velocity in each rock type tends to remain approximately constant with depth, and it may even decrease if the geothermal gradient is locally high.

NAFE and DRAKE (1963) investigated the relationship between *P* velocity and density by plotting these against each other for water-saturated sediments and sedimentary rocks (Fig. 2.28). The observations cluster near a curve and measurements on igneous and metamorphic rocks extend this

Table 2.2 Average densities of coarse-grained igneous rocks and eclogites. These average density estimates have been taken from tables in the *Handbook of physical constants* (CLARK, 1966) and are based on compilations made by R. A. Daly, Francis Birch and S. P. Clark.

	Number of samples	Mean density $(kg\ m^{-3})$	Range $(kg\ m^{-3})$
Granite	155	2670	2520–2810
Granodiorite	11	2720	2670–2790
Syenite	24	2760	2630–2900
Quartz diorite	21	2810	2680–2960
Diorite	13	2840	2720–2960
Gabbro and norite	38	2980	2720–3120
Peridotite	3	3230	3150–3280
Dunite	15	3280	3200–3310
Eclogite	10	3390	3340–3450

Table 2.3 Compressional wave velocities in rocks. Taken from a table compiled by Frank Press in the *Handbook of physical constants* (CLARK, 1966) and mainly based on measurements made by Francis Birch and Gene Simmons.

	Number of samples	Mean density $(kg\ m^{-3})$	Mean velocity $(km\ s^{-1})$ 100 MPa	Mean velocity $(km\ s^{-1})$ 1000 MPa
Granite	10	2643	6·13	6·45
Granodiorite	mean of locality	2705	6·27	6·56
Quartz diorite	2	2852	6·44	6·71
Gabbro and norite	3	2988	7·02	7·24
Dunite	5	3277	7·87	8·15
Eclogite	4	3383	7·52	7·87
Greywacke	2	2692	5·84	6·20
Slate	1	2734	5·79	6·22
Amphibolite	1	3120	7·17	7·35

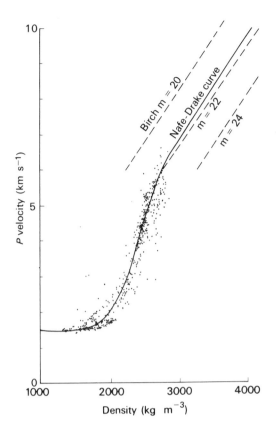

Fig. 2.28 Plot of P velocity against density observed for a wide selection of water-saturated sediments and sedimentary rocks, extended for hard rocks. BIRCH's (1961a) empirical relationship between density and P velocity as a function of mean atomic weight m is also shown on the diagram. Partly redrawn from NAFE and DRAKE (1963), *The sea*, Vol. 3, p. 807, Interscience Publishers.

curve without offset. The Nafe-Drake curve enables us to estimate density of shallow rocks from the P velocity subject to an error of about $100 \, \mathrm{kg \, m^{-3}}$, and it has been widely used in the joint interpretation of gravity and seismic refraction surveys.

BIRCH (1961a) went one stage further by determining an empirical relationship between density, P velocity and chemical composition as represented by mean atomic weight. He used laboratory observations of P velocity at $1000 \, \mathrm{MPa}$ to ensure that the pores had collapsed so that the relationship would be valid beneath a few kilometres depth. He found an approximately linear relationship between velocity V_p and density ρ for a mean atomic weight m as follows (Fig. 2.28):

$$\rho = a(m) + bV_p$$

where $a(m)$ is a constant depending on mean atomic weight and b is another constant. This linear relationship applies approximately to rocks of differing composition and to phase transformations such as that of gabbro to eclogite. Birch speculated that it would also apply to simple compression. The constant b was determined as 3.05 and the constant $a(m)$ for $m = 21$ was found to be -1.87.

ANDERSON (1967a) and several later workers have expressed Birch's law in the form

$$V_p = A(m)\rho^\lambda$$

where the constant $A(m)$ is a function of mean atomic weight and the experimentally determined value of λ is about 1.5. In practice, the linear and power law formulations are almost equivalent when applied over a limited range of variation but the power law form has some theoretical basis when applied to solids undergoing compression. A similar relationship, expressable in linear or power law forms, is observed to apply approximately to the bulk sound velocity $V_\phi = \sqrt{k/\rho}$ (WANG, 1968). Birch's law and its extensions are useful in interpreting seismic velocity variations within the crust and deeper levels of the Earth, but they need to be treated with some caution as they are empirical rather than theoretical relationships. LIEBERMANN and RINGWOOD (1973) found experimentally that the velocity-density relationships across phase transitions can deviate widely from the predictions of Birch's law. Furthermore, the law fails at or near partial melting when the decrease in seismic velocity is much greater than that predicted from the change in density.

In interpreting variations of elastic wave velocities within the Earth, the main problem is to sort out whether they are caused by (i) the normal effect of temperature and pressure gradients on rock of uniform chemical and mineralogical composition, or (ii) change in chemical composition, or (iii) phase changes affecting the mineralogy but not the chemical composition. Certain guidelines can help the interpretation. (i) First order discontinuities, which should be recognizable on ability to give near-normal incidence reflections, are almost certainly caused by change in chemical composition. This is because phase changes affecting multi-component systems involving solid solution, as most rocks are, would be expected to be spread over a discrete interval of depth. (ii) The results of experimental petrology can give information on the mineral assemblages stable at given temperature, pressure and water vapour pressure. (iii) The empirical relations above make possible the prediction of density change accompanying a given change in seismic velocity for the different possible interpretations. (iv) Knowledge of Poisson's ratio determined jointly from P and S velocities (p. 38) and the occurrence of seismic anisotropy (p. 42) may place constraints on mineralogical composition. (v) Gross earth data, such as its mean density or moment of inertia or free oscillation periods, may place further constraints on allowable density distributions. This last guideline is more applicable to mantle and core than to the crust.

Composition of the upper crust

Table 2.4 shows estimates of the mean composition of (i) the crystalline basement rocks of a continental shield, (ii) the rocks forming young mountain ranges, and (iii) acidic and basic igneous rocks. The table shows that the mean composition of the topmost part of the continental crust is between that of acidic and basic igneous rocks, and more specifically it is between that of granodiorite and quartz diorite.

It was shown above (p. 57) that the crystalline basement rocks are on average about 100–150 $kg\,m^{-3}$ denser than granite, and that their mean density is about 2750–2800 $kg\,m^{-3}$, with variation from region to region. This density is what would be expected for a rock with average composition lying between granodiorite and diorite.

The observed P velocity of the upper crust (5·9–6·3 $km\,s^{-1}$) agrees with the experimental determinations on granites at 100 MPa confining pressure, and taken at its face value this suggests a granitic composition for the upper crust. This is in marked disagreement with the above-mentioned estimates of chemical composition and density which suggest a more basic composition than granite. The reason for the lower than expected P velocity is probably partly that slates and greywackes are abundant in the basement; these rocks possess densities which are typically higher than 2700 $kg\,m^{-3}$ but they have distinctly low P velocities.

A further problem relating to the upper crust is the interpretation of the postulated low velocity zone between 8 and 11 km depth or thereabouts, if it exists. Mueller and Landisman suggested that the decrease in velocity with depth is caused by the effect of the steep near-surface temperature gradient outweighing the effect of pressure below a depth of 5 km. The inferred discontinuous increase of velocity at the base of the low velocity channel is interpreted as the boundary between

Table 2.4 Estimates of the composition of parts of the topmost crust compared with acid and basic igneous rocks. (1) and (3)–(4) are taken from POLDERVAART (1955); (1) gives an average for the Canadian Shield estimated by Grout and (3) is an estimate of the mean composition of the upper crust in Norway based on the sampling of glacial clays made by Goldschmidt; (5) and (6) are taken from NOCKOLDS (1954). (2) is taken from EADE and others (1966) and is based on direct sampling of 200 000 sq. miles of the New Quebec area. H_2O and CO_2 have been excluded from the figures given.

	Canadian Shield		Glacial clays from Norway	Average sediment from young fold mountains	Average silicic igneous rock	Average mafic igneous rock
	(1)	(2)	(3)	(4)	(5)	(6)
SiO_2	63·9	65·8	62·1	58·6	69·2	48·6
TiO_2	0·8	0·5	0·8	0·6	0·5	1·8
Al_2O_3	17·0	16·4	16·6	12·9	14·7	15·7
Fe_2O_3	2·4	1·5	7·3	2·9	1·7	2·8
FeO	3·0	3·0	—	2·2	2·2	8·1
MnO	—	—	—	—	0·1	0·2
MgO	1·8	2·3	3·5	4·3	1·1	8·7
CaO	4·1	3·4	3·2	14·2	2·6	10·8
Na_2O	3·7	4·1	2·2	1·4	3·9	2·3
K_2O	3·1	2·9	4·1	2·7	3·8	0·7
P_2O_5	—	—	0·2	0·1	0·2	0·3

'granitic' rocks above and dioritic rocks below, this interface being known to German seismologists as the Förtsch discontinuity. Alternatively, the low velocity zone can be interpreted as a compositional change to more granitic rocks at depth (BOTT, 1961). This latter idea is incorporated into the crustal model of SMITHSON and DECKER (1974) who suggest that the upper crust is divisible into (*i*) a surface zone of intermediate metamorphic rocks containing granitic intrusions grading down into (*ii*) a more felsic migmatite zone.

Composition of the lower crust

The lower crustal layer, where it can be distinguished from the upper crust, is recognized by a *P* velocity of greater than about 6·5 km s^{-1} but less than 7·6 km s^{-1}. We seek possible mineralogical and chemical explanations of this highly variable and little-known part of the continental crust.

Simple increase in velocity with confining pressure cannot explain velocities as high as 6·7 km s^{-1} in the lower crust. *Either* the chemical composition is more basic than that of the upper crust, *or* the stable mineral assemblage differs from that of the upper crust.

It used to be thought that the lower crust is formed of basalt or gabbro. The observed *P** velocity of 6.7 km s^{-1} matches the experimentally determined velocity of gabbro. It was also assumed that basalt magma was formed by local fusion of the layer. It is now known that the upper mantle is the main source of basalt magma, and therefore the petrological justification for a crustal basaltic layer no longer exists.

Important evidence on the mineralogical and chemical composition of the lower crust has been provided by an investigation of the stable mineral assemblages of rocks of basaltic composition without water at pressures up to 3000 MPa and temperatures between 1250 and 1500 K (GREEN and RINGWOOD, 1967; RINGWOOD and GREEN, 1966; ITO and KENNEDY, 1971; RINGWOOD, 1975).

These are:

	Rock type	Stable mineral assemblage
Low pressure	basalt gabbro pyroxene granulite	plagioclase pyroxene(s) $\pm$ olivine $\pm$ spinel
Intermediate	garnet granulite	garnet pyroxene(s) plagioclase
High pressure	eclogite	garnet pyroxene $\pm$ quartz

The effect of increasing the confining pressure at a fixed temperature is to convert gabbro or basalt into eclogite through a transitional stage of garnet granulite. The stability fields for rocks having the composition of the quartz-tholeiite variety of basalt are shown in Fig. 2.29. The field boundaries have been extrapolated to lower pressures and are compared with possible geothermal gradients. Green and Ringwood found that the transition to eclogite would occur at slightly lower pressure for rocks of olivine basalt composition. These experimental results imply that under dry conditions it is eclogite, not gabbro, which is the stable form of rocks of basic composition throughout the *whole* of the normal continental crust. Eclogite has a P velocity of over 8·0 km s^{-1} which would place it in the upper mantle by definition. Garnet granulite has a velocity of 7·5–8·0 km s^{-1}. Thus the interpretation of the lower crust as a layer of basic rock (gabbro or basalt) or its high pressure form under dry conditions appears to be decisively ruled out, provided that the equilibrium assemblage is attained and the extrapolation of the experimentally determined field boundaries is correct.

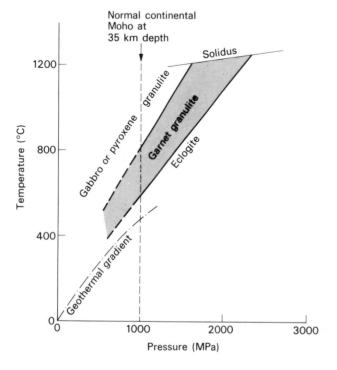

Fig. 2.29 The stability fields for rocks having the composition of the quartz-tholeiite variety of basalt, showing also an average continental geothermal gradient. Adapted from RINGWOOD (1975), *Composition and petrology of the Earth's mantle*, p. 25, McGraw-Hill. Used with permission of McGraw-Hill Book Company.

These results raise the further problem of why gabbro and basalt do occur at and near the Earth's surface, where they are apparently metastable. The reason is that they crystallize from magma at high temperature and low pressure in the stability field of gabbro. Cooling occurs relatively rapidly to low temperatures and the metastable assemblage is frozen in. At low temperature in the upper crust, the rate of phase reaction from gabbro to eclogite is probably too slow for significant change to occur during the geological time scale. On the other hand, at temperatures of 550 to 850 K appropriate to the lower crust, one would expect the mineral reactions to occur fast enough for equilibrium to be established over a period of 100 My or even much shorter.

If the old idea of a basaltic layer is demonstrably wrong, what are the rocks of the lower crust which exhibit P velocities in the range 6·5 to 7·4 km s^{-1}? One possibility is that the lower crust is relatively dry and that it is formed of rocks of between acidic and intermediate composition which have undergone high pressure modification. The experimental work supports the idea that granite and diorite undergo high pressure modification to the granulite facies at temperatures and pressures appropriate to the lower crust. RINGWOOD and GREEN (1966) have discussed the possible mineral assemblages which might occur in such rocks (Table 2.5), and have calculated the density and P velocity in them. The table shows that velocities in the range 6·7–7·4 km s^{-1} could readily be explained in terms of high pressure forms of metamorphic rocks having a bulk granodioritic to dioritic composition. The exact velocity would depend on both temperature and composition, leaving scope for wide variation from region to region. A gradational velocity-depth distribution would normally be expected between the upper crust and the lower crust marked by high pressure mineral assemblages. However, a sharp boundary may occur in some regions where high grade granulites are unconformably overlain by lower grade metamorphic rocks, such as beneath north-west Scotland and its shelf (SMITH and BOTT, 1975).

Another possible interpretation of the lower crust is that it is 'wet' and that admixed basaltic rocks occur as amphibolite, which is the stable form in the presence of substantial water vapour pressure below about 750 K. The typical mineral assemblage would be: amphibole, plagioclase, epidote and iron-rich garnet. The P velocity of amphibolite is about 7·0–7·6 km s^{-1} and the lower crustal velocity would be somewhat lower if amphibolites were admixed with more silica-rich

Table 2.5 Estimated mineral assemblages, densities and compressional wave velocities for typical acid to intermediate rocks at low and high pressures. Taken from RINGWOOD and GREEN (1966); the 'diorite' has 60·4% SiO_2 and the 'granodiorite' has 65·6% SiO_2. The mineral assemblages are given by weight and the high-pressure assemblages represent probable mineralogies at 3000 MPa and 1100° C.

	'Diorite'		'Granodiorite'	
	Low pressure	High pressure	Low pressure	High pressure
Quartz	14·8	20·2	21·7	27·8
Orthoclase	7·7	7·8	13·0	12·8
Plagioclase	56·5	—	52·0	—
Clinopyroxene	5·8	47·2	2·3	41·6
Hypersthene	11·0	—	7·7	—
Garnet	—	8·4	—	4·6
Kyanite	—	15·6	—	12·5
Ore minerals	4·2	0·7	3·3	0·7
Density (kg m^{-3})	2830	3200	2780	3070
P velocity (km s^{-1})	6·6	7·6	6·4	7·3

rocks. Metamorphic and geochemical evidence suggests that the typical lower crust is more likely to be dry than wet, supporting an interpretation as granulite rather than amphibolite-rich lower crust. Yet another possibility is that the lower crust may be formed of the aluminium-rich rock gabbroic anorthosite consisting of feldspar and subordinate pyroxene with minor garnet, quartz and kyanite (GREEN, 1970). Under normal lower crustal temperature and pressure conditions, such a rock would be expected to have a density of about $3150\ \mathrm{kg\ m^{-3}}$ and a P velocity within the range $6\cdot8$–$7\cdot4\ \mathrm{km\ s^{-1}}$. Interest has been added to this suggestion by the discovery that the Moon has such a crust.

Thus the lower continental crust probably consists of a heterogeneous mixture of metamorphic and subordinate igneous rocks having a bulk composition similar to that of diorite or andesite, or perhaps slightly more silicic than diorite. DEN TEX (1965) suggested on geological grounds a mixture of about 55 % acid to intermediate granulites, 40 % basic granulites and eclogites and 5 % ultrabasic rocks. SMITHSON and DECKER (1974) suggested an overall andesitic composition made up of a heterogenous mixture of pyroxene granulites, dioritic and quartz dioritic gneiss, anorthosite, pyroxene granitic and syenitic gneiss and intrusive rocks. It is possible that amphibolite may locally dominate the lower crust in atypical localities where water is present in sufficient abundance.

Nature of the continental Moho

The Moho beneath the continents has most commonly been interpreted in recent years as a compositional boundary separating the silica rich rocks of the crust from the ultrabasic rocks of the mantle. An alternative hypothesis which has been widely discussed is that the Moho marks a phase transition between gabbroic rocks of the lower crust and eclogite forming the topmost mantle.

There are now cogent reasons from experimental petrology and seismology against interpretation of the continental Moho as the gabbro-eclogite transition. Firstly, Ringwood and Green have convincingly shown that eclogite is the stable form at conditions of temperature and pressure normally prevailing in the lower crust; therefore the Moho cannot be explained by a downward transition from gabbro to eclogite. Secondly, it is shown in Chapter 4 (p. 187) that a variety of the ultrabasic rock peridotite is the most likely composition for the upper mantle, although pockets of eclogite may occur. This is supported by the convincing demonstration that P_n varies with direction beneath West Germany (BAMFORD, 1977) which can be explained by a peridotite with abundant orientated olivine crystals but not by an eclogite. Thirdly, the transition from basalt to eclogite takes place through an intermediate granulite mineral assemblage with smoothly varying properties over a depth interval of about 15 km depending on geothermal gradient (Fig. 2.29). Such a gradational transition is incompatible with the seismic evidence indicating a relatively sharp Moho typically occurring over 2–5 km at most in stable continental regions.

Another argument against general interpretation of the continental Moho as a phase change, applicable to certain specific regions, depends on lack of correspondence between Moho depth and temperature as inferred from heat flow (BULLARD and GRIGGS, 1961). The garnet granulite–eclogite boundary would be expected to be about 6 km deeper for every 100 K hotter. The average heat flow over a large area of eastern Australia is nearly twice the average for the Precambrian shield of western Australia (Fig. 7.5), implying that the temperature at 30–40 km depth is likely to be at least 200 K hotter beneath eastern Australia. If the Moho beneath Australia is caused by the phase transition, then it should be at least 10 km deeper beneath the east than the west. However, crustal thickness is about the same beneath the two regions and it certainly does not differ by more than 5 km. A similar difficulty faces the phase transition hypothesis when applied to western U.S.A., where the Moho at about 50 km depth beneath the Great Plains is estimated to be cooler than at about 30 km depth beneath the Basin and Range province (PAKISER, 1963).

The overwhelming weight of evidence is against generally interpreting the continental Moho as the basalt/eclogite phase transition, although such an explanation of the Moho may have local application to a few exceptional regions. The most reasonable general interpretation consistent with our knowedge of the lower crust and uppermost mantle is that it marks the boundary between intermediate granulites of the lower crust and olivine-rich ultrabasic rocks of the mantle. A gradation over about 2 km or more is to be expected as a result of chemical diffusion over long periods and occasional tectonic activity.

2.7 Continental drift

The good correspondence between the opposite coastlines of the Atlantic Ocean has been commented on since it was first recognized by Sir Francis Bacon in 1620. The idea of continental drift, however, was first taken seriously by geologists as a result of the work of TAYLOR (1910), BAKER (1911), WEGENER (1912) and DU TOIT (1937). In particular, it was Alfred Wegener, the German meteorologist and Greenland explorer, who first developed the hypothesis in detail, seeking evidence from widely differing disciplines. One of the new mainstays of Wegener's argument was the Permo-Carboniferous glaciation which affected South America, South Africa, Australia and India, suggesting that these continental land masses were grouped around the south pole at that time. Wegener suggested that during the Upper Palaeozoic there was a single large continental mass which he called Pangaea. This mass broke into fragments which tended to drift away from the pole and towards the west during the Mesozoic and Tertiary.

Wegener suggested that centrifugal force acting on the relatively highstanding continents would cause them to migrate towards the equator. Tidal attraction of the Sun and Moon and precessional effects could cause them to move westwards. These mechanisms are quite inadequate and received severe criticism. Wegener bowed to these criticisms in the 1928 edition of his book *Die Entstehung der Kontinente und Ozeane* and inclined towards the modern theory of sub-crustal convection currents as the mechanism.

The historical development of the theory of continental drift has been reviewed by TARLING and TARLING (1971) and by HALLAM (1973). The four main stages are as follows:

(i) Up to about 1910, early speculation was based almost entirely on the similarity of opposite coastlines across the Atlantic Ocean.

(ii) Between 1910 and 1955 a large amount of geological evidence supporting continental drift was assembled, particularly in the southern hemisphere by DU TOIT (1937) and others, but the theory was not widely accepted by geologists working in the northern hemisphere who had not seen the evidence at first hand.

(iii) Between 1955 and 1960 palaeomagnetism introduced a new quantitative approach which appeared to confirm that the continents had drifted in much the same way as suggested by Du Toit. The theory became much more widely accepted although there were still many notable opponents.

(iv) Since 1961, the verification of the sea-floor spreading and plate tectonic hypotheses has spectacularly confirmed the occurrence of continental drift and has enabled the history of Mesozoic and Tertiary continental movements to be worked out in detail. The main focus of controversy has now moved to Palaeozoic and earlier continental drift.

It is now recognized that there were two main continental masses during most of the Devonian and Carboniferous period. The southern continent *Gondwanaland* consisted of South America, Africa, Madagascar, India, Australia and Antarctica. The northern continent *Laurasia* consisted of North

America, Greenland, Europe and Asia excepting India. These two continental masses collided in the Hercynian orogeny to form the single supercontinent *Pangaea* which was in existence during most of the Permian and Triassic. The supercontinent started to break-up in late Triassic time and the fragments representing the present-day continental masses have separated from each other during the Mesozoic and Tertiary as the Atlantic and Indian Oceans have been formed between them.

The geological evidence

A short summary of the main geological arguments which support continental drift is given here. A useful early account is given in the book by DU TOIT (1937).

The *fit of the continents* across the Atlantic has been a longstanding argument for drift. When examined in detail, the coastlines themselves do not fit together particularly closely, but this is because they do not mark the edges of the regions underlain by continental crust. As shown by CAREY (1958), an excellent fit is obtained between the edges of the continental shelves of South America and Africa provided that the Niger delta is regarded as a post drift feature. There is also a good fit across the North Atlantic provided that Iceland is regarded as oceanic in origin and that Rockall is a microcontinent. The fit across the Atlantic calculated by computer is shown in Fig. 2.30. The Gondwanaland fragments around the Indian Ocean have been fitted together by SMITH and HALLAM (1970) and in several subsequent reconstructions (e.g. NORTON and SCLATER, 1979) but some uncertainties still remain, particularly concerning the original locations of India and Madagascar.

The next stage is to investigate the *fit of the geological features*, including fold mountain belts, major faults, sedimentary basins, the structure and age pattern of the basement, dyke swarms, and the stratigraphy of the sedimentary rocks. If these also fit together, the argument for drift becomes greatly enhanced. Examples include the excellent match of Caledonian and Hercynian fold belts across the North Atlantic, and the fitting together of the Samfrau orogenic belt of Gondwanaland. Further, the Upper Palaeozoic successions of rocks in the Gondwana continents are closely similar to each other but are strikingly different in lithology and fauna from rocks of the same age in Laurasia. In general, there is a convincing fit of the geological features across the Atlantic and between the fragments of ancient Gondwanaland.

Palaeoclimatology is the branch of geology which aims at reconstructing past climates. It provides some of the most convincing geological evidence for the drift hypothesis. The methods available include the use of:

(*i*)   past glacial deposits to indicate arctic climates;
(*ii*)   evaporites suggesting high temperature and low precipitation;
(*iii*)   bauxite which is formed by tropical or subtropical weathering;
(*iv*)   reef-deposits suggesting tropical or subtropical climates;
(*v*)   dune bedding indicating past wind directions;
(*vi*)   direct temperature measurement by oxygen isotope studies.

Caution is needed in interpreting some of the evidence, and also in attempting to deduce palaeolatitudes from the evidence because climatic conditions over the globe may have been severely different at times in the past. The results are most convincing when the global pattern of climate at a given time can be assembled. For instance, during the Permo-Carboniferous time the Gondwana continents were all affected by widespread glaciation at the same time as reef deposits, coal and evaporites were being formed in Britain and U.S.A. There can be little doubt that the Gondwana countries were nearer to the south pole, and Britain and U.S.A. were nearer to the

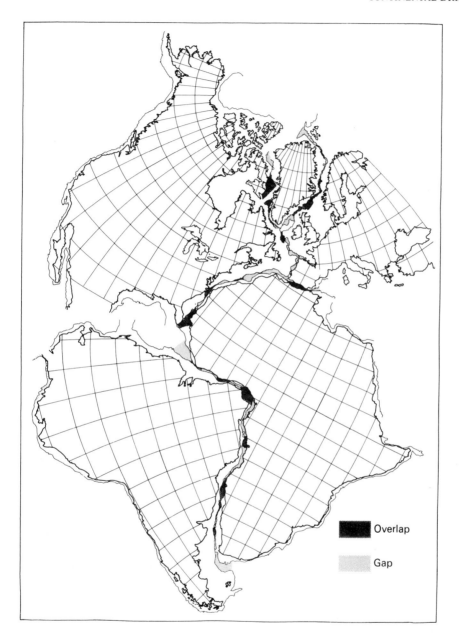

Fig. 2.30 Fit of all the continents around the Atlantic, obtained by least squares fitting at the 500 fathom contour. Redrawn from BULLARD and others (1965), *Phil. Trans. R. Soc.*, **258A**, opp. p. 49.

equator than they are now. The good agreement between the results obtained from palaeo-magnetism and palaeoclimatology supports the general reliability of both methods.

The *past and present distributions of animals and plants* have been used widely as an argument for continental drift. For instance, the Gondwana continents formed a single floral province from Lower Devonian to Lower Jurassic, which is in marked contrast to the present plant distribution in former Gondwana fragments, each of them now forming an independent floral province with few species in common. Strong support for the Pangaea supercontinent comes particularly from the fossil reptiles which migrated by land routes and dispersed throughout all the linked continents during the Permian and Triassic, aided no doubt by the exceptionally low sea level of this period. At no other time in geological history has this been possible.

Palaeomagnetism, a method of testing whether the continents have drifted

Palaeomagnetism is the study of the geomagnetic field during the geological past. It makes use of the permanent magnetization, generally known as natural remanent magnetization (NRM), which may be picked up by a rock when it is formed or during some later geological event. The results of palaeomagnetic studies are of great importance in two ways. They show up important properties of the geomagnetic field, such as reversal of polarity, which are not apparent in the relatively short historical record of it (p. 250). They also provide a method of determining the palaeolatitude and palaeoazimuth direction at suitable locations during the geological past, thus providing an important tool for determining how continents have moved relative to each other and to the equator during geological time. Good general accounts of palaeomagnetism are given by CREER (1970), TARLING (1971), McELHINNY (1973) and IRVING and DUNLOP (1982).

An outline of the palaeomagnetic method is as follows. Orientated specimens of rocks from a locality of known age are collected. The direction of their remanent magnetization is then measured using an astatic or spinner magnetometer. Spinner magnetometers are now interfaced to microprocessors enabling large numbers of samples to be rapidly measured and processed. The recent introduction of superconducting magnetometers has been a most significant step, enabling very weakly magnetized specimens to be studied and experiments in rock magnetism to be advanced. After certain precautions have been taken (see below), the measured direction of magnetization for a sample is used as an estimate of the direction of the magnetic field (known as the ambient field) when the rock was formed or when its magnetization was picked up. The secular variation is approximately averaged out by measuring several samples from slightly different geological horizons for each locality. The results are usually plotted on a stereographic projection such as is shown in Fig. 2.31 and a statistical estimate of the true mean and its circle of confidence are obtained using appropriate statistical methods developed by R. A. Fisher.

The remanent and induced magnetization of rocks are effectively caused by relatively small fractions of minerals which are ferromagnetic (in the general sense of the term), magnetite being the commonest type. Rocks may pick up a natural remanent magnetization through several processes including the following. *Thermo-remanent magnetization* (TRM) is the type acquired by igneous rocks when they cool through the Curie point of the magnetic mineral, which is usually about 850 K or less, and well below the melting point of the rock. This is almost always in the direction of the ambient field, but its polarity may be normal or reverse depending on that of the main geomagnetic field at the time. When the magnetic minerals cool below the blocking temperature, the remanent magnetization becomes permanently frozen in. *Depositional remanent magnetization* (DRM) occurs in some sediments as a result of alignment of the magnetic grains in the direction of the ambient field, with some depositional bias, as deposition occurs. *Chemical remanent magnetization* may occur after a rock has been formed by the production of new magnetic minerals through chemical reactions associated with diagenetic alteration, weathering or metamorphism. The newly formed magnetic minerals pick up remanent magnetization in the direction of the field at the time. Another type is *viscous magnetization* (VRM) which may be picked up when a rock lies in a weak magnetic field at relatively low temperature for a very long period of time. As more

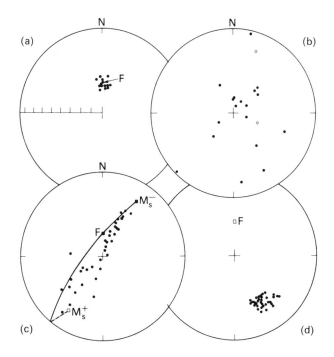

Fig. 2.31 Examples of palaeomagnetic observations from a single locality. Directions of positive inclination are shown as solid circles and negative inclination as open circles. F is the present geocentric axial dipole field.
   (**a**) Payette formation (Neogene sediments), Idaho, U.S.A., showing palaeomagnetic directions consistent with present field;
   (**b**) Arikee formation (Miocene sediments), South Dakota, U.S.A., which are widely scattered;
   (**c**) Triassic marls from Sidmouth, England, showing scatter along great circle between the stable normal and reverse directions for the Triassic and the present field, caused by unstable magnetization (VRM), reproduced from CREER (1957);
   (**d**) Tertiary lava flow, Australia, showing consistent reversed directions differing significantly from present field.
Redrawn from IRVING (1964), *Paleomagnetism*, p. 55, John Wiley.

experimental work in rock magnetism is done, the complexity of the rock magnetization processes is increasingly recognized.

If a remanent magnetization can be retained without change through a long period of geological time, it is said to be *hard*. If it is easily lost or its direction is changed, then it is said to be *soft*. VRM is usually soft. The palaeomagnetic techniques used to remove the unwanted soft components of magnetization are known as 'washing'. Two standard ways of washing are: (*i*) demagnetizing the rock in an A.C. field of decreasing intensity, which removes the soft components but does not affect the hard magnetization; and (*ii*) heating the rock to an appropriate temperature, thereby removing the components of magnetization picked up at low temperature but leaving the high temperature component.

The mean direction of the remanent magnetization at a geological locality is used to estimate the ancient latitude (using the formula $\tan\theta = \frac{1}{2}\tan I$) and the ancient north direction on the assumption that the magnetic field of the past, averaged over a few thousand years to remove the secular variation, can be approximately represented by a geocentric dipole aligned along the rotation axis. This is a basic assumption of palaeomagnetism which is probably not infallible, but is consistent with the predictions of the dynamo theory of origin of the field (p. 257) and is experimentally supported by the reasonably good agreement between palaeomagnetic and palaeoclimatological estimates of ancient latitudes. Other problems of the method include the need to obtain accurate dating of the rocks and to ensure that the measured magnetization was actually acquired at the dated age of the rock. In view of these problems and the complexity of the magnetization process, surprise is sometimes expressed at the success of the early applications of palaeomagnetism to the testing of the continental drift hypothesis.

The basic palaeomagnetic approach to the testing of continental drift is to determine the palaeolatitude and azimuth for each continental mass at as many points in the geological time scale as possible. Three quantities, however, are needed to determine the past position and orientation without ambiguity, such as latitude and longitude of a fixed point and orientation of a fixed direction. Palaeomagnetism yields estimates of two of these quantities to an accuracy which is typically about 5°, but it cannot tell us the longitude. This limitation can be partially overcome by using other indications of the relative positions of continents in the past, such as the fit between continental margins, the fact that they cannot overlap, and reconstruction of the sea-floor spreading history (Chapter 3).

Palaeomagnetic results obtained from a single continent such as South America can be presented pictorially in two basic ways as shown in Fig. 2.32. If it is assumed that the continent has remained fixed in position, then the position of the poles can be estimated from the ancient dip and declination; a convenient way of showing this is to construct an *apparent polar wandering curve*, which marks the migration of one of the poles relative to the fixed continent through geological time (Fig. 2.32b). The alternative method is to assume that the Earth's axis of rotation has remained fixed and to plot the past positions of the continent, assuming values of the longitude (Fig. 2.32a). The palaeomagnetic results for a single continent which has not been internally deformed can be validly interpreted in either way or as a combination of both of them.

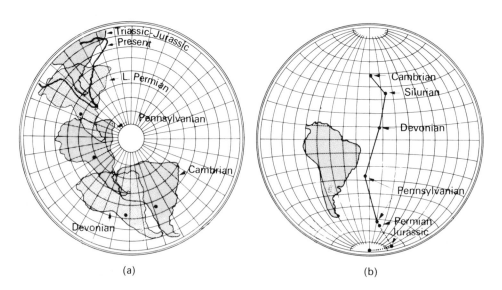

(a)   (b)

Fig. 2.32 Palaeomagnetic results for South America since the Cambrian shown by (a) past palaeolatitudes and orientations assuming a fixed south pole, and (b) a polar wandering path assuming a fixed continent. Note that later work (e.g. Fig. 2.33) has modified the detail of the inferred continental drift or polar wandering. Redrawn from CREER (1965), *Phil. Trans. R. Soc.*, **258A**, 29.

The next stage is to compare the results obtained from several continents. A convenient way of doing this is to plot the apparent polar wandering curves for several continents on the same diagram, as shown in Fig. 2.33. If all the continents had a common polar wandering curve, then we should know that they had remained stationary relative to each other and that the pole had wandered during geological time. Figure 2.33 shows that this is not so. The polar wandering curves diverge from each other going back in time. This means that the continents must have moved relative to each other. The pole may or may not have wandered — the palaeomagnetic evidence is not able to decide this issue.

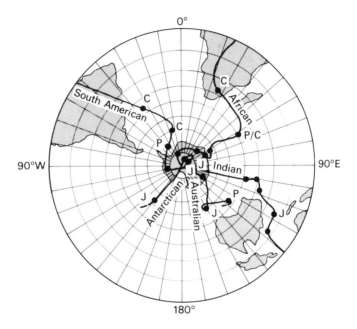

Fig. 2.33 Apparent polar wandering paths for the Gondwana fragments now forming the southern continents. J – Jurassic, P – Permian, C – Carboniferous. Based on data given by CREER (1970).

It was RUNCORN (1956) who first recognized that the apparent polar wandering paths for Britain and North America deviate from each other going back to the Triassic, suggesting that America has subsequently been displaced westwards relative to Europe by about 20°. Soon afterwards palaeomagnetic results became available from the southern hemisphere and it became apparent that the Gondwana fragments must have drifted substantially relative to each other and to the Laurasian continents since the late Palaeozoic (IRVING and others, 1963; CREER, 1965). Although the sampling and measurement techniques used in this early work do not meet modern standards, nevertheless the broad conclusion that the continents have drifted relative to each other since the early Mesozoic has subsequently been substantiated. The early palaeomagnetic work was instrumental in swinging geological opinion towards acceptance of the drift hypothesis.

Because of the ambiguity concerning longitude and the lack of precision, palaeomagnetism is not now the best method of determining the relative continental movements since the Palaeozoic. Much more accurate reconstructions can be made using the evidence of oceanic magnetic anomalies and fracture zones as described in Chapter 3. The continental palaeomagnetic results, however, are broadly consistent at most points with reconstructions based on sea-floor evidence and they also make it possible to place the ancient continents at their correct latitude. Palaeomagnetism has also been useful in recognizing the rotation of small continental fragments such as Spain and Italy. The early work on such continental fragments was carried out in the Mediterranean (ZIJDERVELD and VAN DER VOO, 1973) but the method has subsequently been applied to other regions where small continental fragments have been involved in orogenic movements.

As the present-day ocean floor is almost entirely of Mesozoic and Tertiary age, the recognition of Palaeozoic and Precambrian continental drift must be based solely on palaeomagnetic and geological evidence. In general, the earlier the period, the less reliable are the inferences based on this evidence. It is now reasonably well established that two main continental masses, Gondwanaland and Laurasia, were in existence for most of the Devonian and Carboniferous. These were separated by a pre-Hercynian ocean which closed during the late Carboniferous Hercynian orogeny to form the supercontinent Pangaea (IRVING, 1979).The most celebrated Lower Palaeozoic ocean is the Iapetus which separated North American and European regions and closed at the end of the Silurian. The collision suture can be traced on excellent geological evidence from the northern Appalachians, through Newfoundland, Britain and Scandinavia (DEWEY, 1969; ROBERTS and GALE, 1978). Although the situation in the Precambrian is much less clear, the differences between the apparent polar wandering paths for the major land masses indicates that there was relative movement between them during the Proterozoic (VAN DER VOO, 1979).

An example of reconstructions based on palaeomagnetic and other evidence is shown in Fig. 2.34. This shows (a) a reconstruction of the two continents Laurasia and Gondwanaland, separated by the Hercynian Ocean, in Lower Carboniferous time, and (b) a reconstruction of the supercontinent Pangaea during the Permo-Triassic.

2.8 Origin of the continental crust

Precambrian time is subdivisible into the *Pre-Archean era* (4600 to 3800 My ago) representing the period of primitive Earth history without presently visible record, the *Archean era* (3800 to 2500 My) characterized by formation of granulitic gneisses of tonalitic average composition and intervening greenstone belts, and the *Proterozoic era* (2500 to 570 My) when geological processes became progressively more like those of the last 500 My. A substantial fraction of the present continental crust had probably been formed by the end of the Archean, although some later additions have occurred on a modest scale. The main processes by which the continental crust was formed are controversial and speculative, as is seen in the compilation edited by WINDLEY (1976). These are now summarized, starting with the well-studied processes of the last 500 My and working back towards the uncertain early history of the Earth.

Andesitic to basaltic magmas are at present being added to continental crust in the mountain ranges of the circum-Pacific belt. Similar igneous activity within oceanic regions forms the relatively thick crust of island arcs which may subsequently be driven against an adjoining continent by subduction of the intervening marginal basin, thus adding to the continental crust by accretion. These processes depend on the formation of oceanic lithosphere at divergent plate boundaries and its subsequent subduction at convergent plate boundaries (p. 130). They have contributed modestly to the volume of continental crust over the last 500 My and probably before. Further minor additions have occurred by extrusion of tholeiitic flood basalts at times of continental splitting and by other continental igneous activity.

Turning to the Archean and early Proterozoic eras, the initial strontium and lead isotopic ratios determined on the gneisses and greenstones are in close agreement with the mantle values of the same age (MOORBATH, 1976). This suggests that the gneisses and greenstones were not derived from a pre-existing sialic crust but by direct differentiation from the mantle at most about 50–150 My before their radiometrically-dated age. The visible part of the Archean crust appears

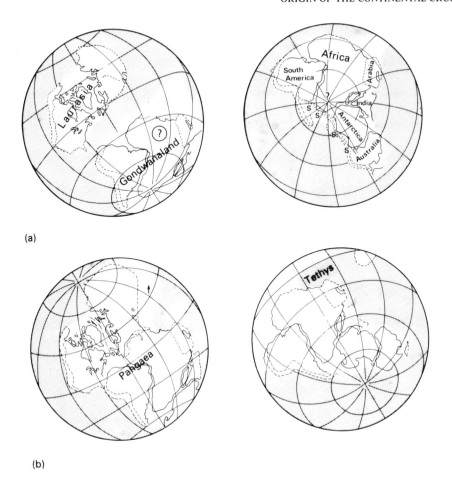

(a)

(b)

Fig. 2.34 Reconstruction of the continental masses (**a**) 340 My ago in the early Carboniferous, and (**b**) 225 My ago in the Permo-Triassic. Redrawn from IRVING (1979), *The Earth, its origin, structure and evolution*, pp. 579 and 581, © Academic Press Inc. (London) Ltd.

to have formed during successive periods of igneous intrusion followed by tectonism and metamorphism. The successive crustal belts were either formed adjacent to each other or plastered against each other to form the Archean continental cores. It is possible that the gneisses represent primitive andesitic volcanism of the circum-Pacific type and that the greenstones were formed by crumpling-up of primitive oceanic marginal basins. However, it is not clear whether the differentiation of crustal material from the mantle occurred as a result of such very vigorous subduction episodes involving primitive oceanic lithosphere or, alternatively, by processes which have no present-day analogues. What is clear is that events during the early, recorded, Precambrian contributed significantly, possibly predominantly, to the formation of our present continental crust.

More speculatively, part of the continental crust may have differentiated from the mantle during the period between the Earth's formation and the oldest known rocks. This may possibly have formed a thin layer covering all the Earth's surface. Some encouragement for this view comes from the knowledge that the lunar crust of anorthositic gabbro and related rocks formed over the first

200 My of the Moon's existence (p. 26). It is believed that a substantial thermal event affected the Earth when the core was formed soon after or even during accretion, possibly causing wholesale or partial fusion of the upper mantle (p. 279). A gabbroic crust which formed in this way might be expected to sink back into the mantle on cooling and possible conversion to eclogite, like the present oceanic lithosphere. But if differentiation gave rise to an anorthositic, dioritic or granitic crust (e.g. SHAW, 1976), this would remain at the surface because of its low density under equilibrium conditions. We have no direct evidence for such a primitive crust encircling or partly encircling the Earth, but it is possible that such rocks form those deeper levels of the continental crust which have not yet been sampled by man. Such a primitive crust would probably have suffered the same intense meteorite bombardment which affected the Moon's surface about 4000 My ago.

Thus the continental crust has formed by differentiation of magma of approximately intermediate average composition from the mantle. The decaying rate of crustal growth is explained by a slight cooling of the Earth over its lifespan, causing a substantial decrease in the rate of heat loss from the interior and the vigour of mantle convection (Chapter 7). According to O'NIONS and others (1979), evolution of the strontium and neodymium isotope ratios observed in basalts at ocean ridges suggests that the continental crust may have differentiated from the upper half of the mantle only, so that the lower half remains primitive. Once formed, the low density prevents large scale recycling of the continental crust into the mantle, in contrast to the oceanic crust and lithosphere (Chapters 3 and 5).

3 The oceanic crust, sea-floor spreading and plate tectonics

3.1 Introduction

One of the important advances in geophysics during the early 1950s was the clear demonstration that fundamentally different types of crust occur beneath oceans and continents. Not only is the oceanic crust much thinner, but the layering is quite distinct. The difference has been highlighted by the discovery that most of the oceanic crust has been formed during the last 200 My, in contrast to the continental crust which goes back over 3500 My.

The oceans, excepting the continental shelves, cover an area of $332 \times 10^6 \, \text{km}^2$ is about 65% of the Earth's surface. The average water depth is 3·8 km. Oceanic regions may conveniently be subdivided into ocean basins, ocean ridges, trenches and continental margins (Fig. 3.1). Some continental margins may be further subdivided into the continental shelf, the continental slope and the continental rise.

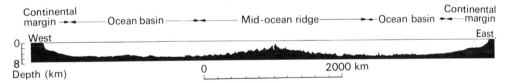

Fig. 3.1 The three major morphological subdivisions of the oceans, illustrated by a profile across the North Atlantic from New England to the Spanish Sahara. Redrawn from HEEZEN (1962), *Continental drift*, p. 237, Academic Press.

The main types of bottom topography of the ocean basins are:

(1) *abyssal plains*, which are smooth plains formed of flat-lying sediments with gradients of less than 1/1000;
(2) *abyssal hills*, with a relief of 50–1000 m and widths of 1–10 km, which are particularly common on the floor of the Pacific and cover more than 80 % of its total area (MENARD, 1964);
(3) *seamounts*, which are circular or oval-shaped underwater mountains rising abruptly from the deeps and having the shape of volcanoes; dredging and magnetic surveys confirm that nearly all seamounts are underwater basaltic volcanoes;
(4) *archipelagic aprons*, which form very smooth areas of the deep seabed near groups of volcanic islands and are made of submarine lava flows smoothed by an overlying veneer of sediment.

The earliest geophysical measurements relevant to the structure of the oceanic crust were the pendulum measurements of gravity made in submarines since 1923 (VENING MEINESZ, 1948; WORZEL, 1965). These showed that the oceans, apart from trenches and island arcs, are in approximate isostatic equilibrium with the continents. This was usually interpreted according to the Airy hypothesis, that the crust is much thinner beneath oceans than continents. Suppose a typical continental crust is 35 km thick and that the densities of crust, mantle and seawater are 2900, 3300 and 1030 kg m^{-3} respectively. If an ocean basin 5 km deep is in equilibrium with it according to the Airy hypothesis, then the basic condition of isostatic equilibrium (p. 50) shows that the oceanic crust would be 6·6 km thick below the seabed. The seismic refraction work described below has shown that the oceanic crust is on average about 6–7 km thick, in good agreement with the predictions of the Airy hypothesis.

Another early method used to investigate the oceanic crust was to study the dispersion of earthquake surface waves which had traversed oceanic paths. Such a study suggested to GUTENBERG (1924) that the oceanic crust is much thinner than the continental crust. The use of Love and Rayleigh dispersion is still of some importance in assessing crustal structure of inaccessible regions, but in the main it has been superseded in oceanic crustal studies by seismic prospecting methods.

3.2 Oceanic crustal structure

Discovery of the oceanic crust

Although early gravity and surface wave studies suggested the presence of a thin crust beneath the ocean basins, our modern knowledge of the thickness and layering of the oceanic crust stems from the adaptation of the seismic refraction method to use in the oceans during the 1950s. The marine refraction methods were pioneered by Dr Maurice Ewing and his co-workers at Lamont-Doherty Geological Observatory and by Dr M. N. Hill at Cambridge. Either separate ships were used for shooting and recording (SHOR, 1963) or the seismic waves were received at buoys and telemetered back to the ship (HILL, 1963). As the oceanic Moho is shallower than beneath continents, the refraction lines needed to be only about 50–70 km long.

The typical pattern of results obtained in such a refraction line is shown in the theoretical model of Fig. 3.2. The oceanic crust was found to consist of three layers. Layer 1 underlies the seabed and the depth to the top of it can be obtained from the ship's precision depth recorder (PDR) or from the wide-angle reflection R_1. Because the P velocity at the top of layer 1 is only slightly greater than that of seawater, the refracted head wave G_1 only rarely occurs as a first arrival. Consequently the velocity of layer 1 could not be found in the usual way, and had to be *either* assumed *or* estimated from the critical distance as determined by the maximum amplitude of R_1. Similarly, the head wave G_2 refracted at the layer 1/layer 2 interface only occurs as a first arrival for a short distance and if layer 2 is thinner than in Fig. 3.2 it may not occur as a first arrival at all. It would then be necessary to use the reflection R'_1 to determine the depth to the top of layer 2, and to assume a velocity for layer 2 for purposes of interpreting the layers below. The two underlying interfaces give rise to easily identifiable first arrivals G_3 and G_4. The water wave D could be used to determine the distance between shot point and receiver.

The results of early crustal refraction lines at sea showed an unexpected uniformity of oceanic crustal structure wherever measurements had been made, with the exception of the crestal regions of some oceanic ridges. The crust beneath the floor of the ocean was subdivided into three layers above the Moho as follows:

	P velocity (km s^{-1})	Average thickness (km)
water	1·5	4·5
layer 1	1·6–2·5	0·4
layer 2	4·0–6·0	1·5
layer 3	6·4–7·0	5·0
- - - - - - - - - - - - - - - - Moho -		
upper mantle	7·4–8·6	

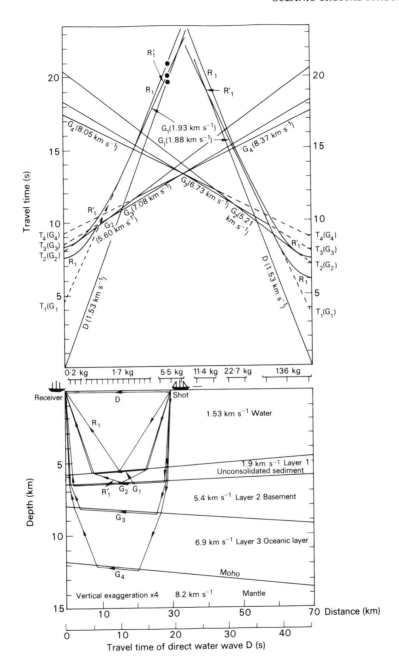

Fig. 3.2 A typical reversed deep-sea refraction station, showing the time-distance graph for refractions, reflections and direct water wave. The ray paths are shown for a 30 km separation of shot and receiver. For further explanation see text. Redrawn from TALWANI (1964), *Marine Geol.*, **2**, 37.

Beneath normal ocean basins, the Moho was found to be about 11 km below sea-level and the oceanic crust to be about 6–7 km thick. An example of the results of early refraction surveys at sea is shown in Fig. 3.3.

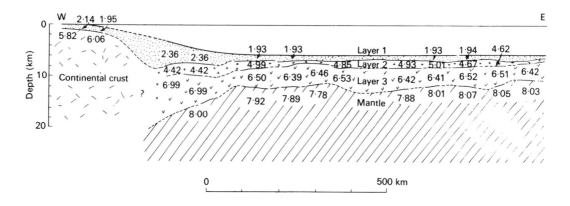

Fig. 3.3 Oceanic crustal structure as determined by refraction lines in the Atlantic Ocean east of Argentine. The line runs from (46·0° S, 60·3° W) on the shelf to (42·7° S, 50·2° W) in the ocean. Redrawn from EWING (1965), *Q. Jl. R. astr. Soc.*, **6**, 19.

Later seismic investigations using more sophisticated techniques have confirmed this major subdivision and have added much detail to knowledge of the internal structure of each of the layers and their variability. Recent work has also shown that the lateral variations in detailed structure are greater than has generally been supposed.

Investigating oceanic crustal structure

During the last two decades, knowledge of oceanic crustal structure has been greatly increased by normal incidence seismic reflection profiling and by use of the airgun-sonobuoy technique. The most recent innovation is the multichannel seismic reflection/refraction experiment using two or more ships. Since 1968 the rocks of the upper crust have also been widely sampled by the Deep Sea Drilling Project.

Seismic reflection profiling as first introduced in oceanic regions was akin to echo-sounding except that a more powerful source of lower frequency content was used to penetrate below the seabed. Small explosive charges were set off at regular interval while the ship was underway, the reflected waves being received by a towed hydrophone. Later on, more efficient and convenient acoustic sources such as the airgun were introduced. A recent development has been the use in oceanic regions of the type of seismic reflection system used for petroleum exploration on the continental shelf, with an array of airguns as the acoustic source and multi-channel digital recording of signals received at separate sections of a long hydrophone streamer towed astern. Such records can be processed by digital computer with great improvement in resolution and penetration.

The reflection records yield the two-way travel-time to the reflecting horizons. The prominence of a given reflector depends on the amplitude of the reflected wave. The relative amplitudes of reflected and incident waves at an interface is called the reflection coefficient r and is given by

$$r = (\rho_1 V_1 - \rho_2 V_2)/(\rho_1 V_1 + \rho_2 V_2)$$

where ρ_i and V_i are the densities and P velocities of the rocks above and below the interface. A large contrast in ρV at an interface means a strong reflection such as is characteristic of the seabed and the layer 1/layer 2 interface. Reflection profiling has been mainly used in the study of layer 1 but deeper information is now being obtained in multi-channel surveys.

The airgun-sonobuoy seismic technique (LE PICHON and others, 1968) has been of great importance in studying the finer details of oceanic crustal structure down to the Moho. In this method, a disposable free-floating sonobuoy with a hydrophone suspended below is launched from the ship which steams away from it firing an airgun at regular intervals. Small explosive charges may be used to extend the line beyond the range of the airgun. The almost stationary sonobuoy telemeters the reflected and refracted seismic arrivals back to the ship. At first the arrivals were displayed on paper records only, but more recent use of magnetic tape recording increases the versatility, including making it possible to study amplitudes. The closely-spaced shots of constant amplitude enable previously unseen layering to be resolved from the refractions and wide-angle reflections they give rise to. Yet another important seismic innovation of recent years is the ocean bottom seismometer or hydrophone station, providing a much quieter environment than the sea surface for recording.

In multi-channel seismic reflection profiling, the average velocity down to a given reflector can be determined by measuring the travel-time t for varying horizontal distance x between shot and receiver. If z is the depth to the reflector and V is the average velocity down to it, then it can be shown by simple geometry that $V^2t^2 = 4z^2 + x^2$ provided that the angle of incidence is small. The gradient of the straight line obtained by plotting t^2 against x^2 is $1/V^2$. By applying this method to a series of reflecting horizons, the interval velocity between each pair of reflectors can be estimated. This method has recently been used with success to determine the velocity structure in layer 1. More commonly, a modified version of the t^2/x^2 method which allows for large angles of incidence has been applied to the wide-angle reflections obtained during airgun-sonobuoy surveys. Insight into the velocity-depth structure, particularly in the vicinity of supposedly sharp interfaces, can also be obtained by use of synthetic seismograms for comparison with sonobuoy and ocean bottom seismic observations.

The most ambitious crustal seismic experiments yet to be carried out at sea are recent multi-channel seismic investigations using two or more ships (e.g. STOFFA and BUHL, 1979). The first type of experiment is the expanding spread profile which aims to determine the velocity-depth structure through the oceanic crust at one precise location. Two ships, one of them firing shots by airgun or explosives and the other receiving on a multi-channel seismic array at least 2·4 km long, steam away from each other at constant speed so that the common depth point remains at the centre of the profile. The results are inverted to give a velocity-depth distribution at the central point by use of a technique known as *tau-p* mapping. In the other type of experiment, two or more ships proceed along the same track at constant offset so as to map lateral variations in crustal structure using both vertical incidence and wide angle reflections from appropriate horizons. If three ships are used, all with multi-channel arrays and firing airguns in turn, common depth point data can be obtained with a continuous aperture of over 25 km, so that interfaces down to the oceanic Moho can together be profiled.

The complete thickness of layer 1 and the top of layer 2 have been sampled at many localities by drilling into the ocean floor. The Deep Sea Drilling Project (DSDP) was initiated in 1968 as the Joint Oceanographical Institutes for Deep Earth Sampling (JOIDES) programme after failure to launch the Mohole project in the early 1960s and successful drilling on the subsided Blake Plateau off Florida during 1965. A phase involving international sponsorship (IPOD) commenced in 1975 and at present the future programme is being reviewed. Up to the time of writing, the specially designed drilling vessel *Glomar Challenger* has been used as the drilling platform. It is possible that a more ambitious programme including drilling widely on the continental margins will take place during the 1980s. The results from the drilling programme are already of great significance to the earth sciences, confirming the predictions of the sea-floor spreading hypothesis and revealing the composition and age of layer 1.

Layer 1 – the oceanic sediments

Starting with the *Challenger* expedition of 1872–6, sampling by dredging and coring and underwater photography have shown that most of the ocean floor is formed of unconsolidated sediments. The pelagic deposits of the deep oceans include brown clays, calcareous and siliceous oozes, and deposits such as manganese nodules formed by precipitation from seawater. Badly sorted sediments derived from the continents and transported by ice are found in polar regions. Terriginous sediments are characteristic of continental margins and these can be transported into nearby ocean basins by turbidity currents. Some transport and redeposition of sediments within the oceans results from cold bottom water currents flowing under the influence of gravity and the coriolis force, forming the 'contourite' type of deposit characteristic of parts of the North Atlantic (JONES and others, 1970).

The layer 1 discovered in the early refraction work was correctly interpreted as consisting of oceanic sediments. The layer was found to be on average about 0·5 km thick, being absent except in local basins near the ridge crests and thickening towards the continental margins. A much more detailed knowledge of the structure, age and composition of layer 1 has emerged since about 1965 as a result of seismic reflection investigations and deep sea drilling.

The most prominent reflecting horizon below the seabed is the interface between layers 1 and 2. This must therefore be a sharp boundary at which there is a substantial change in acoustic properties. It is found that the interface is consistently rougher than the seabed (Fig. 3.4). The surface topography of layer 2 may be up to 1 km in amplitude and the wavelength of the undulations is typically 10–20 km. The rough character of the interface extends beneath the abyssal plains. Smaller scale roughness of the layer 1/layer 2 interface is indicated by the hyperbolic reflections caused by scattering from small irregularities, a feature which may be diagnostic of oceanic crust.

Within layer 1 in the American basin of the North Atlantic (Fig. 3.5), three prominent reflecting horizons were discovered soon after the reflection method had been introduced to the oceans (EWING and others, 1966). Related horizons have been found in other parts of the Atlantic. *Horizon A* is a prominent and persistent interface occurring about 300 metres below the ocean bed; it cuts out against the layer 1/layer 2 interface at the flank of the mid-Atlantic ridge. It can be underlain by a stratified zone up to 500 m thick but usually less. It is somewhat smoother than the seabed and much smoother than the layer 1/layer 2 interface. The horizon is recognized in the South Atlantic. *Horizon β* marks the upper surface of a deeper set of stratification which closely overlies the layer 1/layer 2 interface. *Horizon B* is a smooth reflector which locally replaces the layer 1/layer 2 interface. The greater resolution of modern reflection methods has enabled further persistent layers to be recognized, such as *horizon X* above horizon A and *horizon A** below it. Within the Pacific Ocean, up to four major subdivisions of the sedimentary layer may overlie the acoustic basement. At the top is an almost ubiquitous upper transparent layer which directly overlies the basement in the eastern Pacific. Below it there is an upper opaque layer which is generally present in the western Pacific. The lower transparent and opaque layers are of more restricted occurrence.

In the American basin the P velocity just below the seabed is $1·47–1·55$ km s^{-1} and thereafter it increases fairly steadily with depth down to horizon A. Both horizons A and $β$ appear to mark discontinuous increases of velocity with depth. The velocity above the acoustic basement is $4·7$ km s^{-1} beneath the rise, decreasing to lower values towards the mid-Atlantic ridge as

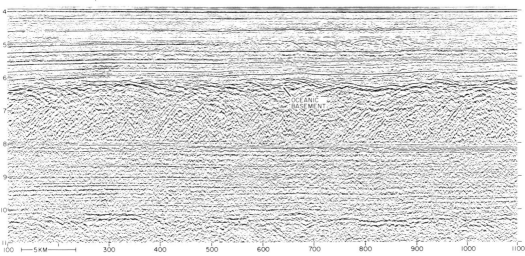

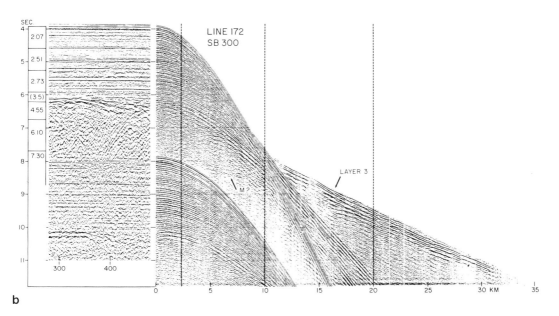

Fig. 3.4

  (a) Multi-channel seismic reflection profile in the Lofoten basin of the North Atlantic (centre of profile at approximately 69·8° N, 10·7° E), showing thicker than average sediments of layer 1 overlying oceanic basement formed by the top of layer 2. The strong reflection at 8 second two-way travel time is the seabed multiple.

  (b) Wide angle reflection/refraction profile obtained using sonobuoy 300 (location shown in (a)), showing the derived velocity-depth structure in km s^{-1} against a section of the normal incidence record. Note the strong primary reflection at about 8 s and 5 km which possibly comes from the Moho.

Both parts of this diagram have been kindly provided by Dr. P. L. Stoffa of Lamont-Doherty Geological Observatory from an unpublished report and have been reproduced with his permission.

Fig. 3.5 The sediment structure of the North American basin of the North Atlantic:

(a) Section from the North American continental margin to the mid-Atlantic ridge, showing the offlap relationship of the sediment layers to the ridge. The location of DSPD drilling site 105 is shown (34° 54′ N, 69° 11′ W). Redrawn from EDGAR (1974), *The geology of continental margins*, p. 244, Springer-Verlag.

(b) The sedimentary layering in the vicinity of DSDP drilling site 105, showing the prominent reflection horizons, the lithological subdivisions and the relationship to the acoustic basement. Redrawn from LANCELOT and others (1972), *Initial reports of the Deep Sea Drilling Project*, 11, p. 946, Washington (U.S. Government Printing Office).

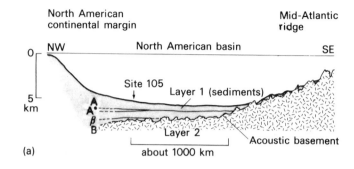

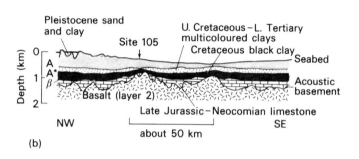

progressively younger horizons cut out against the basement. With knowledge of the interval velocities, accurate depths can be computed. Figure 3.5 shows that the sedimentary layer is about 5 km thick beneath the continental rise and thins to nothing on the flank of the mid-Atlantic ridge.

The stratigraphical succession of sediments in the American basin of the North Atlantic has been investigated during legs 1, 2 and 11 of the Deep Sea Drilling Project, as summarized by EDGAR (1974) and shown in Fig. 3.5. The earlier drill holes identified horizon A as an early to middle Eocene chert interval but later drilling showed that the horizon still remains acoustically prominent even where chert is sparse or absent. Horizon A* marks the boundary between early Tertiary volcanic clays and underlying black clays. Horizon β marks the base of the black clays, which are underlain by Jurassic to early Cretaceous limestones. Horizon B has been identified as the sediment-basalt contact in regions where the layer 1/layer 2 interface is anomalously smooth.

More generally, drilling and seismic profiling has shown that the sediments of layer 1 appear to overlap onto all the ocean ridges, with the oldest sediments overlying the acoustic basement at greatest distances from the ridges and progressively younger horizons cutting out against the basement towards the ridges (Fig. 3.5). Individual units may also thin away from the continental margins because of decreasing sedimentation rates. Thus sediments are thin or absent from the ocean ridges and are locally 5 km or even thicker beneath parts of the continental rise bordering the Atlantic Ocean. The oldest sediments found in the main ocean basins do not go back beyond the Jurassic, in contrast to the much longer age span of sediments found on continental crust. Another remarkable feature displayed by the oceanic sediments, contrasting with typical continental sequences, is the almost complete lack of faulting, folding or tilting shown by them. Thus the oceanic sedimentary layer differs from continental sediments in lithology, average age and tectonic style. Most of the features of layer 1 are explicable in terms of the sea-floor spreading hypothesis described later in this chapter.

Layer 2 – the oceanic volcanic layer

Layer 2 of the oceanic crust extends down from the acoustic basement below the sediments to the level where the *P* velocity reaches 6·4 km s^{-1} or above. The layer is normally between 1 and 2·5 km thick and the *P* velocities range widely between 3·4 and 6·2 km s^{-1}. These velocities can be matched by those of either basaltic lavas or consolidated sediments, both interpretations having adherents during the early 1960s. On the basis of sampling and the strong magnetization required to account for oceanic magnetic anomalies, it is now recognized that the layer is predominantly formed by basalts although some sediment is locally caught up within the lavas.

Layer 2 crops out extensively at the seabed at the ocean ridges where it has been widely sampled by dredging. It has also been investigated in great detail by a wide variety of modern methods including submersibles in a few selected regions of the ocean ridge system, such as the short segment of the crest of the mid-Atlantic ridge at 37°N studied during project FAMOUS (Franco-American Mid-Ocean Undersea Study). The layer has been penetrated to shallow depth in many DSDP drill holes in the ocean basins and on the flanks of the ridges. The sampling shows that the top of layer 2 is formed of basaltic pillow lavas and lava debris of olivine-tholeiite variety. Olivine tholeiite is a type of basalt whose composition, when expressed in terms of certain standard minerals, contains olivine and the calcium-poor pyroxene hypersthene. The average composition of the abyssal tholeiites does not vary much between ocean ridges and ocean basins or from one ocean to another, except where anomalously thick crust is developed such as below Iceland. Local variation is much more pronounced and is mainly attributable to magma fractionation at shallow depth.

Results from numerous seismic wide angle reflection and refraction experiments using the airgun-sonobuoy technique have suggested that layer 2 can be divided into three subdivisions (HOUTZ and EWING, 1976). Layer 2A, with *P* velocity of about 3·6 km s^{-1}, is the uppermost division. Its occurrence is restricted to the ocean ridges and to pockets in the ocean basins particularly near eruptive centres. Layer 2A only occasionally yields observable refraction arrivals, its presence normally being recognized where the acoustic basement is shallower than the underlying layer 2B refractor. Layer 2A is thickest beneath the ridge crests, where it reaches 1·5 km on the mid-Atlantic ridge and 0·7 km on the East Pacific rise. It thins from the ridge crests towards the ocean basins, generally becoming undetectable by present seismic means at a distance of a few hundred kilometres from the crest as it apparently passes laterally into the underlying layer 2B. Drilling has shown that layer 2A consists of porous rubbly basalt and that it may include some pockets of trapped sediment. The region of the ridges where layer 2A is present is characterized by hydrothermal circulation of seawater through the crust as recognized by the heat flow pattern (p. 288). It is probable that the porous, unconsolidated nature of the layer makes this circulation possible.

Layer 2B underlies layer 2A at the ocean ridges but normally forms the top of layer 2 beneath ocean basins. According to Houtz and Ewing, layer 2B gives rise to clear refractions and wide-angle reflections, indicating a relatively sharp upper boundary even when overlain by layer 2A. The *P* velocity varies between about 4·8 and 5·5 km s^{-1} and the thickness is estimated to be about 0·7 to 1·5 km. Such a layer might be formed of consolidated basaltic lavas which have velocities within this range. The postulated underlying layer 2C, where detected, has a more uniform velocity of 5·8 to 6·2 km s^{-1} and an estimated average thickness of 1 km. The high velocity may signal the presence of a significant proportion of intrusive basic igneous rocks admixed with lavas.

A radically different concept of oceanic crustal structure, and of layer 2 structure in particular, has recently been proposed (KENNETT, 1977; LEWIS, 1978; SPUDICH and ORCUTT, 1980). This is based

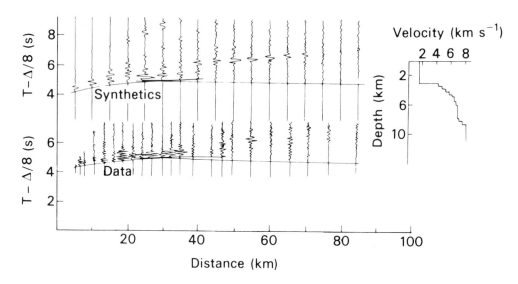

Fig. 3.6   An example of the velocity-depth structure of the oceanic crust obtained by use of synthetic seismograms. This refraction line was obtained on crust 2 My old on the Cocos plate at about 13·5° N, 103·2° W. The stacked records (below) are compared with the best-fitting synthetic records (above), both being shown in the form of reduced travel-time graphs. The velocity-depth distribution corresponding to the best-fitting synthetics is shown to the right. Redrawn from LEWIS and SNYDSMAN (1979), *Tectonophysics*, **55**, 97.

on linear inversion of the travel-time data to obtain allowable limits on the velocity-depth distribution, followed by use of synthetic seismograms to determine those models within the allowable range which are most consistent with the observed amplitudes and waveforms. The results suggest that the oceanic crust is better interpreted in terms of continuous variation of velocity with depth rather than by a series of discrete layers each of which has a constant velocity. According to these models, layer 2 is characterized by rapid increase of velocity with depth, with velocity gradient averaging 1 to 2 s^{-1}, in contrast to a much more uniform velocity within the underlying layer 3 (Fig. 3.6). Substantial lateral variation in the velocity-depth structure of layer 2 is indicated, such as the occurrence of low velocities at the top of the layer at ocean ridges. The presence of some fine sub-layering, as postulated by Houtz and Ewing, is not ruled out but is regarded as being at the best only marginally resolvable by present explosion seismology. This new concept of a gradational layer 2 is in good agreement with the petrological model of the oceanic crust discussed later in the chapter, with pillow lavas grading down into a dyke complex. The increase in velocity away from the ocean ridge crests can be mainly attributed to sealing of the cracks and pores.

Layer 3 – the main oceanic crustal layer

Layer 3 forms the main thickness of the oceanic crust. The top of the layer gives rise to well-defined refractions which occur as first arrivals, but wide-angle reflections are absent suggesting that the boundary with layer 2 is gradational. Modelling by synthetic seismograms further suggests that the boundary is not a steep gradient, but marks the transition from the steep velocity-depth gradient of layer 2 above to the almost uniform velocity of the upper part of layer 3 (Fig. 3.6). Layer 3 is commonly between 3 and 7 km in thickness and measured P velocities within it range between 6·4 and 7·7 km s^{-1}. A statistical study made by CHRISTENSEN and SALISBURY (1975) suggests that the layer may thicken from about 3 km at ridge crests to about 5 km a few hundred kilometres away, beyond where no further thickening has been detected. In contrast to layer 2, the velocity in the upper part of layer 3 (6·7 to 7·0 km s^{-1}) does not appear to change significantly with the age of the crust (LEWIS, 1978).

The structure of the basal part of layer 3 is still controversial. According to SUTTON and others (1971), widespread occurrence of a 7·0 km s^{-1} high velocity basal layer is observed in the Pacific Ocean, although it cannot be detected everywhere. CHRISTENSEN and SALISBURY (1975) subsequently subdivided the lower oceanic crust into layer 3A (6·5–6·8 km s^{-1}) and layer 3B (7·0–7·7 km s^{-1}). They recognized two distinct types of lower oceanic crust which on average are as follows: Type 1, 1 km of 6·4 km s^{-1} above 5 km of 7·1 km s^{-1}; Type 2, 3 km of 6·8 km s^{-1} above 3 km of 7·5 km s^{-1}. The use of synthetic seismograms, however, has thrown doubt on the universal existence of a high velocity basal layer 3B. SPUDICH and ORCUTT (1980) pointed out that certain arrivals previously attributed to such a basal layer may be better explained in terms of reflections from the Moho. A relatively uniform velocity throughout layer 3, or a slight increase of velocity with depth, can satisfactorily explain many of the observed profiles without recourse to a high velocity basal layer above the Moho. LEWIS (1978) has even cited evidence for a shallow low velocity layer at the bottom of layer 3 on the Cocos plate.

The composition of layer 3 has been the subject of continuing controversy. It was originally regarded as gabbroic but HESS (1962, 1965) alternatively suggested that it consists of partially serpentinized peridotite derived by hydration of the rocks of the uppermost mantle. The reaction

$$\text{olivine} + \text{water} \rightarrow \text{serpentine}$$

can only occur below about 770 K and requires an adequate source of water. In proposing the hypothesis of sea-floor spreading, Hess suggested that the 770 K isotherm beneath the crests of ocean ridges occurs about 6 km below the seabed and that olivine above this level is converted to serpentine. As the newly formed crust is carried laterally by the spreading process, the temperature falls and the supply of water becomes inadequate for further reaction. Thus the Moho becomes frozen in as the boundary between peridotite below and its hydrated derivative serpentinite above. The observed range of P velocities in layer 3 is consistent with 20% to 60% serpentinization, the rock with the highest serpentine content having the lowest velocity. The serpentinite hypothesis has more recently been advocated by BOTTINGA and ALLEGRE (1973) on the grounds that the oceanic crustal temperatures must be too low for the formation of gabbro from a magma, but this argument is obviated now that the thermal significance of hydrothermal circulation in the oceanic crust is recognized (p. 288).

Most marine geologists now regard layer 3 as formed of gabbroic rocks or their metamorphosed derivatives, metabasites. CANN (1968) postulated amphibolite but later suggested it to be formed of isotropic gabbros above and cumulate gabbros below, possibly partially metamorphosed (CANN, 1974). A similar model has been proposed by CHRISTENSEN and SALISBURY (1975), with metabasites or hornblende gabbros forming the upper part (layer 3A?) and cumulate gabbros and associated ultrabasic layered rocks derived by crystal settling forming the lower part (layer 3B?). These hypotheses are all linked to concepts of formation of the oceanic crust which are discussed later in the chapter.

A diagnostic test between the serpentinite and gabbro hypotheses can be applied by use of Poisson's ratio which can be determined for layer 3 rocks from the observed ratios of P to S velocities (CHRISTENSEN, 1972). Poisson's ratio for serpentinites varies between 0·34 and 0·38, being on average about 0·33 for a peridotite which is 50% serpentinized. For gabbroic rocks it lies between about 0·25 and 0·30. The observed values for layer 3 range between about 0·27 and 0·29, clearly indicating a gabbroic rather than serpentinite composition. In further support of the gabbro hypothesis, DSDP site 334 about 1° W of the mid-Atlantic ridge crest at 37° N penetrated gabbros and peridotites of cumulate type at anomalously shallow depth beneath a thin lava layer (HODGES and PAPIKE, 1976).

A wide variety of basic and ultrabasic igneous rocks have been dredged from the ocean ridges and fracture zones. These include abundant hornblende gabbros possibly derived from layer 3A and cumulate gabbros possibly from layer 3B. Serpentinites are also commonly found. How can these be accounted for if they are not derived from the crust? The most likely explanation is that they represent hydrated blocks of the mantle which have formed in well-fractured localities where water can penetrate down and faulting can uplift them to the surface.

The oceanic Moho

The oceanic Moho occurs at an average depth of about 7 km below the seabed and 11 km below sea-level. The average value of P_n is 8·15 km s^{-1} except near the ridge crests where it is 7·6 km s^{-1}. Clear P_mP reflections are observed in some localities (EWING and HOUTZ, 1969). The Moho is marked by a steep velocity gradient at the base of layer 3 and it may locally approximate a sharp boundary. Modelling by synthetic seismograms indicates a variable transition zone averaging 1 km in thickness but possibly varying between 0 and 3 km (SPUDICH and ORCUTT, 1980). Present evidence cannot resolve further fine detail.

HESS (1964) suggested that olivine crystals in the topmost sub-oceanic mantle may be orientated as a result of the sea-floor spreading process he postulated. This would cause anisotropy in the P_n velocity because the compression wave velocity in olivine depends on direction; the lowest value occurs parallel to the β-crystallographic axis which typically occurs perpendicular to the banding in olivine rich rocks. Hess consequently examined the refraction profiles in the Pacific made by Rait and Shor to see if such anisotropy could be detected. He found that P_n systematically varied with direction as predicted, ranging from 8·0 km s^{-1} in a north-south direction parallel to the ocean ridge crest to 8·6 km s^{-1} in an east-west direction parallel to the fracture zones. These measurements were made in the vicinity of the Mendocino fracture zone off the west coast of U.S.A. and later measurements (RAIT and others, 1971) confirmed the widespread occurrence of anisotropy in the Pacific with an average velocity difference of about 0·3 km s^{-1}. According to CHRISTENSEN and SALISBURY (1975) the mean mantle velocity observed parallel to the ridge crests is 8·0 km s^{-1} and that perpendicular to them is 8·3 km s^{-1}.

Such variation of P_n with direction can only readily be explained if the topmost sub-oceanic mantle contains a substantial amount of olivine. This indicates that the composition is peridotitic rather than eclogitic. If layer 3 is gabbroic, then the Moho must be a compositional boundary. However, it is possible that a layer of serpentinite accretes to the base of layer 3 away from ocean ridges, the Moho then being a hydration boundary. More detailed petrological discussion of the oceanic Moho is deferred until later in the chapter. Whatever the details, it is clear that the continental and oceanic Mohos are fundamentally different types of boundary.

3.3 Oceanic magnetic anomalies and sea-floor spreading

Oceanic magnetic anomalies

Oceanic magnetic anomalies were originally discovered by use of a fluxgate magnetometer with the sensing element towed two or more ship-lengths behind to be removed from the ship's magnetic field. Proton precession magnetometers are now normally used. Both types of instrument measure the total-field magnetic anomalies to an accuracy of about $1\gamma (= 10^{-9}$ tesla). The measurements are easy to make and since the mid 1950s most of the oceanic regions have been surveyed in sufficient detail to reveal the main pattern of magnetic anomalies.

A most important discovery was made when the magnetic anomalies off the west coast of North America were first mapped in some detail (RAFF and MASON, 1961; MASON and RAFF, 1961). A quite unexpected pattern of north-south orientated strip-like positive and negative magnetic anomalies was found to dominate the whole region surveyed. The pattern of anomalies may be depicted by showing the positive strips in black, as in Fig. 3.7. The continuity of the strips is interrupted by the

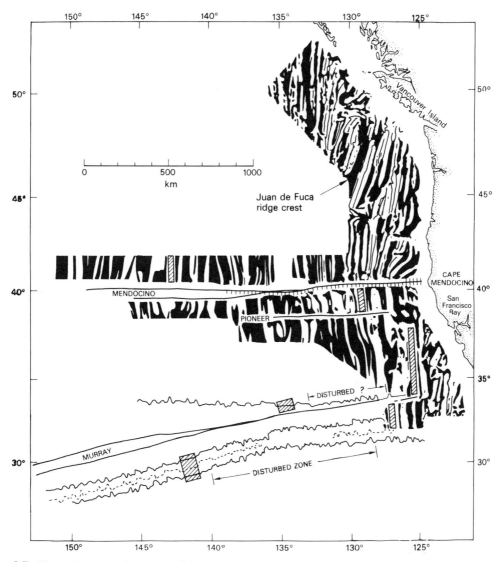

Fig. 3.7 Magnetic anomaly lineations off the west coast of North America. Positive anomalies are shown in black. Note the offset of the magnetic anomaly pattern at the fracture zones and the change in the amount of offset along the Murray fracture zone. After MENARD (1964), *Marine geology of the Pacific*, p. 46, McGraw-Hill.

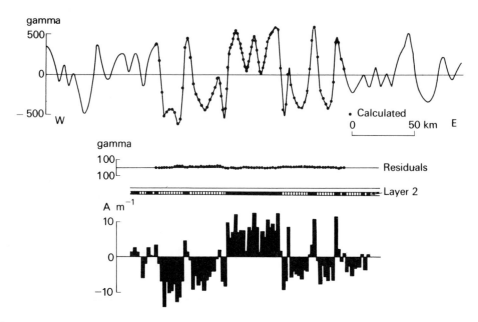

Fig. 3.8 Interpretation of a magnetic anomaly profile across the Juan de Fuca ridge, north-eastern Pacific, in terms of a distribution of magnetization within layer 2, obtained using a matrix inversion method. Layer 2 has been subdivided into two-dimensional rectangular blocks, and the magnetization of each block has been computed and is shown graphically below. The residuals are the observed minus calculated values of magnetic anomaly. Note the symmetry of the anomalies and of the interpreted magnetization distribution about the centre of the ridge (which is at the centre of the profile). Redrawn from BOTT (1967b), *Geophys. J. R. astr. Soc.*, **13**, 320.

east-west fracture zones, where they are apparently displaced laterally by up to more than 1100 km. The individual strips are about 10–20 km wide and the peak-to-peak amplitudes reach up to $1000\,\gamma$ (Fig. 3.8).

Later work has shown that strip-like magnetic anomalies are typical of oceanic regions, although the pattern is not everywhere quite as regular as off the west coast of North America. Furthermore, it has been found that the strips run parallel to the crests of ocean ridges and that the pattern of anomalies is symmetrical about the crests (Fig. 3.8), allowing for lack of symmetry in the shapes of anomalies caused by the local magnetic field direction not being vertical in general. These anomalies are in strong contrast with continental magnetic anomalies, which show no such simple and regular pattern and are generally of much smaller amplitude.

These large amplitude magnetic anomalies with steep gradients must be caused by highly magnetic rocks at relatively shallow depth. The sediments of layer 1 are effectively non-magnetic and cannot contribute significantly to the anomalies. An interpretation of a profile across the Juan de Fuca ridge in the north-eastern Pacific (Fig. 3.7) in terms of lateral variations of magnetization within layer 2 is shown in Fig. 3.8. In the model, the positive anomalies are underlain by magnetization in the direction of the Earth's field and the negative anomalies are underlain by reverse magnetization; however, a layer of uniform magnetization could be added without affecting the agreement so that the anomalies could equally well be explained by alternate strips of more and less strong magnetization. A contrast in magnetization of the order of $20\,\mathrm{A\,m^{-1}}$ is needed to explain the observed amplitudes. If one attempts to interpret such large amplitude anomalies in terms of magnetization within layer 3 alone, it is found that excessively strong and highly irregular patterns of magnetization are needed (BOTT, 1967b). It is therefore clear that the anomalies are caused mainly by strong lateral variations in magnetization within layer 2 although magnetization within layer 3 may also contribute. This is consistent with the basaltic composition of layer 2, as basalt or intrusive rocks of basic composition are the only common rocks which are sufficiently magnetic to cause such large and consistent anomalies. The origin of the strips of strongly contrasting magnetization remained obscure until they could be related to sea-floor spreading by the Vine-Matthews hypothesis.

The idea of sea-floor spreading

The modern concept of the origin and development of the oceans dates from the suggestion of HESS (1962) and DIETZ (1961) that new oceanic lithosphere, including the crust, forms beneath the crests of ocean ridges by the progressive upwelling of mantle material from greater depths. The ocean floor spreads laterally in both directions to accommodate the newly formed lithospheric material (Fig. 3.9). It was suggested that this is the process by which continental drift occurs (p. 77). North America and Europe, for instance, gradually drifted apart as new oceanic crust was formed between them producing the Atlantic Ocean over the last 200 My or thereabouts. On the other hand, sea-floor spreading in the Pacific must be related to sinking of oceanic lithosphere at or near the active continental margins around forming the circum-Pacific belt.

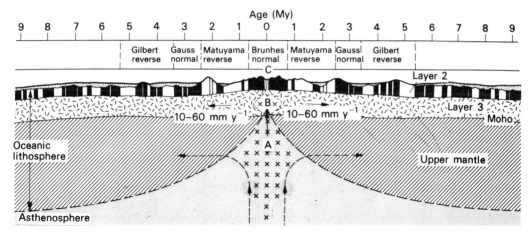

Fig 3.9 The sea-floor spreading hypothesis and the related process of formation of new oceanic lithosphere by upwelling and of new oceanic crust by separation of the magma fraction, as the lithosphere on either side spreads laterally at about 10–60 mm y^{-1}:
  A marks the rise of basalt magma to form the crust, with the residual material being left to form the mantle part of the oceanic lithosphere;
  B marks the magma chamber which cools to form layer 3;
  C marks the rapid cooling of basalt magma to form the pillow lavas and dykes of layer 2.
In layer 2, normal magnetic polarity is shown black and reverse polarity is shown white, using the geomagnetic time scale described later in the chapter. Vertical exaggeration is about 10:1 for a spreading rate of 30 mm y^{-1}.

The spreading ocean floor can be regarded as the uppermost visible part of a system of mantle convection, as discussed in Chapter 9. Whatever the flow pattern at greater depth, an upwelling current beneath the ridge crests brings hot asthenospheric material almost to the surface as it turns to flow horizontally in both directions and cools to form the oceanic lithosphere. As pressure is released during the upward flow, partial fusion starts at a depth probably between 150 and 30 km and the upwelling mantle material becomes increasingly enriched in a basalt magma fraction (Fig. 3.9). A small fraction of the upwelling material separates to the top where it cools rapidly to form new oceanic crust. The remainder is pushed laterally as it cools more slowly to form the mantle part of the oceanic lithosphere which eventually attains a thickness of about 100 km.

Hess originally suggested that a mush of upper mantle rock, basaltic magma and water separates to form the crust, the magma rising to the top to form layer 2 and the water and residual solid ultrabasic material reacting to form a layer 3 composed of serpentinite. The prevalent view now is that the crust forms entirely from the magma fraction which rises to the top beneath the ridge crest because of its low density. Part of the magma is quenched rapidly to form layer 2 and the remainder cools more slowly to form the gabbroic rocks of layer 3. If partial fusion in the upwelling material starts at 100 km depth, then the magma fraction needed to form a 6 km thick crust is 6 % and the rock forming the mantle part of the lithosphere will be depleted throughout by removal of the basalt fraction. On the other hand, if partial fusion does not start until 30 km depth, then only the uppermost 30 km of the oceanic lithosphere will be depleted having lost a 20 % basalt fraction.

The Vine-Matthews hypothesis

Spectacular support for the sea-floor spreading hypothesis has come from the study of oceanic magnetic anomalies. The first important step was taken by VINE and MATTHEWS (1963). They suggested that the alternating strips of positive and negative magnetic anomaly are caused by underlying blocks of layer 2 alternatively magnetized in the normal and reverse directions of the Earth's magnetic field (Figs 3.8 and 3.9). They pointed out that this interpretation follows almost as a corollary from the combination of the ideas of sea-floor spreading and of periodic reversal of the Earth's magnetic field as had been established shortly before by palaeomagnetic observations (p. 251). As new crust is formed in the crestal zone of an ocean ridge by igneous processes, shortly after solidification it cools through the Curie point and picks up a strong component of permanent (remanent) magnetization in the direction of the ambient field. This swamps the induced magnetization, so that blocks magnetized when the field was reversed produce magnetic anomalies of opposite sign. The newly magnetized block is then split and forced apart to make room for fresh injections of new crust (Fig. 3.9).

According to the Vine-Matthews hypothesis, the past history of reversals of the Earth's main magnetic field is fossilized in the oceanic crust at least as far back as the late Mesozoic. The oceanic layer 2 has acted as a magnetic tape, recording the polarity of the magnetic field as new crust is formed at the crest of ridges (VINE, 1966). This crustal 'tape-recorder' can be replayed by observing the magnetic anomalies along profiles crossing the ocean ridges. The record is distorted by noise, mainly produced by variation in the depth to the upper surface of layer 2. Provided the noise does not swamp the signal, study of oceanic magnetic anomalies can be used to study the past character of the main magnetic field and to provide a tool for dating the formation of oceanic crust.

Before the oceanic magnetic anomalies could be used in these ways, it was necessary to verify the Vine-Matthews hypothesis by showing that the observed pattern of magnetic anomalies near ridge crests is consistent with an independently observed history of magnetic reversals over the last few million years. Such a palaeomagnetic time-scale of reversals was first obtained by COX and others (1964) who compiled the measured remanent magnetization directions of lavas from various land localities which had also been reliably dated by the potassium-argon method. This time-scale went back 4 My and included two main periods (epochs) of normal and two of reverse magnetization. According to the revision of NESS and others (1980), the scale is as follows:

Epoch	Magnetic polarity	Age range (My)
Brunhes	normal	0·72–present
Matuyama	reverse	2·47–0·72
Gauss	normal	3·40–2·47
Gilbert	reverse	–3·40

This time scale includes four short periods of normal polarity during the Matuyama reverse epoch and two of reverse polarity during the Gauss normal epoch. These are called the Jaramillo (0·9 My), Gilsa (1·66–1·87 My), Olduvai (2·0 and 2·1 My), Kaena (2·9 My) and Mammoth (3·1 My) events.

Shortly after the discovery of this time-scale, palaeomagnetic measurements on deep-sea cores gave further confirmation of the scale and allowed it to be accurately tied in to the faunal zones of the oceans (OPDYKE and others, 1966; HAYS and OPDYKE, 1967). Cores from high latitudes are most convenient to use because the magnetic field is nearly vertical so that accurate polarities can be obtained without need to orientate the core. Results from some antarctic cores are shown in Fig. 3.10, illustrating the excellent agreement between the different cores and with the faunal zones based on microfossils. Deep-sea cores have the advantage over the use of lavas in that the record is known to be continuous and relative ages can be resolved much more accurately. The excellent agreement between the results obtained using deep-sea cores and lava successions on land demonstrated beyond reasonable doubt that the geomagnetic field has reversed at irregular intervals over the last 5 My. A magnetic time-scale had been firmly established going back about 5 My and a new method of stratigraphical correlation had been introduced. Subsequently the record of reversals based on deep-sea cores has been extended back 20 My (OPDYKE and others, 1974; THEYER and HAMMOND, 1974).

Using this palaeomagnetic time-scale of reversals, PITMAN and HEIRTZLER (1966) and VINE (1966) independently computed the theoretical magnetic anomaly profiles which would be produced near ocean-ridge crests for various spreading-floor rates. The agreement with the observed magnetic anomaly profiles was found to be excellent provided that appropriate spreading rates were chosen, as shown for the Juan de Fuca ridge and the East Pacific rise out to 4 My age in Fig. 3.11. The excellent agreement between the theoretical and observed patterns of anomalies out to 4 My age on all the profiles studied confirms the validity of the Vine-Matthews hypothesis and thus verifies the sea-floor spreading hypothesis. The comparisons also show that the magnetic anomalies produced by the irregular topography of layer 2 do not swamp the anomaly pattern caused by reversals of polarity of the magnetic field. The magnetic anomalies can therefore be used to work out the average spreading rate over the last 4 My. The early results were as follows:

Ridge	Spreading rate (mm y^{-1})	Reference
Juan de Fuca (46° N)	29	Vine
East Pacific rise (51° S)	44	Vine
N.W. Indian Ocean (5° N)	15	Vine
South Atlantic (38° S)	15	Vine
Pacific-Antarctic	45	Pitman and Heirtzler
Reykjanes	10	Pitman and Heirtzler

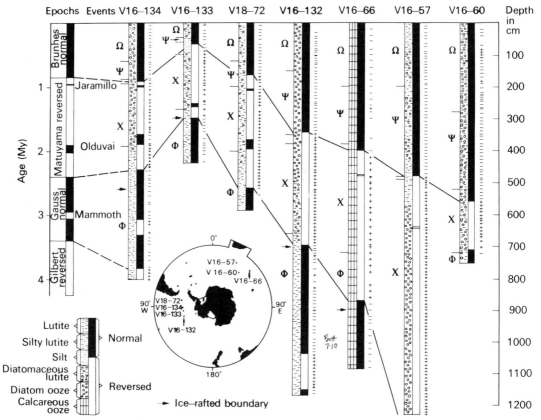

Fig. 3.10 Correlation of the magnetic stratigraphy in seven cores from the Antarctic, from locations shown in the inset. The magnetic polarity of the individual specimens is shown to the right of each section, minus signs indicating normal and plus signs indicating reversed polarity. Boundaries between the fossil zones are shown to the left of each section. Reproduced from OPDYKE and others (1966), *Science, N. Y.*, **154**, 350.

The rate of crustal separation across a ridge is twice the spreading rate. Applying this to the Atlantic Ocean, the present spreading rates are consistent with the formation of the ocean during the Mesozoic and Tertiary. Thus the sea-floor spreading hypothesis explains quantitatively how continental drift has taken place.

Establishing the geomagnetic time scale and dating the ocean floor

Once the Vine-Matthews hypothesis had been verified back to an age of 4 My, the way was open to use oceanic magnetic anomalies to construct a geomagnetic time scale of reversals going much further back in time. Soon after, both PITMAN and HEIRTZLER (1966) and VINE (1966) showed that a single sequence of normally and reversely magnetized blocks going back about 10 My could explain the magnetic anomalies across a wide selection of ocean ridge crests. Fig. 3.12 shows the good agreement obtained for the profiles across the Pacific-Antarctic and Reykjanes ridges provided that appropriate spreading rates are assumed. Later work has shown that the good agreement between the oceans extends back to the Jurassic (HEIRTZLER and others, 1968; LARSON and PITMAN, 1972).

The practical problem of constructing a geomagnetic time scale of reversals is to know how the spreading rate has varied in the past and whether there are any significant breaks in spreading

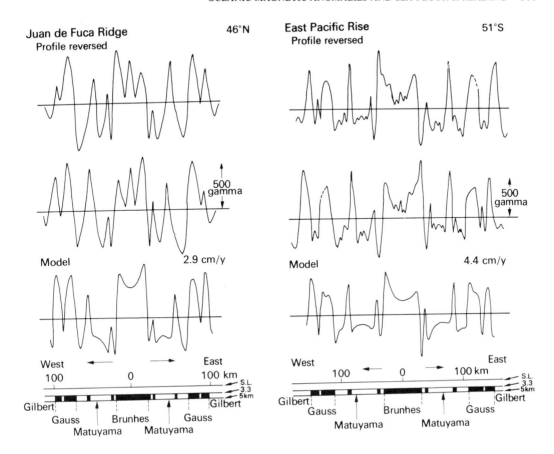

Fig. 3.11 Observed magnetic anomaly profiles across the Juan de Fuca ridge and the East Pacific rise compared with *above* the profile in the reverse direction to demonstrate the symmetry about the ridge crests, and *below* the theoretical magnetic anomaly according to the Vine-Matthews hypothesis assuming the palaeomagnetic time-scale of reversals and an appropriate spreading rate. It is assumed that the magnetic blocks are confined to layer 2. Black denotes normal magnetization, unshaded denotes reverse. Redrawn from VINE (1966), *Science, N. Y.*, **154**, 1409.

history. When the spacings of the magnetic anomaly patterns of different oceans are compared, it is apparent that the spreading rates of most of them must have varied. Any time scale therefore needs to be calibrated by independent dating of the reversals at a selection of points. The calibration was first done by dating the oldest sediments above layer 2 encountered in deep sea drill holes but the detection of reversal sequences in thick limestone successions on land accurately datable by fossils is now being increasingly used to provide a more accurate calibration of the scale.

HEIRTZLER and others (1968) originally constructed a geomagnetic time scale going back 80 My by assuming that the spreading rate in the South Atlantic has remained constant during the formation of the ocean. They did this by matching the observed magnetic anomaly profiles across the ocean with theoretical profiles computed for an oceanwide sequence of normally and reversely magnetized blocks forming layer 2, symmetrical about the ridge crest. They obtained an acceptable fit by trial and error. The resulting spatial sequence of magnetized blocks of opposite polarity could then be converted into a geomagnetic time sequence of reversals provided that the spreading rate over the last 80 My has been the same as that over the last 4 My. On this time scale,

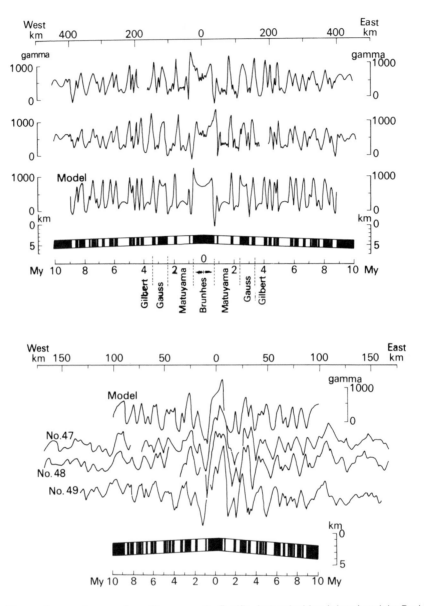

Fig. 3.12 Observed magnetic anomaly profiles across the Pacific-Antarctic ridge (*above*) and the Reykjanes ridge (*below*) compared with computed models according to the Vine-Matthews hypothesis with the same sequence of normally and reversely magnetized blocks, but different spreading rates of 45 and 10 mm y^{-1} respectively. The Pacific-Antarctic profile is shown in the reverse direction at the top to indicate the symmetry. Reproduced from PITMAN and HEIRTZLER (1966), *Science, N. Y.*, **154**, 1166.

the prominent normal polarity epochs were numbered 1 to 32, going back in time. The time scale had to be regarded as a working hypothesis until independent calibration could be obtained.

The Heirtzler time scale was first independently checked during the third leg of the Deep Sea Drilling Project when holes were drilled at sites stretching across the South Atlantic at 30° S as shown in Fig. 3.13 (MAXWELL and others, 1970). The lavas at the top of layer 2 were too badly

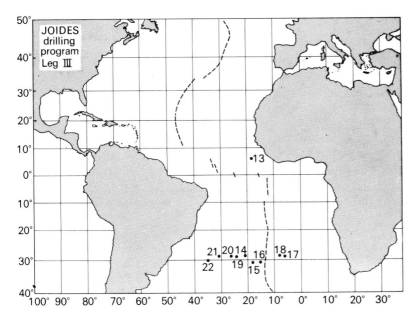

Fig. 3.13 Drilling sites occupied during leg 3 of the Deep Sea Drilling Project. Redrawn from MAXWELL and others (1970), *Science, N. Y.,* **168**, 1048.

altered to give reliable radiometric ages and the best that could be done was to date the oldest sediments above layer 2 using fossils, giving a slight underestimate of the true crustal age depending on the time gap before the first sediments were deposited on the new crust. Remarkable results were obtained. The age of the oldest sediments was found to show a closely linear increase with distance from the ridge crest (Fig. 3.14) and agreed within a few million years with the age of the crust inferred from the Heirtzler time scale (Table 3.1).

These results from leg 3 of the Deep Sea Drilling Project were of exceptional scientific importance. They gave the first direct confirmation of the predictions of the Vine-Matthews

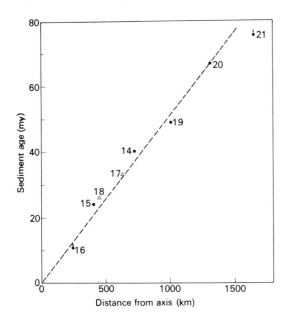

Fig. 3.14 Age of the oldest sediments at leg 3 drilling sites of DSDP plotted as a function of distance from the axis of the mid-Atlantic ridge. Numbers next to points indicate site numbers (Fig. 3.13), the triangles showing sites on the eastern flank. Arrows at sites 16 and 21 indicate possibility of older sediments directly above basalt which were not recovered. Redrawn from MAXWELL and others (1970), *Science, N. Y.,* **168**, 1055.

Table 3.1 South mid-Atlantic ridge drilling sites (MAXWELL and others, 1970).

Site No.	Sediment thickness (metres)	Hypothesized magnetic anomaly age (My)	Sediment age (My)	Distance from ridge axis (km)
16	175	9	11	191
15	141	21	24	380
18	178	—*	26	506
17	124	34–38*	33	643
14	107	38–39	40	727
19	141	53	49	990
20	72	70–72	67	1270
21	131†	—	76†	1617

* Location of these sites within the characteristic magnetic anomaly pattern is uncertain.
† Basement rock not reached at site 21.

hypothesis and of the theory of sea-floor spreading. Moreover, they confirmed the reasonable accuracy of the Heirtzler geomagnetic time scale and gave confidence in the use of the scale back to an age of 80 My with an accuracy of about ±5 My. The results show that spreading in the South Atlantic Ocean appears to have been a continuous process over the last 80 My. This must also apply to the Pacific and Indian Oceans where the same sequence of magnetic anomalies has been observed, although with some shifts of the spreading axis. This suggests that there are no significant breaks in spreading in the main oceans, such as have been recognized in the Red Sea by GIRDLER and STYLES (1974). However, the relatively uniform spreading rate in the South Atlantic confirmed by the drilling suggests that the changes in the spreading rates for the Pacific and Indian Oceans must be real.

The geomagnetic time scale was extended back to about 160 My by LARSON and PITMAN (1972) on the basis of a set of older magnetic lineations observed in three separate regions of the western Pacific. These are best known as the Hawaiian lineations. The same sequence of magnetic anomalies is also observed in the Atlantic near the North American and North African margins, where they are known as the Keathley lineations. The time scale was constructed by assuming a uniform spreading rate in the Pacific and was calibrated using two DSDP holes, one in the Pacific and one in the Atlantic near Bermuda. Conspicuous marker levels in the time scale were numbered M1 to M22, going back in time.

An updated version of the reversal time scale going back to an age of 160 My is shown in Fig. 3.15. This incorporates the time scale of NESS and others (1980) for the last 100 My. The four calibration points on this time scale are as follows. A radiometrically determined age of 3·40 My marks the base of the Gauss normal polarity epoch (MANKINEN and DALRYMPLE, 1979). A polarity reversal at 10·30 My has also been radiometrically dated. An age of 54·9 My is assigned to the base of anomaly 24 on the inference that it coincides with the Palaeocene-Eocene boundary, but it is possible that this point may be up to 5 My younger. The base of anomaly 29 is fixed at 66·6 My on the basis of palaeontological dating of the reversal sequence observed in an Upper Cretaceous to Palaeocene limestone succession studied near Gubbio in Italy (ALVAREZ and others, 1977). The time scale is interpolated between these four fixed points and the scale is extrapolated between 67

Fig. 3.15 The geomagnetic polarity time scale for the last 160 My, based on NESS and others (1980) back to 90 My age and LARSON and HILDE (1975) for the earlier part of the scale. Periods of normal polarity are shown in black and the numbers assigned to prominent magnetic anomalies are shown to the right of each column. Note that the scale from 80 to 160 My age is condensed by a factor of two. Calibration points back to 90 My age are marked by arrows.

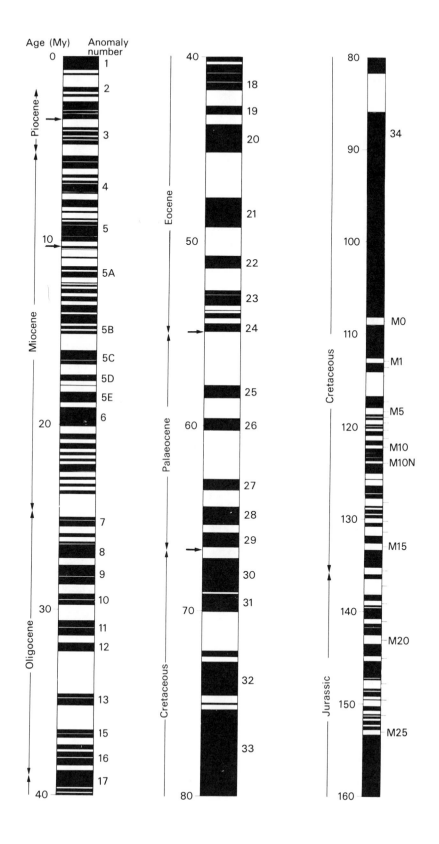

Age (My)   Anomaly number

and 86 My ages on the assumption that the spreading rate in a selected part of the North Pacific was constant between 55 and 80 My (LABRECQUE and others, 1977). Beyond 100 My, the revised scale of LARSON and HILDE (1975), based on the Hawaiian lineations and calibrated using several DSDP holes, has been incorporated in Fig. 3.15. In general, the time scale is probably accurate to a few million years at worst but undoubtedly it will be further updated as improved calibration is obtained. It should be noted that the resolution deteriorates beyond 23 My age as a result of more poorly defined magnetic anomalies and lack of supplementary evidence from cores.

The practical use of the time scale to date observed magnetic lineations depends on the irregular character of the pattern of reversals. For instance anomaly 5, which is about 10 My old, marks a change in the character of the reversal pattern. Anomaly 31 and the sequence 21 to 24 provide easily recognizable markers. Particularly striking are the two long periods of normal polarity in the Cretaceous (86 to 108 My) and Upper Jurassic (153 My back) which give rise to conspicuous quiet magnetic zones where large amplitude magnetic anomalies are lacking. Where the identification is less obvious, it may be necessary to compute the theoretical anomaly pattern for selected parts of the time scale for comparison with the observed pattern. The identification of the anomalies is most difficult where the lineations are north-south orientated at equatorial latitudes, as rocks horizontally magnetized along the strike direction do not produce detectable magnetic anomalies. Once marker anomalies have been identified, they can be mapped on the ocean floor by magnetic survey and the history of sea-floor spreading can be worked out as far back as the mapped horizons are available. Most of the floor of the major oceans has now been dated in this way and it has been found that the oldest parts only go back to the Jurassic. Thus the history of continental drift since the early Mesozoic can now be worked out in detail, although discussion of this is outside the scope of the book.

Magnetization of the oceanic crust

A simple pattern of crustal magnetization was assumed during the testing of the Vine-Matthews hypothesis and the establishment of the geomagnetic time scale. The steep magnetic anomaly gradients observed indicated that a significant part of the anomaly must be caused by layer 2 magnetization and it was assumed that uniformly magnetized normal and reverse polarity zones extended over the thickness of layer 2 with vertical boundaries separating them. However, the observed anomalies could equally well be explained by a thinner layer of more strongly magnetized rocks or by a composite source including layer 3. A direct estimate of the magnetization was made by TALWANI and others (1971) by running profiles along the magnetic strips parallel to the crest of the Reykjanes ridge. They assumed that the layer causing the magnetic anomalies is uniformly magnetized and has a flat bottom. Variations in magnetic anomaly along the magnetic strips must then be attributed to the topography on the upper surface of layer 2. By correlating the observed magnetic anomaly with the sub-sediment bathymetry, they were able to show that the magnetization below the central Bruhnes anomaly is $30\,A\,m^{-1}$ and that it decreases to $12\,A\,m^{-1}$ at 75 km from the ridge crest and to about $7\,A\,m^{-1}$ at anomaly 5 which is 100 km from the crest. Using a profile across the ridge, they then used these magnetization values to show that the magnetic layer is 400 m thick if uniform magnetization is assumed. This suggested that the upper part of layer 2, rather than the whole of the layer, is the main source of the magnetic anomalies. A similar result was obtained over the Gorda rise off northern California by ATWATER and MUDIE (1973) where results obtained using a deep-towed magnetometer indicated a magnetization of about $8\,A\,m^{-1}$ and a magnetic layer 500 m thick.

Some major problems arising out of magnetic measurements made on DSDP cores have been reviewed by HALL (1976). These cores show a much more complex pattern of magnetization than that inferred by interpretation of magnetic surveys. Three types of magnetization are recognized in the cores. Induced magnetization is relatively unimportant and can be ignored. Stable remanent magnetization picked up when the rock cooled at its time of formation gives rise to the observed magnetic polarity zones. Viscous remanent magnetization picked up more recently is also significant and helps to confuse the pattern. Stable remanent magnetization is found to be dominant in the pillow basalts near the top of layer 2 but the relative importance of stable and viscous magnetization in the massive basalts is unclear. The measurements also indicate that the directions and polarities of magnetization vary unexpectedly within a single drill hole. In some (but not all) instances this may be the result of drilling at or near a polarity transition zone. Another puzzling feature is that the average magnetization measured appears to be significantly lower than the values inferred from magnetic survey interpretation, suggesting that the magnetic layer may be thicker than 500 m, possibly extending down into layer 3. Thus the drilling shows a lot of confused local detail whereas the magnetic surveys reveal the broad pattern. On the other hand, the sample measurements show that magnetization may be reduced with time by reaction between basalt and cold seawater, in agreement with deductions from magnetic surveys. On balance, it is clear that much further work is needed on drill cores before the nature of the oceanic crustal magnetization is properly understood.

The magnetic anomaly profiles can be used to estimate the width of the transition between adjacent polarity zones within the oceanic crust. According to ATWATER and MUDIE (1973), the transition in the late Tertiary oceanic crust of the Gorda rise occurs over about 1·7 km. Wider transition zones have been inferred for older crust in the Pacific, suggesting that the transition zone may broaden with age. The reversal of the geomagnetic field is known to occur sufficiently rapidly to produce an effectively sharp boundary. Thus the finite width of the transition zone may result from either a finite width of the injection zone at the ridge crest or deep-seated crustal contributions to the anomalies or both.

A two-layered model of oceanic crustal magnetization has been proposed by CANDE and KENT (1976). The strongly magnetized upper layer in the model comprises the uppermost 500 m of layer 2 and this contributes about 75 % of the amplitude of the observed magnetic anomalies. This is underlain by a thicker, more weakly magnetized layer roughly corresponding to layer 3 which produces the remaining 25 % of the anomalies. The polarity boundaries in the rapidly quenched upper layer are effectively vertical but in the more slowly cooling lower layer they dip outwards from the ridge crest. This outward dip on the boundaries between magnetic blocks in the lower layer overcomes a misfit between the polarity patterns on opposite sides of ridges which occurs in some regions when vertical boundaries are used. It also accounts for the apparent broadness of the transition between polarity zones.

3.4 Ocean ridges

It was shown above that new oceanic crust is probably formed beneath the crestal zones of oceanic ridges. The ocean ridges are thus of great importance in understanding the origin of the oceans and the formation process of oceanic crust.

Bathymetry and geology

The ocean ridge system forms the largest uplifted linear surface feature of the Earth (Fig. 3.16). Intensive study of the ridge system dates from about 1956, when Ewing and Heezen found that the belt of shallow focus earthquakes following the crest is continuous and that the individual ridges form an interconnected world-wide system. It has a total length of 65 000 km and individual ridges are typically about 1000 to 4000 km wide. The crests rise about 2–3 km above the average depth of

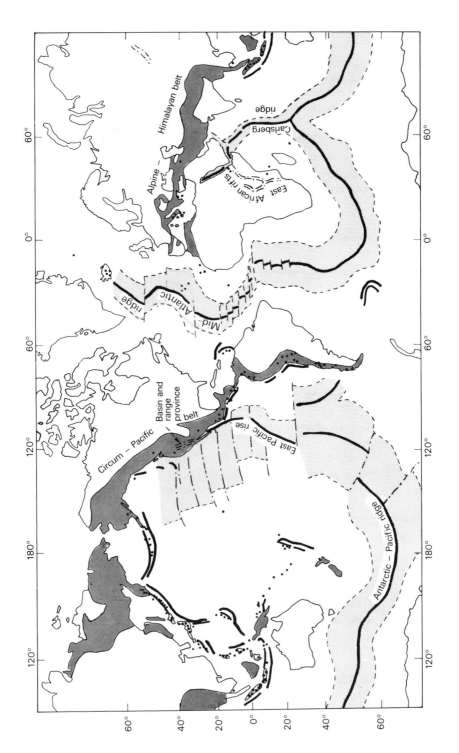

Fig. 3.16 Map showing the Earth's major surface features. Tertiary and recent mountain ranges and island arcs are shown in dark tint. Ocean trenches are shown as thick black lines. Ocean ridges and their crests are also shown. Major active or recently active volcanoes are marked as black dots. Compiled from various sources.

the adjacent ocean basins. On a broad scale the ridges are gently arched uplifts but locally the topography can be extremely rugged (Fig. 3.1). The seamounts forming the sub-bottom tend to be elongated parallel to the crest. Transverse fracture zones, discussed later in the chapter, form prominent topographical features crossing certain parts of the ridges.

In the Indian and Atlantic Oceans, where the areal extent is being increased by sea-floor spreading, the crests of the ridges lie close to the median line of the ocean. The ridges are relatively narrow and rugged, and a prominent deep trench known as the median rift is typically present along the crest. In the FAMOUS region of the mid-Atlantic ridge a 50 km long segment of the median rift valley lying between two small fracture zones (36·5° to 37° N) has been studied in great detail (HEIRTZLER and VAN ANDEL, 1977, and succeeding papers in the April-May numbers of the Bulletin of the Geological Society of America, volume 88). The median valley here is about 30 km wide. The peaks on either side are about 1300 m deep and the rift-floor reaches between 2500 and 2800 m deep. There is commonly an inner rift floor about 1–4 km wide which is flanked by a series of fault-controlled terraces but elsewhere the floor is wider and the terraces less conspicuous. The floor of the inner valley is formed of a series of low hills and depressions elongated parallel to the valley, with evidence of intense fissure igneous activity and small scale faulting and fracturing. The active volcanism suggests that the inner rift floor is the site of current formation of new crust and that it may be underlain by a magma chamber below. The normal faults marking out the central valley, the terraces and the rim mountains probably provide the mechanism by which the crustal blocks are progressively uplifted eventually to form the rim mountains as sea-floor spreading carries them laterally (Fig. 3.17).

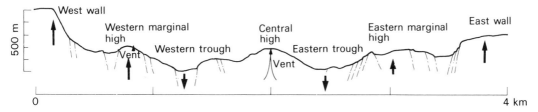

Fig. 3.17 Diagrammatical cross section of the inner rift valley of the mid-Atlantic ridge in the FAMOUS region (about 36° 50′ N), based on several dive traverses using submersibles and showing the main structural units. No vertical exaggeration. Redrawn from BALLARD and VAN ANDEL (1977), *Bull. geol. Soc. Am.*, **88**, 524.

In contrast to the mid-Atlantic ridge, the East Pacific rise is located at the east side of the shrinking Pacific Ocean. It is much wider and less rugged than the other ridges discussed above and it has no prominent trench along the crest of it. These contrasts between the Atlantic and Pacific types of ridge are probably fundamental, reflecting the differences in spreading rate or tectonic setting of these two types of ocean.

The East Pacific rise passes into North America at the Gulf of California and the extension of the Carlsberg ridge of the Indian Ocean passes into the Gulf of Aden, where it divides, one branch continuing up the Red Sea and the other passing into the East African rift system (p. 63). Both western North America and East Africa are regions where plateau uplift of about 1·5–2 km has occurred during the Tertiary.

Rock samples dredged from the ridges show that basalt of the abyssal olivine-tholeiite type is by far the commonest rock. It commonly occurs as pillow lava. Gabbro, serpentinite and other igneous rock types, including some metamorphosed varieties, are also found, particularly along the scarps of fracture zones. The islands along the crest are almost all basaltic volcanoes, although St Paul rocks lying on an east-west fracture zone on the equatorial mid-Atlantic ridge is formed of coarse-grained ultrabasic rocks including serpentinite. Iceland is the largest 'outcrop' of an ocean ridge and it is formed of basalt and other igneous types of late Tertiary to Recent age. Recent volcanic activity occurs in the central graben, and the rocks become progressively older towards the east and west. Recent volcanic activity is typical of the whole crestal region of ocean ridges and must account for the major proportion of all terrestrial volcanism; in contrast, continental volcanism is relatively insignificant.

Broad structure of crust and upper mantle beneath ridges

The pioneer submarine gravity measurements made by VENING MEINESZ (1948) showed that the mid-Atlantic ridge crest at 40°N is in approximate isostatic equilibrium. Subsequently, surface ship gravimeter traverses across the ridges have confirmed that they are in approximate isostatic equilibrium, although local topographic features are uncompensated and cause the gravity profiles to be irregular. Over the East Pacific rise and the mid-Atlantic ridge (Fig. 3.18), there are small positive free air anomalies over the axial region and slightly negative anomalies on and beyond the flanks. This would be expected for relatively deep isostatic compensation. There may be some small deviation from perfect equilibrium at the ridge crest resulting from the tectonic processes occurring below.

Seismic refraction measurements of crustal structure across the East Pacific rise show that the layers 2 and 3 are apparently broadly continuous across the crest, apart from a narrow magma chamber beneath the crest (p. 118). The crust beneath the crest tends to be slightly thinner than that

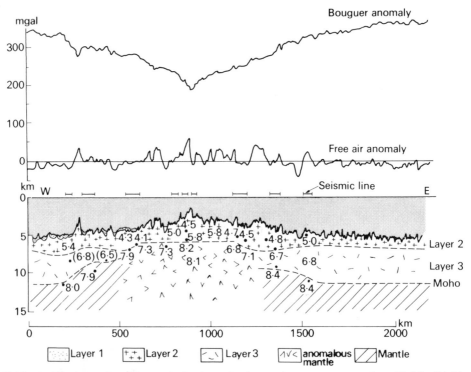

Fig. 3.18 Gravity anomalies and seismically determined crustal structure across the mid-Atlantic ridge, from (36·6°N, 48·3°W) to (25·5°N, 29·8°W) approximately. The Bouguer anomalies were obtained by filling up the sediment basins and the oceans with rock of density 2600 kg m⁻³. *P* velocities shown in km s⁻¹. Note that more recent investigations suggest that layer 3 is continuous across the ridge crest. Redrawn from TALWANI and others (1965), *J. geophys. Res.*, **70**, 343.

beneath the adjacent ocean basins, and the topmost mantle velocity beneath the crest is anomalously low. Across the mid-Atlantic ridge (Fig. 3.18), layer 1 rocks are present only in intermontane basins. Layer 2 continues uninterrupted across the crest. The older refraction work suggested that layer 3 becomes confused with the topmost mantle beneath the crestal region where P velocities are highly variable, but more recent work (WHITMARSH, 1975; FOWLER, 1976) has emphasized the continuity of layer 3 and the Moho across the crestal region in the FAMOUS area. However, the crust does not thicken – if anything it is slightly thinner beneath the ridge than beneath the adjacent ocean basins.

The seismic refraction evidence conclusively rules out the idea that the isostatic compensation beneath ridges is caused by crustal thickening. It must be caused by an anomalously low density region in the underlying upper mantle. On the assumption that the low density rocks are related to the region of anomalously low P velocity underlying the Moho at the ridge crest, TALWANI and others (1965) constructed models of the deep structure beneath the mid-Atlantic ridge satisfying both the gravity and seismic observations, using the empirical Nafe-Drake curve (Fig. 2.28) to relate density and velocity. One of their models is shown in Fig. 3.19. A difficulty here is to understand how such a large density contrast of -250 kg m^{-3} can occur in the uppermost mantle. An alternative model (Fig. 3.20) in which a density contrast of -40 kg m^{-3} extends to a depth of 200 km has been suggested by KEEN and TRAMONTINI (1970). Although these models cannot yield a unique interpretation, they do indicate two undisputable features: (*i*) a substantial thickness of anomalously low density rocks in the upper mantle is needed to explain the gravity anomalies, and (*ii*) the zone of anomalously low density rocks appears to widen downwards from the Moho.

Other important features of the deep structure beneath ocean ridges which are treated in more detail later in the book are summarized here. *Firstly*, an extensive region of low P velocity probably reaches down to a depth of the order of 200 km in the upper mantle beneath ocean ridges. This is mainly based on evidence from Iceland (p. 157) which is admittedly an anomalous region; here the delay of P arrivals from distant earthquakes and the surface wave dispersion characteristics indicate that a P velocity as low as 7·4 km s^{-1}, in contrast to normal average upper mantle values above 8·0 km s^{-1}, needs to extend down to a depth of at least 200 km to account for the observations. The S wave velocity anomaly is probably even more pronounced. This anomalous region of low seismic velocities may be similar to that observed to occur beneath the East African rift system (p. 63). *Secondly*, the ocean ridges are regions of high heat flow despite lowering of the observed values by hydrothermal circulation of seawater in the crust (p. 284). This high heat flow and the occurrence of volcanism at the ridge crests indicate that temperatures must reach the melting point of basalt at shallow depths beneath the ridges. *Thirdly*, surface wave dispersion studies in the Pacific region show that the lithosphere thickens away from the ridge crest where it is a few kilometres thick at the most, reaching about 35 km thickness at 15 My spreading age and 70 km at 50 My (p. 159). Thus the upper mantle beneath the ridges can be regarded as an 'intrusion' of hot asthenospheric material into the cooling and spreading lithosphere. *Fourthly*, the ocean ridge crests are associated with a world-circling belt of shallow-focus earthquakes for which focal mechanism studies indicate horizontal deviatoric tension perpendicular to the ridge crests (p. 131). This is consistent with the observed patterns of fissuring and normal faulting observed at ridge crests.

The deep structure of ocean ridges needs to be interpreted within the framework of the sea-floor spreading hypothesis. The anomalous deep structure should then be explicable in terms of the following two processes: (*i*) the upwelling of hot mantle material from below about 100 km depth, involving partial fusion followed by separation of the magma fraction to the top to form the crust

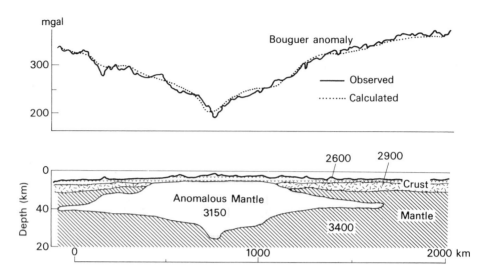

Fig. 3.19 A possible density model of the crustal and upper mantle structure beneath the mid-Atlantic ridge along the profile of Fig. 3.18. The densities (in kg m^{-3}) assigned to the various layers in the model are shown. Redrawn from TALWANI and others (1965), *J. geophys. Res.*, **70**, 348.

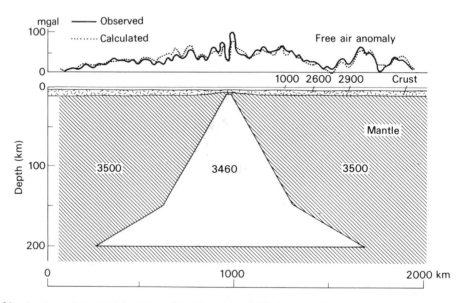

Fig. 3.20 An alternative type of density model to that of Fig. 3.19 of the crustal and upper mantle structure beneath the mid-Atlantic ridge, computed for a gravity profile crossing the ridge at 46° N. Densities are shown in kg m^{-3}. Redrawn from KEEN and TRAMONTINI (1970), *Geophys. J. R. astr. Soc.*, **20**, 487.

while the depleted material from which the magma has been removed goes to form part or all of the mantle part of the lithosphere; and (*ii*) cooling of the upwelled material as it spreads laterally in both directions away from the spreading axis. The horizontal scale of the cooling lithosphere greatly exceeds its vertical thickness so that cooling effectively takes place by vertical flow of heat. According to SCLATER and FRANCHETEAU (1970), the broad topography of ocean ridges relative to ocean basins can be attributed to the contraction of the lithosphere as it cools on spreading (p. 285). If the ocean ridge topography is caused by such cooling, then the width of each ridge should be proportional to the spreading rate. This is accurately borne out by observations which show that the sea depth relative to the ridge crest is proportional to $t^{\frac{1}{2}}$ for the first 70 My of spreading, where t is the age of the sea floor (DAVIS and LISTER, 1974). The bathymetric depth d in metres is then given by the empirical relationship

$$d = 2500 + 350 \, t^{\frac{1}{2}}$$

where t is measured in million years (PARSONS and SCLATER, 1977). This relationship readily accounts for the relative widths of the fast spreading East Pacific rise and the slow spreading mid-Atlantic ridge. The theory of the cooling oceanic lithosphere is further discussed following p. 284.

Within the framework of the spreading and cooling oceanic lithosphere, the anomalously low density region in the upper mantle which supports the elevation of the ridges can be attributed to three possible sources as follows. *Firstly*, the simple thermal contraction of the hot upwelled material beneath the ridge crests as it cools while spreading laterally should give rise to a relatively low density region beneath the ridges. It can be shown that such thermal contraction should give rise to a bathymetric profile proportional to $t^{\frac{1}{2}}$ as long as heat is not added to the base of the lithosphere (p. 287). In order to assess this contribution, let us assume that the average temperature down to 100 km below the Moho is 500 K hotter beneath the ridge crest than beneath the adjacent ocean basins. Taking the average density to be $3300 \, \mathrm{kg \, m^{-3}}$ and the volume coefficient of thermal expansion to be $3 \times 10^{-5} \, \mathrm{K^{-1}}$, the average mantle density down to 100 km beneath the ridge crest is $50 \, \mathrm{kg \, m^{-3}}$ less than beneath the ocean basins. Assuming isostatic equilibrium, this would support a ridge elevation of 2·2 km relative to the basin. This effect thus appears to provide an adequate explanation for most of the observed ridge topography. *Secondly*, the presence of a partially fused fraction within the upwelling material reduces the overall density. If the magma fraction is 1 % then the reduction in density would be about $6 \, \mathrm{kg \, m^{-3}}$. Extending over a 100 km depth range, this would support an additional increment of ridge elevation of 0·25 km. The effectiveness of this source of ridge elevation is difficult to assess as the magma fraction is unknown, this depending on the rate at which the magma migrates from its source at depth into the crust. *Thirdly*, the high temperature beneath ridges may give rise to a low density assemblage of minerals in the upper mantle. The most significant effect may be the presence of a lens-shaped body of low density plagioclase pyrolite (p. 187) in the mantle above about 40 km depth beneath ridges. This high temperature and low pressure mineral assemblage may be up to $70 \, \mathrm{kg \, m^{-3}}$ lower in density than the normal pyroxene pyrolite model of the upper mantle. Extending over a depth range of 20 km, this could isostatically support an additional ridge elevation of about 0·7 km. Such a lense resembles the low density region in the Talwani model (Fig. 3.19) except that the density contrast is smaller. One difficulty confronting this hypothesis is that plagioclase may not be able to develop in significant amounts if the mantle at these shallow depths is barren after removal of the fused fraction following partial melting.

Thus thermal contraction of the spreading oceanic lithosphere probably accounts for most of the elevation of ocean ridges relative to the basins. Partial fusion and solid-solid phase transitions may contribute to a lesser extent. According to this assessment, the Talwani model places too great emphasis on the lens of low density material at shallow mantle depth. The model of Keen and Tramontini provides a closer approximation to a realistic density distribution but some elements of both these simple models are probably present.

Birch's relationship between P velocity and density of solid rocks (p. 71) does not appear to apply to the anomalous upper mantle region beneath ocean ridges. The observed anomaly in P velocity beneath Iceland is much greater than the predictions of Birch's law based on the density anomaly would suggest. This severe deviation from Birch's law suggests that the low P velocities are mainly attributable to partial fusion rather than to the thermal effects on the solid rocks which predominantly cause the gravity anomalies. One may speculate that the upwelling material beneath ridge crests has significantly lower seismic velocities than the normal asthenosphere because of the extensive partial fusion.

Structure beneath the ridge crest

Ideas on the origin of oceanic crust which developed during the early 1970s (see section 3.5) suggested that a narrow magma chamber, which passes laterally into layer 3 on either side, should be present beneath ocean ridge crests. Evidence for such a magma chamber beneath the spreading axis of the East Pacific rise was obtained by ORCUTT and others (1976) from a seismic refraction line along the crest at 9° N using ocean bottom seismometers. They discovered a pronounced low velocity zone starting at 2 km below the seabed and extending down towards the Moho where P_n was observed to be $7{\cdot}7\,\mathrm{km\,s^{-1}}$ (Fig. 3.21). Above the low velocity zone, the P velocity increases downwards from $5{\cdot}4$ to $6{\cdot}7\,\mathrm{km\,s^{-1}}$. The low velocity zone is apparently absent in crust $2{\cdot}9$ My old and an almost normal crustal profile was found at 5 My spreading age. Similar results were obtained by ROSENDAHL and others (1976) in the same region using two ships and sonobuoys. They found that the P_n arrivals are significantly delayed over the crustal horst. Assuming a P velocity of

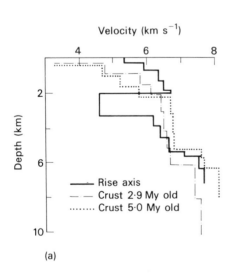

(a)

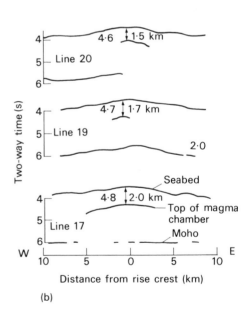

(b)

Fig. 3.21 Evidence for a magma chamber beneath the crest of the East Pacific rise at about 9° N:
  (a) Crustal velocity-depth models along the rise axis and on young oceanic crust on the west flank obtained by refraction lines using ocean bottom seismometers. The well-developed low velocity zone beneath the rise axis, which is absent from the other profiles, is interpreted as a magma chamber. Redrawn from ORCUTT and others (1976), *Geophys. J. R. astr. Soc.*, **45**, 317.
  (b) Line drawings of prominent reflectors obtained along three multi-channel seismic reflection profiles across the crest of the rise at between 9° and 10° N. The prominent reflector at about 0·5 s two-way travel time beneath the seabed is interpreted as the top of the crustal magma chamber and the deeper reflector is interpreted as the Moho. Average velocity down to the top crustal reflector shown in km s^{-1}. Redrawn from HERRON and others (1980), *Geophys. Res. Lett.*, **7**, 991.

5 km s^{-1} in the low velocity zone, they inferred that a magma chamber extended between depths of 2 and 5·5 km below the seabed, wedging out to zero thickness in both directions at 5 to 10 km from the spreading axis. Further evidence for the presence of a crustal magma chamber beneath the ridge crest at this location was obtained by HERRON and others (1980) using multi-channel seismic reflection profiling. A strong reflector 1·5 to 2·0 km below the seabed was interpreted as the top surface of a magma chamber between 2 and 8 km wide and the delay in the associated reflections from the Moho was attributed to the lowering of P velocity caused by the presence of the magma.

The search for a magma chamber beneath the median valley of the slower spreading mid-Atlantic ridge has been less conclusive. FRANCIS and PORTER (1973) used the limited range of earthquakes detected on an ocean bottom seismometer sited in the median valley at 45° N to suggest that the velocity decreases with depth in the crust starting about 2 km beneath the seabed. They interpreted this as evidence of a possible magma chamber beneath. WHITMARSH (1975) detected a shallow intrusion zone beneath the median valley at 37° N in the FAMOUS area but FOWLER (1976) found that propagation of shear waves across the crest in this region precluded the existence of any sizeable magma chamber at shallow depth. These results are conflicting but they do not necessarily rule out the presence of a small magma chamber about 1–2 km wide occupying the deeper part of layer 3 beneath the spreading axis.

Why is it that some ridges are rugged and rifted while others are smooth and lack a conspicuous median valley? Commonly the fast spreading ridges are the smooth ones and the slow spreading ridges are rifted, but there are exceptions such as the slow-spreading Reykjanes ridge which is rugged but lacks a median valley. The answer to this puzzle depends on knowledge of the mechanism of formation of the median rift.

One possibility is that low density material is added to thicken the crust as it spreads laterally from the median valley towards the flanking rim mountains, but there is no observational evidence for substantial accretion to the crust here. Another possibility is that the rifted ridges are periodically inflated and deflated producing mountains and valleys successively at the spreading centre; this would mean that the present time is a general period of deflation which seems implausible. Perhaps the most satisfactory explanation of the median valley is that the zone of upwelling hot asthenospheric material beneath the ridge crests exerts a viscous drag on the adjacent 'walls' of newly formed oceanic lithosphere on either side, thereby slightly uplifting the flanking lithosphere while depressing the region above the apex of the upwelling zone (SLEEP, 1969; OSMASTON, 1971). The process of uplifting the floor of the median valley into the flanking hills appears to take place by movement on a series of inward dipping normal faults in response to tension in the crust produced by the low density upwelling material. CANN (1974) suggested that the existence of a median rift valley is thus primarily related to the viscosity of the upwelling material rather than to spreading rate, a low viscosity corresponding to a smooth crest and a higher one corresponding to a rifted crest because of the increased shearing stress.

3.5 Origin of the oceanic crust

According to the sea-floor spreading theory as described above, oceanic crust is newly formed in a narrow zone beneath the central axis of ocean ridges. The new crust is formed from the basaltic magma fraction developed in the underlying upwelling mantle. The magma fraction rises to the top because of its low density and then it solidifies to form the crust. The upper part cools rapidly to form the lava-dyke complex of layer 2 and the lower part cools more slowly to form the gabbroic layer 3. The processes by which this takes place are examined in this section. There have been two main approaches to the modelling of the crust-forming processes, petrological modelling aided by observations on ophiolite suites and thermal modelling of the cooling of the magma. Both yield a similar picture of the crust-forming process.

The basic petrological model of CANN (1970, 1974) postulates that a linear magma chamber exists at the level of layer 3 beneath the ridge crest. Above the magma chamber, there is an upper zone of rapidly quenched lavas which goes to form layers 2A and 2B, and a lower zone of dykes which can probably be equated with layer 2C. Layer 3 is formed by crystallization of the magma in the axial chamber. This occurs in two ways. An upper layer of isotropic gabbro forms by crystallization from the roof downwards and a lower layer of cumulates is formed by settling of crystals onto the floor of the chamber. These two sublayers may possibly be identified with layers 3A and 3B respectively, assuming these to exist. According to this model, the magma chamber is widest at the depth of the layer 3A/layer 3B junction and narrows upwards and downwards. The lower part of the cumulate zone may on average be an ultrabasic rock rich in olivine with a seismic velocity indistinguishable from that of the underlying mantle. This gives rise to the contrasting concepts of the petrological Moho marking the base of the magma chamber beneath the spreading axis and the seismological Moho occurring within the cumulate pile at the level where the P velocity first reaches about 7.8 km s^{-1}. According to the petrological model, the newly formed igneous rocks may suffer metamorphism as a result of equilibration of the mineral assemblages at raised temperature in the presence of abundant seawater.

A strikingly similar picture of oceanic crustal structure has been derived more directly by study of ophiolite complexes found in orogenic fold belts. These complexes consist of layered sequences of ultrabasic and basic igneous rocks of characteristic type associated with radiolarian cherts and flysch deposits. The commonly occurring association of serpentinite, pillow lava and radiolarian chert is often referred to as the Steinmann Trinity, after the geologist who originally recognized it. The origin of the ophiolite complexes puzzled geologists until it was suggested that such rocks occurring in Cyprus in the Troodos mountains may represent a sequence of oceanic crust and topmost mantle thrust up to the Earth's surface during intense orogenic deformation (GASS, 1968). If this interpretation is correct, as is now widely accepted, then ophiolite sequences provide a cross section of oceanic crust available to be seen on land, yielding detail unavailable to geophysical study at sea.

As an example of an ophiolite sequence, the downward succession of rocks forming the Blow-me-down ophiolite massif of the Bay of Islands complex in Newfoundland (WILLIAMS, 1973; SALISBURY and CHRISTENSEN, 1978) is as follows:

0·0–1·0 km   Pillow basalts with minor intercalations of red chert, dykes common, metamorphosed to greenschist facies in lower part (layer 2A/2B)

1·0–1·5 km   A transitional zone of fine-grained dolerite dykes in greenschist metamorphic facies, brecciated probably by contact with seawater at high pressure, gradational boundaries above and below

1·5–2·6 km   Sub-vertical sheeted dolerite dykes emplaced against each other without intervening country rock and averaging 0·5 m wide, metamorphic facies passing down from greenschist to epidote-amphibolite, fairly sharp lower boundary of zone (layer 2C)

2·6–3·8 km   Relatively coarse-grained gabbro, with greenschist metamorphic facies overprinting epidote-amphibolite facies (layer 3A)

3·8–5·3 km   Predominantly pyroxene gabbro with cumulate layering developing towards the base

5·3–6·4 km   Interlayered anorthositic olivine gabbro, troctolite and plagioclase peridotite, probably formed mainly by crystal settling (layer 3B)

6·4–10·8 km   Beneath a 50 m thick transitional zone of crushed rocks, partially serpentinized peridotites of dunite and hartzburgite type predominate, the complex being separated from the underlying country rock by a thrust fault at its base (represents uppermost mantle)

Most levels in the complex have been profoundly influenced by the presence of abundant water (probably seawater) but the serpentinization near the base probably occurred during or after tectonic emplacement on land. The complex as a whole bears a remarkable resemblance to the independently derived petrological model of Cann. Tentative identification of the oceanic seismic crustal layering is shown above. Although the agreement here is excellent, the identification of all ophiolite complexes with oceanic crust is less straightforward—for instance several complexes such as that of Cyprus are much thinner than normal present-day oceanic crust.

KIDD (1977) has taken the petrological model one stage further by making computer simulations of the process of dyke emplacement and lava formation at the spreading centre and comparing these with observations on a selection of ophiolite complexes. He found that the central zone of the sheeted dyke complex is entirely formed of dykes, indicating 100% extension. The boundary between the dykes and the underlying gabbro is relatively sharp and the dykes show about 10% more margins chilled one way than the other way indicating that more than 10% of them intruded into the middle of earlier dykes before they had cooled. Both these observations, when compared with the simulations, indicate that the dykes must be almost all intruded within a narrow zone not more than about 50 m wide. Kidd suggested that most of the dykes reach the sea floor on emplacement and that the extruded lavas do not flow for more than 2 km on either side of the injection zone, being rapidly quenched by seawater. The lava pile thus builds up to its full thickness over about 2 km of lateral sea-floor spreading from the injection zone, the base of the lavas being progressively dropped by normal faulting and tilting to accommodate the later additions to the pile while allowing the seabed to remain approximately flat.

The commonly observed symmetry of sea-floor spreading about the injection zone is perhaps best explained by the preferential emplacement of new dykes up the hot central axis of the spreading zone. Asymmetrical spreading may arise either if the cooling of the dyke injection zone by seawater circulation occurs more rapidly on one side than the other, or if the spreading axis migrates by jumping laterally at intervals.

Further insight into the process by which oceanic crust forms can be obtained by thermal modelling of the cooling magma chamber beneath the spreading axis (SLEEP, 1975; KUSZNIR and BOTT, 1976). Kusznir and Bott assumed that seawater penetration rapidly cools the lavas of layer 2 but does not penetrate significantly below this level at the ridge crest. It is then possible to model the shape of a steady state magma chamber which cools in pace with the spreading rate by crystallization at its roof to form gabbro. Numerical methods were used to enable the release of the heat of fusion to be taken into account. These computations showed that a magma chamber should exist provided that the spreading rate exceeds about 5 mm y^{-1}. For slower spreading rates, or even for faster rates if significant seawater cooling extends below the layer 2 level, instantaneous dyke-like solidification of the magma would be expected to occur right down to the Moho. The computations show that the roof of the magma chamber dips steeply outwards below the axis and that its slope decreases laterally away from the axis. Crystal settling to form a cumulate layer at the base of the magma chamber has the effect of narrowing the magma chamber and flattening its roof, its greatest width occurring at the junction between the gabbros which solidify at the roof and the cumulates. Assuming that the magma chamber is 5 km in vertical extent below the axis and that cumulates form the lower third of layer 3, then the computations predict that the half-width of the magma chamber should be slightly above 1 km for 10 mm y^{-1} spreading rate and about 11 km for 60 mm y^{-1} rate. The computations also indicate that for faster spreading rates the thicknesses of the dyke complex and fine grained gabbros are reduced but that the underlying zone of cumulates is thickened.

There is excellent agreement between the petrological and thermal models of oceanic crustal formation and with observations of the rock succession in ophiolite complexes (Fig. 3.22). Central to these models is the presence of a magma chamber and an associated narrow zone of dyke injection above the magma chamber where cooling is rapid. Without such a magma chamber, it is difficult to understand how layer 3 could form except as a dyke complex. As described earlier, recent seismic investigations have detected such a magma chamber of the expected width beneath the crust of the East Pacific rise in localities which have been investigated (p. 118).

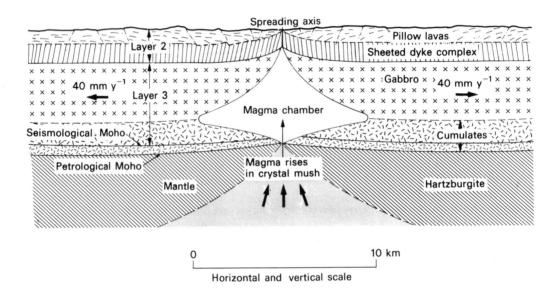

Fig. 3.22 Diagrammatical cross section of the structure of the crust and topmost mantle beneath an ocean ridge crest, drawn without vertical exaggeration. Layer 2 structure is based on the model of KIDD (1977) and shows the tendency of the pillow lavas to dip inwards and the dykes to dip steeply outwards resulting from rotation near the spreading axis. The shape and size of the magma chamber is based on the thermal model of KUSZNIR and BOTT (1976).

Why is it that crust of almost identical thickness is produced at fast-spreading and slow-spreading ridges? This does not depend on the crust forming process but rather on the amount of magma available. This in turn depends on the depth at which partial fusion commences in the upwelling region of the mantle beneath. If the underlying temperature regimes are the same, then partial fusion will start at the same depth beneath each type of ridge and the same total magma fraction will be produced. Hence the crustal thickness will be approximately the same. However, the thermal modelling does suggest that there may be small contrasts in the crustal layering between the fast-spreading and slow-spreading ridges. For instance, a fast-spreading ridge would be expected to produce a thicker cumulate layer but a thinner sheeted dyke zone.

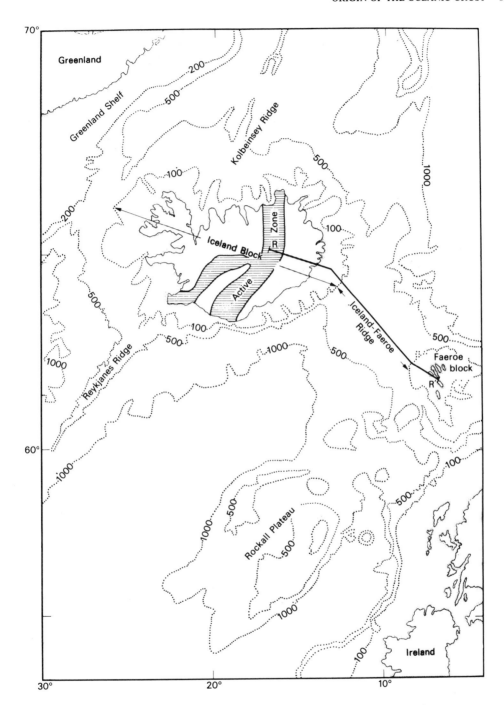

Fig. 3.23 Map of part of the north-eastern North Atlantic, showing upstanding regions of shallow bathymetry. The Rockall Plateau and the Faeroe block are interpreted as parts of a microcontinent underlain by continental crust. In contrast, the Icelandic transverse ridge, which includes the Iceland block and the Iceland-Faeroe ridge, is interpreted as an aseismic transverse ridge formed of anomalously thick oceanic crust. After BOTT (1974), *Geodynamics of Iceland and the North Atlantic area*, p. 35, D. Reidel Publishing Company.

The main features of the oceanic crustal layering are thus established within 2–5 My of formation, but two significant modifications occur subsequently. *Firstly*, layer 2A becomes transformed into the upper part of layer 2B on the flanks of the ridges, the transition occurring in the region where seawater circulation ceases to be a significant thermal factor (p. 289). This probably occurs as a result of plugging of the pores and cracks by deposition of minerals. *Secondly*, layer 3 appears to thicken from 3 to 5 km average thickness during the first 30 My of spreading history. This may possibly be explained by further gabbroic intrusion into the layer as it spreads, or by partial serpentinization of the topmost mantle rocks reducing their seismic velocity.

3.6 Icelandic type crust

The oceanic crust, in comparison with the continental crust, is remarkably uniform in structure and thickness over most of the oceanic regions, despite some small regional variation. An exception, however, to this uniformity is posed by certain regions of upstanding topography within the oceans which are underlain by anomalously thick crust. Some such regions of shallow bathymetry are proved microcontinents (p. 210) such as the Rockall Plateau (Fig. 3.23) and the Seychelles Bank, but others may be underlain by oceanic crust of anomalously great thickness. Perhaps the best known region of this type includes Iceland and the associated Icelandic transverse ridge of shallow bathymetry extending from eastern Greenland to the Faeroe Islands. This feature forms an aseismic transverse ridge crossing the North Atlantic and intersecting the active ocean ridge at Iceland itself (Figs 3.23 and 3.24).

The crust beneath Iceland is substantially thicker than the normal oceanic crust (PÁLMASON, 1971). This is why Iceland stands above sea-level in contrast to the normal submerged parts of the ocean ridge system. The upper crust varies between 1 and 9 km in thickness and it can be subdivided into two or three sub-layers with velocities ranging between 2·7 and 5·1 km s^{-1}. These upper layers are formed by lava flows which crop out at the surface and associated minor intrusive rocks. These sub-layers might together be regarded as a thick representative of the oceanic crustal layer 2, but formed sub-aerially rather than below water. This is underlain by the main crustal layer having an average P velocity of about 6·5 km s^{-1} which is probably equatable to layer 3 of the oceanic crust. The boundary with the underlying 7·2 km s^{-1} layer occurs at 8–9 km depth beneath south-western Iceland and at 14–15 km depth beneath south-eastern and northern Iceland. It is not immediately clear whether the 7·2 km s^{-1} layer forms the top of an unusually low velocity upper mantle or is a thick lower crustal layer, but teleseismic delay times and surface wave dispersion studies suggest that it extends down to about 200 km depth, favouring the former interpretation.

The crust beneath the now-submerged Iceland-Faeroe ridge is even thicker than that beneath Iceland (Fig. 3.24). There are at least two upper crustal layers (3·2 to 5·8 km s^{-1}) above a 6·7 km s^{-1} refractor at about 7 km depth, this probably being equivalent to the top of layer 3. The Moho occurs at a depth of about 30–35 km, being at comparable depth to that beneath the continents. This crust resembles that beneath Iceland except that it is thicker and the comparable velocities are rather higher.

How did this thick 'Icelandic type' crust beneath Iceland and the Iceland-Faeroe ridge originate? Two fundamentally different views have been expressed regarding the origin of the aseismic ridge. BELOUSSOV and MILANOVSKY (1977) took the view that Iceland is underlain by continental crust, this implying that Greenland and Norway have not drifted apart but rather that the ocean here resulted from transformation of continental into oceanic crust with resulting subsidence. The alternative view which is generally accepted is that Greenland and Europe have

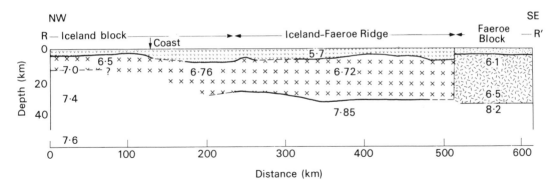

Fig. 3.24 Crustal structure of the Iceland-Faeroe ridge and its junctions with the neighbouring Iceland and Faeroe blocks along line RR' (Fig. 3.23) as determined by seismic refraction surveys. *P* velocities are shown in km s⁻¹. After BOTT and GUNNARSSON (1980), *J. Geophys.*, **47**, 226.

drifted apart as a result of the formation of the North Atlantic by sea-floor spreading. The thick crust beneath the aseismic ridge thus represents an unusually voluminous differentiation of crustal material from the underlying mantle, this at first causing the extensive early Tertiary continental volcanism on the continental borderlands and later being concentrated at the spreading axis.

If it is accepted that the Icelandic transverse ridge has formed by sea-floor spreading, then its formation must represent exceptionally strong differentiation of crustal material from the underlying mantle, producing crust of up to six times the normal oceanic thickness. WILSON (1963) suggested that such aseismic ridges and chains of volcanic islands originate from hot spots in the underlying mantle. Subsequently, this hypothesis has been widely applied to the Icelandic transverse ridge. The hotter mantle means that partial fusion in the upwelling material below the ridge crest starts deeper and therefore produces larger quantities of magma. Whether the hot mantle results from a mantle plume rising from near the core-mantle boundary or is the result of a major convective overturn occurring just prior to the split of Greenland from Europe is a matter of controversy.

3.7 Fracture zones and transform faults

Fracture zones are prominent topographical features on the sea-floor which cross ocean ridges, apparently displacing their crest laterally (Fig. 3.25). They are typically long linear depressions with associated parallel uplifted blocks, often but not always perpendicular to the ridge crest. They appear to form some of the straightest features on the Earth's surface but in reality they are found to be arcs of small circles on the surface. The steep marginal scarps of the troughs form excellent targets for dredging, normal oceanic crustal rocks being found in some hauls but from others there is abundant evidence of additional metamorphism and shearing. Large elongated masses of serpentinite commonly underlie the uplifted blocks, these probably having been faulted up from the topmost mantle.

One of the important discoveries resulting from the magnetic anomaly surveys in the north-eastern Pacific was the large lateral displacements of the magnetic anomaly pattern across the Mendocino, Pioneer and Murray fracture zones (Fig. 3.7). The Mendocino and Pioneer faults show a combined left-lateral displacement of the magnetic anomaly strips amounting to about 1400 km, and the right-lateral displacement at the Murray fault is 680 km at the western end and

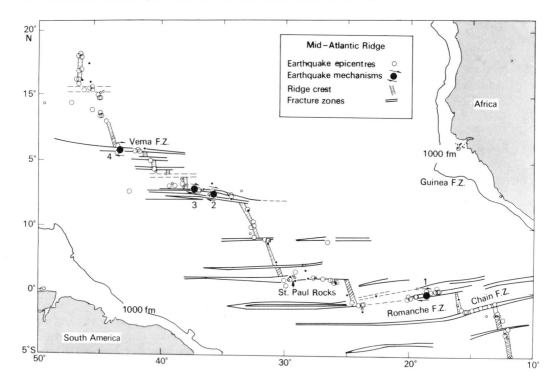

Fig. 3.25 Fracture zones offsetting the ridge crest along the equatorial section of the mid-Atlantic ridge, showing epicentres of earthquakes observed to occur during the period 1955–1965. Focal mechanism solutions are shown for four of the fracture zone events, indicating that the sense of the motion is consistent with transform faulting but not with fault displacement of the ridge crest. Note that the epicentres cluster (1) along the crest, and (2) on the fracture zones between adjacent portions of the crest, but not beyond. After SYKES (1967), *J. geophys. Res.*, **72**, 2137.

only 150 km at the eastern end (VACQUIER, 1965). The problem is to understand how such large horizontal movements can affect adjacent blocks of the oceanic crust, and how this lateral displacement can change drastically along the length of a single fault such as the Murray fracture zone. A similar problem is posed by the great continental strike-slip faults, such as the San Andreas fault of California and the Great Glen fault of Scotland. These faults must die out somewhere because they do not continue right round the Earth.

A simple yet profound solution as to how these large strike-slip faults can terminate has been suggested, within the framework of the sea-floor spreading hypothesis, by WILSON (1965). He suggested that they terminate at the ends of mobile belts, which they meet, commonly, but not necessarily, at right angles. The lateral displacement on one side of the fault is taken up either by formation of new crust along a terminated segment of ocean ridge or by crustal shortening along a terminated segment of mountain range or ocean trench. Wilson called this newly recognized class of strike-slip faults by the name *transform fault*. The concept of transform faults also explains the long-standing problem as to how mobile belts can be terminated. It leads to the idea that mobile belts are linked by transform faults into an interconnected network which subdivides the Earth's surface into a series of 'rigid' plates which undergo relatively little internal deformation. This is the basic idea of plate tectonics.

Transform faults were grouped by Wilson into six basic classes depending on the type and orientation of the two mobile belts they join (Fig. 3.26). The three possible types of junction at one end are (*i*) an ocean ridge, (*ii*) a compression feature joined from the concave side, and (*iii*) a compression feature joined from the convex side. Each of the six classes can be further subdivided into left-lateral (sinistral) and right-lateral (dextral) types.

Fig. 3.26 Diagram illustrating the six possible types of dextral transform fault:
(**a**) ridge to ridge;
(**b**) ridge to concave side of arc;
(**c**) ridge to convex side of arc;
(**d**) – (**f**) the three possible types of arc to arc connection.

The lower part of the figure shows the same faults at a later stage of development, with the now inactive parts marked as dashed lines. Note that the direction of motion in (**a**) is in the opposite sense to that required to offset the ridge. Redrawn from WILSON (1965), *Nature, Lond.,* **207**, 344.

The oceanic fracture zones are by far the commonest type of transform fault. Most of these connect segments of ocean ridge crest at both ends, thus being of ridge-to-ridge type. They vary in scale from the small fracture zones which offset the ridge axis by a few kilometres, such as those seen in the FAMOUS area of the Atlantic Ocean, to the major ones of the north-eastern Pacific (Fig. 3.7) and the equatorial Atlantic (Fig. 3.25). These fracture zones all appear to play an important part in the economy of the sea-floor spreading process although their role and occurrence differs somewhat between the slow-spreading oceans of Atlantic type and the fast-spreading oceans of Pacific type, as discussed in turn below.

Best known in the Atlantic is the series of parallel fracture zones which displaces the crest of the mid-Atlantic ridge in equatorial latitudes (Fig. 3.25). These used to be interpreted as *left-lateral* transcurrent faults displacing a once-continuous crest. However, Wilson interpreted them as a series of *right-lateral* transform faults related to the opening up of the South Atlantic Ocean and

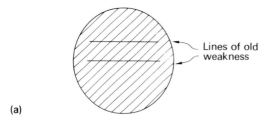

(a)

Lines of old weakness

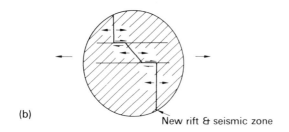

(b)

New rift & seismic zone

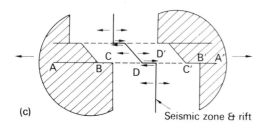

(c)

Seismic zone & rift

Fig. 3.27 Diagram illustrating three stages in the drifting apart of two continents such as South America and Africa, showing how transform faults have developed and played their part in the movement. The seismic activity would be mainly along the heavy lines. Redrawn from WILSON (1965), *Nature, Lond.,* **207**, 345.

the separation of South America from Africa (Fig. 3.27). The apparent lateral displacement of the crest is not a true offset, but it is the consequence of the shape of the original split between the continents. The shape of the original split must be maintained by the mid-ocean ridge if the new ocean floor and adjacent continents remain undeformed and the spreading axis forms crust symmetrically and does not migrate laterally on one side relative to the other. Under these conditions, the spreading rate must be the same on both sides of each fracture zone. Between the intersections of the ridge axis on either side with the fracture zone, the crustal blocks on either side move laterally relative to each other at twice the spreading rate, in the opposite sense to the displacement of the axis. This is supported by earthquake mechanism studies which confirm the nature and sense of movement (e.g. Fig. 3.25). Beyond the spreading axis on each side, the fractures cease to be active faults and the two adjacent plates move together, the topographic features being preserved until they become buried by sediment. In the equatorial Atlantic region, the initial split of the continents occurred partly by transform faulting producing offset or sheared continental margin segments and partly by separation at the new ridge axis producing rifted margin segments.

Opening oceans such as the Atlantic may be terminated at their extremities by active or fossil transform faults separating oceanic from continental crust. Wilson interpreted the northern end of the mid-Atlantic ridge in this way. During the early Tertiary, the ridge split south of Greenland, one branch passing on each side of it. The now extinct western branch terminated against the postulated Wegener transform fault passing between Greenland and Ellesmere Island. The still active eastern branch terminates against the De Geer fault which crosses from north Norway to Greenland, passing just south-west of Spitsbergen, with an active section north-west of Spitsbergen. A similar major transform fault, now extinct, permitted the separation of the Falkland plateau from the present south-eastern African margin at the south end of the early-forming Atlantic Ocean. These faults show how a spreading sea-floor can abut against less mobile parts of the lithosphere.

Yet another role of the transform fault is seen in the north-eastern part of the North Atlantic, a region where there have been repeated lateral jumps of the spreading axis producing, for instance, the present asymmetrical location of the ridge axis north of Iceland. If a segment of an initially continuous spreading axis jumps laterally, then transform faults must connect the newly jumped segment to the adjacent ridge sections. If, at a later stage, the ridge becomes continuous again, the extinct fracture zone will remain visible until eventually buried by sediment. The Jan Mayen fracture zone is a relict of this type. Small fracture zones may also develop where spreading is oblique to the ridge axis, with the consequent formation of short ridge segments normal to the spreading direction separated by short fracture zones; such a situation developed temporarily along the obliquely spreading Reykjanes ridge between about 40 and 10 My ago (VOGT and AVERY, 1974).

Some of the fracture zones of the north-eastern Pacific differ from those of the Atlantic in that the spreading rate may change across them, thus accounting for instance for the varying offset of the magnetic anomaly pattern across the Mendocino and Murray fracture zones (Fig. 3.7). This can only occur where one of the junctions between spreading axis and transform fault also connects directly or indirectly to another mobile belt. Such a situation developed when the active continental margin of western North America overrode the East Pacific rise causing the observed time-varying offset. In this region the San Andreas transform fault crossing continental crust now joins the termination of the East Pacific rise in the Gulf of California to the short Juan de Fuca ridge off Vancouver Island and the active margin has ceased to exist.

Oceanic fracture zones, interpreted as transform faults, must separate regions of oceanic crust of differing age on opposite sides. As the depth of the sea-floor depends on the age of the cooling and subsiding oceanic lithosphere (p. 117), the fracture zones would be expected to mark an overall change in bathymetric depth from one side to the other. Such a change in depth is observed to occur, and it is particularly conspicuous where the fracture zone separates relatively young crust of large age difference such as occurs across the Mendocino fracture zone (Fig. 3.7).

Transform faults, as originally defined by Wilson, are plate boundaries at which crust is neither created nor destroyed, the relative movement between opposite sides being parallel to the strike of the fault. This definition may not be strictly true for all oceanic fracture zones. There is evidence that a small amount of compression, or alternatively extension with new crustal formation (leaky transform fault), may occur along certain fracture zones. However, the effect is probably sufficiently small to be negligible in terms of gross plate movement. Thus transform faults in general and oceanic fracture zones in particular are of great importance in defining the directions of relative movement between adjacent plates, as is discussed in the next section.

3.8 Plate tectonics

The concept of plate tectonics has evolved from the earlier hypotheses of continental drift, sea-floor spreading and transform faults. The overall concept was proposed almost simultaneously by

McKENZIE and PARKER (1967) and by MORGAN (1968) and has been outlined in detail by LE PICHON and others (1973). The basic idea is that the outermost strong shell of the Earth which forms the lithosphere suffers strong deformation only along relatively narrow linear mobile belts. The mobile belts and the interconnecting transform faults divide the lithosphere into a series of 'rigid' plates which do not undergo any significant internal stretching, folding or distortion but which move relative to each other, the motion being taken up at the plate boundaries. Most of the global release of tectonic and seismic energy is thus concentrated at the boundaries between plates. The relative motion between plates is further governed by the geometric constraints applying to motions of rigid shells on the surface of a sphere.

It is a well established facet of isostatic theory that the weak asthenosphere is overlain by a relatively strong lithosphere (or tectonosphere) which is on average about 80–100 km thick. This rheological model is supported by much modern observational, experimental and theoretical evidence as described in Chapter 8, although there is still difficulty in unambiguously locating the intervening boundary. Another important basis of plate tectonics is that the plates of lithosphere are capable of transmitting stress over large horizontal distances without buckling (p. 314).

The three basic types of plate boundary are as follows:

(i) *Divergent* (or *constructive*) *boundaries*, where new lithosphere is produced at the crests of ocean ridges;

(ii) *Convergent* (or *destructive*) *boundaries*, where the surface is being destroyed as two plates approach each other;

(iii) *Transform faults* (or *conservative boundaries*), where plates move laterally relative to each other and lithosphere is neither produced nor destroyed.

The structure of divergent boundaries (ocean ridges) and the nature of transform faults have been described earlier in the chapter. Convergent boundaries are of two main types. The first type develops where oceanic lithosphere occurs on one or both sides of a plate boundary which coincides with the axis of an ocean trench. Here oceanic lithosphere is recycled into the mantle. The sinking tongue of lithosphere forms a subduction zone which dips at about 45° away from the ocean and beneath the adjacent island arc or Andean mountain range as described in Chapter 5. The oceanic lithosphere can sink because its average density is higher than that of the underlying asthenosphere. The second type of convergent boundary forms where there is continental lithosphere on both sides of the boundary. Because of its relatively low density, the thick continental crust cannot be subducted in significant amounts. Consequently, a collision type mountain range, such as the Alps (Chapter 2) or Himalayas, is produced at the boundary. The relationship between the three types of plate boundary, as they occur in the Pacific region, is shown in Fig. 3.28.

Plate tectonics provides a geometrical explanation of how sea-floor spreading and continental drift can take place on the surface of a nearly spherical Earth without deformation of the ocean-floor or the continents except at the well-known mobile belts. It relates most of the Earth's primary tectonic activity including continental drift and the formation of ocean ridges, fold mountains, trenches, island arcs, plateau uplifts and rift valleys to the processes of sea-floor spreading and subduction. Most of the geological implications of plate tectonics follow from purely geometrical reasoning without need to refer to the underlying cause of the movement of plates. In fact, discussion of the mechanism is deferred entirely until later chapters, particularly Chapter 9.

Earthquakes and plate tectonics

Seismology has been of very great importance in the initial establishment and confirmation of the plate tectonic theory (e.g. ISACKS and others, 1968), and has continued to be of importance in the

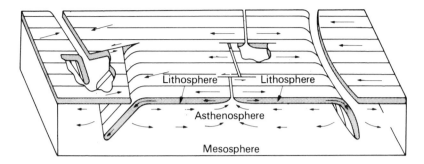

Fig. 3.28 Block diagram illustrating schematically the sea-floor spreading and plate tectonic hypotheses, and more particularly the relationship between ocean ridges, island arc-trench systems and transform faults of ridge-to-ridge type in the Pacific region. The arrows indicate the relative motion between adjacent blocks, and the return flow which is here assumed to occur within the asthenosphere. Adapted from ISACKS and others (1968), *J. geophys. Res.*, **73**, 5857.

study of processes at plate boundaries. Earthquake focal mechanisms are discussed in detail in Chapter 8 and the contribution of seismology to the evaluation of the structure and processes at subduction zones is described in Chapter 5.

The Earth's main earthquake activity would be expected to be concentrated at the plate boundaries where the vigorous tectonic activity occurs. This is borne out by the world map of earthquake epicentres (Fig. 3.29), on which an interconnecting network of foci marks out the plate boundaries. Most of the known earthquakes are concentrated in these belts. The lack of activity in the extensive plate interior regions is equally spectacular, although a few plate interior earthquakes do occur. Each type of plate boundary is associated with a characteristic type of seismicity and focal mechanism, which agrees well with the expected type of faulting. The belt is widest and extends to greatest depths around the Pacific and along the Alpine-Himalayan mountain ranges and is relatively narrow along the crests of the ocean ridges.

The narrow linear belt of shallow focus earthquakes follows the crest of the ocean ridge system along its entire length and extends along the East African rift system. Most of the earthquakes are of small to moderate size, reaching up to about magnitude 6. The total release of energy is negligible in comparison with that of the circum-Pacific and Alpine-Himalayan belts. SYKES (1967) showed that the epicentres occur in two main settings as follows (Fig. 3.25). Firstly, both isolated events and also earthquake swarms accompanying volcanic activity occur in the immediate vicinity of the ridge crest. These very shallow events are predominantly associated with normal faulting indicative of crustal tension, with the inferred extension direction perpendicular to the axis of the ridge as would be expected. Secondly, epicentres along fracture zones tend to cluster between the offset sections of the crest and are relatively rare beyond these locations. Earthquake mechanism studies indicate strike-slip movement on steeply dipping fault planes (Fig. 3.25), with the sense of the motion agreeing with the expected transform fault motion. Thus the seismological evidence supports the interpretation of the fracture zones as transform faults rather than as wrench faults displacing the ridge crest. FRANCIS (1968) has further shown that the total energy released by the fracture zone events is probably at least 100 times as great as that released by the crestal events and that the magnitude-frequency distributions differ between the two types of event.

The convergent plate boundaries are marked by the broad circum-Pacific and Alpine-Himalayan earthquake belts, which include shallow, intermediate and deep focus events, the

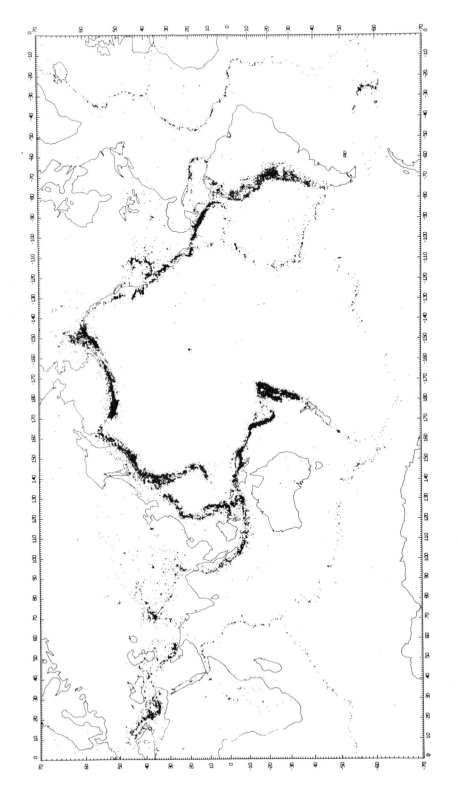

Fig. 3.29 World-wide distribution of earthquake epicentres (0–700 km depth) for the period 1961–1967, as compiled from the U.S. Coast and Geodetic Survey records. After BARAZANGI and DORMAN (1969), *Bull. seism. Soc. Am.*, **59**, in pocket.

deepest foci being just over 700 km deep in the Tonga trench region. Most of the global release of seismic energy takes place in the large earthquakes of these two belts. In the circum-Pacific belt, shallow focus events predominate on the continental side of the trenches, but other shallow, intermediate and deep focus events cluster on a plane dipping on average about 45° from the trenches beneath the associated island arc or mountain range. These approximately mark the upper surface of the sinking oceanic lithosphere, which is also characterized by its low seismic attenuation (high Q). The pattern of earthquake mechanisms is more complicated than that at ridges but it is fully consistent with the inferred relative motions of the converging plates. A fuller discussion is given in Chapter 5.

Thus the seismological evidence on distribution of earthquakes and their focal mechanisms has given strong support to the plate tectonic theory. This evidence outlines the plate boundaries with a clarity not otherwise attainable. It supports the idea that ocean ridges are extension features where new lithosphere is formed and that fracture zones are transform faults related to sea-floor spreading. Seismology has also enabled the tongues of sinking lithosphere to be identified and mapped in some detail.

Determining the relative motion between plates

Any conceivable displacement of a conformable spherical cap on a spherical surface can be produced by rotation about an appropriate axis passing through the centre of the sphere. Such a displacement can be completely specified by one of the two poles where the rotation axis cuts the surface of the sphere and by the angular rotation about the axis needed to cause the displacement. Similarly, the relative movement of two plates is defined by the pole of rotation and the angular velocity of rotation. One of the most important basic problems in plate tectonics is to determine the present-day instantaneous pole of rotation and relative angular velocity between the various pairs of plates on the Earth's surface. The 'instantaneous' values refer to those averaged over the last 3 to 5 My, which is the shortest interval over which accurate velocities can be determined.

Three main types of observation are normally now used to determine the relative motion between two plates. *Firstly*, the spreading rate at divergent plate boundaries can be determined from the spacing of the oceanic magnetic anomalies on either side of the ridge crest. As spreading can take place obliquely to the ridge crest, the strike of the magnetic anomalies is also required. Relative to the axis of rotation, the spreading rate is maximum at the equator and falls off with increasing latitude θ as $\cos \theta$. Thus the location at which the spreading rate is measured is also required. *Secondly*, the most precise method available for determining the local direction of relative motion between adjacent plates is by use of the trend of transform faults, the movement being parallel to the fault trace. Transform faults therefore lie along lines of latitude, which are small circles, relative to the pole of rotation. A great circle drawn perpendicular to such a small circle must therefore pass through the pole of rotation. If two or more great circles can be constructed from transform faults along a plate boundary, then their intersection will give the pole of rotation. *Thirdly*, the local slip direction between two plates can sometimes be determined by studying the focal mechanism of suitable earthquakes, as explained in Chapter 8 (p. 320). This method is not as satisfactory as the use of transform fault trends, but it has the advantage that it can be applied at most convergent plate boundaries.

As ocean ridges are typically intersected by numerous transform faults, the relative motion of the plates at a divergent boundary can be fully determined from the spreading rates and transform fault trends. The situation is less satisfactory at convergent boundaries where the relative velocity of the adjacent plates cannot be measured directly and where the slip direction usually determined from earthquakes can be in significant error. It is therefore necessary to determine the relative motion at most convergent boundaries indirectly. This is possible because the instantaneous rotation between two plates can be treated as a vector, the direction being along the rotation axis and the magnitude being proportional to the angular velocity of relative motion. Suppose that the relative motion between plates A and B is represented by the vector **AB** and that between plates B and C by vector **BC**. Then the relative motion between the plates A and C is given by the sum of these two vectors. This method can be extended to determine the relative motion between any numbers of plates, provided that there are suitably distributed divergent boundaries to make the solution fully determined. There are about twelve major plates forming the Earth's surface and there are sufficient divergent plate boundaries to enable the relative motions to be satisfactorily determined between most of these. This method, however, only applies to the small instantaneous rotations and more involved techniques described by LE PICHON and others (1973) need to be used if finite rotations are to be studied.

One of the earliest analyses of present-day plate motions was carried out by LE PICHON (1968), who subdivided the Earth's surface into six main plates. He used fracture zones and spreading rates to compute the relative motion of the plates separated by divergent plate margins. He then computed the relative motions of plates at the convergent plate boundaries assuming that the Earth's surface area remains constant. Le Pichon avoided specifying the spreading rates across the ridges in the southern parts of the Indian Ocean, because otherwise the problem would have been overdetermined. However, his computations yielded estimates of the spreading rates across these ridge portions which agreed excellently with the observed rates.

More recent analyses of present-day plate motions, such as those of CHASE (1978) and MINSTER and JORDAN (1978), make use of all the available information on slip direction and spreading rate. All the data is simultaneously inverted to yield estimates of the poles and relative rates of rotation between all pairs of plates. As the number of observations used greatly exceeds the number of unknown parameters, the problem is overdetermined and a solution is obtained by minimizing the sum of squares of the residuals. The observations can be appropriately weighted depending on their reliability prior to inversion. The analyses of Chase, and of Minster and Jordan, both assume that the Earth's surface is divisible into twelve major plates, but Chase included the Phillipine but not the Caribbean plate whereas Minster and Jordan included the Caribbean but not the Phillipine plate. Chase made use of 90 spreading rate estimates from the ocean ridges, 69 measurements of oceanic transform fault trend, and 101 earthquake slip vectors. The poles and rates of rotation between selected pairs of plates obtained by Chase are shown in Table 3.2 and the relative motion of adjacent plates at selected points on the global network of plate boundaries is shown in Fig. 3.30. The misfit between observed and calculated spreading rate is generally about 1mm y^{-1} and only exceptionally exceeds 5 mm y^{-1}. The misfits in transform fault orientation are about 1° to 12° and those of earthquake slip vectors are mostly less than 15°. Despite using a somewhat different data base and inversion procedure, Minster and Jordan obtained results in good general agreement with those of Chase.

The global analysis (Table 3.2 and Fig.3.30) enables the relative motion across plate boundaries to be calculated where it cannot be measured directly. The results show that India and Asia are converging on each other in the Himalayan region at a rate of about 50 mm y^{-1} in a north-south direction. The convergence of Europe and Africa at the Mediterranean-Alpine plate boundary is at about 8 mm y^{-1}. The ocean floor of the Nazca plate is being subducted beneath South America at about 100 mm y^{-1} and comparable subduction rates occur along the western margins of the Pacific Ocean. Subduction occurs oblique to the trend of the Aleutian arc and dextral strike-slip motion of about 55 mm y^{-1} is predicted to occur along the San Andreas transform fault.

The creation of new ocean floor at the Pacific ocean ridges is not keeping pace with the rate of

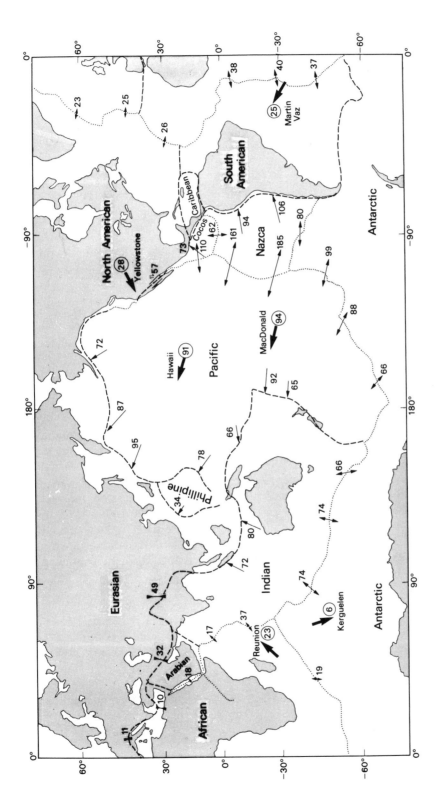

Fig. 3.30 The main plates forming the Earth's surface, showing divergent plate boundaries as a dashed line and convergent and transform fault boundaries as a dotted line. The directions and rates (in mm y^{-1}) of the relative motion, as computed by CHASE (1978), are shown at selected points on the plate boundaries. The computed relative motions of the plates at six selected hot spots relative to the mean hotspot reference frame are also shown (in mm y^{-1}).

Table 3.2 Poles of rotation and rates of rotation for the relative movement between selected pairs of adjacent plates (after CHASE, 1978). Seen from above the pole, the plate first named moves anti-clockwise with respect to the second plate. South latitudes and west longitudes are negative.

Plate pair	Pole of rotation		95% confidence ellipse for pole (deg)		Rate of rotation $(10^{-7}$ deg $y^{-1})$ with 95% confidence limit
	lat. (deg)	long. (deg)	semi major axis	semi minor axis	
African–Antarctic	−6·6	−35·7	12·5	6·5	1·77 ± 0·34
African–Eurasian	29·2	−23·5	4·5	2·0	1·42 ± 0·52
African–North American	80·0	71·7	10·1	2·2	2·58 ± 0·24
African–South American	63·9	−34·3	7·9	2·4	3·60 ± 0·24
Arabian–African	34·9	19·2	10·2	3·5	3·71 ± 1·07
Cocos–Pacific	39·7	−107·9	2·9	2·0	21·33 ± 1·97
Eurasian–North American	53·7	137·3	6·1	2·2	2·29 ± 0·15
Indian–Antarctic	17·4	32·1	4·1	3·4	6·79 ± 0·23
Nazca–Pacific	50·9	−87·0	4·3	2·9	16·85 ± 0·68
North American–Pacific	48·2	−72·3	2·6	2·1	8·64 ± 0·43
Pacific–Antarctic	−66·2	96·5	2·1	1·8	10·05 ± 0·27
Pacific–Indian	−62·0	174·3	2·2	1·8	12·72 ± 0·49

destruction of ocean floor at the subduction zones of the circum-Pacific belt. Eurasia and the Americas are slowly converging onto the Pacific Ocean at a rate of about 20–40 mm y^{-1}. Thus the Pacific is a shrinking ocean, in contrast to the expanding Atlantic and Indian Oceans which are increasing in area. At the present rate, the Pacific Ocean would vanish in about 300 My, although collisions between the surrounding continents would probably stop the process before this occurred completely. The global pattern indicates that the plates containing parts of the Atlantic and Indian Oceans tend to be growing in size whereas those containing parts of the Pacific are decreasing in size. A further consequence is that the mid-Atlantic and Indian Ocean ridges are progressively moving away from each other.

Many more detailed inferences of great geological interest can be made from the relative plate motions and by tracing them back into the past. Examples include the evolution of the various types of triple junction where three plate boundaries join and the effects of the subduction of an ocean ridge as has occurred during the late Tertiary off the western coast of North America with subsequent development of the San Andreas transform fault system. Discussion of these aspects is outside the scope of the book.

Direct measurement of relative plate motion

The strain rate in the vicinity of certain plate boundaries situated on land, such as the San Andreas transform fault and the central rift in Iceland, has already been measured in some detail by ground based geodetic surveying methods. The pattern of relative motion, strain accumulation and strain release has been found to vary along the San Andreas plate boundary. Furthermore, the strain associated with the fault extends into the adjacent plate interiors for a distance of the order of 100 km. For these reasons, ground based surveying techniques are of great use in investigating the local pattern of motion within about 20 km of the boundary but are of limited value in determining the overall relative motion between the adjacent plates. Accurate measurement of distance over a much longer baseline is required.

Two methods which use space technology are now available for measuring accurately the relative motion of plates over baselines of several thousand kilometres in length. Both of these methods use an extra-terrestrial reference point to determine the distance apart of points on the Earth's surface. One method is based on laser ranging to the Moon, or to a satellite in a high enough orbit to avoid perturbations of terrestrial origin. Very short laser pulses are transmitted by a fixed or mobile station and strike a retroreflector on the Moon or the artificial satellite which returns the pulses to the sending station. The distance is determined from the travel time of the pulse after corrections for the various perturbing factors have been made. By making a suitable number of observations at two sites on the Earth's surface, their distance apart can be measured with a repeatability of a few centimetres. The alternative method is known as very long baseline interferometry (VLBI). This makes use of distant radio sources (quasars) which transmit an irregular signal. The difference in travel time of the wave train received at two terrestrial radio telescopes can be determined by correlation methods. The distance between the two stations can again be determined to a precision of a few centimetres.

Preliminary determinations of the relative motion across the San Andreas fault has already been made by successive satellite laser ranging measurements in 1972, 1974 and 1976 (SMITH and others, 1979). The two measuring sites were situated at Quincy in northern California and at Otay Mountain near San Diego which are about 900 km apart. The preliminary results indicate a shortening of the baseline by 90 ± 30 mm y^{-1} which is somewhat higher than the dextral strike slip of about 55 mm y^{-1} predicted by plate tectonics. The VLBI technique is also being used to measure the relative motion across the San Andreas fault and results should shortly become available (NIELL and others, 1979).

The precision of both methods has been increasing as the measuring and correcting techniques improve, the uncertainty now being of the order of 30 mm. Extensive programmes for the measurement of relative plate motions using these methods are planned for the 1980s and it is to be anticipated that results of great importance will be obtained during the next ten to twenty years. As well as investigating contemporary plate motions, in contrast to those averaged over the last few million years, the measurements will eventually make it possible to determine whether the plates move smoothly and uniformly or irregularly. It will also be possible to determine, in association with ground based methods, how the motions vary from plate boundary to plate interior.

Estimating absolute plate motions

The absolute motions of lithospheric plates can only be determined if their movements can be related to a stationary reference frame which is fixed deep within the Earth's interior. This idea appeared to have a sound basis in the early 1970s when it was believed that the lower mantle is too stiff to convect, but it is less plausible now that the lower mantle is inferred to be convecting (p. 346). Nevertheless, it now seems probable that the lithospheric plates themselves form the upper boundary layer of the main convecting system (p. 352) and that the rate of convective flow in the lower mantle is consequently much slower on average than the plate motions at the surface. Under such circumstances, the concept of absolute plate motion relative to the deep mantle may still have some approximate validity over short time intervals.

There have been several attempts to suggest a reference framework for absolute plate motions. The most widely used framework is based on the occurrence of linear chains of volcanoes and aseismic ridges which vary systematically in age of formation along their length. One of the best known examples is the Hawaiian-Emperor seamount chain which extends as a ridge for more than 4000 km within the Pacific plate interior. The Hawaiian ridge has been formed over the last 20 My and the volcanoes along it vary systematically in age. The magma source appears to have migrated east-south-east at a rate of about 90 mm y^{-1}. According to WILSON (1963) and MORGAN (1971) the migrating volcanic activity is attributed to the motion of the plate above a stationary hot spot which continues to supply magma over a period of some tens of million years and is assumed to be anchored to the fixed framework in the lower mantle. According to this hypothesis, the source of the Hawaiian volcanism is stationary and the Pacific plate is moving at a rate of about 90 mm y^{-1} in a west-north-west direction. MORGAN (1971) suggested that the upwelling from the deep mantle takes the form of a narrow plume.

The various hot spots distributed over the Earth's surface would be expected to remain stationary relative to each other if the hot spot hypothesis is correct. The observed relative motion between hotspots averages less than about 5 mm y^{-1} which is over an order of magnitude less than their average motion relative to the plates they penetrate. Their apparent directions of migration are also consistent to better than 10°. Exceptions to the good consistency are shown by the Easter Island and Iceland hot spots. In general the hot spot hypothesis provides a plausible working hypothesis for motions over the last 10 My or thereabouts but the consistency breaks down going further back into the past. The present absolute plate motions related to the hot spot framework, as inferred by CHASE (1978), are shown in Table 3.3 and selected values are shown in Fig. 3.30.

Table 3.3 Motions of plates relative to the mean hotspot frame of reference (after CHASE, 1978). Conventions as in Table 3.2.

Plate	Pole of rotation		95% confidence ellipse for pole (deg)		Rate of rotation (10^{-7} deg y^{-1}) with 95% confidence limit
	lat. (deg)	long. (deg)	semi major axis	semi minor axis	
African	31·8	−61·3	18·0	10·5	1·97 ± 0·47
Antarctic	58·3	−144·6	15·2	7·3	1·46 ± 0·81
Arabian	40·4	−7·3	14·3	7·0	4·87 ± 0·95
Cocos	22·8	−117·0	5·3	2·8	14·64 ± 2·43
Eurasian	18·5	−108·7	30·1	29·4	1·09 ± 0·52
Indian	29·8	31·6	6·7	4·8	6·59 ± 0·42
North American	−36·8	−70·7	15·6	11·0	2·51 ± 0·54
Nazca	36·0	−94·3	9·8	5·7	8·71 ± 1·18
Pacific	−64·7	106·8	3·8	3·3	8·79 ± 0·70
South American	−70·7	−131·3	15·8	10·9	2·32 ± 0·66

LLIBOUTRY (1974) inferred that the lithosphere as a whole should not be rotating relative to the lower mantle. This suggests that an alternative absolute motion framework can be determined from the average rotation of the lithospheric plates. Other suggestions which partly stem from the hot spot hypothesis are that the African plate (BURKE and WILSON, 1972), or the Caribbean plate (JORDAN, 1975), can be regarded as stationary. As shown by MINSTER and JORDAN (1978), these other frameworks are approximately but not exactly consistent with the hot spot framework.

No great reliability should be placed on inferred absolute plate motions. However, the rough consistency obtained indicates that the velocities of flow within the deep mantle are probably much slower than the motions of lithospheric plates. This supports the idea that the plates are driven by forces acting on their edges rather than by the drag of underlying convection currents which move faster than the plates (p. 360).

Plate tectonics as a unifying theory in geological sciences

Over the last twenty years, a number of quite independent lines of evidence from geology, palaeomagnetism, oceanic geophysics and seismology have converged to provide an overwhelming case for the occurrence of continental drift within the framework of sea-floor spreading and plate tectonics. Continental drift is now understood to occur as a result of known tectonic processes at plate boundaries without causing any significant distortion of the vast majority of the Earth's surface.

Plate tectonics has also provided for the first time a unified explanation of the origin of most of the Earth's primary surface features. *Ocean ridges* are related to formation of new lithosphere and their elevation is caused by the high underlying temperatures. *Rift valley systems* may be continental extensions of ocean ridges and may mark lines of incipient continental splitting. *Ocean trenches, island arcs* and *Andean-type mountain ranges* occur at convergent boundaries where oceanic lithosphere is being subducted and *collision mountain ranges* at convergent margins with continental crust on both sides. *Large strike slip faults* and oceanic *fracture zones* are transform faults related to lateral movement of the adjacent plates. *Passive margins* occur where oceanic and continental crust adjoin within a plate interior and they mark the original line of continental splitting. The structure of these major features is described earlier in this chapter or elsewhere in the book (Chapters 2 and 5).

Plate tectonics has been a unifying principle in other branches of geology. For instance, many types of igneous rock can be related to processes at plate margins. The genesis of sedimentary rocks and the past distributions of flora and fauna have become much better understood within the framework of evolving continents and oceans. Plate tectonics has thus led to deeper understanding in almost all branches of geology and lithospheric geophysics, although the danger remains of applying it in too uncritical a way.

The theory of plate tectonics has been developed without need to refer to the underlying driving process in the mantle, which will be discussed in Chapter 9. However, it is worth mentioning at this stage that the sea-floor spreading and subduction processes must have a profound effect on the mantle. Suppose that the average rate of plate separation is 56 mm y^{-1} along 60 000 km of ocean ridge and that the lithospheric plate which is recycled into the mantle is 80 km thick. Then about $2 \cdot 7 \times 10^8$ km^3 of oceanic lithosphere of surface area $3 \cdot 36 \times 10^6$ km^2 is recycled into the mantle every million years, the average residence time being 92 My. At the present rate of recycling, the whole of the upper mantle and transition zone down to 700 km depth would be overturned over about 1100 My and the whole of the mantle over about 3300 My. It will be shown in Chapter 7 that this recycling process is the main mechanism of escape of heat from the deep interior of the Earth.

4 The mantle

4.1 Introduction

The mantle is the largest of the three major subdivisions of the Earth, forming 83 % by volume and 69 % by mass. Its upper boundary is the Moho and it is separated from the core by the Gutenberg discontinuity at a depth of about 2886 km.

As mentioned in Chapter 1, the mantle may be subdivided into three zones on the basis of the elastic wave velocity distribution with depth. These are:

Zone B	21–370 km	upper mantle
Zone C	370–720 km	transition zone
Zone D	720–2886 km	lower mantle

The boundaries between the zones are marked by a change in the velocity-depth gradient and their depths cannot be fixed exactly. The depth of 21 km for the top of the upper mantle is an average value.

Our knowledge of the mantle has been much improved as a result of the 'Upper Mantle Project' and its successor the 'Geodynamics Project'. The Upper Mantle Project was initiated at the Helsinki meeting of the International Union of Geodesy and Geophysics in 1960, when it was resolved that particular attention of earth scientists should be devoted to the outer 1000 km of the Earth. Since then, many important discoveries have been made. The Inter-Union Commission on Geodynamics was established in 1970 to administer the activities of the Geodynamics Project; this project has emphasized the importance of processes in the mantle in the development and origin of the Earth's surface features. The Geodynamics Project has now been superseded by a new international programme on the geodynamics and evolution of the lithosphere.

This chapter aims to describe the seismological and electrical properties of the mantle and the interpretation of them in terms of the composition. The discussion of the thermal and non-elastic processes which occur within the mantle is deferred to later chapters.

4.2 Seismological methods of investigating mantle structure

Body waves

The use of body waves to investigate the structure of the mantle leans heavily on the classical Herglotz-Wiechert method of determining the velocity-depth distribution (Chapter 1). This method assumes radial symmetry. Then the travel-time T of a body wave travelling between two points on the surface separated by an angular distance of Δ (Fig. 4.1) is related to the velocity-depth function by the following equation, which is derived by applying the methods of differential geometry to the ray paths (BULLEN, 1963):

$$\Delta = 2p \int_{r_p}^{r_a} r^{-1} (\eta^2 - p^2)^{-\frac{1}{2}} dr,$$

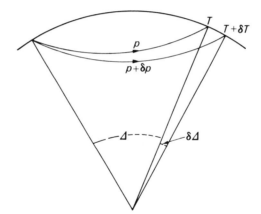

Fig. 4.1 Properties of a seismic ray.

where $v(r)$ = velocity as a function of radius r, r_a = radius of Earth's surface, r_p = radius at depth of greatest penetration of ray, $\eta = r/v(r)$ and p = a ray parameter relating to the angle of launching of the ray (it can be shown that $p = dT/d\Delta$ provided there is radial symmetry). This equation determines T as a function of Δ if the velocity-depth distribution is known. It also enables the variation of amplitude with Δ resulting from geometrical spreading to be estimated for a given velocity distribution, because it tells us how much Δ changes for the cone of rays from p to $p + \delta p$ – a relatively small change would correspond to a high amplitude and vice versa. In practice, T is known as a function of Δ and we wish to use the equation to determine the velocity-depth function. Because the unknown function $v(r)$ occurs under the integral sign, this involves solution of an integral equation. It can be shown to reduce to Abel's integral equation and a unique solution can be obtained by numerical methods provided that $dv/dr < v/r$ over the appropriate range of depth. BULLEN (1963) gives the theory of the solution.

This method was used by Jeffreys and Gutenberg to derive the velocity-depth distribution for P and S waves through the mantle. When applied in greater detail, however, there are certain characteristics of the velocity-depth distribution in the mantle which may preclude unambiguous interpretation of the time-distance observations for body waves. The more important difficulties are described below, and methods by which some of them can be overcome are given.

Firstly, the Herglotz-Wiechert method fails when the velocity decreases with depth more rapidly than v/r. Other methods of investigation (see below) have revealed just such a velocity decrease with depth in the upper mantle for S waves and locally for P waves. Hence this classical method of seismology is of limited value for investigating upper mantle structure. Under this condition the solution of the integral equation is indeterminate for the velocity-depth function in the low velocity zone and below it. A typical time-distance graph showing this is given in Fig. 4.2(a). The characteristic feature caused by the low velocity zone is the shadow zone where no rays emerge at the surface. The recognition of a shadow zone points to the existence of a low velocity zone but it does not remove the ambiguity in interpreting the velocity-depth distribution.

If both the source of the waves and the receiving station are situated above a low velocity layer, all that can be done is to assume a velocity distribution across the layer possibly based on other evidence such as surface waves; the underlying velocity distribution can then be determined uniquely.

A *second* type of difficulty inherent in the Herglotz-Wiechert method is caused by sudden increases in the velocity gradient with increasing depth. These are known as second order

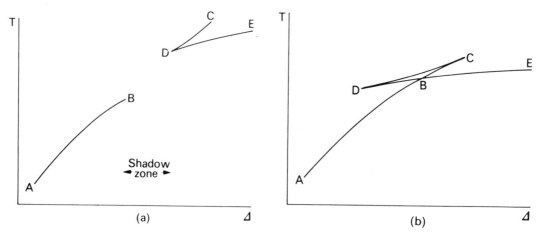

Fig. 4.2 Time distance (T-Δ) graphs illustrating the effects of (**a**) a low velocity zone, and (**b**) a sudden increase of velocity with depth causing triplication.

discontinuities. They cause a triplication of the time-distance graph as shown in Fig. 4.2(b). If the complete curve including the segments BC, CD and DB is known, then the velocity-depth distribution can be deduced without ambiguity using the Herglotz-Wiechert method; but if the segment ABE alone is known the uniqueness is lost. The complete curve is difficult to observe using conventional seismological stations, because late arrivals closely following the first arrival tend to be masked. The problem of fully defining the time-distance curve in the vicinity of a triplication can be overcome to some extent by making use of seismological array stations (see below).

A *third* difficulty arises from lack of spherical symmetry within the Earth. Allowance can be made for the spheroidal shape of the Earth and for variations in crustal structure. But it is now known that there are considerable lateral variations in P and S velocities within the upper mantle and possibly also at greater depths. It is usually assumed that these lateral variations are averaged out in broad velocity-depth distributions.

This difficulty can be turned to advantage. Use can be made of systematic deviations of travel-time from the values predicted by standard tables to show up regional velocity variations within the mantle. 'Rays' from distant earthquakes emerge steeply through the upper mantle and crust. If the travel-times for a series of distant earthquakes covering a range of azimuths are combined, then any systematic deviation from the predicted arrival times should reflect abnormality in the velocity structure of the underlying crust or upper mantle. Correction can usually be applied for the crust, leaving the contribution from the upper mantle.

Fourthly and lastly, observational errors cause uncertainty in the velocity-depth structure deduced from body waves. Important improvement can be gained through use of artificial explosions including nuclear shots, for which the time and exact position of the source are accurately known. Modern instrumentation, including array stations and the much improved world-wide network, has also substantially improved accuracy in the recognition and timing of phases.

The problem of observational errors is much more serious for S than for P. It still remains more difficult to recognize the onset of S, despite modern improvements. S is also affected by a pronounced low velocity channel in the upper mantle. Consequently investigation of the S velocity-depth distribution in the upper mantle and transition zone rests heavily on the use of surface wave dispersion.

Phased seismological array stations

An important recent innovation in seismology has been the introduction of phased array stations. These consist of arrays of individual seismometers recording one (or more) components of ground motion, spread over the ground in an appropriate pattern. The output from each individual seismometer is recorded on a separate track of magnetic tape at a central installation, making it possible to apply versatile processing methods using digital computers, either at the time of recording using on-line computers or at a later date. Phased array stations were originally established to aid the detection of underground nuclear explosions, but they also provide us with an important seismological tool for investigating the Earth's interior.

The first phased array station, now dismantled, was built by the United Kingdom Atomic Energy Authority (UKAEA) in 1961 at Pole Mountain in Wyoming. The UKAEA seismology group under the leadership of Dr. H. I. S. Thirlaway has built four permanent array stations at Eskdalemuir in south Scotland, Yellowknife in Canada, Warramunga in Australia and Gauribidanur in India. These stations essentially consist of two lines of short-period vertical seismometers which cross at right angles, as shown in Fig. 4.3. A particularly large and sophisticated array station incorporating seismometers of different type was built in Montana, U.S.A. This was known as LASA, standing for Large Aperture Seismic Array, and it incorporated a series of sub-arrays which are deployed as shown in Fig. 4.3(b). Another large array station known as NORSAR has been set up in south Norway.

The versatility of a phased array station depends on the ability to apply time delays to the recordings from individual seismometers before combining outputs in various ways. This makes it possible to steer the array to search for seismic signals coming from a specified direction. The delays are applied to individual seismometers so that the wave arrivals from the specified azimuth and dip direction reinforce each other when their signals are combined, while arrivals from other directions are as far as possible suppressed. By repeating this operation for a series of different directions, it is possible to determine the delays which give the strongest reinforcement, and thus to determine the direction of approach of the signal both in azimuth and in dip. If a fast enough computer is available, this can be done in real-time, i.e. as fast as the event is recorded.

The simplest use of an array station is to improve the clarity of seismic signals by increasing the signal-to-noise ratio. However, the opportunity to steer the array enables the direction of approach of a wave to be determined. As the broad velocity-depth distribution in the Earth is well known, this makes it possible to locate the position of a given event. It also enables us to measure $dT/d\Delta$ directly, because this quantity is given by the velocity of the wavefront across the array (Fig. 4.1). Array stations can also be used to separate interfering signals coming from different directions. An example is shown in Fig. 4.4, in which LASA has been used to extract the long period record of an Argentine earthquake which was received at the same time as an event from the Kuriles. As will be seen below, results of processing array records have been of considerable importance in unravelling the velocity structure of the mantle.

Surface waves

Two types of surface elastic wave can propagate in the presence of a free surface such as the Earth's surface. These are Rayleigh and Love waves, named after the scientists who predicted their existence. Early in the history of instrumental seismology, it was observed that both Rayleigh and Love waves are generated by earthquakes and that the resulting wave-trains are dispersed. The study of the dispersion of long-period surface waves is of fundamental importance in assessing the S velocity structure of the upper mantle.

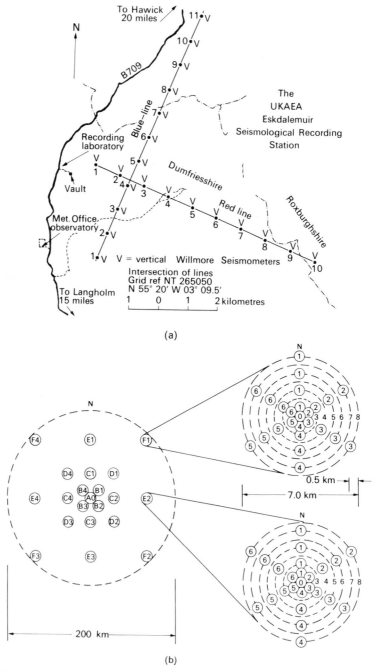

Fig. 4.3

(a) A map showing the geometry of the United Kingdom Atomic Energy Authority phased seismological array station at Eskdalemuir, south Scotland, Redrawn from TRUSCOTT (1964), *Geophys. J. R. astr. Soc.*, **9**, 61.

(b) The geometry of the Large Aperture Seismic Array (LASA) in Montana, U.S.A. This consisted of 21 sub-arrays each consisting of 25 short-period vertical seismometers, and a three-component long-period set of seismometers at the centre of each sub-array. Redrawn from CAPON and others (1969), *Geophysics*, **34**, 306. Used with the permission of the Society of Exploration Geophysicists.

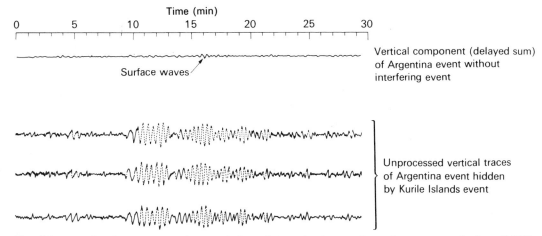

Fig. 4.4 Use of LASA to suppress long-period interfering teleseism. Redrawn from CAPON and others (1969), *Geophysics*, **34**, 317. Used with the permission of the Society of Exploration Geophysicists.

Surface waves which are sensitive to the structure of the upper mantle have periods ranging from 30 s to 600 s. Modern improvements in the design of long-period seismographs have made it possible to record such long-period surface waves.

Rayleigh waves are the only type of surface wave which can occur in a uniform elastic half-space (Fig. 4.5a). Particle displacement is confined to the vertical plane containing the direction of propagation. The amplitude of the displacement decreases with increasing distance from the free surface. For Poisson's ratio of 0·25 the velocity of propagation is $0·92\beta$ for all wavelengths, where β is the S wave velocity. The motion of a particle at the free surface is a retrograde ellipse with its major axis vertical, the ratio of the axes being about 1·47.

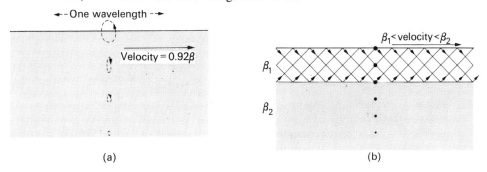

(a)   (b)

Fig. 4.5 Sketches to illustrate the propagation of surface waves.
  (a) *Rayleigh waves*, showing how the particle motion changes with depth from the free surface; particle motion is within the plane of the diagram. S velocity is represented by β.
  (b) *Love waves*, which can be represented by the constructive interference of rays in the upper layer which are repeatedly reflected between the surface and the interface (at supercritical incidence); particle motion is perpendicular to the plane of the diagram.

The Rayleigh wave train becomes 'dispersed' if the elastic moduli and density vary with depth below the free surface. A wave of given period T and wavelength λ travels with a velocity dependent on the distribution with depth of P and S velocity and density, being particularly sensitive to the S velocity at a depth of about $0·4\lambda$. Consequently, the velocity of propagation depends on wavelength, the longer wavelengths sampling the properties over a greater depth range. Rayleigh waves of period 20 s and greater are thus sensitive to the upper mantle beneath continental regions.

The simplest type of structure which can propagate Love waves is a uniform layer with one free surface and the other surface in contact with a uniform half-space, such that the S wave velocity in the layer (β_1) is less than in the half-space (β_2) (Fig. 4.5b). Particle motion is perpendicular to the direction of propagation (as for S waves). Within the half-space the amplitude decreases exponentially with distance from the boundary. Such Love waves are dispersed and the phase velocity varies from β_1 for the very short wavelengths to β_2 for the very long wavelengths. They are closely similar to waveguide waves in radar, quantum theory, etc. Unlike Rayleigh waves, Love waves are not affected by the sea. Love waves can occur in more complicated structures provided velocity increases initially with depth; the character of the dispersion curve reflects the layering.

Suppose we have a dispersed train of waves as shown in Fig. 4.6, containing a packet of waves having a spread of wavelengths close to λ (for simplicity we could consider just two wavelengths, $\lambda + d\lambda$ and $\lambda - d\lambda$). If the velocity depends on wavelength, then the waves travel in a packet as shown, but the individual peaks and troughs travel with a different velocity from the packet itself.

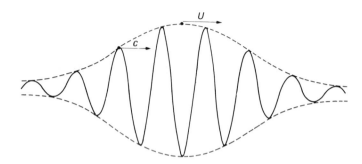

Fig. 4.6 A gaussian wave packet (which is dispersed), illustrating the meaning of *phase velocity c* which is the velocity of the individual wave peaks, and *group velocity U* which is the velocity of the envelope of the wave packet.

The peaks and troughs travel with the *phase velocity*, which is the velocity with which an unmixed wave would travel. The packet travels with the *group velocity* which represents the velocity with which the energy is transmitted. The group velocity U is related to the phase velocity c by the equation $U = c - \lambda dc/d\lambda$, λ being the wavelength. If the phase velocity dispersion curve is known, the group velocity dispersion curve can be obtained by differentiation. The reverse process requires an integration and introduces an arbitrary constant. An example of a dispersed train of Rayleigh waves is shown in Fig. 4.7.

The phase velocity of a given wavelength in a surface wave train can be determined by measuring the velocity of a single crest or trough of the appropriate wavelength as it passes between two long period seismograph stations located along a path of propagation, or more generally by noting its passage across an array of three seismograph stations. This is known as the time correlation method. An alternative method, known as the Fourier phase method, uses Fourier analysis of the record at each station to isolate specific frequencies before calculating their velocity of passage between the stations. Both these methods yield a phase velocity dispersion curve appropriate to the structure of the region between the stations. A group velocity dispersion curve can be obtained by observing the passage of a dispersed wave-train originating from an earthquake of known focus and time of origin as it passes a seismograph station. The average group velocity for the region between the earthquake epicentre and the station for a given period is d/t, where d is the distance along the great circle between the epicentre and the station and t is the time between the event and the passage of the wave. The group velocity determination can sometimes be improved by applying band pass filtering to the records at one or more stations. Modern methods of determining group

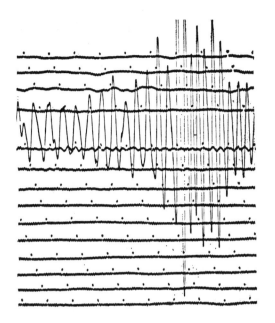

Fig. 4.7 East-west component of the long-period record observed at College, Alaska, of an earthquake in the Himalayan region on October 21, 1964, Timing marks on the record are at one-minute intervals. The part of the record shown here gives a good example of a dispersed train of Rayleigh waves. The record is not confused by Love waves because the seismograph is orientated along the direction of propagation of the waves. Note the large amplitude arrivals of about 20 s period towards the end of the train, which are known as the Airy phase. By courtesy of U.S. Coast and Geodetic Survey.

and phase velocity are mostly variants on the above basic methods but make use of digital filtering techniques. A useful review is given by KOVACH (1978).

The method of interpretation is to compare observed phase or group velocity dispersion curves with theoretically computed curves for assumed models of the Earth. Before the computer era, theoretical curves could only be constructed for relatively simple models such as one or two plane layers overlying a uniform half-space. These were adequate for early investigations of crustal structure such as distinguishing oceanic and continental crusts, but not for application to the mantle. A method for computing dispersion curves for multi-layered models using matrix methods to apply the boundary conditions between layers was given by HASKELL (1953). This method is readily applicable to computers. It can be modified to take into account the Earth's curvature. The computer adaptation of Haskell's method enables theoretical dispersion curves to be calculated for a realistic model of the mantle.

The practical inverse problem is to obtain a model of the underlying structure which satisfies the observed dispersion curve. This could be done by simple trial and error, but the fitting process can be greatly speeded-up and made semi-automatic by use of partial derivatives. These specify the changes to the group or phase velocities caused by an incremental change in the P or S velocity or the density of an individual layer in the multi-layered model, and can be calculated by extending Haskell's method. Having computed the partial derivatives for each layer, the non-linear inverse problem is sufficiently well behaved for the methods of linear algebra to be used to adjust the model and obtain an improved fit. If necessary, the process can be repeated until an acceptable fit is obtained. In practice, it is usual to relate P velocity (Rayleigh waves only) and density to S velocity by empirical relationships and to adjust the S velocities to obtain the fit, thus avoiding ambiguity in the final model.

The Love wave dispersion curve depends on the rigidity modulus and density of each layer, or more conveniently on the S velocity and density. The curve is several times more sensitive to S velocity than to density. Unique interpretation of observed results cannot be obtained unless either density or S velocity is assumed. In dealing with the upper mantle, it is usual to assume the density and make use of the dispersion to derive an S velocity distribution. An acceptable S distribution must satisfy the observed body wave travel-times for S, and if agreement is not reached the density assumptions would need to be changed. Rayleigh wave dispersion is dependent on P and S velocities and on density, although the waves which penetrate the upper mantle are most sensitive to S velocity. The very long-period Rayleigh waves and their counterpart, the spheroidal free oscillations, are more strongly dependent on P velocity and density; as the P velocity distribution in the lower mantle is known fairly well, potentially they can give information on the density of the lower mantle and even of the core.

The Rayleigh and Love waves discussed above belong to the fundamental mode, which means that there are no nodal planes at which the displacements are zero. Higher mode surface waves, for which one or more nodal planes occur, are recognizably excited by some earthquakes. They are an important tool for investigating the S velocity structure of the upper mantle.

The pattern of surface wave propagation becomes more complicated if the rocks of the upper mantle are significantly anisotropic. In particular, there would be some modification to the pattern of particle motion and some coupling between the Love and Rayleigh modes (CRAMPIN, 1977). Thus surface waves have potential for recognizing and studying bulk anisotropy in the upper mantle.

Free oscillations

It is known that free natural vibrations of the Earth are excited by earthquakes. They were first convincingly recognized after the Chile earthquake of May 22, 1960. There are two types, the *torsional vibrations* in which the periodic displacement is everywhere perpendicular to the radius vector, and the *spheroidal vibrations* which involve radial and tangential displacement. Each type can be subdivided into an indefinite number of modes, which depend on the disposition of the nodal surfaces. Three types of nodal surface can occur (Fig. 4.8): (*i*) concentric spherical surfaces; (*ii*) systems of concentric cones with apices at the centre; and (*iii*) equally spaced diameters which intersect the surface at two poles. The fundamental modes have a node of type (*i*) only at the centre.

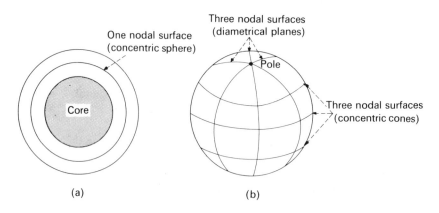

(a)   (b)

Fig. 4.8 Diagram to illustrate the three possible types of nodal surface for free vibration of the Earth.
(a) shows a single spherical surface (first overtone) for torsional vibrations, the core-mantle boundary and the free surface being antinodes. Torsional vibrations are restricted to the crust and mantle, but the core participates in spheroidal vibrations.
(b) shows three conical and three diametrical nodal planes appropriate to torsional or spheroidal vibrations. Note that the position of the pole is arbitrary.
The torsional mode incorporating all seven nodal planes shown would be referred to a $_1T_6^3$ and the corresponding spheroidal mode would be $_1S_6^3$. In general, the $_lT_n^m$ and $_lS_n^m$ modes possess l spherical, m diametrical and $(n-m)$ conical nodal surfaces.

The observation of free vibrations depends on use of instruments sensitive to ultra-long period oscillations. Strain seismometers, which measure strain in place of ground displacement or velocity, respond to both torsional and spheroidal vibrations. Earth tide gravimeters can detect the spheroidal vibrations. The various modes are recognized on the trace by carrying out a power spectrum analysis. If a given mode is sufficiently strongly excited, it appears as a peak on the power spectrum, the position of the peak giving an estimate of the period. The power spectral densities for strain seismometer records of the Alaskan and Chilean earthquakes are shown in Fig. 4.9. The fundamental modes and a few higher overtones are most strongly excited in these records.

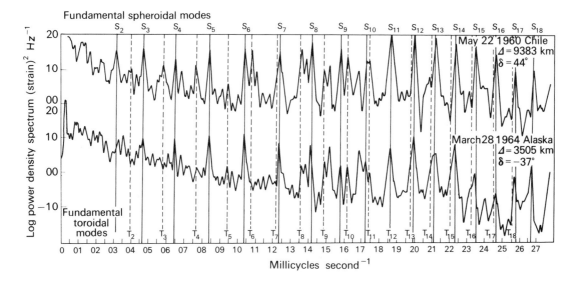

Fig. 4.9 Power spectral density of the Alaskan and Chilean earthquakes recorded on a strain seismometer at Isabella, California. The angle δ is the deviation of the great circle path from the axis of the strain seismometer. Redrawn from SMITH (1966), *J. geophys. Res.*, **71**, 1187.

The spheroidal vibrations are equivalent to the standing wave set up by equal trains of Rayleigh waves travelling in opposite direction round the Earth. The torsional vibrations are equivalent to interfering Love wave trains. The period of a given mode of free vibration can be re-expressed as the phase velocity of the equivalent surface wave. Thus the observation of the periods of the free oscillations enables the Love and Rayleigh wave phase velocity dispersion curves to be extended from about 300 s to 2576 s (torsional) and to 3229 s (spheroidal). The longest periods correspond to the second order fundamental modes, and the higher order modes correspond to progressively shorter periods.

Solving the inverse problem in geophysics

Geophysics is concerned with the determination of the physical properties of the Earth's interior. This can be accomplished within limits by fitting theoretical models to the observations from seismology, geodesy, geomagnetism, etc. Obtaining a model to fit a set of observations is known as solving the inverse problem. The corresponding forward problem, which is usually much simpler to solve, is the process of computing the theoretical values of the observable quantities appropriate to a specified model. Several theoretical difficulties arise in the solution of the inverse problem. All that can be done here is to give a brief glimpse of these problems so that the limitations which apply to most models can be recognized.

It has been known for a long time that the solution to certain inverse problems in geophysics suffers from ambiguity, when more than one model can be fitted to a set of observations. For instance, an infinite number of density distributions within the Earth could be fitted to the Earth's gravitational field at external points, even if this were perfectly known. Furthermore, only a finite number of observations can be made and thus at best the model can only be expressed in terms of the same number of independent parameters. It is therefore necessary to incorporate assumptions and simplifications into the model to ensure that the parameters describing the model are uniquely obtained.

A more sophisticated but related difficulty is that of determining the resolution and precision of a solution, taking into account the errors of observations. The important formal theory for such analysis has been established by BACKUS and GILBERT (1967, 1968, 1970) and described by PARKER (1977). Each parameter of a model can at best be regarded as a weighted average value over a finite region or depth range. The weighting function is known as the averaging or resolving kernel. The spread is the measure of the width of a resolving function, usually its standard deviation. A wide choice of resolving kernels is generally possible, each set being associated with a corresponding set of standard error values for the parameters, calculable from the errors of observations. In general, the narrower (or more 'delta-like') the spread of the kernels, the larger the standard errors of the parameters, and vice-versa. There is thus a 'trade-off' between resolution and precision in solving the inverse problem. A satisfactory solution is one in which the errors are acceptably small without the spread becoming unacceptably wide. The method of Backus and Gilbert enables an acceptable compromise to be reached. It is also very useful in the design of realistic models, prior to inversion. Both the error and the spread are often shown in presenting the results of inversion.

The simplest but most tedious method of solving the inverse problem is by trial and error, involving repeated solutions of the forward problem. In practice, the inverse problem can normally be solved by numerical methods using computers without need for repeated interventions by the investigator. One method is to compute the effect on each observable quantity of perturbing each of the model parameters by an incremental amount. If the problem is linear, the methods of linear algebra can be used to obtain the best-fitting model directly. If it is non-linear, then linear methods can often be used to obtain successively improving fits. A variety of other numerical methods can be used to solve the non-linear inverse problem, including Monte Carlo techniques and various non-linear optimization methods.

4.3 Distribution of P and S velocities in the mantle

The P and S velocities are the most accurately known physical properties of the mantle. Over most of the mantle, they are probably known to better than 1 %. The average radial distribution forms the basis for subdivision of the mantle. The radial and lateral variations provide the framework for discussion of the physical and chemical properties.

The radial distribution of P and S velocity

Early determinations of the P and S velocity distributions with depth in the mantle, such as those of JEFFREYS (1939a) and GUTENBERG (1959), were based on Herglotz-Wiechert inversion of travel-time data, with a low velocity zone in the upper mantle being incorporated in the Gutenberg model on the basis of a shadow zone (Fig. 4.2a). Jeffreys based his inversion on the tables of travel-times against angular distance compiled by JEFFREYS and BULLEN (1940), which are often referred to as the

JB tables. Several sets of improved travel-time tables are now available, such as those of HERRIN (1968). However, in recent determinations of the radial velocity distributions of the mantle, inversion of free oscillation data is used in conjunction with the body wave data. The body waves resolve detail in the upper mantle and elsewhere whereas the free oscillations yield much more accurate average values over depth ranges of a few hundred kilometres. Such inversion of gross earth data has the additional advantage that the density distribution is obtained directly, albeit with rather poor resolution.

ANDERSON and HART (1976) derived a radial distribution of seismic velocities and density within the Earth as a whole by inverting 400 modes of free oscillation of the Earth. The resolution for seismic velocities is about 100–400 km so that the fine detail above 700 km depth cannot be resolved. Such detail was incorporated by using as the starting model the *P* and *S* velocity distributions beneath the western United States obtained by HELMBERGER and WIGGINS (1971) and HELMBERGER and ENGEN (1974) using travel times, amplitudes and waveforms of body wave arrivals, as described later. The model was further improved by HART and others (1977) by allowing for the effects of attenuation on free oscillation periods (Fig. 4.10). As the fine detail of the final model is strongly influenced by the starting model, the velocities down to about 800 km depth are a compromise between the broad average global structure and the detailed structure beneath mid-western United States.

An alternative approach of considerable practical value is to construct a simple reference model of the Earth based on parameterization. The Earth is subdivided into a series of spherical shells within each of which the seismic velocities and density are assumed to vary smoothly with depth as a simple linear, quadratic or cubic polynomial function of earth radius. Discontinuities in the functions or their gradients occur at the boundaries between adjacent shells. Thus for example the mantle above the low velocity zone, the region between the 400 and 670 km steps, the outer core and the inner core can be treated as separate shells. The coefficients of all the polynomial representations are then determined by inverting the data consisting of free oscillation periods, travel-times of body waves, and the mass and moment of inertia of the Earth. The first parametrically simple earth model was published by DZIEWONSKI and others (1975). More recently, DZIEWONSKI and ANDERSON (1981) have obtained a Preliminary Reference Earth Model (PREM) using 1000 free oscillation periods, 500 travel-times of body waves and 100 attenuation values for free oscillations. This model is compared with that of HART and others (1977) and on a model based solely on body waves in Fig. 4.10. The greatest difference between the models is in the upper mantle and transition zone, and agreement is good in the lower mantle. An interesting feature of PREM is that anisotropy was allowed for in the inversion. The best fit was obtained if the horizontal components of *P* and *S* velocities are 2 to 4% faster than the vertical components down to 220 km depth in the mantle, although anisotropy was not found at greater depths. Average values down to 220 km are shown in Fig. 4.10.

The upper mantle: The topmost mantle is referred to as the 'lid' of the low velocity zone and it is commonly regarded as forming the lower part of the lithosphere. According to most radial distributions, the seismic velocities remain constant or increase slightly with depth across this zone, which averages about 60 km in thickness. A more detailed and somewhat different velocity-depth structure of the topmost mantle is revealed by seismic refraction experiments in both continental and oceanic regions. The average value of P_n determined by crustal refraction experiments is about $8 \cdot 15 \text{ km s}^{-1}$ but there is also evidence for fine structure below the Moho from seismic refraction lines which extend beyond 300 km where P_n dies out (HIRN and others, 1973; FUCHS, 1979). Large amplitude arrivals are observed with apparent velocities ranging between $8 \cdot 3$ and $8 \cdot 7 \text{ km s}^{-1}$. The simplest interpretation is that these are refracted arrivals from steep velocity gradients or

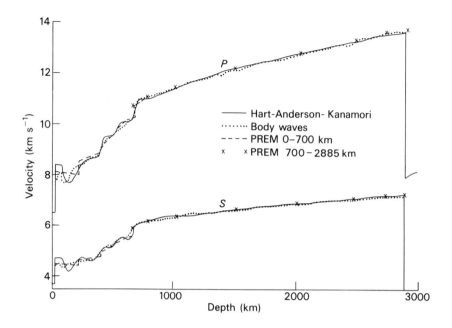

Fig. 4.10 Some recent P and S velocity-depth distributions in the mantle, showing the distribution of HART and others (1977) and the Preliminary Reference Earth Model of DZIEWONSKI and ANDERSON (1981) which are both based on the inversion of gross earth data including body waves and free oscillations, and a solution based on the inversion of body waves only. The body wave distributions are taken from BURDICK and HELMBERGER (1978) for P down to about 750 km depth, HELMBERGER and ENGEN (1974) for S down to about 750 km depth, and SENGUPTA and JULIAN (1978) for P and S below 750 km depth.

discontinuities within the topmost mantle, possibly with intervening low velocity layers between them. For instance, HIRN and others (1973) observed along a profile across France two segments named P_I and P_{II} having apparent velocities of 8·2 and 8·5 km s^{-1} and these were interpreted in terms of refractors at 55 and 80 km depths respectively. This evidence may indicate fine layering with strong velocity contrasts within the uppermost mantle but the possibility that the arrivals are partly or wholly caused by lateral inhomogeneity should not be ruled out.

According to the average radial Earth models, there is a low velocity zone for P and S extending between depths of about 80 and 220 km. The presence of a low velocity zone in the upper mantle was first suggested by GUTENBERG (1926) on the basis of the observed rapid decrease of amplitude of P arrivals with increasing distance from the epicentre out to about 15° (1600 km). Beyond this distance, the amplitude increases by more than a factor of ten. Gutenberg attributed this to the shadow zone effect (Fig. 4.2a) caused by a low velocity zone for P starting at about 80 km depth. He subsequently found that the shadow zone is better developed for S waves, implying that the low velocity zone is more pronounced for S than for P (GUTENBERG, 1948). These early amplitude studies were important in showing up the existence of the low velocity zone but using only shallow focus sources it was impossible to uniquely define the velocity distribution across the zone. This difficulty was partially overcome by GUTENBERG (1953) by using intermediate and deep focus earthquakes to give a direct estimate of the velocity at the focus, although such a method is of very restricted geographical application.

The existence of an average low velocity channel for S in the upper mantle has been amply confirmed by surface wave studies. In contrast to body waves, these can sample the average structure of the upper mantle relatively easily. Low velocity zones can be detected although the fine detail cannot be resolved. DORMAN and others (1960) were first to apply the method to investigate the average global *S* velocity structure of the upper mantle. They used Rayleigh waves up to 250 s period and found that a global low velocity channel for *S* in the upper mantle is needed to explain the observed group velocity dispersion curves. Subsequent work has supported this conclusion but has shown that the low velocity zone is not well developed beneath continental shield regions.

The presence of a global low velocity channel for *P* is less easy to establish as surface waves are insensitive to *P*. The best evidence comes from the *P* shadow zone and more recently from the use of synthetic seismograms to model the waveform of *P* arrivals. Using this method, BURDICK and HELMBERGER (1978) found a low velocity *P* zone beneath western United States between about 70 and 160 km depths with a minimum velocity of 7·70 km s^{-1} at 100 km depth, and GIVEN and HELMBERGER (1980) found a low velocity zone between 140 and 230 km depths beneath northwestern Eurasia. In contrast, the earlier work of BRUNE and DORMAN (1963) found no evidence for such a zone beneath the Canadian Shield although a slight decrease of the velocity of *P* with depth could not be ruled out. On balance, the evidence points to the existence of a global low velocity channel for *P* but it is probably not as well developed as the *S* low velocity zone, despite the contrary indication in the velocity distributions of HART and others (1977) which depended on initial assumptions. It may not be universally present.

The base of the low velocity zone is probably marked by a steep velocity gradient. LEHMANN (1962) had originally suggested that there is a small discontinuous increase in velocity at a depth of about 220 km beneath Europe and North America. BURDICK and HELMBERGER (1978) used waveform analysis of *P* arrivals to show that a steep velocity gradient between depths of about 150 and 175 km fits the data better for western United States. Below the lower boundary of the low velocity zone, velocities apparently increase steadily towards the top of the transition zone, detail being unresolvable at present.

As pointed out earlier, there is evidence for gross anisotropy down to about 220 km depth, in that *P* and *S* velocities appear to be about 2 to 4% higher in horizontal directions than in the vertical direction. Such anisotropy is indicated by the anomalously high *S* velocity values yielded by inversion of free oscillation data. Transverse anisotropy may also occur.

The transition zone is marked by a rapid increase in *P* and *S* velocity with depth, contrasting with the relatively small velocity gradients in the upper mantle above it and in the lower mantle below. It was originally recognized in the Jeffreys velocity-depth distribution as a second order discontinuity at 400 km depth where the velocity-depth gradient abruptly steepens. This used to be referred to as the 20° discontinuity.

New insight into the fine structure of *P* across the transition zone has come from careful studies of recordings in the range 10° to 30° from shallow earthquakes and nuclear explosions. The most detailed studies have been made in the western and central United States because of the frequent occurrence of suitable sources and the good seismic instrumentation including array stations. Thus NIAZI and ANDERSON (1965) used the Tonto Forest array station in Arizona to determine directly the apparent velocity $dT/d\Delta$ across the array for 70 earthquakes with epicentres between 10° and 30° from the station. By assuming the velocity-depth distribution down to 225 km, they were able to apply the Herglotz-Wiechert technique to the $dT/d\Delta$ observations to obtain the velocity distribution down to 800 km depth. They found that the rapid increase in *P* velocity with depth occurs in two steps. Each of the two steep rises in velocity gives rise to triplication of the travel time curve which can be recognized convincingly from the $dT/d\Delta$ observations (Fig. 4.11) and each causes a focussing of the rays as shown in Fig. 4.12.

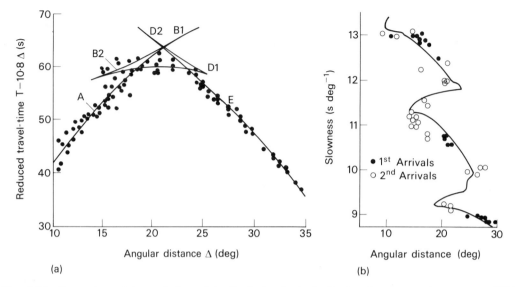

Fig. 4.11 Comparison of the predictions of the P velocity-depth distribution of BURDICK and HELMBERGER (1978) with travel-time and apparent velocity observations from the western United States:

    (**a**) Reduced travel-time curve. Segment A represents arrivals which do not penetrate down to the transition zone. The retrograde branches B1–B2 and D1–D2 represent rays returned from the 400 km and 650 km discontinuities respectively and segment E represents arrivals penetrating below the transition zone.

    (**b**) Observed and calculated slowness ($dT/d\Delta$) values. Note that slowness is the reciprocal of apparent velocity. Adapted from BURDICK and HELMBERGER (1978), *J. geophys. Res.*, **83**, 1703.

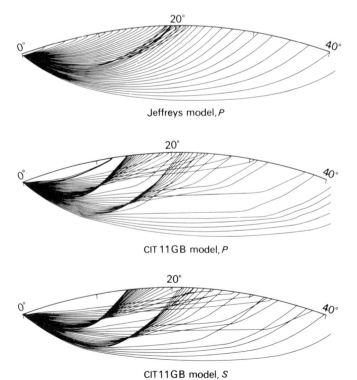

Fig. 4.12 Ray paths for some older velocity-depth distributions in the upper mantle and transition zone, showing the large amplitudes at 20° associated with the Jeffreys P model, and the concentration of rays at about 14° and 21° caused by the steep velocity-depth increases at about 400 and 700 km depths in CIT11GB models of JULIAN and ANDERSON (1968). Redrawn from JULIAN and ANDERSON (1968), *Bull. seism. Soc. Am.*, **58**, 348 and 351.

A further advance in the study of *P* and *S* velocity-depth distributions across the transition zone and across the overlying low velocity zone (see earlier) has come from the use of synthetic seismograms to study amplitude and waveform by Helmberger and his colleagues. This involves a tedious process of trial and error but it yields better resolution of some of the detail than that obtained using travel-time and $dT/d\Delta$ data only and it also gives better depth control as the velocity distribution across the low velocity zone above is determined rather than being assumed. WIGGINS and HELMBERGER (1973) applied this method to observations from nuclear events at the Nevada test site and several earthquakes and derived a model of the *P* velocity across the underlying transition zone with sharp discontinuities at 430 and 650 km and a discontinuity in slope at about 550 km depth. In a more recent analysis, BURDICK and HELMBERGER (1978) used long-period as well as short-period *P* observations in the waveform matching. They found a rather simpler structure gave the best fit to the observations, with sharp discontinuities at 400 and 670 km depths but without any discontinuity or steep gradient at 550 km (Figs 4.10 and 4.11). Using the same technique, GIVEN and HELMBERGER (1980) have determined an almost identical structure for the transition zone beneath north-western Eurasia, with a 5 % jump in *P* velocity at 420 km depth and a second discontinuity with a 4 % increase in velocity at 675 km depth. The Anderson-Hart velocity-depth distributions within the Earth used the earlier Wiggins-Helmberger distribution in their starting model; consequently the 500 km step in the Anderson-Hart distributions (Fig. 4.10) should be treated with some reserve.

The first indication of a double step in the *S* velocity-depth distribution between 350 and 700 km depth came from surface wave dispersion (TOKSÖZ and others, 1967) but the resolution of surface waves and free oscillations is now regarded as inadequate for studying the fine structure of the transition zone. The best method is to model the travel times, apparent velocities, amplitudes and waveforms of *S* arrivals out to 30° from the epicentre, as described above for *P*. HELMBERGER and ENGEN (1974) used this method to determine the *S* velocity structure down to 1000 km depth beneath the United States based on horizontally polarized *S* waves. This model, named SHR14, has steep increases in velocity starting at about 370 and 600 km depths and a further short but sharp increase at 500 km depth (Fig. 4.10). There are several obvious inconsistencies between this model and the *P* model T7 shown in Fig. 4.10 and the next step will be to obtain mutually consistent *P* and *S* distributions.

The lower mantle extends from the base of the transition zone to the core-mantle boundary which occurs at an estimated average depth of 2886 ± 3 km (p. 233). Below 720 km depth, the *P* and *S* velocities in the mantle rise gently with depth until the bottom 200 km where the gradients decrease and may even reverse. The velocity-depth profiles based on inversion of free oscillation periods give the most accurate average levels but the resolving width is of the order of 300–400 km so that local detail is obscured. Body wave studies give better resolution with a spread of the order of 70 km but are affected by lateral inhomogeneity in the lower mantle. Body wave studies have benefitted by use of array stations to determine $dT/d\Delta$ directly. The lower mantle body wave distributions shown in Fig. 4.10 were obtained by SENGUPTA and JULIAN (1978) by studying travel times of body waves from deep-focus earthquakes. These give high quality signals which are less biassed by heterogeneity near the source than those from shallower events. The two main current controversies concern the nature and depth of radial inhomogeneities and the velocity distributions in the lowermost 200 km of the mantle.

Most of the evidence for small irregularities in the lower mantle radial velocity-depth distributions comes from $dT/d\Delta$ measurements on *P*. For instance, WRIGHT and CLEARY (1972) used data obtained at the Warramunga array station in Australia to infer a prominent region of low gradient starting at 800 km depth, with other less prominent low gradients starting at 1070, 1260,

1750 and 2460 km and high gradients starting at 1160, 2180 and possibly 2700 km depths. On the other hand, the distributions of Sengupta and Julian show regions of high gradient which they regard as possibly significant at 2400 and 2600 km only. The smaller irregularities on their distributions are regarded as being below the limits of resolution and thus probably of no real significance. One feature of interest is that the local variation appears to be more pronounced for P than for S.

There is a well-established decrease in the P and S velocity-depth gradients in the lowermost 200 km of the mantle. This zone was named D'' by Bullen. It has been further suggested by BUCHBINDER (1971) and by some others that the P velocity actually *decreases* quite sharply with depth below about 2750 to 2800 km depth and a similar velocity reversal has also been suggested for S. The P distribution of Sengupta and Julian shows a slight decrease of velocity in the lowermost 60 km but the S distribution shows no reversal. The existence of such velocity reversals, however, remains an open question because of doubts of the applicability of geometrical ray theory near the core-mantle boundary.

Lateral velocity variations

About twenty five years ago hardly anyone suspected that the radial velocity structure of the mantle varied from region to region. It is now known that there are substantial lateral variations in the P and S velocity structure of the upper mantle and that some smaller lateral variations occur at greater depths. The evidence comes from three main sources, namely seismic refraction surveys revealing variations in P_n, surface wave studies which highlight variations in the upper mantle low velocity zone for S, and anomalies in the travel-times of P and S arrivals which are mainly caused by advance or delay in the upper mantle.

Perhaps the most spectacular lateral variation occurs beneath active margins and island arcs where oceanic lithosphere is recycled, but discussion of this is deferred to Chapter 5. Here the emphasis is given to the broad regional variations in upper mantle structure.

Upper mantle beneath continents: Crustal seismic refraction surveys have shown that the P velocity just below the Moho ranges from about $7{\cdot}5\,\mathrm{km\,s^{-1}}$ beneath certain active tectonic regions to $8{\cdot}6\,\mathrm{km\,s^{-1}}$ beneath certain stable regions. The variation of P_n beneath U.S.A. is sufficiently well known to produce a map of it. This was done by HERRIN and TAGGART (1962) who used the extensive recordings of P_n from the GNOME nuclear explosion in New Mexico to determine interval velocity from one station to another. Their results agree well with local refraction survey determinations of P_n and an updated version of their map is shown in Fig. 4.13. The map shows that P_n is substantially lower beneath the western United States than beneath the central and eastern parts of the country; within western United States it varies from one structural province to another (PAKISER, 1963).

It is a more difficult problem to determine how the overall velocity-depth structure of the upper mantle varies from region to region. The inversion of P and S body wave data incorporating waveform analysis is a tedious process and can only be applied to a limited number of regions close to suitable seismic sources. Local variations in the P velocity structure have been most successfully studied beneath North America. LEHMANN (1964) used travel times of P towards the north-west and north-east from the GNOME nuclear explosion set off in a salt dome in New Mexico to show that the P velocity structure of the upper mantle differs strikingly between the western tectonic region and the eastern stable region of the United States. The P low velocity zone is pronounced in the west where it starts just below the Moho; in the central and eastern regions it is much less conspicuous and hardly exists at all beneath the Canadian Shield. Subsequent work using more sophisticated techniques has confirmed this large contrast in P velocity structure between the western and eastern

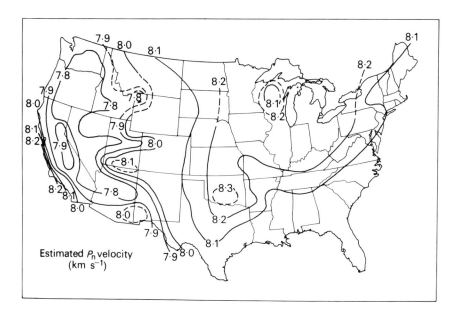

Fig. 4.13   Estimated P_n velocity beneath the United States based on a flat earth approximation. After TUCKER and others (1968), *Bull. seism. Soc. Am.*, **58**, 1256.

regions. Further afield, GIVEN and HELMBERGER (1980) have shown that the low velocity *P* zone is much less well developed beneath north-western Eurasia than beneath the western United States.

A more widely applicable method of determining regional variations in the mean value of *P* (or *S*) in the upper mantle is to make use of systematic deviations of arrival times at seismic stations from the values predicted by standard tables which assume radial symmetry. Where rays arrive obliquely, a correction needs to be applied to yield the travel-time anomaly of a vertically incident ray. The travel-time anomaly which is independent of the direction of arrival of a ray at a station can then be attributed to the velocity structure in the underlying upper mantle rather than to deeper irregularities. As shown in Fig. 4.14, relative delays of the order of 2–3 seconds between two stations imply very large contrasts in upper mantle velocity-depth structure. CLEARY and HALES

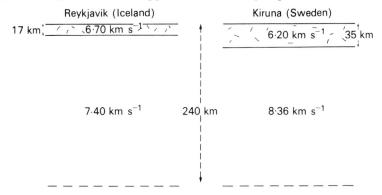

Fig. 4.14   Differences in upper mantle *P* velocity structure beneath Reykjavik and Kiruna which would explain the differences in *P* arrival times, relative to the Jeffreys-Bullen tables, at these two stations.

(1966) used the travel-times predicted by the Jeffreys-Bullen tables as a standard to show that the anomalies in P arrival time for the United States range from -1 to $+1$ seconds. The arrivals are early in the central and eastern parts and late in the western part, as would be expected. Such large residuals can only arise from large variations in the average P velocity in the upper mantle beneath. If the velocity anomalies are spread over a vertical range of 400 km, then the average P velocity over this range must differ by about 0.4 km s^{-1} between west and east to account for the 2-second differences. This is a substantial difference. According to HALES and DOYLE (1967), the anomalies in S are much larger than those in P, ranging from about -4 to $+3$ seconds; early arrivals occur in the east and centre, and late arrivals in the west, as for P. The observed ratio of S delay to P delay is 3.72 ± 0.43. To explain the 7-second difference, very large variations in the average S velocity structures are needed.

Broad regional variations in the S velocity structure of the upper mantle are also revealed by surface wave studies, albeit with rather low resolution. The main problem is that fundamental mode surface wave paths long enough to yield suitable dispersion curves usually cross two or more types of regional structure. DZIEWONSKI (1971) has shown, however, that dispersion data from a suitable number of composite paths can be analysed to separate out the 'pure-path' dispersion curves appropriate to each type of region. He used this method to show that continental tectonic regions have a much better developed low velocity zone for S than shield regions where it is only marginally developed if at all. According to Dziewonski's simple models, the average value of S down to 200 km depth is about 0.3 km s^{-1} lower beneath tectonic regions than beneath shields, but surprisingly it is 0.2 km s^{-1} higher between 200 and 400 km depths. Subsequent work (Fig. 4.15) has broadly confirmed this pattern and has made it possible to make more detailed analyses of continental regions (e.g. PATTON, 1980).

To summarize, the low velocity layer is best developed beneath tectonic regions where P and S velocities are on average the lowest, whereas the zone is least well developed and the velocities are highest beneath shields. Lateral variation of S appears to be much greater than that of P.

Upper mantle beneath oceans: It was pointed out in Chapter 3 (pp. 98 and 115) that P_n is anomalously low beneath ocean ridge crests and that there is significant anisotropy in oceanic P_n

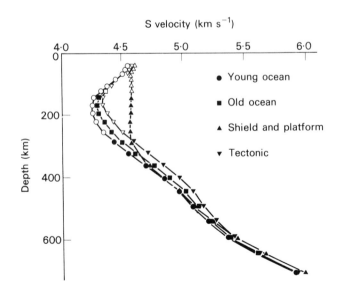

Fig. 4.15 Regional S-velocity models for the upper mantle based on inversion of great-circle Rayleigh wave phase velocities. Open symbols denote slight confidence. Redrawn from LÉVÊQUE (1980), *Geophys. J. R. astr. Soc.*, **63**, 36.

attributable to orientation of olivine crystals in relation to the spreading direction. Otherwise, most of the available evidence on regional variation in the upper mantle beneath oceans comes from surface wave dispersion and study of *S* delay times. The upper mantle *S* structure is found to vary systematically with the age of the overlying ocean floor.

The regional upper mantle structure beneath the Pacific and Atlantic Oceans has been studied by surface wave dispersion, particularly that of Rayleigh waves of 20 to 200 second period. In most studies, the 'pure-path' method of Dziewonski described above has been used to isolate the dispersion curves for a series of age zones of the ocean floor such as 10–20 My (Fig. 4.16). The dispersion curve for each age zone is then interpreted in terms of a model of upper mantle *S* velocity structure. Most interpretations have assumed constant velocities for the mantle part of the lithosphere and for the underlying asthenosphere, and the dispersion curves have then yielded an estimate of the depth to the lithosphere-asthenosphere boundary appropriate to each age zone. The results indicate that the lithosphere thickens with age from about 30 km at 5 My to about 100 km at over 100 My. FORSYTH (1977) also determined lateral variations of *S* velocity in the upper mantle, which he found to increase from $4 \cdot 31 \, \text{km s}^{-1}$ in young lithosphere to an average of $4 \cdot 55 \, \text{km s}^{-1}$ in old lithosphere and from $4 \cdot 10$ to $4 \cdot 25 \, \text{km s}^{-1}$ in the underlying asthenosphere. These studies are not yet able to resolve variations in depth to the base of the asthenosphere which is assumed to be at a constant depth of 170 km. Nor can detailed structure beneath the ridge crests be resolved.

An alternative method is to study the *S* delay times associated with shallow focus earthquakes at ridge crests and within oceanic parts of plate interiors. DUSCHENES and SOLOMON (1977) used this technique to show that the *S* delays decrease systematically with the age of the ocean floor by 6 ± 1

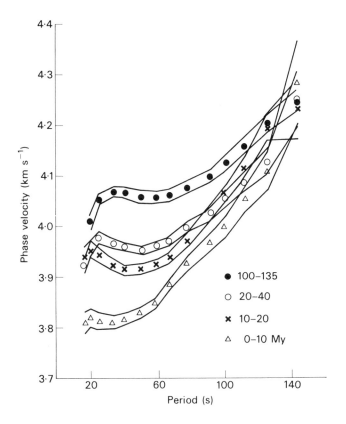

Fig. 4.16 Pure-path Rayleigh wave phase velocity dispersion curves for sea floor of varying age (shown in My). The solid lines are one standard deviation away from the observed velocities. Redrawn from FORSYTH (1977), *Tectono-physics*, **38**, 101.

seconds from ocean ridges to ocean floor 100 My old. These results are consistent with the surface wave dispersion models but the S delays at the ridge crests also reveal a region of exceptionally low S beneath, which is outside the resolution of surface wave studies. Duschenes and Solomon suggested a theoretical model of S velocity structure in the uppermost 100 km of the sub-oceanic upper mantle based on thermal modelling of a wet peridotitic mantle. This model, which is shown in Fig. 4.17, is in good general agreement with the surface wave dispersion and S delay times. Duschenes and Solomon also found that the delays are systematically 2 seconds less for events in the Indian Ocean region, which they attribute to the rapid northward movement of the ocean ridges of the Indian Ocean relative to the underlying transition zone and lower mantle.

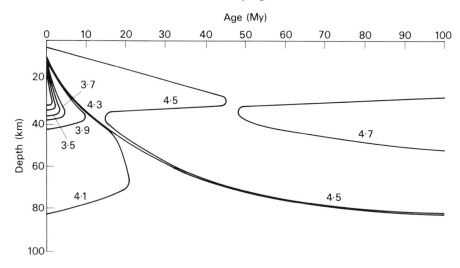

Fig. 4.17 Contours of S velocity as a function of depth and sea-floor age in the oceanic lithosphere and upper asthenosphere based on a theoretical wet peridotite model and in approximate agreement with the observed S residuals. The 4·3 km s^{-1} contour marks the boundary between the seismological lithosphere and the underlying asthenosphere. Redrawn from DUSCHENES and SOLOMON (1977), *J. geophys. Res.*, **82**, 1995.

Contrast between sub-oceanic and sub-continental upper mantle: While regional variation of upper mantle structure has been established within continental and within oceanic regions, the comparison between them is more controversial. The problems concern the magnitude of the difference in S velocity structure and whether the differences are limited to the uppermost 200 km of the mantle or extend to much greater depths.

The earliest indication that the average S velocity structure of the upper mantle differs substantially beneath continents and oceans came from surface wave dispersion (DORMAN and others, 1960; AKI and PRESS, 1961). This work suggested that the S low velocity channel is better developed beneath oceans than continents, either because it starts at shallower depth or because of lower velocities within it. Subsequent surface wave studies, including those using the improved 'pure-path' technique (DZIEWONSKI, 1971), have generally supported this conclusion (Fig. 4.15). Certain studies, however, suggest that the differences in structure vanish at about 250 km depth (e.g. OKAL, 1978) whereas others indicate differences extending to greater depth (e.g. LÉVÊQUE, 1980). This problem will probably be eventually resolved by use of higher mode surface waves (overtones) which are more sensitive to the velocity structure between depths of 200 and 1000 km. Preliminary results already indicate that small differences of around 1 % may extend to 1000 km depth and that the difference in vertical one-way *S* travel time between continents and oceans is of the order of 1·7 seconds.

An alternative approach is to use *S* travel-time anomalies. Use has been made of phases of type ScS_n, which are steeply incident *S* waves which have been multiply reflected at the core-mantle boundary *n* times (Fig. 4.19). By determining the difference between travel times of phases reflected *n* and $(n-1)$ times, the local delay in the upper mantle associated with the selected reflection point can be determined. Using this technique to compare reflection points in the western Pacific with those within continental regions, SIPKIN and JORDAN (e.g. 1980) found that there is substantial delay of *S* associated with the old ocean of the western Pacific in comparison with average continental regions. Taking the different crustal structure into account, they found that the one-way *S* delays averaged about 5 seconds more for the western Pacific than for continental shield regions and 4 seconds more than average continental regions. In contrast, OKAL and ANDERSON (1975) found that the older oceanic regions such as the western Pacific have travel-time residuals similar to those of continental shields. The surface wave data suggests that the true situation probably lies somewhere between these two extreme interpretations.

It seems reasonable to accept that the average *S* velocities are greater in the upper mantle below continents than below oceans, the contrast in one-way travel-time being about 1–3 seconds. The exact magnitude of the difference and the depth to which it extends are still problematical. The differences in *P* structure are hardly known, although the comparison of *P* delay times between Iceland and north Europe (Fig. 4.14) suggests that the average oceanic *P* in the upper mantle is lower beneath oceans than continents but that the differences are much less than for *S*.

Transition zone and lower mantle: It is generally accepted that the large lateral variations of seismic velocity in the upper mantle substantially decrease below about 300 km depth. Lateral variation within the transition zone may be insignificant or alternatively small systematic variations of up to 1 % may occur. The largest lateral variation probably occurs where the oceanic lithosphere is subducted. It will be shown later in the chapter that the transition zone is caused by a series of pressure-induced and temperature sensitive phase transitions which cannot vary in depth on a regional scale by more than a few kilometres without causing much larger global gravity anomalies than are observed. Thus large lateral variations would not be expected to affect the transition zone, except locally where penetrated by sinking lithosphere or rising hot plumes of convection currents.

Large scale lateral heterogeneity in the seismic velcoity pattern of the lower mantle has been detected by studying anomalies in travel times. DZIEWONSKI and others (1977) investi-gated such heterogeneities for *P* by subdividing the mantle into five spherical shells and determining velocity anomalies within each shell in terms of spherical harmonic coefficients by analysis of nearly 700 000 travel-times. They found that the region above 1100 km depth is dominated by perturbations of less than 5000 km wavelength but longer wavelength perturbations were significantly resolved below 1500 km depth. The velocity perturbations, which are of the order of ± 0.03 km s^{-1} amplitude, are interpreted as being greatest in the lowermost 700 km of the mantle.

Smaller scale heterogeneity in the lowermost 200 km of the mantle is also indicated by scattering of *PKP* and associated phases which travel through the outer core. This scattering gives rise to precursors as described in Chapter 6 (p. 235). Much of this scattering probably takes place at the core-mantle boundary but some of it probably takes place as a result of lateral heterogeneity in the lowermost 200 km of the mantle. Such irregularity is to be expected if this region marks a lower thermal boundary layer (p. 349).

4.4 Radial distribution of density, elastic moduli and allied properties

Prior to the 1970s, the radial density distribution within the Earth's interior was derived from the P and S velocity distributions by assuming an appropriate relationship between seismic velocity and density and by using other constraints to fix the change in density across prominent boundaries such as the core-mantle interface. In particular, any acceptable model of the density distribution must reproduce the observed values of the Earth's mass and moment of inertia. Further constraint on allowable density distributions can be provided by assuming a specific petrological model for the upper mantle and computing the density at the relevant temperatures and pressures. The determination of the periods of a sufficient number of the Earth's free oscillations has now opened up the possibility of determining the density and seismic velocity distributions together by inversion. However, the resolution of the resulting density distribution is poor and it is still necessary to use the earlier methods to obtain detail on a scale of less than 400 km in the upper mantle and transition zone or less than about 1000 km in the lower mantle.

Knowledge of the P and S velocities at a specified depth within the Earth gives two equations connecting three important physical quantities, namely density and the bulk and rigidity moduli. A further equation connecting them is needed if they are to be determined uniquely from the velocities. In the homogeneous shells of the Earth which are subject solely to adiabatic self-compression, the Adams-Williamson equation provides the additional relationship needed. This equation relates increase in density with depth caused by adiabatic self-compression under hydrostatic pressure to the seismic velocities. It is derived as follows. A seismic parameter ϕ which can be computed from the P and S velocities is defined as follows:

$$\phi = k/\rho = V_p^2 - \tfrac{4}{3}V_s^2$$

For adiabatic compression $k = \rho \, dp/d\rho$ by definition, where p is pressure. If the pressure is hydrostatic then $dp/dr = g\rho$ where g is acceleration due to gravity. Eliminating p and k gives

$$d\rho/dr = g\rho/\phi$$

which is the Adams-Williamson equation. Step-by-step integration of this equation makes it possible to compute the density distribution and gravity throughout a homogeneous shell of the Earth provided their values are known at the top. The following modified form of the equation can be applied to regions where the temperature distribution is not adiabatic:

$$d\rho/dr = g\rho/\phi + \alpha\rho\tau$$

where τ is the difference between actual and adiabatic temperature gradients and α is the coefficient of thermal expansion.

The Adams-Williamson equation has been applied justifiably to the lower mantle (D) and the fluid outer core (E). It has been used for the upper mantle (B) with less justification. It cannot be used for the crust because of variation in composition, or for the mantle transition zone (C) because of the occurrence of major phase transitions. The Adams-Williamson approach cannot determine increases of density across transition zones and discontinuities; some of these jumps in density can be determined by using the constraints of the Earth's mass and moment of inertia, but it is still necessary to make some further assumptions to obtain a unique model of density.

Models of the Earth's radial density distribution have been presented by several workers. Each of these models depends on a separate set of assumptions. Some of the better known models are described briefly below.

Bullen has developed two basic types of model in a series of papers culminating in a book

(BULLEN, 1975). Models of type A are based on the following assumptions. The density just below the Moho is taken as $3320\,\mathrm{kg\,m^{-3}}$, which is appropriate for a rock of dunitic composition. The Adams-Williamson equation is used for regions B, D and E, and the outer core is taken to have zero rigidity modulus. The density distribution across the mantle transition zone (C) is assumed to be a quadratic function of radius, tied to the density values at the top and bottom of the region and to the density-depth gradient at the top of D. The constraints imposed by the Earth's mass and moment of inertia determine the increase in density across C and at the core-mantle boundary. The density at the Earth's centre is arbitrarily defined. The original model A was tied to an old value of the Earth's moment of inertia of $0.3335\,Ma^2$ and assumed an unacceptably high density at the Earth's centre of $17\,300\,\mathrm{kg\,m^{-3}}$. The improved model A″ is based on a better estimate of the moment of inertia of $0.3308\,Ma^2$ and incorporates a much more acceptable central density of $12\,510\,\mathrm{kg\,m^{-3}}$.

One implication of the models of type A is that the bulk modulus k and its pressure gradient dk/dp are found to change only slightly across the core-mantle boundary despite the change in composition. This led Bullen to formulate the 'compressibility-pressure hypothesis', that k is a smoothly varying function of pressure, irrespective of composition or state, below about 1000 km depth. Models of type B are based on use of the compressibility-pressure hypothesis for the lowermost 200 km of the mantle (D″) and below the outer core (F and G); the Adams-Williamson equation was used for D′ and E. The original model B had unacceptably high density values at the Earth's centre and in the upper mantle, stemming from the poor value of the moment of inertia used in the earlier models. The more recent model B_2 which is based on the improved value of the moment of inertia is in fairly close agreement with model A″, verifying the approximate validity of the compressibility-pressure hypothesis.

Models of type A and B have now been superseded, but their method of construction is informative. The most recent models in this series are of type HB (HADDON and BULLEN, 1969). Model HB_1 (Table 4.1) has been obtained by modifying model A″ to satisfy the periods of the fundamental modes of the free oscillations then available.

BIRCH (1964) alternatively used his empirical relationship between P velocity, density and mean atomic weight (p. 71) for regions B and C in preference to applying the Adams-Williamson equation to B. Below the mantle transition zone, he used similar assumptions to those of the Bullen models of type A. He used P and S distributions similar to those of Jeffreys except in the upper mantle where a low velocity zone was incorporated and at the inner-outer core boundary where a smoother distribution was adopted. Models I and II were constructed assuming the density just below the Moho to be 3425 and $3320\,\mathrm{kg\,m^{-3}}$ respectively. Model II closely resembles Bullen's model A″, the maximum discrepancy amounting to only $120\,\mathrm{kg\,m^{-3}}$. Yet a further type of density model was presented by CLARK and RINGWOOD (1964) based on an assumed petrological model for the upper mantle but otherwise based on a similar approach to that of Bullen and Birch.

A density distribution obtained with the seismic velocities by inversion of the free oscillation data is incorporated in the Anderson-Hart model of Earth structure (p. 151). However, the resolution achieved by the inversion is relatively poor for density, the averaging length being about 400 km in the upper mantle and transition zone and about 1000 km in the lower mantle. The more detailed features of the resulting density distribution (Table 4.1 and Fig. 4.18) carry through from the starting model for the inversion, which is based on use of Birch's approach as described above. This density model is thus a compromise between a Birch type model based on the body wave structure beneath western U.S.A. and a broad distribution consistent with the free oscillation data.

The density distribution is least well known in the upper mantle. Here the relationship between seismic velocities and density is obscured by solid phase transitions and proximity to melting

Table 4.1 Some estimates of the density distribution of the mantle and core, after HART and others (1977), HADDON and BULLEN (1969), DZIEWONSKI and ANDERSON (1981) for Preliminary Reference Earth Model (PREM), and CLARK and RINGWOOD (1964). Densities shown in kg m^{-3}.

	Depth (km)	Hart & others Model QM2	Haddon & Bullen Model HB$_1$	PREM	Clark & Ringwood Pyrolite model
	60	3520	3332	3377	
	100	3390	3348	3373	3285
	200	3330	3387	3362	3404
	300	3380	3424	3484	3481
	400	3700	3775	$\begin{cases} 3543 \\ 3724 \end{cases}$	3556
	500	3720	3925	3850	3682
	600	3970	4075	3976	3906
Mantle	670	$\begin{cases} 4040 \\ 4380 \end{cases}$	$\begin{cases} 4150* \\ 4200* \end{cases}$	$\begin{cases} 3992 \\ 4381 \end{cases}$	
	800	4510	4380	4461	4360
	1000	4600	4538	4580	
	1400	4770	4768	4806	
	1800	4990	4983	5018	
	2200	5230	5188	5222	
	2600	5450	5387	5421	
————	2886	$\begin{cases} 5520 \\ 9970 \end{cases}$	$\begin{cases} 5527^+ \\ 9927^+ \end{cases}$	$\begin{cases} 5566^+ \\ 9903^+ \end{cases}$	
Outer core	3000	10140	10121	10073	
	4000	11310	11383	11321	
	5000	12060	12130	12085	
————	5156	$\begin{cases} 12120 \\ 12300 \end{cases}$	12197$^\times$	$\begin{cases} 12166^\times \\ 12764^\times \end{cases}$	
Inner core	6371	12570	12460	13088	

* At 650 km depth for HB$_1$.
$^+$ At 2878 km depth for HB$_1$ and 2891 km depth for PREM.
$^\times$ At 5121 km depth for HB$_1$ and 5149·5 km depth for PREM.

temperature. An important controversy arose from the Monte Carlo inversion of gross earth data by PRESS (1970) which suggested that the average mantle density down to 150 km depth has the otherwise unacceptably high value of about 3500 kg m^{-3} and that there is an underlying low density zone centred at about 300 km depth. WORTHINGTON and others (1972) later showed that these anomalous features arose from unduly narrow constraints applied to S between 300 and 600 km. There remain substantial differences between the various density models. Perhaps the most satisfactory models for the upper mantle are of the type constructed by CLARK and RINGWOOD (1964); these assume a pyrolite or eclogite composition and take into account self-compression, thermal expansion and experimentally determined phase transitions. They predict a small downward decrease of density of about 10 kg m^{-3} between the Moho and about 80 km depth where thermal expansion outweighs the effects of self-compression and phase transitions. The Bullen-Haddon models do not include a low density zone but they do not conform to recent upper mantle S distributions. On the other hand, the Anderson-Hart model, using Birch's relationship to connect density and P velocity based on the anomalous western United States region, shows a

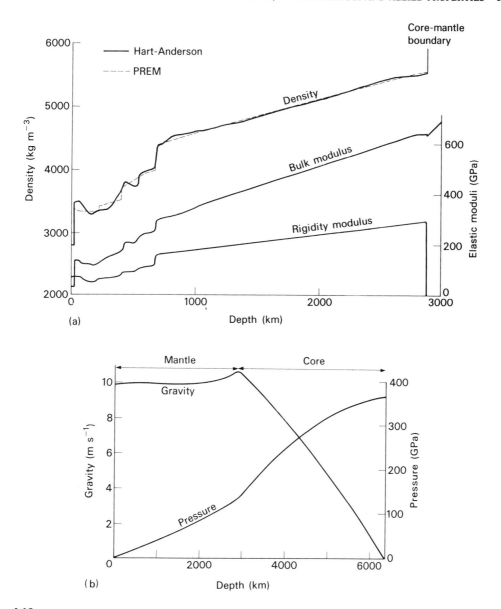

Fig. 4.18
(a) The density, bulk and rigidity moduli distributions in the mantle shown as a function of depth. The density distributions are taken from HART and others (1977) and DZIEWONSKI and ANDERSON (1981) and the elastic moduli from ANDERSON and HART (1976).
(b) Gravity and pressure distributions in the mantle and core after ANDERSON and HART (1976).

decrease of density of 200 kg m^{-3} between the Moho and 150 km depth with anomalously high values of nearly 3500 kg m^{-3} for the topmost mantle. This model probably exaggerates the low density zone by a factor of ten and yields too high a density above it. On present evidence, the density is on average about 3340 kg m^{-3} just below the Moho and rises slightly with depth to about

80 km depth. It then decreases by 10 to 20 kg m^{-3} down to 100 km depth below which it rises gradually with depth down to the base of the upper mantle.

The steep rise in velocity and density with depth across the mantle transition zone cannot be attributed solely to homogeneous self-compression but results mainly from solid phase transitions (p.193). The increase in density across the zone, other than due to self-compression, is about 650 kg m^{-3}. The Anderson-Hart model incorporates three steep downward increases in density with depth reflecting the P velocity distribution used; as pointed out earlier (p. 155), the existence of the middle step at 525 km depth is now regarded as doubtful. Across most of the lower mantle (D'), the available evidence indicates that density increases with depth in conformity with the Adams-Williamson equation, but in the lowermost 200 km (D'') the density still increases with depth whereas the seismic velocity gradients decrease or even reverse.

Once the density distribution within the Earth has been specified, the value of gravity can be computed at each depth and the pressure-depth distribution evaluated without complication. The rigidity and bulk moduli can be computed from the P and S velocity and the density, using the basic formulae for P and S velocities (p. 6). These distributions are shown graphically in Fig. 4.18.

4.5 The anelasticity of the crust and mantle

Introduction

Free elastic vibrations within the Earth do not continue indefinitely. The strain energy is progressively dissipated as heat through imperfections of elasticity often collectively referred to as solid friction. This effect is known as anelasticity. It occurs in all solids and is observed at indefinitely small strains. The anelastic properties of the Earth are of interest in connection with the damping of seismic waves, the earth tides and the Chandler wobble.

Anelasticity causes free vibrations to decay in amplitude. The effect on forced vibrations such as earth tides is to cause strain to lag behind the applied stress. Anelastic damping is conveniently measured by the quality factor Q. The most convenient definition for seismological purposes is that $2\pi Q^{-1}$ is equal to the fraction of the total strain energy dissipated per cycle. The logarithmic decrement, $\ln(A_1/A_2)$ where A_1 and A_2 are successive maximum amplitudes, is approximately equal to πQ^{-1} provided that Q is fairly large. In dealing with forced vibrations, Q^{-1} is the tangent of the phase angle between the applied stress and the strain. The alternative definitions of Q are not exactly equivalent for seismic waves, but are close enough for our purposes.

The seismic damping within the Earth is normally expressed in terms of Q_α for compression waves and Q_β for shear waves. Alternatively, the damping can be expressed in terms of Q_k for purely compressional deformation and $Q_\mu (= Q_\beta)$ for pure shear. These values of Q are likely to be frequency dependent to a greater or lesser extent. Anelastic damping affects surface waves and free oscillations. Q_R denotes the anelastic attenuation of Rayleigh waves or spheroidal free oscillations of given period, and Q_L that of Love waves and torsional free oscillations. Such values of Q are frequency dependent because the longer periods penetrate to greater depths within the Earth, quite apart from any inherent frequency dependence. Q_L is a function of the distribution with depth of Q_β over the region affected by the vibration, and Q_R is similarly a function of Q_α and Q_β.

Damping of seismic waves

Observations of the decay with time of seismic waves and free oscillations can be used to estimate the anelastic properties of the Earth's interior within the period range 1 second to 1 hour. The straightforward use of body waves to determine the average radial distribution of Q_α and Q_β through the mantle is not practicable because of large variations in amplitude produced by low velocity zones and second order discontinuities, and because of lateral inhomogeneity. A more satisfactory method is to invert the measured decay of surface waves and free oscillations of varying period and to use selected body wave data to provide additional constraints.

A particularly useful constraint is provided by nearly normal incidence S waves which have been repeatedly reflected between the core-mantle boundary and the Earth's surface (Fig. 4.19). The reflection coefficient at both boundaries can be taken as unity for S waves and the geometrical spreading factor can be accurately calculated. By comparing the amplitudes of successive reflections, such as ScS_1 and ScS_2 or ScS_2 and ScS_3, the average radial value of Q_β for the mantle can be estimated. The observations show some significant regional variation, yielding a mean value of about 285 (ANDERSON and HART, 1978), the uncertainty being about 40. By using deep focus earthquakes and comparing the amplitudes of ScS and the later arriving phase $sScS$ which is first reflected from the surface, the average value of Q_β between 600–700 km depth and the surface is found to be about 165. Similarly, PcP and related phases can be used to determine the average radial value of Q_α in the mantle, but with lesser accuracy as the reflection coefficient at the core-mantle boundary is less well determined. Another type of constraint provided by body waves is to measure the ratio of travel time T to apparent Q of waves travelling through the mantle. In obtaining such constraints from the attenuation of body waves, it is desirable to carry out spectral analysis in order to isolate waves of specific period.

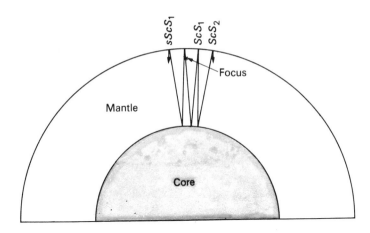

Fig. 4.19 Ray paths used for estimating Q in the mantle from body wave amplitudes.

In measuring Q for surface waves, it is first necessary to isolate waves of specific period by spectral analysis of the wavetrain. The decay in amplitude can then be measured using two stations along a great-circle path. Alternatively, a single seismograph station can be used to measure the attenuation of surface waves which pass round the Earth more than once. Accurate corrections can be applied for geometrical spreading and for other effects such as the ellipticity of the Earth. This enables Q to be determined for each selected period in the surface wave spectrum. At longer periods, the decay of the free oscillations can be measured by analysing strain seismometer or gravimeter records at successive periods of time after excitation. The values of Q_L across the spectrum, obtained from Love waves and torsional oscillations, can then be inverted to give an estimate of the Q_β radial distribution through the mantle provided that Q_β can be assumed to be independent of frequency. Once Q_β is known, Q_α can be obtained with lesser accuracy by inverting the values of Q_R obtained for Rayleigh waves and spheroidal oscillations. One of the shortcomings of the present seismic velocity and attenuation models of the mantle is that these are not independent of each other although they have been obtained by separate inversions. In this respect, a recent simultaneous inversion of surface wave phase velocity and attenuation of Love waves has been carried out for western U.S.A. (LEE and SOLOMON, 1978). It is to be hoped that this new and more satisfactory approach will be applied to the whole of the mantle soon.

Figure 4.20 shows a recent model of the radial distribution of Q_β through the Earth, obtained by ANDERSON and HART (1978) by inverting data from surface waves with period greater than 65 s and free oscillations and using body wave data as constraints. The general picture which emerges is as follows (Fig. 4.20). Values of Q are high in the crust but fall steeply to a minimum averaging about 85 between depths of 80 and 150 km. The effect of the low Q zone is seen in the low values of Q_L for Love waves of period 60 to 100 seconds. Q_β then rises steeply below the low Q zone, rising to a peak in the lower mantle between 1800 and 2700 km depth, the value here being rather poorly constrained. A pronounced region of low Q in the lowermost 200 km of the mantle is suggested on several grounds, notably by the increase in T/Q_α for mantle P waves travelling beyond epicentral distances of 80° and by the fall-off in Q_R for low order spheroidal oscillations of period above about 1000 seconds. The model is based on the assumption that Q is frequency independent over the period range 65 seconds to 1 hour, although this assumption is loosely tested during the construction of the model and found to be acceptable within present limitations.

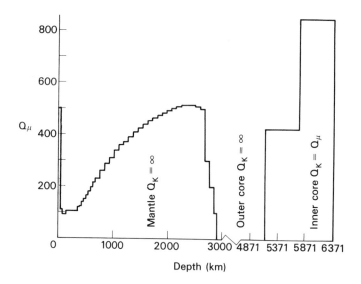

Fig. 4.20 The Q model SL8 for the Earth, redrawn from ANDERSON and HART (1978), *J. geophys. Res.*, **83**, 5878.

There are distinct differences between the inferred distributions of Q_β down to at least 300 km beneath oceans and continents. Beneath the Pacific Ocean, Rayleigh wave attenuation yields an average Q_β of about 170 down to 70 km depth below which there is a low Q zone with minimum value of about 70 at 150 km depth (MITCHELL, 1976). Body wave studies indicate that remarkably high Q occurs at least locally in the uppermost oceanic lithosphere and that attenuation of S is particularly strong beneath ocean ridge crests. Stable continental regions lack evidence for any well developed low Q zone in the underlying upper mantle and continental Q values are generally higher than those beneath oceans (PATTON, 1980). Pronounced low Q zones, however, do occur in the upper mantle beneath the tectonic regions of America and Eurasia.

The low Q (high attenuation) zone in the upper mantle is usually interpreted in terms of proximity to the melting temperature. It is often regarded as being equivalent to the asthenosphere, with the overlying high Q zone forming the lithosphere. The simultaneous inversion of surface wave velocity and attenuation carried out by LEE and SOLOMON (1978) shows that the low shear velocity and low Q zones in the upper mantle are probably coincident.

The observed attenuation of free oscillations is best satisfied if almost all the seismic loss within the Earth occurs in shear with only a minor loss of energy occurring in compression. The small compressive loss is believed to occur almost entirely either in the inner core or in the upper mantle, probably the latter (DZIEWONSKI, 1979). Thus $Q_k^{-1} = 0$ throughout most of the crust and mantle. Under such circumstances, it can be shown that

$$Q_\alpha/Q_\beta = \tfrac{4}{3}(V_p/V_s)^2$$

so that the value of Q_α is a factor of 2·4–2·7 times greater than Q_β throughout the mantle. The evidence also indicates that Q is frequency independent or only weakly dependent for periods ranging from about 1 hour to 10 seconds but that Q increases with decreasing period for body waves of period between 10 and 1 seconds.

Damping of the Chandler wobble

It was pointed out in Chapter 1 that the Chandler wobble is a free oscillation of 435 day period which affects the Earth's axis of rotation and which may be excited by the redistribution of mass associated with large earthquakes. The damping of the Chandler wobble provides an indication of Q of the mantle for oscillations of 14 month period.

The decay time of the Chandler wobble probably lies between about 20 and 120 years. The corresponding values of Q_w for the wobble lie between 50 and 300, the best present estimate being about 100 (SMITH and DAHLEN, 1981). Three processes may contribute to the damping, these being (*i*) dissipation of energy in the oceans by frictional slip between the seawater and the seabed, (*ii*) anelastic dissipation in the crust and mantle, and (*iii*) dissipation at the core-mantle boundary by electromagnetic, viscous or topographical coupling. The contributions from the oceans and the core-mantle boundary are probably small and the main dissipation thus occurs within the mantle.

The relationship between Q_w for the wobble and Q_β for the mantle at 14 month period is not obvious because most of the energy involved is in the rigid body rotation rather than deformation. Consequently it has been widely supposed that the ratio Q_w/Q_β is about 10, but SMITH and DAHLEN (1981) have now shown that this ratio is probably between 1·83 and 1·62 depending on whether the pole tide is ignored or not. If it is assumed that all the dissipation occurs within the mantle and that Q is constant with depth, this yields an estimate of 175 for Q_β at 14 month period. This is slightly lower than the value of about 300 obtained for seismic waves and free oscillations, suggesting a slight dependence of Q on frequency at periods within the range 1 hour to 14 months. JEFFREYS (1958) suggested that the frequency dependence of damping can be modelled by the modified Lomnitz law of transient creep. This implies that Q should be proportional to f^α where f is frequency of the oscillation and α is a constant. SMITH and DAHLEN (1981) obtained a value of α within the range 0·09–0·15, although they could not rule out the possibility of frequency independence.

Jeffreys used the modified Lomnitz law to extrapolate the rheological properties determined from the damping of seismic waves and the Chandler wobble to deformation of the mantle at much longer period. He found that the mantle would be too stiff to convect and later used this as an argument against continental drift and plate tectonics. The more general opinion, however, is that the extrapolation from 14 month to about 10 My period on the basis of an empirical relationship is unjustifiable and invalidated by other evidence.

The mechanism of seismic damping

Many experimental observations of anelasticity in metals and rocks have been made. The observed values of Q for rocks are mostly within the range 50–1000 (BRADLEY and FORT, 1966). In both metals and rocks Q appears to be nearly independent of frequency up to strains of about 10^{-6}–10^{-5}, and it is not strongly dependent on temperature. Q increases with confining pressure, suggesting that at low pressures pores play an important part in anelastic mechanisms. The dominant anelastic process under experimental conditions is grain boundary sliding, causing frictional losses. This process would be inhibited by pressure below a few hundred kilometres depth, suggesting that the experimental results are not relevant to the anelasticity of the mantle.

The values of Q in the mantle are probably mainly influenced by the temperature and pressure, with mineralogical composition and presence of volatiles having a subordinate effect. A variety of mechanisms have been suggested, including various grain boundary effects, stress-induced ordering of defects in crystals, damping caused by vibration of dislocations, ferromagnetic interactions, thermal anelasticity and flow of partially melted material through intergranular spaces. The most widely canvassed mechanism is that of grain boundary relaxation losses (ANDERSON, 1967b; ANDERSON and HART, 1978). This is probably caused by stress-induced migration of defects at the crystal boundary rather than by frictional sliding at the boundary. An oscillating stress field such as is produced by passage of a seismic wave causes fluctuations of the crystal defects at the boundary with consequent loss of energy. This mechanism is thermally activated so that it is strongly temperature dependent, particularly at higher temperatures, but it is approximately independent of strain amplitude. The mechanism would be expected to be frequency dependent, the maximum absorption of energy occurring when the applied stress has the same frequency as that of formation of defects. In practice, it is likely that a range of frequencies of defect formation will occur so that the value of Q appears to be nearly frequency independent over the seismic band. Calculations show that these frequencies are likely to occur in the seismic band, so that the mechanism has theoretical justification. Furthermore, it predicts loss of energy in shear rather than in pure compression, in good agreement with the observations.

4.6 Temperature-depth distribution in the mantle

Simple extrapolation of the near-surface geothermal gradient can only be used to infer the temperatures down to a depth of a few tens of kilometres at the most. Such extrapolation fails at greater depths because of the presence of heat sources and the occurrence of heat transfer mechanisms other than simple thermal conduction. Two other types of approach are useful in inferring the temperature-depth distribution in the mantle. Firstly, some indication of temperature at certain specific depths can be obtained by geothermometry, that is by inferring the temperature from measurable physical properties, such as the depth of a phase transition whose pressure-temperature relationship has been experimentally or theoretically determined. Secondly, theoretical temperature-depth distributions can be computed provided that the thermal properties, heat sources and mechanisms of heat transfer are known; such an approach is important in determining the upper mantle temperatures. The resulting temperature-depth profile for the mantle is much more uncertain than the seismic and related properties, although there have been some significant improvements in knowledge over the last few years.

The material of the mantle is known to be in the solid state apart from some local melting in the upper mantle. The melting temperature therefore places an upper limit on the temperature at any depth. As the rock forming the mantle contains several mineral species, the temperature of the solidus at which partial fusion commences differs from that of the liquidus at which fusion is complete. The melting temperature can be determined experimentally at depths relevant to most of the upper mantle, but at greater depths it is necessary to extrapolate extensively. Such extrapolation applied to simple crystalline substances is based mainly on development of the Lindemann law of melting, which postulates that melting occurs when the mean square amplitude of vibrations of the atoms exceeds a certain fraction of the lattice spacing. ROSS (1969) used statistical mechanics to generalize the Lindemann law for various types of interatomic force. He found that the melting temperature T_m of crystalline substances for which the interatomic repulsion between pairs of atoms follows an inverse nth power law is given by

$$T_m/T_o = (1 - \Delta V/V_o)^{-\frac{1}{3}n}$$

where $\Delta V/V_o$ is the fractional reduction in volume at the applied pressure and T_o is the melting temperature at zero pressure. This equation probably applies to silicate minerals of the mantle and to iron, the value of n for iron being estimated as 8·4. Various empirical formulae have also been used to predict melting temperatures within the Earth. One of the earliest to be applied was Simon's relationship between melting temperature and pressure which is

$$P = a((T/T_o)^c - 1)$$

where a and c are constants determinable by experiment. This formula has been widely applied to estimating the melting temperature of iron at core pressures. It has now been shown to be satisfactory for Van der Waals solids such as solid helium and some organic compounds, but not for metals or silicates for which it predicts too high melting temperatures at high pressure. A more satisfactory empirical relationship is that of KRAUT and KENNEDY (1966), which states that

$$T_m/T_o = 1 + b\Delta V/V_o$$

where b is a constant. This relationship forms the first two terms in the expansion of the Ross-Lindemann equation given above. Thus on theoretical as well as experimental grounds, it tends to overestimate the melting point of ionic substances such as the minerals of the mantle. It applies accurately to the alkali metals but its status for iron is unclear.

In the upper mantle, the known melting point of forsterite as a function of pressure gives an upper limit on the liquidus whereas the experimentally determined commencement of partial fusion of peridotite gives an estimate of the solidus temperature. Forsterite melts at 2160 K at the surface, rising by 1·5 K km^{-1} with increasing depth. Basalt forms by partial fusion of dry peridotite at about 1350 K at the surface, rising to 1800 K at 100 km depth. These temperature-depth profiles are related to the change in density on melting and the specific latent heat of fusion by the Clapeyron equation (p. 275). However, if there is a small amount of water present in the peridotite, say 0·1%, then the solidus is substantially lowered below about 75 km depth, particularly between 100 and 150 km depth where it is between about 1300 and 1400 K. Olivines and pyroxenes are believed to undergo drastic high pressure breakdown in the transition zone (p. 193) and because of this the solidus and liquidus cannot be estimated by simple extrapolation of the upper mantle distributions. The liquidus curve for the lower mantle has been estimated by KENNEDY and HIGGINS (1972) assuming a periclase (MgO) composition and using the Kraut-Kennedy equation to extrapolate to high pressure from the melting point and its theoretically derived pressure gradient at atmospheric pressure. Their solidus curve (Fig. 4.21) is based on the assumption, which has some

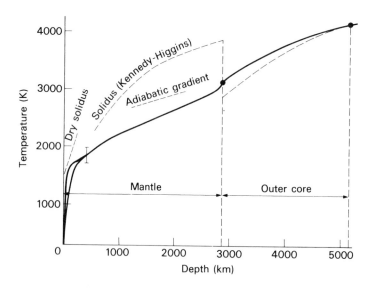

Fig. 4.21 A temperature-depth model for the mantle and outer core, based on STACEY (1977). The estimate at 400 km depth is based on the olivine-beta phase transition. The upper mantle distributions are shown in more detail in Fig. 4.22.

theoretical justification, that the gap between liquidus and solidus should widen with increasing pressure.

Comparing the near-surface geothermal gradient with the melting point curves shows that the temperature-depth gradient must decrease substantially below about 50–80 km depth, otherwise there would be wholesale melting of the mantle which is known not to occur. The average temperature gradient beneath continents and oceans is about 24 K km^{-1}. If this continued downwards unmodified, then the temperature at 100 km depth would be about 2650 K, which exceeds the melting point of olivine at that depth. There must be a considerable reduction in the temperature-depth gradient below about 60 km depth. The reasons for this fall-off in the gradient are discussed in Chapter 7 where it is attributed to heat transfer by convection below the lithosphere, near-surface concentration of radioactive heat sources particularly in continents, and transient cooling of the lithosphere particularly below oceans. The seismic low velocity and low Q channel of the upper mantle is commonly regarded as the region of the mantle where the melting point of basalt is most closely approached, if not reached. This suggests that the temperature-depth gradient becomes equal to the melting point gradient of basalt (solidus of peridotite) of about 3 K km^{-1} at some depth between about 100 and 200 km, and that below about 200 km it is somewhat less than this.

Most, if not all, of basalt magmas must form by partial fusion in the upper mantle, as temperatures in the crust are too low. Beneath Hawaii, earthquake foci associated with eruption of basalt occur between 45 and 60 km depth (EATON and MURATA, 1960) suggesting that the magma is formed at a depth of about 60 km. Thus the temperature at about 60 km depth below Hawaii is probably about 1500–1550 K. Using earthquakes in this way to detect the source of basalt magma is one reliable method of determining the actual temperature at a point in the upper mantle, but this method is limited to regions of active volcanism.

A more widely applicable method of estimating upper mantle temperatures down to about 200 km depth, albeit only at certain specific times in the past, is by pyroxene geothermometry and geobarometry. This method, pioneered by BOYD (1973), depends on the study of equilibrium mineral assemblages in chunks of solid rock from the upper mantle carried up to the surface in rising magma. Boyd used the garnet peridotite inclusions found in kimberlite 'pipes' of late

Cretaceous age in northern Lesotho. Most of these inclusions are a type of peridotite known as garnet lherzolite (p. 185) containing the orthopyroxene enstatite, the clinopyroxene diopsidic augite and garnet. The method assumes that these mineral phases were in equilibrium with each other at the source prior to the igneous activity which brought them up to the surface fast enough for the equilibrium assemblage not to be modified. Since the emplacement of the kimberlite, the assemblage has been frozen in at the cool temperatures near the surface. Solid solution of enstatite in diopside is sensitive to temperature but not to pressure, the ratio of Ca/(Ca + Mg) in the diopside being used by Boyd to estimate temperature with experimental results providing the calibration. Knowing the source temperature, the pressure was then estimated by means of the aluminium content of enstatite in equilibrium at the source with garnet. As the inclusions come from a range of depths, Boyd was able to infer the Cretaceous geotherm beneath Lesotho prior to the magma forming event. His interpretation of the geotherm shows two segments. The shallower segment extends from about 130 to 170 km depth with the inferred temperature rising downwards from 1170 to 1290 K. The deeper segment, extending from 180 to 220 km depth, is steeper with the temperature increasing downwards from 1470 to 1670 K. The shallower segment probably represents the true geotherm, within the errors of the method. The anomalously steep deeper segment may arise because of shear heating or from heating associated with the magma source. It is unlikely to be a true feature of the geothermal gradient at this depth. Pyroxene geothermometry has been applied to other regions where kimberlites occur with similar results. MACGREGOR (1974) used the method on inclusions in the alkali basalts of Hawaii to indicate a temperature of about 1170 K at 50 km depth. Subsequent work has suggested the need for some modification of the original method, such as a full thermodynamical treatment to take into account the other components such as ferrous iron (POWELL, 1978). However, the temperature estimates made by Boyd and other early workers are thought to be substantially correct but the pressure estimates are probably too high, overestimating the depth by about 30 km. Allowing for this, the measurements on kimberlite inclusions indicate a temperature beneath typical shield regions of about 1250 K at just over 100 km depth and those on alkali basalts indicate that the oceanic upper mantle is significantly hotter at equivalent depth.

The temperature at the top of the mantle transition zone can be approximately estimated by using experimental data on the phase transition from olivine to the higher pressure beta phase (p. 193), corresponding to the uppermost steep rise in velocity with depth at about 400 km depth. For olivine of estimated mantle composition containing about 90 % forsterite and 10 % fayalite, the experimentally determined median point of the transition is 11·4 GPa (1 gigapascal = 10 kbar) at 1270 K (see RINGWOOD, 1975). Estimates of the pressure-temperature gradient of the transition range from 3·0 to 4·8 MPa K^{-1}, indicating that the temperature of the transition at 400 km depth would be between 1700 and 1970 K. This estimate should be treated with caution as it ignores the possible influence of other phases such as pyroxenes on the location of the seismic step. Furthermore, it is critically dependent on the estimated depth of the step, with an increase in its depth of only 10 km being associated with a temperature estimate almost 100 K higher. It is shown below that there are substantial lateral variations of temperature in the upper mantle. These temperatures must converge together by 400 km depth, because otherwise there would be variations in depth of the transition which would produce larger gravity anomalies than those observed (p. 199).

Thus far we have two or three fixed temperature points spanning the upper mantle. Derivation of the complete temperature-depth distribution must be based on assumptions about the distribution of heat sources and the mechanisms of heat transfer. To discuss such distributions, we need to anticipate some of the results given in Chapter 7. *Firstly*, about 40 % of the continental heat flow comes from decay of long-lived radioactive isotopes in the crust whereas the contribution from the thin oceanic crust is negligible; as oceanic and continental heat flow averages are about equal, this means that the upper mantle should be significantly hotter beneath oceans than beneath continents. *Secondly*, most of the oceanic heat loss is now known to come from cooling of the oceanic lithosphere as it spreads away from the ocean ridges, whereas the cooling of the continental lithosphere occurs on a longer time-scale and contributes a smaller fraction of the heat flow. *Thirdly*, heat escapes through the continental and oceanic lithospheres by thermal conduction whereas the main mechanism of heat transfer below the lithosphere is probably by convection and diapiric upwelling; this accounts for the occurrence of a much lower temperature-depth gradient below the lithosphere than within it. The temperature differences in the upper mantle beneath continents and oceans differ not only because of the concentration of the heat sources in the continental crust, but also because of the markedly different styles of heat escape through them. The upper mantle beneath continents is probably at least 100–200 K cooler on average than that beneath oceans; the continental shield regions have the coolest upper mantle and the ocean ridge regions the hottest.

Prior to the 1970s, the mantle was widely assumed to be static and non-convecting. The well-known temperature-depth distributions of CLARK and RINGWOOD (1964) were based on this assumption, which is no longer acceptable. Upper mantle temperature-depth distributions based on a more recent assesment of the Earth's thermal regime are shown in Fig. 4.22. The oceanic

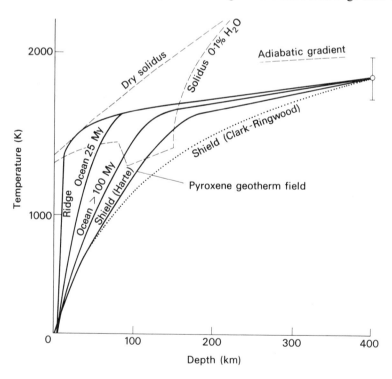

Fig. 4.22 Estimated regional temperature-depth distributions for the upper mantle, showing the probable range of variation. The oceanic distributions are based on the progressive cooling of the lithosphere as it ages. The continental shield distributions are those of CLARK and RINGWOOD (1964) based on a conduction solution and HARTE (1978) based on pyroxene geotherms and upper mantle convection. The estimate at 400 km depth is based on the inferred temperature of the olivine-beta phase transition. The solidus is after RINGWOOD (1975).

geotherms are based on the progressive cooling of the oceanic lithosphere as it spreads laterally from the ridge crest towards the deep ocean basins, with a much shallower gradient in the underlying asthenosphere where convection is assumed to occur. The shield geotherm shows a shallowing of the gradient through the crust where radioactive heat sources occur and a further shallowing below the base of the lithosphere at 150–200 km depth where the dominant heat transfer mechanism changes from conduction to convection. The shield geotherm of Clark and Ringwood is shown for comparison; it is notably cooler between 100 and 300 km depths and lacks the sharp fall-off in temperature-depth gradient at the base of the lithosphere. Figure 4.22 shows that substantial lateral temperature differences occur in the upper mantle, particularly above 200 km depth, the sub-oceanic mantle generally being significantly hotter than that beneath the continents. These differences, however, become negligible by 400 km depth.

Going deeper into the mantle, it is normally assumed that the temperature rises with increasing depth slightly more steeply than the adiabatic gradient. The adiabatic gradient arises as follows. If a thermally isolated volume of rocks is subjected to change in pressure, the temperature also changes. The slope dT/dp is called the adiabatic gradient. Within the Earth it is normally expressed in terms of depth rather than pressure. Thermal convection can only occur if the actual temperature-depth gradient exceeds the adiabatic gradient by a critical amount. The adiabatic gradient is related to the coefficient of volume expansion α, the specific heat capacity at constant pressure c_p and the density ρ by

$$dT/dp = \alpha T/c_p\rho.$$

Within the Earth, $dp/dr = -g\rho$, where r is the radius and g is gravity. Putting $g = 10\,\mathrm{m\,s^{-2}}, \alpha = 1·5 \times 10^{-5}\,\mathrm{K^{-1}}$ (p. 275), $T = 2500\,\mathrm{K}$ and $c_p = 1·27\,\mathrm{kJ\,kg^{-1}\,K^{-1}}$, these equations give $-dT/dr = 0·3\,\mathrm{K\,km^{-1}}$ in the lower mantle. This is unlikely to be in error by more than about 30%. The adiabatic gradient is likely to be significantly steeper across the transition zone. However, taking this gradient below 400 km depth and assuming the temperature at this depth is at least 1720 K, we can see that the temperature at the core-mantle boundary at 2886 km depth must exceed 2450 K. On the other hand, the solidus curve for the lower mantle (Fig. 4.21) places an approximate upper limit of 3850 K.

A more precise estimate of the temperature at the core-mantle boundary can be obtained indirectly. This is based on the assumption that the inner-outer core boundary marks the solidus of the Fe-FeS eutectic (p. 247). The temperature at this boundary was estimated by STACEY (1977) to be about 4150 K by using the Lindemann law to extrapolate to higher pressure the experimental results on this eutectic obtained by USSELMAN (1975a). The temperature-depth distribution in the outer core is assumed to be the adiabatic gradient, which can be computed with reasonable accuracy. Stacey thus estimated the temperature at the core-mantle boundary to be 3157 K. If this is correct, then the average temperature-depth gradient through the mantle below 400 km appears to exceed the adiabatic gradient by 80% at the maximum, which is plausible and consistent with the occurrence of convection (p. 344).

4.7 Electrical conductivity of the mantle

The electrical conductivity distribution down to about 1500 km depth can be investigated by studying the short-period variations of the Earth's magnetic field ranging from less than a second to several years. The spectrum of the secular variation gives some indication of the conductivity of the lower mantle.

The variations in the Earth's magnetic field with period of less than a year originate outside the solid Earth, but they include a secondary internal component caused by induced currents flowing in the crust and mantle. The currents are known as telluric currents.

The most conspicuous short-period variation is the diurnal variation. A typical record of diurnal (or daily) variation is shown in Fig. 4.23(a). The same pattern tends to be repeated with some variation from day to day. If a record covering a period of many days is subjected to spectral analysis, it is found that the dominant period is one day and that there are conspicuous harmonics of period 12 hour, 8 hour and so on, as shown in Fig. 4.24. The diurnal variation is believed to be caused by the interaction of the conducting ionospheric layers of the upper atmosphere with the main magnetic field. The ionospheric layers move in response to solar and lunar tidal forces, the main diurnal tide probably being thermally driven; this causes a varying pattern of horizontal current loops in the ionosphere as the conducting layers cut the magnetic lines of force. The diurnal variation is mainly produced by two large current loops in the ionosphere at about 100 km height on the side of the Earth which faces the Sun. Longer period variations of tidal origin include (i) a 27 day period and its harmonics which may be related to tides caused by the Moon's orbital motion, and (ii) an annual period. A variation of 11 year period is associated with the sunspot cycle.

Other short-period variations are caused by interaction between the geomagnetic field and the plasma stream known as the solar wind which originates from the Sun (Fig. 4.25). The solar wind divides and flows round the Earth, confining the magnetic field to the magnetosphere. The

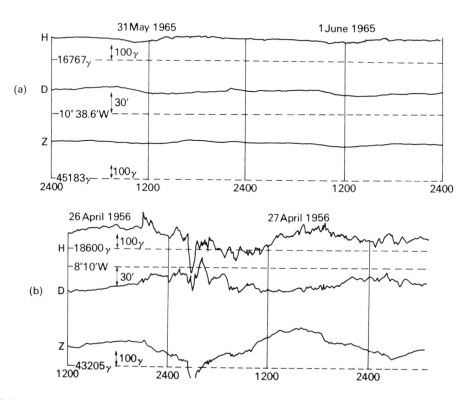

Fig. 4.23 Magnetic observatory records showing short-period variations in horizontal field (*H*), declination (*D*) and vertical field (*Z*). (**a**) is a typical quiet day variation at Eskdalemuir observatory, south Scotland, and (**b**) is a magnetic storm at Abinger. (**a**) redrawn from the original record, and (**b**) redrawn from BULLARD (1967), *Q. Jl R. astr. Soc.*, **8**, 149.

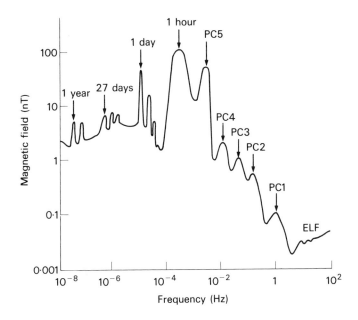

Fig. 4.24 The amplitude spectrum of the short-period variations in the horizontal geomagnetic field indicating the peaks used in magnetic induction research. PC1, etc., refer to micropulsations. The 11 year period is not shown. Redrawn from SERSON (1973), *Phys. Earth planet. Interiors*, **7**, 313.

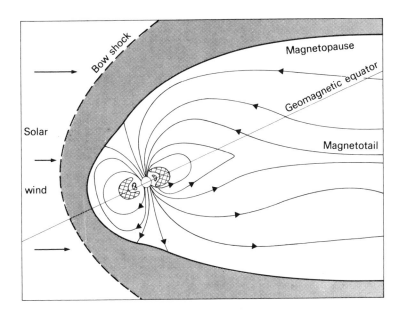

Fig. 4.25 The Earth's magnetosphere, after HUTTON (1976), *Rep. Prog. Phys.*, **39**, 496.

interaction produces a bow shock which confines the field on the sunward side to a region of about ten earth radii and stretches it out on the opposite side to form the magnetotail. *Magnetic storms* are caused by large and sudden increases in the energy density of the solar wind. This distorts the magnetosphere and causes injection of high energy particles into it. The record of a magnetic storm is shown in Fig. 4.23(b). This commences with a sudden increase in the horizontal field which occurs almost instantaneously over the Earth's surface, followed by a depression of the field which may last several days yielding relatively long-period harmonic components. Superimposed are shorter period irregular variations of about 10 to 100 minute period which are prominent during the first few hours of the storm and which are of more restricted areal extent. *Micropulsations* of 0·2 second to 10 hour period are also caused by interaction between the solar wind and the magnetosphere.

The magnetic variations can be used to estimate the conductivity within the outer parts of the Earth because the strength of the induced telluric currents depends on the electrical conductivity distribution. The depth to which the currents penetrate depends on the period of the variation – short period variations only penetrate to shallow depths and longer periods penetrate deeper. The depth of penetration of an oscillation of period T above a uniform half-space of conductivity σ is given by $(T/4\pi^2\sigma)^{\frac{1}{2}}$. Thus the whole spectrum of external magnetic events extending from one second period to a year or more is useful for probing the conductivity of the Earth down to about 1000–1500 km depth. The method depends critically on the possibility of separating the external and internal parts of the field by spherical harmonic analysis or by other methods. By determining the relative amplitude and phase difference of the external and internal components of the variation of given period, it is possible to estimate a weighted mean value of the conductivity down to the depth of penetration of the currents. By carrying out such an analysis for a range of different period variations, a crude conductivity-depth distribution can be estimated down to the depth penetrated by the largest period variation available for analysis, using trial and error modelling or inversion. Contrasting approaches are used to determine the broad radial variation of conductivity between the upper mantle and 1000–1500 km depth and the more detailed conductivity-depth distributions down to about 300 km depth which show large regional variations.

Radial distribution of electrical conductivity

The main feature of the conductivity-depth distribution in the mantle revealed by global studies is a downward increase of conductivity by one or two orders of magnitude between depths of 300 and 1000 km. This has been determined by the following method, which is described with reference to the diurnal variation and its harmonics but equally well applies to other variations of more than a few hours period. The average diurnal variation for quiet days is obtained for the vertical and horizontal components of the magnetic field from magnetic observatories over the world. The records for each observatory are Fourier analysed into simple harmonic variations of 24, 12, 8 and 6 hour period. A spherical harmonic analysis is then carried out on the worldwide spread of data for each time period to separate the internal and external parts of the variation. Alternatively, the separation can be carried out by integral methods. Thus the amplitude ratio and phase difference between external and internal parts of each period of variation can be determined. The results can be used to estimate the conductivity distribution as a function of radius down to the depth of maximum penetration.

CHAPMAN (1919) was the first person to make this analysis on the diurnal variation. For the 24 hour period variation he found the external to internal amplitude ratio to be 2·8 and the phase difference to be $-13°$ and he also obtained results for the higher harmonics. He interpreted his results in terms of a uniformly conducting 'core' having a conductivity* of $3·6 \times 10^{-2}\,\mathrm{S\,m^{-1}}$ surrounded by a non-conducting shell 250 km thick. This model fails to satisfy the longer period variations associated with magnetic storms. A series of models constructed to explain both the daily

---

* Conductivity is the reciprocal of specific resistance (resistivity); the SI unit is siemens per metre; $1\,\mathrm{S\,m^{-1}} = 1\,\mathrm{Ohm^{-1}\,m^{-1}}$.

and storm time variations were proposed by LAHIRI and PRICE (1939). Within the 'core' the conductivity was taken to vary as a power of the radius, and a thin conducting layer representing the oceans was also incorporated. These models show a downward increase in conductivity at about 600 km depth to at least $1\,\mathrm{S\,m}^{-1}$. BANKS and BULLARD (1966) found that the annual variation yields an estimate of $2\,\mathrm{S\,m}^{-1}$ at 1300 km depth.

More recent analyses made by BANKS (1969, 1972) indicated a rise in conductivity from just below $10^{-2}\,\mathrm{S\,m}^{-1}$ above 300 km depth to about $1\,\mathrm{S\,m}^{-1}$ at 700 km depth. PARKER (1971) re-analysed the 1969 data of Banks by Backus-Gilbert inversion to obtain a smaller increase in conductivity from 10^{-1} to $1\,\mathrm{S\,m}^{-1}$ between depths of 200 and 1000 km, with a resolving length of about 300 km. LARSEN (1975) used electromagnetic fields observed on Hawaii to obtain a closely similar distribution to that of Parker applicable to the Pacific region, with an average conductivity of about $10^{-1}\,\mathrm{S\,m}^{-1}$ down to 200 km depth increasing to about $0.9\,\mathrm{S\,m}^{-1}$ at 700 km depth, but he also found a local high conductivity at 330–380 km depth which is not present in other models. Thus there are discrepancies of an order of magnitude between estimates of the conductivity above 300 km depth but there is general agreement that the conductivity rises steeply from 300 to 700 km to a value of about $1\,\mathrm{S\,m}^{-1}$.

An approximate estimate of the conductivity of the lower mantle can be obtained from the secular variation, which is believed to involve variations in the magnetic field at the core-mantle interface. The shortest variation which is observed at the Earth's surface is slightly less than four years. If it is assumed that shorter period variations do occur but are prevented from penetrating to the surface by the relatively highly conducting lower mantle then the conductivity can be roughly estimated. DUCRUIX and others (1980) applied this method to a large secular variation impulse occurring in the northern hemisphere in the late 1960s to estimate that the conductivity of the lowermost mantle is about $100\,\mathrm{S\,m}^{-1}$ and that the average value over the bottom 500 km does not exceed this. Comparable results were also obtained by use of the 11 year cycle of the external field. An interpretation of the radial conductivity distribution of the mantle is shown in Fig. 4.26.

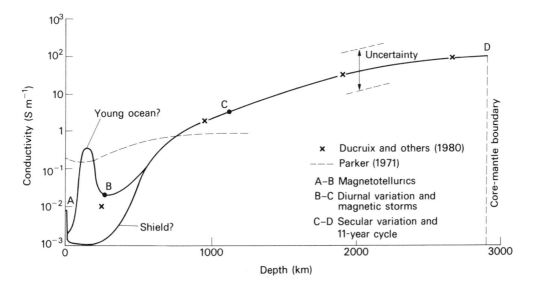

Fig. 4.26 Estimated distribution of electrical conductivity in the mantle showing the range of likely regional variation in the upper mantle and the high values of the lower mantle.

The steep rise in electrical conductivity approximately coincides with the mantle transition zone. It will be shown later (p. 184) that the mantle down to about 400 km depth consists mainly of silicates amongst which magnesium-rich olivine is predominant. In the lower mantle the stable phases are probably perovskite and the oxides of magnesium and iron. The increase in conductivity can probably be best explained in terms of the effects of temperature and phase transition on olivine and related minerals.

Silicate minerals such as olivine are practically insulators at room temperature but become semiconductors as the temperature is raised. Three types of conduction can occur in them. *Impurity semi-conduction* is caused by the presence of foreign atoms with a misfitting valency in a crystal lattice; these produce either excess electrons or 'holes' (missing electrons) which migrate through the lattice when an e.m.f. is applied. *Electronic semi-conduction* is caused by movement of free electrons raised into the conduction band through thermal agitation. *Ionic semi-conduction* is caused by movement of ions, as in an electrolyte. The observed conductivity is the sum of the three types of conductivity, although one type is usually dominant.

Each type of conductivity is thermally activated, so that

$$\sigma = \sigma_0 e^{-E/kT}$$

where σ is the conductivity, T is the temperature, σ_0 and E are constants (which may depend on pressure) and k is Boltzmann's constant. The most rapid increase in conductivity with temperature occurs when $T = \frac{1}{2}E/k$. Much below this temperature the conductivity is small relative to σ_0 and above $T = E/k$ it asymptotically approaches its maximum possible value σ_0. The activation energy E controls the temperature range at which the conductivity increases strongly with temperature.

In olivine, impurity semi-conduction ($\sigma_0 = 10^{-4} - 10^{-2}\,\mathrm{S\,m^{-1}}$) is probably dominant up to about 900 K and this is probably the main conduction process in dry rocks of the crust and topmost mantle. Above about 1000 K electronic semi-conduction increases strongly and swamps the impurity conduction. Above about 1400 K ionic conductivity becomes dominant at low pressure, but as it is strongly inhibited by pressure it is probably unimportant below 400 km depth. However, ionic conductivity is likely to be predominant in magma and in weakened lattices near the melting point, and it may therefore be an important factor in raising the level of conductivity in the upper mantle as explained later.

Electronic semi-conduction is probably the predominant conduction process in the transition zone and lower mantle. The conductivity of olivine depends on pressure, temperature and oxygen fugacity, but temperature is the most important influence. Experimental measurements at zero pressure show that the conductivity rises by a factor of about 100 from $10^{-3} - 10^{-2}\,\mathrm{S\,m^{-1}}$ at 1900 K to $10^{-1} - 1\,\mathrm{S\,m^{-1}}$ at 2270 K. A further downward increase in conductivity may result from the phase transitions. AKIMOTO and FUJISAWA (1965) observed a hundredfold increase in the conductivity of the iron olivine fayalite as it undergoes phase transition to spinel at 770 K (Fig. 4.27). By analogy, a similar increase in conductivity with depth might be expected as magnesium-rich olivine breaks down to higher pressure phases within the transition zone. It seems probable that the actual increase in conductivity between 300 and 1000 km depths is the combined effect of temperature and phase transitions.

Conductivity of the crust and upper mantle – regional variations

Global analysis of short-period geomagnetic variations fails to resolve the conductivity-depth structure of the uppermost 200 km of the Earth because of the large lateral variation. Other techniques applicable to individual regions are needed. The two most widely used techniques are geomagnetic deep-sounding and the magnetotelluric method.

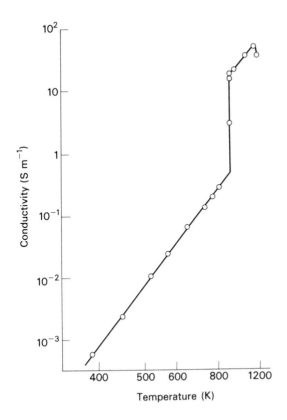

Fig. 4.27 Electrical conductivity of the iron-olivine fayalite (Fe_2SiO_4) at 4380 MPa confining pressure as a function of temperature. Adapted from AKIMOTO and FUJISAWA (1965), *J. geophys. Res.*, **70**, 447.

Geomagnetic deep-sounding makes use of temporary three-component magnetometer stations which are deployed as an array over the region to be studied. This method has been reviewed by FRAZER (1974). The main difficulty is in separating the internal and external parts of the magnetic variations recorded. Usually either the source field has to be assumed or semi-quantitative methods of interpretation are used. These include measurement of the ratio of horizontal and vertical field variations of specific period and their expression in terms of vectors which indicate where the anomalous telluric currents are flowing.

In the magnetotelluric method of CAGNIARD (1953), the variation of one (or both) components of horizontal magnetic field is recorded continuously at a station. At the same station, the electric field component perpendicular to the recorded magnetic field is measured by placing probes in the ground and recording variations of the e.m.f. between them. The variation in electric field is caused by the induced currents. Direct measurement of the electric field makes it possible to separate the internal and external parts of the horizontal magnetic field variations, provided that it is assumed that the field variations are uniform over a much wider extent than the depth penetrated by the induced currents and that the underlying conductivity distribution is a function of depth alone. From the records, the ratios of amplitudes of the electric and magnetic fields are determined for variations ranging from a period of a few seconds up to a day or more, depending on the penetration required. These ratios can then be interpreted in terms of a conductivity-depth profile. In practise, it is desirable to measure all three components of the magnetic field and two components of electric field, so that the redundant data can be used to check the validity of the assumptions. The magnetotelluric method has now been adapted to operate satisfactorily on the ocean floor (FILLOUX, 1977).

The electrical conductivity structure of the crust and upper mantle is now reviewed, starting at the surface and working downwards for the continents and oceans in turn. The electrical conductivity of sea-water is about $4\,S\,m^{-1}$ and for water-saturated sedimentary rocks it ranges between 10^{-3} and $1\,S\,m^{-1}$. The conductivity of dry continental crustal rocks beneath the sediments is about 10^{-6} to $10^{-3}\,S\,m^{-1}$. The outer layers of the Earth are poorly conducting apart from the more highly conducting skin formed by the oceans and sediments, and some anomalous local conductors within the crust.

Magnetotelluric measurements indicate that in non-anomalous stable continental regions the electrical conductivity of the lower crust and upper mantle down to 80 km or more is of the order of $10^{-3}\,S\,m^{-1}$. In some regions such as Massachusetts there is indication that the conductivity increases to about $10^{-2}\,S\,m^{-1}$ at about 80 km depth but in shield regions the low conductivity region extends much deeper. Much higher conductivities of about $0.1\,S\,m^{-1}$ occur at fairly shallow depth beneath active tectonic regions such as western North America (Fig. 4.28) (GOUGH, 1974) and the East African rift zone (BANKS and OTTEY, 1974). The high conductivity is attributable to high temperature and may be related to the presence of a melt fraction as described below.

Magnetotelluric and geomagnetic deep-sounding measurements have recently been made on the floors of the Pacific and Atlantic Oceans. These reveal the presence of a well-resolved high conductivity zone below the oceanic lithosphere and approximately corresponding to the asthenosphere. As would be expected, the depth to the highly conducting layer increases with the age of the overlying lithosphere. OLDENBURG (1981) has applied Backus-Gilbert inversion to the data from the Pacific Ocean obtained by FILLOUX (1977) and by others. He showed that the conductivity reaches a maximum value of about $0.1\,S\,m^{-1}$ at depths of 70, 120 and 180 km for lithospheric ages of 1, 30 and 72 My respectively. Regions of lower conductivity occur above and below the conducting zone but their detail is not well resolved. Magnetotelluric measurements in the neovolcanic region of Iceland reveal a conducting crust and uppermost mantle ($\sigma \sim 0.03\,S\,m^{-1}$) extending downwards from a depth of about 4 km with a more highly conducting layer at the base of the crust which may represent a magma chamber (THAYER and others, 1981).

The anomalously high electrical conductivity of about $0.1\,S\,m^{-1}$ observed below the oceanic lithosphere and in active tectonic regions cannot be satisfactorily explained by electronic conduction in solid rocks. This would require a temperature of at least 1950 K, which is unrealistic at 100 km depth. The best explanation is that a melt fraction is present in the highly conducting rocks. Basalt magma has an electrical conductivity 100 to 1000 times greater than that of peridotite under the same conditions. SHANKLAND and WAFF (1977) have shown that quite a small melt fraction of about 5 % within periodotite can give rise to a conductivity of about $0.2\,S\,m^{-1}$ at about 1650 K provided that it forms an interconnecting network within the rock. The observed magmatic activity in continental tectonic regions and at ocean ridge crests supports this interpretation. A further implication is that a melt fraction is probably present in the sub-oceanic asthenosphere out to at least 70 My age of the overlying ocean floor. A lack of similar high conductivity beneath stable continental regions may suggest a lack of significant partial fusion.

Local conductivity anomalies

One well known type of magnetic variation anomaly occurs near the margin of the oceans. Here there are large lateral variations in conductivity down to 5 km depth because sea-water is more than 10 000 times a better conductor than crustal rocks. Because of this, strong fluctuating telluric

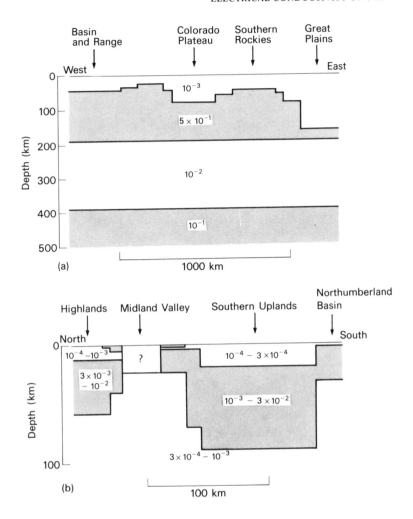

Fig. 4.28 Two-dimensional regional electrical conductivity distributions showing (**a**) lateral variation in the upper mantle high conductivity zone beneath western United States, and (**b**) shallower variations in conductivity beneath south Scotland. Electrical conductivity values are shown in units of $S\,m^{-1}$. (**a**) is redrawn from GOUGH (1974), *J. Geomagn. Geoelect., Kyoto,* **26**, 107 and (**b**) is redrawn from HUTTON and others (1981), *Phys. Earth planet. Interiors,* **24**, 85.

currents flow in the oceans in response to the short-period magnetic variations. Near the edge of an ocean, the strongest currents tend to flow parallel to the coast. The magnetic field vector associated with an electric current is perpendicular to the plane containing the current and the point of observation. Because the depth of oceans is relatively small compared with the width of most continental shelves, the magnetic vector on the adjacent continent associated with the oceanic telluric currents would be expected to be nearly vertical. Such strong variations in the vertical component are indeed observed within 50–100 km of the coasts. The high conductivity of the sea-water does not completely explain the anomalous magnetic variation near some continental margins. For instance, PARKINSON (1962) has shown that the horizontal field variations near the Australian coast are larger than could be caused simply by the ocean; a deep-seated change in conductivity associated with the transition from a more highly conducting sub-oceanic upper mantle to the more weakly conducting sub-continental upper mantle contributes to the observed variations.

Local anomalies in geomagnetic variations within continental interior regions can arise as a result of underlying disc-like regions of high conductivity or by channelling of currents induced elsewhere in the Earth along a local conductor. A good example of the latter type is associated with a long, narrow belt of high electrical conductivity stretching from south-eastern Wyoming to the edge of the Canadian Shield in Saskatchewan. This is probably caused by conducting rocks within the metamorphic basement. CAMFIELD and GOUGH (1977) have suggested that it marks an ancient fracture zone in which the high conductivity is caused by graphitic schists and other hydrated conducting rocks. Scotland is a region where there are lateral variations in crustal conductivity which cause complicated magnetic variation anomalies (Fig. 4.28) (HUTTON and others, 1981). Low conductivity occurs in the crust and upper mantle below the Caledonian foreland region of north-western Scotland. There are regions of high conductivity at shallow crustal depths beneath the Midland Valley and northern England and in the lower crust beneath the Highlands and the Southern Uplands. Other well known magnetic variation anomalies occur in the Canadian Arctic, western North America, Germany and Japan.

Some anomalies in geomagnetic variation are caused by the high conductivity of sea-water. A few, such as one in north Germany, are caused by highly conducting sedimentary rocks such as salt deposits. Some anomalies are caused by highly conducting rocks within the metamorphic basement, such as graphite schists. Some are probably caused by the presence of trapped water within the crust. Others may be caused by anomalously high temperatures in the underlying lower crust and upper mantle which may or may not be associated with a melt fraction or magma chamber. Discovering the origin of a specific anomaly often involves use of other geological and geophysical investigations.

To summarize section 4.7 (Fig. 4.26), the oceans and waterlogged sediments locally form a highly conducting layer at the Earth's surface. The conductivity of the normal underlying crystalline rocks of the lithosphere is about 10^{-3} S m^{-1} and results from impurity semi-conduction. Local anomalous regions occur within the continental crust for a variety of reasons. Beneath stable continental regions the conductivity may increase downwards to about 10^{-2} S m^{-1} at 80 km or deeper as a result of electronic semi-conduction at the higher temperature. Beneath the oceans the upper mantle conductivity is an order of magnitude higher than beneath stable continental regions, reaching a maximum of about 10^{-1} S m^{-1} within the asthenosphere as a result of a small melt fraction. High conductivity is also characteristic of the upper mantle below continental tectonic regions. The conductivity increases by one or two orders of magnitude between depths of about 300 and 700 km, this being the combined effect of higher temperature on electronic semi-conduction and phase transitions. The conductivity is about 1 S m^{-1} at 700 km depth and rises downwards to about 100 S m^{-1} at the bottom of the mantle.

4.8 Composition of the mantle

Composition of the upper mantle

The P velocity just below the Moho is typically about 8·0–8·2 km s^{-1}. The two common rock types which have comparable P velocity at the temperature and pressure appropriate to the Moho are (*i*) the ultrabasic rock *peridotite*, and (*ii*) *eclogite* which is the high pressure form of the basic rock gabbro. Peridotite is predominantly composed of olivine which may be accompanied by orthopyroxene, clinopyroxene, garnet and/or spinel. The main types are as follows (RINGWOOD, 1975):

Dunite	olivine
Hartzburgite	olivine, orthopyroxene, $\pm$ spinel
Lherzolite	olivine, ortho- and clinopyroxene, $\pm$ spinel
Garnet harzburgite	olivine, orthopyroxene, garnet
Garnet lherzolite	olivine, ortho- and clinopyroxene, garnet

The olivine is about 90% forsterite and 10% fayalite, and garnet is characteristic of high pressure conditions. Eclogite is composed of garnet and clinopyroxene which may be accompanied by subordinate orthopyroxene, quartz, kyanite, plagioclase or olivine. Table 4.2 shows the contrast in composition between peridotite and eclogite. The main mineralogical distinction is that peridotites contain abundant olivine and less than 15% garnet whereas eclogites contain little or no olivine but normally have more than 30% garnet. The presence or absence of abundant olivine is diagnostic.

Table 4.2 Composition of rock types relevant to the upper mantle.

	Dunite (anhydrous) [1]	Spinel lherzolite nodules (Hawaii) [2]	Garnet lherzolite nodules (South Africa) [3]	Pyrolite [4]	Eclogite nodules [5]
SiO_2	41·3	44·8	46·5	45·1	45·3
TiO_2	trace	0·2	0.3	0·2	0·5
Al_2O_3	0·5	3·0	1·8	4·6	15·3
Cr_2O_3	–	0·3	0·4	0·3	0·1
Fe_2O_3	1·2	1·6	} 6·7	0·3	2·1
FeO	5·9	8·3		7·6	9·9
MnO	0·1	0·1	0·1	0·1	0·3
MgO	49·8	38·2	42·0	38·1	12·7
CaO	trace	2·5	1·5	3·1	10·3
Na_2O	trace	0·4	0·2	0·4	1·4
K_2O	trace	0·1	0·2	0·02	0·3
H_2O	–	0·6	–	–	1·7

1 From GREEN and RINGWOOD (1963) giving average composition;
2 Average of eleven analyses given by KUNO (1969);
3 Average of nine analyses of nodules from kimberlites given by CARSWELL and DAWSON (1970) (water free);
4 Average water-free mantle pyrolite quoted by RINGWOOD (1975);
5 Average of ten eclogite nodules from kimberlites quoted by O'HARA and others (1975).

Two criteria based on the observed seismic velocities suggest that peridotite is the dominant rock of the topmost mantle (RINGWOOD, 1975). (*i*) The values of Poisson's ratio calculated from P_n and S_n (p. 38) range between about 0·24 and 0·27 which is consistent with peridotite but falls short of the few experimentally determined values for eclogite, which exceed 0·30. (*ii*) Oceanic values of P_n vary with direction, averaging about $8\cdot0\,km\,s^{-1}$ parallel to ridge crests and $8\cdot3\,km\,s^{-1}$ perpendicular to them (p. 98). Anisotropy of P_n is observed beneath Germany but is not characteristic of all continental regions (p. 42). Such anisotropy, where it occurs, is best explained by the preferred orientation of olivine crystals. Measured P velocities in individual olivine crystals vary between extreme values of $9\cdot87$ km s^{-1} in the [100] direction to $7\cdot73$ km s^{-1} in the [010] direction (VERMA, 1960) but a lesser anisotropy is characteristic of peridotites because of the scatter of crystal orientation and presence of other minerals. In contrast eclogite shows much less directional variation of P velocity as orthopyroxene is much less anisotropic than olivine and garnet is isotropic.

Another physical property of the uppermost mantle upon which limits can be set is the density. Peridotites have densities within the range 3250 to 3400 kg m^{-3} whereas those of fresh eclogites are between 3400 and 3600 kg m^{-3} (RINGWOOD, 1975). WOOLLARD (1970) estimated the density contrast between the crust and mantle to lie between 370 and 420 kg m^{-3} by correlating surface elevation against seismically determined depth to the Moho, assuming isostatic equilibrium. He also estimated the mean density of the crust to be 2870–2890 kg m^{-3}, implying that the density of the uppermost mantle lies between 3240 and 3310 kg m^{-3}. Even allowing for increase of density with depth within the crust, the mantle density below the Moho is unlikely to exceed 3400 kg m^{-3}. This is consistent with peridotite but not with eclogite. On the other hand, inversions of gross earth data yield rather higher estimates of density just below the Moho, but as pointed out earlier in the chapter (p. 163) such inversions have poor resolution and the high density value at the top of the mantle is probably the spurious effect of too deep a low density zone below.

Further indications of the composition of the upper mantle come from rock types which are believed to have been brought up from the mantle by tectonic or igneous processes. For instance, certain types of peridotite and serpentinite bodies now found at the surface probably represent slices of the uppermost sub-oceanic mantle which have been tectonically emplaced in the crust. Not all peridotites originate in this way. Some are known to have formed by crystal accumulation in a magma chamber, possibly followed by movement of the crystal mush. But other groups, notably the Alpine-type peridotites and those at the base of ophiolite sequences probably come from the uppermost mantle and have been tectonically emplaced in the crust during an orogeny. Other peridotites which have been serpentinized are found in oceanic fracture zones and these probably represent hydrated mantle rocks which have been uplifted to the seabed by block faulting. All these types probably come from the uppermost sub-oceanic mantle and cannot be regarded as typical of the upper mantle. Those in fracture zones represent young oceanic lithosphere whereas the other types come from older lithosphere which has been carried to a convergent plate boundary by sea-floor spreading and has subsequently been obducted into the crust by thrusting. These peridotites are of hartzburgite composition with some local occurrences of dunite. Hartzburgite and dunite are types of peridotite which are deficient in aluminium and cannot therefore give rise to significant further amounts of basalt by partial fusion. This is consistent with the idea that the uppermost sub-oceanic mantle has lost its available fraction of basalt as a result of partial fusion below the ridge crest. OXBURGH and PARMENTIER (1977, 1978) have pointed out that such depleted mantle material is about 60 kg m^{-3} less dense than undepleted mantle. They further suggested that masses of such depleted mantle may break loose from the sinking oceanic lithosphere at subduction zones, rising because of their low density to form part of the upper mantle beneath the adjacent continent. This would give rise to a compositional difference between sub-oceanic and sub-continental upper mantle.

Direct information on the composition of rather deeper parts of the upper mantle comes from the peridotite fragments known as olivine nodules which occur widely in alkali basalts. Some of these may have formed by crystal accumulation in a magma chamber but others with more uniform mineralogy probably represent fragments of the mantle brought up from depths of between 40 and 100 km. Their composition varies between dunite and lherzolite, the average being spinel lherzolite. Such nodules occur on oceanic islands such as Hawaii, as well as on the continents, indicating that the sub-oceanic lithosphere grades down from an uppermost barren hartzburgite zone into spinel lherzolite. Other fragments of the sub-continental mantle are brought up from even greater depths in kimberlite, which is a type of mica peridotite which occurs in small pipes and is much sought after because some of them contain diamonds. The fragments are mostly garnet lherzolite but a few of them are eclogite. It was pointed out earlier in the chapter (p. 173) that these fragments probably represent the sub-continental mantle material over the depth range 130 to 220 km.

In summary, the physical properties of seismic velocities and density and the supposed samples from the mantle both indicate a dominantly peridotitic composition. The topmost zone of the sub-oceanic mantle is probably formed of hartzburgite which passes down into spinel lherzolite. Beneath the continents, spinel lherzolite gives way at about 80 km depth to garnet lherzolite with some segregations of olivine-free eclogite. It is possible that there is an accumulation of depleted hartzburgite in the sub-continental upper mantle at fairly shallow depth.

The pyrolite model

What is the composition of the bulk of the upper mantle? It may possibly be close to that of the least depleted garnet lherzolite nodules found in kimberlite pipes, particularly those richest in CaO, Al_2O_3 and Na_2O. RINGWOOD (1975) has pointed out that even these must have suffered a small degree of partial fusion.

To overcome this difficulty, RINGWOOD (1962) postulated that the undepleted upper mantle consists of a mixture of dunite representing fully depleted mantle and basalt representing the melt fraction. This mixture produces a hypothetical type of peridotite named pyrolite by Ringwood. As originally proposed, this consists of three parts of average anhydrous dunite and one part of average basalt. RINGWOOD (1975) re-estimated the composition of pyrolite using a variety of more sophisticated approaches, such as combining 17% or primitive oceanic tholeiite with 83% of barren hartzburgite, or 99% of least fractionated lherzolite with 1% of nephelinite, with the result shown in Table 4.2.

GREEN and RINGWOOD (1963) found that pyrolite can crystallize in four different mineral assemblages under conditions of temperature, pressure and water vapour pressure appropriate to the upper mantle. These are as follows:

Ampholite	olivine, amphibole, accessory chromian spinel (3250–$3280\,\mathrm{kg\,m^{-3}}$)
Plagioclase pyrolite	olivine, plagioclase, enstatite, clinopyroxene, accessory chromite ($3240\,\mathrm{kg\,m^{-3}}$)
Pyroxene pyrolite	olivine, aluminous enstatite, aluminous clinopyroxene, spinel (3300–$3320\,\mathrm{kg\,m^{-3}}$)
Garnet pyrolite	olivine, pyrope garnet, pyroxene(s) ($3370\,\mathrm{kg\,m^{-3}}$)

The densities are stated for surface temperature and pressure. According to Green and Ringwood, these same assemblages would be expected for a wide range of possible upper mantle compositions ranging from basalt-dunite ratios of 1:1 to 1:10.

Pyroxene pyrolite is the mineral assemblage which should normally occur down to about 70 km depth. The mineralogy is equivalent to that of an aluminium-rich spinel lherzolite. At greater depths it would be expected to give way to the higher pressure form *garnet pyrolite*. According to the model, garnet pyrolite should be the normal stable assemblage present between depths of about 80 and 350 km. *Ampholite* would occur in low temperature, low pressure, water-rich environments. As the uppermost mantle has probably been dehydrated, this assemblage is likely to be of restricted occurrence if present. *Plagioclase pyrolite* is stable down to 35 km depth at temperatures higher than about 900 K and at low water vapour pressure. This assemblage may possibly occur in the uppermost mantle beneath regions of high heat flow although it cannot form from barren hartzburgite just below the oceanic Moho which lacks sufficient aluminium to form plagioclase.

Plagioclase pyrolite is about 70 kg m^{-3} lower in density than pyroxene pyrolite so that its presence may contribute to the low density of the upper mantle below high heat flow regions.

Origin of magma in the upper mantle

Most igneous rocks, with the possible exception of some granites and granodiorites, form from magma produced by partial fusion in the upper mantle. The most abundant source underlies ocean ridges where an average volume of about 20 km^3 of basaltic magma of olivine-tholeiite composition rises each year to form new oceanic crust. Another important source of magma is at subduction zones where the dominantly andesitic magmas of the calc-alkaline series are produced. A third type of magma source in the upper mantle underlies plate interiors, producing a variety of magma types such as alkali basalts of Hawaii and kimberlites of continental shield regions. Overall, basalt is by far the commonest type of magma formed in the upper mantle. It is the normal product of partial fusion of fertile peridotite.

The specific latent heat of fusion of basalt is about 400 kJ kg^{-1} so that a large amount of heat is required to cause significant melting. Partial melting on the observed scale probably cannot occur simply as the result of a general rise of temperature caused by thermal conduction in a static mantle. More drastic processes of shorter duration are required. Probably the most important cause of partial fusion is the progressive lowering of the melting temperature as pressure is reduced in upwelling mantle material. This may occur in a rising convection current or during isolated diapiric upwelling. When a mass of rock rises fast enough to prevent exchange of heat with its surroundings, it cools slightly as its temperature follows the adiabatic gradient, which is about 0.3 K km^{-1} in the mantle (AB in Fig. 4.29). It meets the steeper melting point curve at B where melting starts. This is the solidus in the upper mantle and it has a gradient of about 3 K km^{-1}. On crossing the solidus, partial fusion progressively occurs increasing the magma fraction as its cooling gradient steepens. Eventually the magma fraction segregates at C and rises to the surface (D), leaving behind an unmelted residuum.

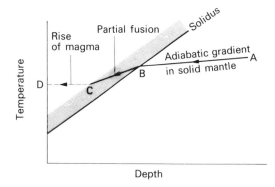

Fig. 4.29 Partial fusion occurring as a result of adiabatic upwelling of mantle material from A to C. Fusion starts at B where the appropriate adiabat crosses the steeper solidus. Magma segregates at C and rises to the surface (D) because of its low density.

The origin of basalt magma in the upper mantle can be understood by reference to Fig. 4.30 which shows a possible scheme for the partial melting of pyrolite containing 0.1% water (RINGWOOD, 1975). The solidus has a minimum temperature trough between depths of about 80 and 160 km which marks the pressure range where water probably remains free, rather than combined in the lattices of hydrous minerals such as amphiboles as occurs above and below the trough. If upwelling occurs along path A, starting either within the trough or below it, an olivine tholeiite melt is in equilibrium with the pyrolite at 20 km depth and shallower. The average melt fraction down to

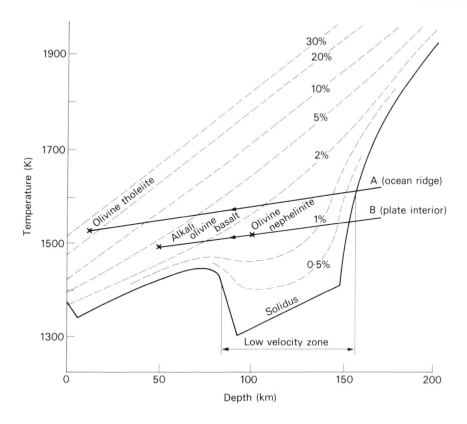

Fig. 4.30 Possible relationships between mantle solidus for 0·1 % water, temperature distribution and degrees of partial melting of pyrolite. Upwelling of material along adiabat A (from below or within low velocity zone) may produce olivine tholeiite at shallow depth in sufficient abundance to account for ocean ridge volcanism. Upwelling along adiabat B could produce plate interior volcanism of various types as a result of a small degree of partial melting. Note that the degree of partial fusion attained and the magma type depends on both the adiabat followed by the upwelling material and on the depth at which the magma segregates. Adapted from RINGWOOD (1975), *Composition and petrology of the Earth's mantle*, p. 155, with permission of McGraw-Hill Book Company.

80 km depth is about 10 %, which on segregation would be sufficient to form the oceanic crust. Thus path A represents a possible mechanism for the formation of ocean ridge basalts. If upwelling proceeds along path B, then alkali olivine basalt would form by segregation of the magma at 50 km depth, whereas other alkali rock types such as olivine nephelinite would form by segregation at greater depth. Diapiric upwelling along path B thus gives a possible explanation of certain types of plate interior volcanism. In reality great complexity may result from other factors such as the presence of CO_2, but Fig. 4.30 illustrates in principle how ocean ridge and plate interior volcanism can occur by upwelling. The production of calc-alkaline magmas at subduction zones is connected with the recycling of oceanic crust back into the mantle and is described in Chapter 5 (p. 217).

A central problem of igneous petrology is to understand how the wide variety of igneous rocks found in the crust can originate. The type of magma produced at source depends on composition, pressure and temperature of the source and the degree of partial fusion which occurs. The composition of the magma may then be modified as it rises towards the surface in one or more stages as a result of fractional crystallization, reaction with wall rock and contamination by foreign inclusions. One or two simple types of source such as the fertile peridotite of the upper mantle or the gabbroic rocks of the subducting oceanic crust (which originally come from the mantle) can produce a wide variety of igneous rock types at the surface.

It has been shown that the oceanic upper mantle is probably compositionally stratified as a result of separation of basalt magma from its uppermost levels and that there may be some compositional difference between the sub-oceanic and sub-continental upper mantles. Otherwise, it has often been assumed that the upper mantle is fairly homogeneous in composition. The geochemistry of igneous rocks now indicates that this is not so, and that there is further lateral heterogeneity in the composition of the upper mantle. These lateral variations are best indicated by the minor element chemistry and by isotopic heterogeneity in Nd, Sr and Pb isotopes. The variations are probably fairly small in terms of major element chemistry. Some of the heterogeneity may be inherited from the primitive Earth and some of it may have developed later by removal of material to form the crust. A more detailed discussion of the evidence and its interpretation is given in the compilation of papers by BAILEY and others (1980).

Interpretation of the seismic velocity pattern in the upper mantle

Modern seismology has revealed some quite unexpected features of the upper mantle, such as the low velocity zone between about 60 and 150 km depths and the large regional variations in P and S velocity structure. Variations in seismic velocity can be caused by one or more of the following effects: (*i*) influence of temperature and pressure on a rock of uniform chemical and mineralogical composition; (*ii*) partial fusion; (*iii*) mineralogical phase transitions; (*iv*) anisotropy; (*v*) chemical inhomogeneity; and (*vi*) presence of open cracks (probably ineffective below 20 km depth).

The most significant seismological feature of the upper mantle is the low velocity zone. This zone approximately coincides with the region of high seismic attenuation (low Q). It seems probable that the low velocity and low Q zones have a common cause as there is experimental and theoretical evidence that strong attenuation should be accompanied by lowering of the elastic moduli, particularly the rigidity modulus. Beneath the oceans and the continental tectonic regions, the low velocity zone may also coincide with a region of high electrical conductivity. The low velocity and low Q zones have often been identified with the asthenosphere and the overlying region with the lithosphere, but this interpretation encounters some problems as described in Chapter 8.

Can the low velocity zone be explained simply by the influence of temperature and pressure on solid rock of uniform composition? Increase in pressure with depth causes a rise in seismic velocity but the geothermal rise in temperature with depth causes a corresponding reduction. If the temperature gradient in the upper mantle is higher than about 8 to 9 K km^{-1} then the P velocity in homogeneous material must decrease slightly with depth as the influence of temperature then slightly outweighs that of pressure (ANDERSON and SAMMIS, 1970). If it exceeds only about 3 to 4 K km^{-1} then the S velocity will decrease with depth. The oceanic geothermal gradient down to about 70 km depth is on average much steeper than 8 K km^{-1} but below 100 km depth it falls off to less than 4 K km^{-1} because otherwise there would be wholesale fusion. This means that a small decrease in velocity with depth down to 70 km and a slight rise below this depth is to be expected in the sub-oceanic mantle. Because of the concentration of radiogenic heat sources in the continental crust, the temperature gradient is probably significantly lower in the upper mantle beneath continents than beneath oceans, giving a possible explanation of the less well developed low velocity zone beneath continents.

Self-compression and thermal expansion, however, are probably not the main explanation of the low velocity zone. The phase transition from pyroxene to garnet pyrolite would be expected to cause a downward increase in velocity of about 0.2 km s^{-1} at 70 km depth sufficient to annul the temperature effect. A further difficulty is that a shallow low velocity zone with gentle gradients above and below would be expected whereas recent seismological results have emphasized relatively sharp upper and lower boundaries to the zone. Nor is the simple temperature effect adequate to account for regional variations in delay times. HALES and DOYLE (1967) found that a difference of temperature of about 1250 K spread over 500 km vertical extent in a homogeneous and solid mantle would be needed to explain the variation of delay times in the United States, which is unrealistic.

The most widely accepted explanation of the low velocity zone in oceanic and tectonic continental regions is that it contains a small melt fraction. The melting point of mantle material is most closely reached between about 50 and 150 km depths. This effect is accentuated by the trough of low melting temperature between 70 and 150 km depths which may be caused by the inability of small quantities of water to combine with minerals within this region (Fig. 4.30). A small melt fraction may remain trapped in this region as it would solidify if it rose towards the overlying higher melting point region. The low melting point trough approximately coincides with the depth range of the sub-oceanic low velocity zone. The influence of a small melt fraction on seismic properties depends critically on whether the melt occupies isolated or interconnected cracks. If the cracks are isolated, then the reduction in velocities and Q would be small. On the other hand, O'CONNELL and BUDIANSKY (1977) have shown that a liquid fraction of less than 1 % occupying an interconnected network of cracks at grain boundaries could account for the observed reduction in seismic velocities and Q value. Such an interpretation is supported by the high electrical conductivity zone underlying the oceanic lithosphere, which appears to require the presence of a melt fraction occupying interconnected cracks (p. 182).

Partial fusion is not the only possible explanation of the low velocity zone. Other thermally activated mechanisms have been suggested. These are related to proximity to melting but do not require actual fusion. One such mechanism is grain boundary relaxation which reduces the rigidity modulus by slip at grain boundaries and increases attenuation by viscous dissipation at the boundaries. Alternatively, experimental and theoretical indications suggest that thermally activated point defects within the grains may possibly give rise to the low velocity zone (SHAW, 1978). In reality, it seems probable that such solid state effects combine with partial fusion to cause the low velocity zone. Partial fusion is probably the dominant mechanism beneath the oceans and in tectonic continental regions but it is possible that the solid state mechanisms are mainly the cause of the shallow low velocity zone for S beneath stable continental regions.

The large lateral variations in the seismic velocity structure of the upper mantle between continents and oceans and within them are probably almost entirely caused by variations in the depth, thickness and intensity of the low velocity zone. This is clearly borne out by the relationship between P and S delay times and density. For instance, HALES and DOYLE (1967) studied the relationship between P and S delay times in the United States (p. 158). They found that the S delay times had a range of nearly 8 seconds and that the observed ratio of S delay to P delay is $3 \cdot 72 \pm 0 \cdot 43$. They pointed out that if Poisson's ratio in the normal and anomalous parts of the upper mantle is the same, then the ratio of S delay to P delay would be $1 \cdot 7 - 1 \cdot 8$, which is significantly different from the observed value. The observed ratio of $3 \cdot 72$ can be accounted for if the rigidity modulus alone changes and the bulk modulus is not affected. This may suggest that the delays are related to partial fusion or approach to it rather than to compositional changes or solid-solid phase transition.

A similar argument can be applied to the P delay of about $1 \cdot 3$ seconds observed in Iceland (p. 157); this needs a reduction of $0 \cdot 6 \, \text{km s}^{-1}$ in P velocity over a vertical depth extent of about 200 km. If Birch's relationship between velocity and density applies, then the corresponding density reduction would be $200 \, \text{kg m}^{-3}$ extending over a vertical extent of 200 km. This would cause a negative gravity anomaly of more than 800 mgal, which is certainly not observed on Iceland, even if allowance for a thin crust is made. Clearly Birch's relationship does not apply, and the simplest interpretation is that there is a melt fraction in the upper mantle beneath Iceland which causes greater reduction in seismic velocity than in density.

Two seismic features of the velocity-depth distribution in the lid above the low velocity zone invite comment. Firstly, there is a general increase in velocity with depth, which is difficult to explain because the normal effect of increasing temperature and pressure should in theory cause a slight decrease in velocity. It is possible that this effect is counteracted by downward gradation towards a garnet pyrolite assemblage or by increase in the eclogite fraction at greater depths. Secondly, thin high velocity layers of 8·3 to 8·7 km s^{-1} have been detected below the Moho in some long lithospheric refraction lines. The abrupt velocity fluctuations with depth which have been postulated are probably too large to explain by compositional variation. If these high velocity layers are real, the best explanation is probably in terms of seismic anisotropy, with the fast [100] axes of olivine crystals preferentially orientated along the observed profile. The intervening lower velocity layers would then represent different patterns of orientation of the olivine crystals. A structural explanation of such layers formed of crystals of varying orientation is not yet available.

The velocity-depth distribution between the low velocity zone and the base of the transition zone is not well resolved. The slight increase in velocities with depth can be attributed to the combined effect of the low temperature gradient and the pressure gradient on garnet pyrolite of uniform composition.

We have seen that proximity to the melting temperature is the best way of explaining the low velocity zone in the upper mantle. Regional variations in the P and S velocity structure are best explained by lateral variations in temperature affecting the depth and intensity of the low velocity zone. This does not rule out other less extreme variations of velocity caused by phase transitions or even chemical inhomogeneity. Temperature and its regional variation, however, seems to be the controlling factor. The temperature has a larger effect than in other regions of the crust and mantle simply because it is in the upper mantle that fusion temperatures can be reached.

Figure 4.31 summarizes some of the main features of the structure and constitution of the upper mantle below oceans and continents.

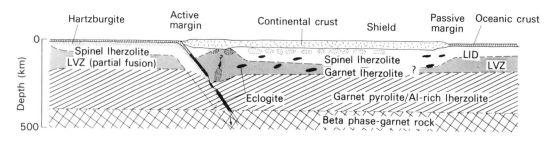

Fig. 4.31   Sketch showing the structure and constitution of the upper mantle beneath continents and oceans.

The transition zone and the lower mantle

BERNAL (1936) first suggested that the '20° discontinuity' may be caused by phase transition under increasing pressure from olivine to spinel, by analogy with two known forms of magnesium germanate (Mn_2GeO_4), one having an olivine structure and the other a spinel structure. Later BIRCH (1952) calculated that the seismic parameter ϕ of the lower mantle, reduced to surface temperature and pressure, is consistent with the presence there of simple oxide phases rather than olivine or pyroxene. This suggested a second stage of break-down of crystal structure in the transition zone to very dense phases such as periclase (MgO), wustite (FeO) and stishovite (SiO_2). The discovery that there are two main steps of steep velocity gradient in the transition zone naturally led to the suggestion (e.g. ANDERSON, 1967c) that the 400 km deep step is predominantly caused by the olivine-spinel transition and that the 650 km deep step is caused by a further stage of break-down to more closely-packed post-spinel phases such as oxides having rocksalt structure. Up to about 1965, interpretation of the transition zone was based on inference from experiments on germanates and germanate-silicate solid solutions and on thermodynamic deductions. The crystal chemistry of germanium compares closely with that of silicon but the phase transitions occur at much lower pressure than those in the corresponding silicates. For instance, in magnesium germanate the olivine spinel transition occurs at 1080 K at atmospheric pressure and at 1210 K at 570 MPa. An important step was made when AKIMOTO and FUJISAWA (1968) observed the transition to spinel in olivines ranging from pure fayalite to a 60% molecular ratio of forsterite. Subsequently, high pressure techniques have improved to such an extent that the transitions in the relevant mantle minerals (magnesium-rich olivines, pyroxenes, garnet) can be experimentally observed at pressures and temperatures spanning the whole transition zone, although accurate delineation of the phase boundaries is yet to be attained at pressures above those at about 400 km depth.

According to the pyrolite model of RINGWOOD (1975), the mantle between about 150 and 330 km depths consists of about 57% by weight of olivine of composition $(Mg_{0.89}, Fe_{0.11})_2 SiO_4$, 17% of orthopyroxene $(Mg, Fe)SiO_3$, 12% of omphacitic clinopyroxene and 14% of pyrope-rich garnet. The transformations affecting olivine are therefore likely to dominate the structure of the transition zone although those affecting pyroxenes and garnet will also be significant. RINGWOOD and MAJOR (1970) first succeeded in transforming olivine of mantle composition; they obtained the unexpected result that olivine with more than about 80% molecular proportion of forsterite transforms to a newly-discovered distorted type of spinel structure which is orthorhombic. This is referred to as the beta-phase. The experimental data indicates that mantle olivine at 1270 K transforms to the beta-phase at 1.18×10^4 MPa. The slope of the transition is 3 MPa K^{-1}, implying that at 400 km depth it should occur at about 1870 K. Experimental evidence indicates that the beta-phase of Mg_2SiO_4 transforms to spinel at a depth of about 500 to 600 km. At even higher pressures, LIU (1976a) has shown that spinel does not break down to all its constituent oxides as had been expected but transforms to a mixture of perovskite ($MgSiO_3$) and periclase (MgO). At 1270 K this was experimentally observed to occur within the pressure range 1.7–2.5×10^4 MPa. This indicates that the 650 km deep seismic discontinuity is probably predominantly caused by the transformation of spinel to perovskite, periclase and wustite.

LIU (1976b) has shown experimentally at 1270–1670 K that the pyroxene enstatite would undergo a similar series of transformations to higher density phases in the transition zone, first breaking down to a mixture of beta-phase and stishovite at about 1.7×10^4 MPa. The beta-phase transforms to spinel at about 2.0×10^4 MPa. As pressure increases further, the spinel and stishovite react to produce $MgSiO_3$ of ilmenite structure and at about 2.5–3.0 MPa the ilmenite transforms to perovskite. These transformations may not actually occur in the mantle because, in presence of aluminium, it is possible that the pyroxenes transform to garnet in the depth range 330–400 km, this being taken into solid solution with the garnet already present at these depths (RINGWOOD, 1975). At pressures corresponding to the deeper parts of the transition zone, the garnet may break down through a complicated series of reactions, possibly first splitting off the calcium perovskite ($CaSiO_3$) and eventually producing magnesium-iron perovskite together with one or more other dense phases carrying aluminium and sodium.

Thus the main features of the mantle transition zone can be interpreted in terms of a pyrolite upper mantle. In contrast, LIU (1980) has shown that the 400 km step could not be explained if the upper mantle composition was eclogite. Putting the evidence together, the main features of the transition zone are tentatively interpreted as follows (Fig. 4.32):

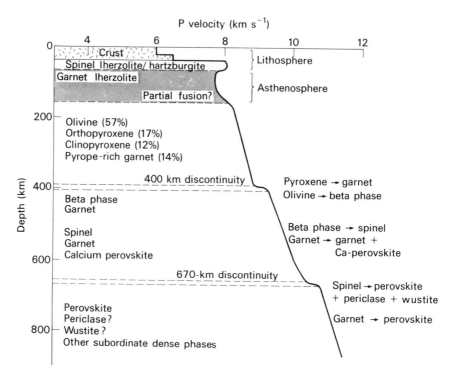

Fig. 4.32   Interpretation of the P velocity-depth distribution of BURDICK and HELMBERGER (1978) for the upper mantle and transition zone in terms of partial fusion and phase transitions.

(*i*)   Between 150 and 350 km depths the predominant mineral phases present are magnesium-rich olivine, enstatite, clinopyroxene and garnet.

(*ii*)   The 400 km deep seismic discontinuity or step is mainly caused by the olivine to beta-phase transformation, but pyroxene may transform to garnet between 330 and 400 km depths thus smearing out the velocity gradient over shallower depths. Below 400 km, the beta-phase and garnet are probably the main constituents present.

(*iii*)   The steep gradient in seismic velocity which may possibly occur at about 550 km depth may be caused by the transformation of beta-phase to spinel and/or by the splitting-off of calcium perovskite from garnet. Below 550 km depth, spinel, garnet and calcium perovskite may represent the main mineral phases present.

(*iv*)   The 650 km deep seismic discontinuity is attributed to the break-down of spinel to perovskite, periclase and wustite. Garnet may also break down in this region to yield perovskite and certain other dense silicate phases containing aluminium and sodium. It is not yet clear whether this discontinuity is caused by simple mineral transformation or reaction, or whether there is an associated compositional change (see below).

It should be emphasized that the details of the mineralogy below about 400 km depth are still controversial.

According to the seismological evidence, the lower mantle, apart from the lowermost 200 km, approximately satisfies the Adams-Williamson equation. Therefore it probably has a relatively homogeneous composition in comparison with the upper mantle and transition zone. Some local minor steepenings of the velocity-depth distribution have been suggested and these may possibly represent slight radial inhomogeneity such as minor phase transitions of a less drastic type than those of the transition zone. The most important outstanding question is whether or not the chemical composition of the lower mantle differs from that of the upper mantle.

Shock wave experiments (e.g. WANG, 1968) and static experiments (SAWAMOTO, 1977) indicate that the density of the lower mantle is probably about 5 % higher than it would be if formed of a mixture of constituent oxides. Most of this discrepancy has been removed now that it is recognized that the dense phase perovskite is probably the most abundant mineral in the lower mantle. However, SAWAMOTO (1977) found that the calculated density and P velocity for a perovskite-periclase mixture are slightly lower than the observed values. The presence of periclase lowers both density and seismic velocity below the observed values. Increasing the iron content would improve the density fit but would increase the velocity discrepancy. Sawamoto suggested that the discrepancy can best be removed by modelling the lower mantle almost entirely as perovskite. It would then have a composition closer to pyroxene $(Mg, Fe) SiO_3$ than to olivine $(Mg, Fe)_2 SiO_4$ implying that the proportion of SiO_2 would be significantly higher in the lower than the upper mantle. However, an explanation not involving a major difference in chemistry between the lower and the upper mantle may yet be possible. Other as yet unknown dense phases may be present. For instance, LIU (1978) showed that periclase may react with aluminium to produce a dense phase of composition $MgAl_2O_4$ having a sodium titanate structure.

The decrease, and possibly reversal, of the seismic velocity-depth gradients in the lowermost 200 km of the mantle is probably best explained by a steep temperature gradient marking the lower boundary zone of mantle convection, but explanation in terms of iron enrichment near the core boundary is also possible.

Chemical evolution of the mantle

The mantle probably formed within a few million years of the Earth's formation when the iron-nickel phase melted out and segregated to form the core, leaving the lower density silicate material behind to form the primitive mantle (p. 244). Consequently, magnesium and silicon are more abundant in the mantle than iron. Subsequently the bulk composition of the mantle has continued to change slightly with time as continental crustal material rich in silicon and aluminium has been irreversibly differentiated from it.

Prior to about 3800 My ago, conditions in the primitive mantle were probably too hot to permit permanent separation of sialic material. The geological evidence (p. 84) indicates that the most active period of differentiation was in the early Precambrian between about 3800 and 2500 My ago. Differentiation has continued at a decreasing pace to the present day. This now occurs by a two stage process, firstly production of oceanic crust by partial fusion of the mantle material and secondly formation of calc-alkaline magmas by partial fusion of subducting oceanic crust and related rocks. The sialic magmas thus produced form new crust at island arcs and add to existing continental crust in Andean type mountain ranges. Taking the present volume of the continental crust to be about 7×10^9 km^3, the average annual production of new continental crust from the mantle over the last 3800 My has been 1·8 km^3 and we can guess that the present rate is about 0·4 km^3 per year.

The mass of the continental crust is only about 0·5 % of that of the mantle. Consequently crustal differentiation has not greatly affected the major element composition of the mantle. Iron and calcium are little changed as they are about equally abundant in crust and mantle. Silicon in the mantle has been depleted by about 1 part in 400 and magnesium has been enriched by about 1 part in 200. Of the abundant elements, aluminium has been most strongly affected, having been depleted by about 1 part in 70. The effect of crustal differentiation has been much more drastic on some less abundant lithophile elements such as uranium, barium, potassium and rubidium. These have been strongly concentrated in the crust and are significantly depleted in the mantle relative to their initial abundance as a result of crustal differentiation.

The chemical evolution of the mantle towards its present state depends on several factors as follows. (i) The primitive mantle may have been homogeneous as is usually assumed or it may have been chemically zoned as a result of inhomogeneous accretion and core formation. (ii) Differentiation of sialic material may have affected the whole of the mantle uniformly or just part of it. The isotope evidence discussed below suggests that the differentiation has probably only affected the upper half of the mantle. (iii) Depleted portions of the mantle from which oceanic and/ or continental crustal material has been separated may have been re-mixed on a broad scale with fertile mantle material, or they may have been segregated to form barren regions. DICKINSON and LUTH (1971) suggested that refractory, depleted parts of the sinking oceanic lithosphere accumulate in the lower mantle whereas OXBURGH and PARMENTIER (1978) suggested that they accumulate in the upper mantle beneath continents. Alternatively, RINGWOOD (1975) suggested that the depleted material lies between depths of about 200 and 650 km. There is clearly no general agreement.

The deepest insight into the geochemical evolution of the mantle has come from the decay products of long-lived radiogenic isotopes (O'NIONS and others, 1979, 1980). The recent study of strontium and neodymium isotopes has been particularly informative and lead isotopes are also useful but less easy to interpret. The principles are illustrated by reference to strontium. Strontium-87 has progressively increased in abundance within the Earth by decay of the long-lived isotope rubidium-87 (p. 11). Other isotopes of strontium are not decay products and their abundances within the Earth have remained constant. Thus the ratio $^{87}Sr/^{86}Sr$ within the whole Earth has progressively increased as a result of the decay of ^{87}Rb. Knowing that Rb/Sr for the bulk Earth is 0·03, it can be shown that the strontium isotope ratio has increased over the Earth's lifespan from its initial value of 0·698 98 to its present bulk value of about 0·7047 (Fig. 4.33). The strontium isotope ratio within the mantle would follow the same course as that of the bulk Earth if during sialic differentiation rubidium and strontium were removed in proportion to their abundances. However, rubidium is a large-ion lithophile element and it is preferentially concentrated into the sialic crust relative to strontium. Rubidium is more than four times more abundant relative to strontium in the crust than in the mantle. Consequently the mantle has been strongly depleted in rubidium relative to strontium as a result of formation of the continental crust and for this reason the $^{87}Sr/^{86}Sr$ ratio within the mantle has increased less rapidly than the gross earth value. As explained in Chapter 1 (p. 12), the strontium isotope ratio occurring at the source of an igneous rock can be determined using the initial isotope ratio method. The strontium isotope ratio characteristic of the source of ocean ridge basalts is determined to be about 0·7028, which indicates that the ratio in the underlying mantle which is yielding new oceanic crust is indeed below that of the bulk Earth.

A similar story emerges from the study of neodymium isotopes. Neodymium-143 is the decay product of samarium-147. Neodymium is preferentially concentrated into the crust relative to samarium. Thus the ratio $^{143}Nd/^{144}Nd$ appropriate to the upper mantle is observed to be higher than the estimated gross earth value. O'NIONS and others (1979) have used the evolution of

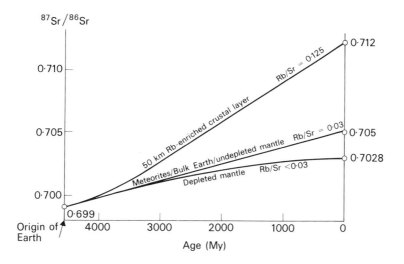

Fig. 4.33 Evolution of the strontium isotope ratio $^{87}Sr/^{86}Sr$ within regions of the Earth as a result of the progressive decay of ^{87}Rb by beta decay, according to a model of O'NIONS and others (1979) in which one half of the mantle differentiates over the Earth's lifespan at a decreasing rate to form the crust. The preferential removal of rubidium relative to strontium from the depleted mantle reservoir into the crustal layer causes the strontium isotope ratio to increase more rapidly in the crustal layer, and less rapidly in the depleted part of the mantle, than in the bulk Earth or in the undepleted part of the mantle. Adapted from O'NIONS and others (1979), *J. geophys. Res.*, **84**, 6096.

strontium and neodymium isotope ratios, and those of lead, to model the geochemical evolution of the mantle. They found that the isotope ratios determined for the source region of ocean ridge basalts is best modelled if the sialic crust has developed by differentiation of one half only of the mantle. This suggests that differentiation has mainly affected the mantle above about 1000 km depth and that the lower mantle has a more primitive composition. Plate interior volcanism may tap this more primitive source but ocean ridge volcanism is normally reworking the upper depleted part of the mantle. If this analysis is correct, it has far-reaching implications for mantle convection, implying that the main convective circulation does not pass through the mantle transition zone – rather there would be separate convection systems in the upper and lower mantle.

4.9 Lateral density variations in the mantle

Lateral variations of density of two main types are known to occur within the mantle. There are substantial lateral density variations of relatively short wavelength in the uppermost mantle beneath ocean ridges and continental plateau uplifts, and longer wavelength features of deeper origin which give rise to the satellite-derived global gravity anomalies.

The combined use of gravity and crustal seismic refraction methods has shown that ocean ridges (p. 115) and continental plateau uplifts (p. 64) are isostatically supported by regions of low density within the underlying upper mantle. These give rise to local negative Bouguer anomalies of about $-150\,mgal$ but the free air anomalies are small because of the isostatic equilibrium. The low density regions occur in the uppermost mantle and are attributable to anomalously high temperature which reduces the density by thermal expansion and by phase transition and partial fusion. The lithosphere is thinned and lateral density variation probably also occurs within the lithosphere and the underlying asthenosphere. These features are not conspicuous on the satellite-derived global gravity anomaly map (Fig. 4.34) because of their short wavelength and the isostatic equilibrium.

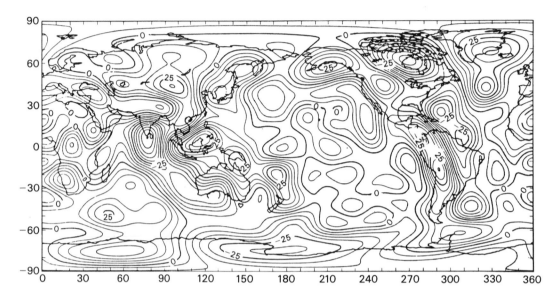

Fig. 4.34   Global free air gravity anomaly map based on spherical harmonic coefficients up to degree and order of sixteen. Redrawn from PHILLIPS and LAMBECK (1980), *Rev. Geophys. space Phys.*, **18**, 45.

The long wavelength variations of the Earth's gravity field have been determined by the study of the orbits of artificial satellites. They can be presented either as a map of deviations of the geoid from a reference ellipsoid (Fig. 1.2) or as a map of gravity anomalies (Fig. 4.34). The shortest wavelength present in Fig. 4.34 is about 2500 km. These global gravity anomalies are almost entirely caused by deep-seated lateral variations of density within the mantle or at the core-mantle boundary.

The satellite-derived gravity anomalies do not appear to be related systematically to the distribution of continents and oceans. This is partly to be expected, because the major surface features are all in approximate isostatic equilibrium so that the gravitational effect of the surface elevation of continents relative to oceans at a point above the Earth's surface is approximately cancelled out by the opposite effect of the root. But in fact a small residual gravity anomaly would be expected above the Earth even for perfect isostatic equilibrium. This is because the compensating mass must have a smaller areal extent than the surface feature as a result of decrease of the Earth's circumference with depth. It is therefore puzzling that the satellite gravity anomalies do not reflect the difference between continental and oceanic regions. A possible explanation stems from the substantial temperature difference between the sub-oceanic and sub-continental upper mantle (p. 174). The hotter sub-oceanic upper mantle should have a lower average density than the cooler upper mantle beneath continents. The isostatic balance between continents and oceans may involve this relatively small upper mantle density contrast in addition to the difference in crustal thickness. This would have the effect of reducing the expected residual gravity anomaly between continental and oceanic regions. A further complication comes from the suggestion by OXBURGH and PARMENTIER (1978) that low density depleted mantle material may accumulate beneath the continents, so that the density differences resulting from temperature and compositional effects tend to cancel out. Whatever the explanation of lack of systematic gravity differences between continental and oceanic regions, it seems clear that the major part of the long wavelength gravity anomalies shown in Fig. 4.34 must originate below the asthenosphere.

The harmonics of the global gravity field of degree less than about five, that is of wavelength greater than 8000 km, may arise partly or wholly from fluctuations in the depth of the core-mantle boundary, at which the density increases downwards by about $4000 \, \text{kg} \, \text{m}^{-3}$. Shorter wavelength components cannot be explained in this way as they would require unacceptably large fluctuations in the depth of the boundary. Some contribution to the longer wavelengths may come from lateral density variation within the lower mantle. Most of the prominent anomalies seen on Fig. 4.34 have a dominant wavelength of 5000 km or less, and these are too sharp to originate below a depth of about 1000 km. They are probably essentially caused by lateral density variation between 200 and 1000 km depth.

Density anomalies within the mantle may be caused by chemical inhomogeneity or by lateral temperature variation. As the mantle below the lithosphere probably lacks finite strength and can flow in response to density perturbations (Chapter 8), the most acceptable explanation is that lateral density variation is a predominantly temperature effect and is related to convective flow. An interesting possibility consistent with a source above 1000 km depth is that the gravity anomalies are partly caused by lateral temperature variation affecting the depth of phase transformations within the mantle transition zone (BOTT, 1971b). The olivine to beta-phase transformation at about 400 km depth involves an increase of density of about $250 \, \text{kg} \, \text{m}^{-3}$. A 10 K rise in temperature causes the transformation to migrate about 1 km deeper. A fluctuation in depth of 1 km amplitude and 3000 km wavelength would give rise to a gravity anomaly at the surface of about 3·5 mgal amplitude. Thus quite small lateral temperature variation of about 60 K affecting the depth of the olivine-beta phase transformation could give rise to global gravity anomalies of the observed type. Temperature variation may also affect the depth of the 650 km transformation but it is not yet known whether this would migrate upwards or downwards in response to a temperature rise.

Perhaps the most conspicuous features on Fig. 4.34 are the positive anomalies which occur around the circum-Pacific belt and a series of negative anomalies which occur adjacent to the positive belt on its outer side. These anomalies contrast with the relatively subdued gravity field of the Pacific Ocean. The positive anomalies occur in the vicinity of the subducting Pacific lithosphere and they may be caused by the sinking of the cool oceanic lithosphere into the mantle. An individual sheet of sinking lithosphere has a high density because of its low temperature. If subduction occurs over a long enough period of time while the position of the sinking slab migrates inwards (as is now happening around the Pacific Ocean), there would be a broad regional cooling of the affected part of the mantle including the transition zone. It is therefore suggested that the positive anomalies bordering the Pacific Ocean are caused by the combined effect of the sinking slabs and the regional cooling over a long period of time. The adjacent negative anomalies include the well-known minimum region of the north-eastern Indian Ocean and a negative anomaly down the eastern side of the American continents. It is possible that these are caused by a complementary upflow of hot material from the lower mantle through the transition zone into the upper mantle, with the downward migration of the olivine-beta phase transition contributing to the resulting anomaly.

A subdued belt of positive gravity anomaly follows the Alpine-Himalayan mountain ranges. This may also be caused by sinking of cool upper mantle material such as must occur beneath a collision mountain range. Positive anomalies occur over the anomalous oceanic regions of Iceland in the North Atlantic and Kerguelen in the southern Indian Ocean but their origin is unclear. Small negative anomalies may be related to imperfect recovery from postglacial loading in the North American and North European shield regions.

5 Continental margins and island arcs

5.1 Introduction

The continental margins are amongst the most spectacular morphological features of the Earth's solid surface, marking the relatively abrupt transition between the continents with their shallow shelves and the deep oceans. The change in elevation across them is primarily the isostatic response of the surface to the junction between thick continental crust and thin oceanic crust. The deep structure associated with margins, however, extends down into the upper mantle and locally reaches the mantle transition zone.

Continental margins are of two main types. *Passive margins* are the seismically inactive type where the adjacent continental and oceanic lithosphere are welded together as parts of the same plate. These margins originated by continental splitting but once formed they ceased to be plate boundaries. Most of the present passive margins are found bordering the relatively young and progressively widening Atlantic and Indian Oceans. Despite their lack of strong seismic activity, passive margins typically undergo considerable subsidence during their evolution. Most of them are covered by thick sedimentary sequences which makes them a prime target for hydrocarbon exploration. By contrast, *active margins* mark the juxtaposition of continental and oceanic lithosphere at convergent plate boundaries. The best development occurs along the eastern edge of the Pacific Ocean bordering the American continents. A deep trench occurs on the oceanward side of the margin and an Andean type mountain range with predominantly andesitic volcanism on the continental side. *Island arcs* are similar to active margins but the convergent plate margin at them separates two portions of oceanic lithosphere. The circum-Pacific belt of island arcs and active margins extends anticlockwise round the Pacific Ocean from Cape Horn to New Zealand and is the scene of the Earth's most vigorous tectonic and seismic activity.

Active and passive margins differ greatly in structure and tectonic activity but a genetic connection can be traced between them. Most passive margins occur along the edges of relatively young oceans which are steadily growing in areal extent by sea-floor spreading. Such oceans cannot widen indefinitely. For instance, the growth of the Atlantic and Indian Oceans must cease shortly after the continents now converging on the Pacific Ocean eventually collide. When an ocean has opened up to its maximum possible size, the passive margins around it must fracture to form new subduction zones if sea-floor spreading is to continue, as in the present Pacific Ocean. The passive margins of the earlier stage of development would thus be transformed into active margins. If further continental splitting occurs at this stage to form a new ocean within the collided continental region, then the large ocean would start to shrink as the continents started to converge onto it, as in the Pacific. Eventually the surrounding continents would be expected to collide and the margins would be caught up in collision mountain building activity. Thus the life history of a typical continental margin starts by birth at the time of continental splitting, continues into maturity as a passive margin, into old age as an active margin, and terminates by death in a collision mountain range. At the present rate of sea-floor spreading, the lifespan is of the order of 300 to 500 My.

For more detailed study, useful collections of papers covering both active and passive margins have been edited by BURK and DRAKE (1974) and by WATKINS and others (1979). The passive margins are reviewed with particular reference to DSDP results by KENT and others (1980). The active margins and island arcs are covered in a review volume edited by TALWANI and PITMAN (1977).

5.2 Passive margins

The passive margins of the Atlantic and Indian Oceans have formed by the progressive break-up of the large Permian and early Triassic land mass called Pangaea which included almost all the major continental areas. The break-up has occurred in a series of stages over the last 200 My or thereabouts. At each stage of splitting, a new complementary pair of passive margins has formed on the opposite sides of the new oceanic region. The oldest margins of the Atlantic Ocean occur on the eastern coast of North America and the western coast of North Africa and were formed in the Jurassic. The youngest ones are those of East Greenland and the Rockall-Norway region which formed during the Palaeocene about 54 My ago. The central and southern parts of the Atlantic Ocean are symmetrical about the mid-Atlantic ridge with a single complementary pair of margins on opposite sides, but the North Atlantic has a more complicated bathymetric structure because of successive margin development resulting from repeated attempts at continental splitting – first the Bay of Biscay, next Rockall Trough, then the Labrador Sea, and lastly the presently widening main branch of the north-eastern North Atlantic.

There are two main types of passive margin. *Rifted margins* are the most widespread and best-studied type and they result from continental splitting at newly formed divergent plate boundaries. *Offset* (or *sheared*) *margins* form by continental separation along transform faults, undergoing an initial stage as a conservative plate boundary. They represent an original offset in the line of continental splitting (WILSON, 1965) and can normally be traced laterally into an oceanic fracture zone marking a continuing offset of the ocean ridge crest. Examples of offset margins include the large marginal offset represented by the Falkland escarpment which forms the southern termination of the Atlantic Ocean and the alternation of short rifted and offset segments along the north coast of the Gulf of Guinea in the equatorial Atlantic.

Structure of passive margins

A typical rifted margin is morphologically subdivisible into continental shelf, continental slope and continental rise. The continental shelf forms the seaward extension of the adjacent continent out as far as the 200 m (100 fathom) depth contour. Flat-lying regions of shelf which have undergone greater subsidence are called marginal plateaus, an example being the Blake plateau off Florida. The continental slope is the relatively steep boundary between the shelf and the deep ocean. Most continental slopes are less than 50 km across, possessing a gradient steeper than 1 in 10. The slope may be cut by submarine canyons through which turbidity currents carry sediments from the shelf down to the abyssal plains. The continental rise forms the gentle slope of the ocean floor from the foot of the slope towards the flat abyssal plains. It may be several hundred kilometres wide or even absent. Offset margins typically have a steeper slope than rifted margins and the rise may not be so well developed.

The main structural features of a rifted margin are summarized in Fig. 5.1. A good example of a fairly typical rifted margin at a mature stage of development is the eastern margin of North America (Fig. 5.2) formed in the Jurassic. In contrast, the north Biscay margin (Fig. 5.3) is an example of a sediment starved margin formed in the Lower Cretaceous. The north Biscay margin

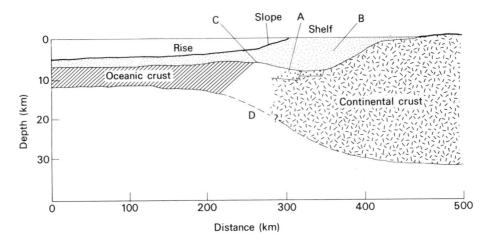

Fig. 5.1 Section across a passive margin showing some of the characteristic features of sediment and crustal structure:

A, pre-split rifting stage sediments;
B, post-split drifting stage sediments;
C, shallow continent-ocean contact masked by sediments;
D, gradational deep crustal transition.
Redrawn from BOTT (1979), *Geological and geophysical investigations of continental margins*, p. 4, American Association of Petroleum Geologists.

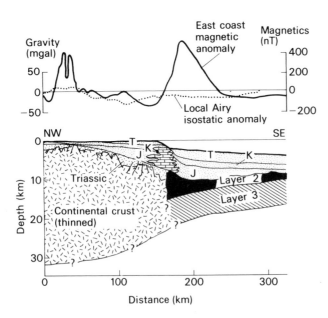

Fig. 5.2 Schematic crustal section across the eastern North American continental margin along a line stretching from Cape Cod, Massachusetts in a south-easterly direction (profile M in Fig. 2.11a), and based mainly on multichannel seismic reflection data. Magnetic and local isostatic gravity anomalies are shown. Note the reef complex beneath the slope. J, Jurassic; K, Cretaceous; T, Tertiary. Redrawn from GROW and others (1979), *Geological and geophysical investigations of continental margins*, p. 77, American Association of Petroleum Geologists.

is of great interest because deep-sea drilling (MONTADERT, ROBERTS and others, 1979) has penetrated into the Jurassic and early Cretaceous sediments pre-dating the formation of the margin and deep geophysical measurements have been made at the same locality. The sediment and crustal structure across rifted margins is discussed below using Figs 5.1 to 5.3 as examples.

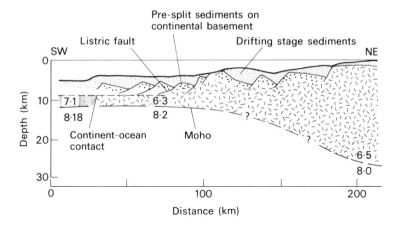

Fig. 5.3   Schematic crustal section across the north Biscay margin (profile N in Fig. 2.11a) showing thin drifting stage sediments of post Aptian age, listric faulting affecting pre-split shallow water Jurassic and lowest Cretaceous sediments overlying continental basement, the continent-ocean contact near the foot of the slope, and extreme thinning of the continental crust beneath the slope. Based on deep sea drilling, seismic reflection and refraction results (velocities shown in km s^{-1}). Modified from DE CHARPAL and others 1978), *Nature, Lond.,* **275**, 710. © 1978 Macmillan Journals Limited.

Passive margins contain some of the thickest known sequences of Mesozoic and Tertiary sediments (Fig. 5.2), locally reaching up to 15 km thickness. The sediments are divisible into two groups, an earlier sequence deposited during a *rifting stage* prior to and during continental splitting and a later one formed during a *drifting stage* subsequent to splitting. The rifting stage sediments appear to have been formed in fault-bounded troughs along the line of the initial continental split but their present great depth of burial makes detailed study difficult. Rift troughs have been identified from seismic reflection records along much of the Atlantic borderlands (BURKE, 1976) but they are not everywhere present. The details of the tectonics and sedimentation of the rifting stage are still enigmatic.

The drifting stage sediments are deposited on passive margins as they progressively sag downwards without obvious indication of faulting. The two main depositional environments are the shelf and the continental rise. These pass into each other beneath the slope where the sediment thickness is near to its maximum. Shallow water sediments are characteristic of the shelf where sedimentation is able to keep pace with subsidence. Shelf sediments may be calcareous or clastic depending on proximity to rivers, climate and other factors. The lithology of the shelf sediments may be important in moulding the slope. Calcareous sediments give rise to steep slopes without canyons, whereas clastic sediments give rise to prograding slope deposits often associated with canyons. In general, thick clastic sediments laid down in deep water dominate beneath the rise. Evaporites may occur at the base of the succession if a newly opened ocean trough had restricted circulation; such evaporites can later give rise to diapirism in the overlying sedimentary succession. If prograding occurs, the position of the slope would be expected to migrate oceanwards as the age of the margin increases. If excessive quantities of sediment are available from a large river, then a delta may form such as those near the mouths of the Niger and Mississippi. On the other hand, some margins have been starved of sediments (see for example Fig. 5.3). Starved margins provide the most suitable locations to study the origin of passive margins and the nature of the crustal transition.

It used to be assumed that the contact between oceanic and continental basement at passive margins occurs about half way down the slope. BULLARD and others (1965), for instance, used the 900 m depth contour as the mark of the contact in their continental reconstruction of the Atlantic. Local misfits in this reconstruction and subsequent geophysical indications show that this criterion is unsatisfactory. It is now known that the position of the slope is more dependent on the subsequent sedimentary history of the margin than on the original location of the split.

Two other criteria have been used successfully in some regions to locate the continent-ocean contact. The edge of the linear oceanic magnetic anomalies should mark the boundary between the oceanic layer 2 and the continental basement. This method works well in some regions, such as along the north-western margin of the Rockall microcontinent where magnetic anomaly 24 of about 54 My age borders the contact on the oceanward side (VOGT and AVERY, 1974). In many other regions, such as the central Atlantic and Rockall Trough, the method may fail because of the occurrence of marginal magnetic quiet zones along the edges of the ocean. These quiet zones probably represent oceanic crust formed during the Upper Jurassic or mid-Cretaceous constant magnetic polarity periods, but parts of them could also represent subsided continental crust. Such a quiet zone, formed during the Jurassic constant polarity period, lies off the east coast of the United States, but here the prominent east coast magnetic anomaly (Fig. 5.2) may possibly be caused by the contrast in magnetization between oceanic and continental basement rocks. Where the magnetic method fails, seismic methods probably offer the best method of locating the contact, but these are difficult to apply where sediments are thick. Where sediments are thin (Fig. 5.3), seismic reflection surveying reveals a much rougher basement on the oceanic side of the contact. Multi-channel reflection lines across some North Atlantic margins such as western Rockall reveal a zone of fan-shaped seaward-dipping reflectors beneath the flat-lying sediments (e.g. ROBERTS and others, 1979, Fig. 12). These may be sub-aerial lavas with some interbedded sediments formed at the rapidly subsiding site of continental splitting just before normal sea-floor spreading became established. Such zones may mark the location of the continent-ocean contact.

The most satisfactory locations for examination of the continent-ocean basement contact are at sediment starved margins such as the north Biscay margin (Fig. 5.3) and those bordering Rockall Trough. The contact at shallow depth appears to be a relatively sharp boundary, rather than gradational. A surprising result at such margins is that the contact occurs at the foot of the slope, or near it. Off south-eastern Greenland, where sediments are relatively thin and the continental crust has been excessively thinned, the contact occurs beneath the rise (FEATHERSTONE and others, 1977). In regions of excessive sedimentation such as the Niger delta, the contact probably occurs beneath the shelf which has been outbuilt onto oceanic crust. Beneath many of the passive margins with thick sediments, the location of the contact remains unknown. At margins where the contact has been located, major faulting is not observed to occur between the continental and oceanic basement.

The deep transition between continental and oceanic crust down to the Moho is best studied by the combined use of the seismic refraction and gravity methods. Seismic methods are not yet sufficiently refined to show exactly how oceanic and continental crust merge together at depth and the gravity method cannot resolve the detail. The normal procedure is to use the seismic refraction stations on either side of the margin to fix the crustal thickness at two or more points, and then the gravity profiles across the margin can be used to deduce the shape of the Moho. The resulting models are very blurred representations of the actual structure at depth, but two important generalizations can be made from them. Firstly, the passive margins appear to be in approximate isostatic equilibrium although not always exactly so. Secondly, the complete transition from continental to oceanic crust at Moho depth takes place over a horizontal distance of 200 km or less (Figs 5.1 to 5.3). At the starved margins (Fig. 5.3), most of the transition in crustal thickness occurs on the continental side of the boundary. It is possible that this situation applies more generally. There is some indication that the oceanic crust does also thicken slightly but significantly towards the crustal contact. Some anomalously high velocity regions have been detected deep within the continental crust at margins.

At greater depth, practically nothing is known of the upper mantle structure beneath passive margins. A thermal contrast would be expected to occur between the relatively hot oceanic lithosphere and the much cooler continental lithosphere and it is also possible that the continental lithosphere differs slightly in composition. The thermal contrast should be most strongly developed at relatively young margins such as that of south-eastern Greenland and should be less conspicuous at older margins such as that of eastern North America. These effects have usually been ignored in the gravimetric modelling of passive margin structure but this must certainly lead to serious error at the younger margins.

In general offset margins differ in structure from rifted margins in having a steep slope with much narrower transition between continental and oceanic crust (SCRUTTON, 1976). Continental basement may occur at shallow depth beneath the shelf edge and slope as a faulted ridge. Offset margins show less evidence of subsidence than rifted margins, indicating a more stable tectonic development.

Tectonics of passive margins

Passive margins of rifted type undergo a history of predominantly vertical tectonic development. The most obvious aspect is the great subsidence which allows thick piles of sediment to accumulate beneath shelf and slope. Less obviously, there is a widening of the deep crustal transition at some stage, involving thinning of the adjacent continental crust. The tectonic history of a typical rifted margin can conveniently be divided into an earlier rifting stage associated with the continental splitting event and a later drifting stage involving flexural downsagging of the margin. The lifespan of a passive margin is eventually terminated when subduction starts. The thick sediments of passive margins may still later become involved in collision mountain building. In this way, shelf sediments become synonomous with what used to be known as geosynclinal deposits.

The rifting stage of development may occur for up to about 50 My prior to the split as well as during it. The normal faulting and graben formation are indicative of crustal tension. Off eastern North America (Fig. 5.2) block faulting occurred during the Triassic with the formation of red beds. At the north Biscay margin (Fig. 5.3) listric normal faulting of step type occurred with an overall extension of about 10 %, but sediments at this stage were only weakly developed. It has been suggested that the East African rift system is an incipient continental split at present undergoing this stage of development. However, the associated crustal doming and extensive continental volcanism of East Africa are not present everywhere along the passive margins. Such intense thermal and volcanic activity is restricted to certain special areas, such as the Greenland-Scotland region of the North Atlantic margins. A more typical situation probably occurs at the north Biscay margin (Fig. 5.3), where DSDP drilling has shown that doming did not occur at the time of the split but rather there was a rapid initial subsidence of about 2 km. This probably occurred because of drastic thinning of the continental crust at the time of splitting. The rifting stage is thus still not well understood but it is nevertheless of great interest because of its relevance to the continental splitting mechanism. This aspect is examined further in Chapter 9.

The drifting stage starts when the new ocean begins to widen by sea-floor spreading. Flexural downsagging of the margin occurs without conspicuous faulting. The resulting sedimentary pile may reach up to 10 km in thickness (Fig. 5.2). These sediments unconformably overlie the faulted sediments of the rifting stage. The rate of subsidence is most rapid at the start of the drifting stage.

Thereafter the rate of subsidence decreases approximately exponentially with a time constant of about 60 My, similar to that of the cooling oceanic lithosphere. The progressive decrease in the subsidence rate is well displayed at the eastern North American margin, where the Jurassic strata are thickest and the Tertiary are thinnest (Fig. 5.2). Unravelling the cause of the drifting stage subsidence has been a prominent geodynamic problem of the last ten years, with some critical evidence coming from the DSDP drilling and associated geophysical measurements at the north Biscay margin (DE CHARPAL and others, 1978). Three main factors apparently contribute to the subsidence. These are sediment loading, thermal contraction of the cooling lithosphere, and crustal thinning.

Isostatic subsidence must occur in response to sediment loading (WALCOTT, 1972). The amount of subsidence in response to a specified load can be calculated if local Airy isostasy is assumed. Suppose that the initial water depth is d and that the densities of seawater, sediment and upper mantle are ρ_w, ρ_s and ρ_m respectively. If the sea is filled up by sediments, then their total thickness must be greater than the initial water depth because of their higher density. The thickness t is given by

$$t = d(\rho_m - \rho_w)/(\rho_m - \rho_s).$$

Putting $\rho_w = 1030 \text{ kg m}^{-3}$ and $\rho_m = 3300 \text{ kg m}^{-3}$, substitution in the formula shows that a sediment thickness of about twice the initial water depth can develop if the sediments have a mean density of 2150 kg m^{-3} and of nearly three times for 2550 kg m^{-3}. A more sophisticated approach is to treat the lithosphere as a thin elastic or visco-elastic plate and investigate its flexure in response to the sediment loading by beam theory. The results from the two approaches do not differ greatly provided that the load is wide in comparison with the thickness of the lithosphere (Fig. 5.4).

Evidently thick shelf successions of sediments cannot be formed in water depths of around 200 m by this mechanism alone, as has been emphasized by the fuller calculations of WATTS and RYAN (1976). The sediment loading effect, however, is of great importance at passive margins in two ways.

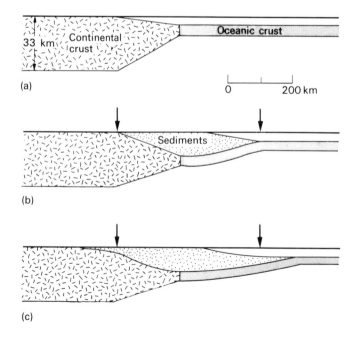

Fig. 5.4 The influence of sediment loading at a passive margin:
(a) The initial situation prior to loading, assuming a pre-existing 200 km wide transition between oceanic and normal continental crust.
(b) The result of local Airy loading by sediments, assuming sediment density of 2450 kg m^{-3} and upper mantle density of 3300 kg m^{-3}.
(c) The result of flexural loading, assuming the lithosphere to have a flexural rigidity of 2×10^{22} N m, with densities as in (b).
Based on WALCOTT (1972) and redrawn from BOTT (1979), *Geological and geophysical investigations of continental margins*, p. 4, American Association of Petroleum Geologists.

Firstly, substantial subsidence can occur where great volumes of sediment are deposited in deep water on the slope and rise to form a delta. Fig. 5.4 shows that up to about 14 km of sediment might be deposited on the rise if a sufficient volume of sediment is available. Such is the situation at the mouth of a large river such as the Niger. Secondly, if subsidence is caused by some other mechanism such as thermal contraction or crustal thinning (see below), then sediment loading up to sea level has the important result of increasing the total amount of subsidence by a factor of between two and three depending on the average sediment density.

The basic idea of thermal subsidence of a passive margin is straightforward (Fig. 5.5). The continental lithosphere near the embryo margin is assumed to be heated and thinned at the time of continental splitting. Thermal expansion reduces the density of the uppermost mantle and crust, causing isostatic uplift similar to that of East Africa. As the new ocean starts to widen, the continental lithosphere at the margin will recover towards its initial elevation with a time constant of around 60 My, this also being about the time constant of cooling of the adjacent oceanic lithosphere. If the crust has been thinned by surficial erosion or by some other process at the time of splitting, then the recovery on cooling will cause subsidence of the shelf below sea level. This mechanism has been quantitatively developed by SLEEP (1971, 1973) assuming that surface erosion thins the crust from above. A smooth exponential decay of subsidence rate with time is predicted and this agrees closely with observations.

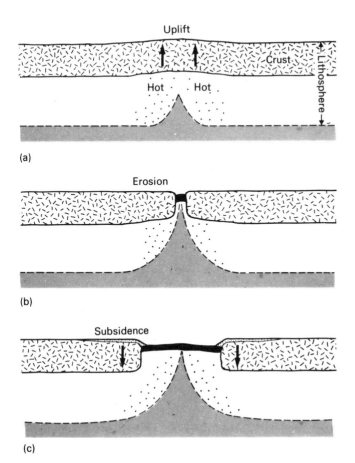

(a)

(b)

(c)

Fig. 5.5 The thermal hypothesis of SLEEP (1971). (a) Uplift following heating and thinning of lithosphere. (b) Initiation of new ocean accompanied by erosion of continental uplifted regions, causing crustal thinning. (c) Subsidence of continental margins as underlying lithosphere cools. Note sharp transition in crustal thickness. Redrawn from BOTT (1979), *Geological and geophysical investigations of continental margins*, p. 5, American Association of Petroleum Geologists.

Sleep's thermal hypothesis based on surficial erosion cannot be the sole cause of regional subsidence at passive margins for the following reasons. The amount of crustal thinning h caused by erosion depends on the initial uplift and the time constants of lithospheric cooling and erosion. Taking a maximum initial uplift of 2 km and an erosional time constant of 50 My, then h is about 4 km. The maximum possible sediment thickness to sea level, including the sediment loading effect, that this will allow is $h(\rho_m - \rho_c)/(\rho_m - \rho_s)$ which must be less than h and is typically about $\frac{1}{2}h$. Thus sediment thicknesses of about 2 km on the shelf can be explained, but unacceptably large amounts of supracrustal erosion would be required to account for sediment thicknesses of 4 to 10 km, such as are observed. Furthermore, according to Sleep a gap of about 50 My must intervene between the onset of spreading and the first marine sediments to allow the erosion to occur, and no such gap exists at margins which have been drilled such as that of North Biscay. Various modifications of the thermal hypothesis based on metamorphism and igneous intrusion affecting the lower crust have been proposed (e.g. FALVEY, 1974), but these also meet difficulty in accounting for large sediment thicknesses.

Despite the reservations stated above, thermal subsidence of passive margins elegantly explains the post-split downwarping provided that substantial crustal thinning can be produced by processes other than surficial erosion. For instance, at the north Biscay margin, where sediment loading is a negligible factor, subsidence of about 2 km that occurred after the formation of the margin is only reasonably explained by cooling of the lithosphere. The problem which is now taken up is to understand how the crustal thinning can occur.

The third main factor which contributes to the subsidence is thinning of the continental crust. Some particularly important evidence on the nature and timing of the crustal thinning has come from the DSDP drilling and associated geophysical investigations of the north Biscay margin (DE CHARPAL and others, 1978). Here, the continental crust thins beneath the slope and is about similar in thickness to the oceanic crust at the crustal contact at the foot of the slope (Fig. 5.3). The DSDP drilling showed that the sediments deposited just after the split were laid down in 2 km depth of water. Rapid crustal thinning at this time must have been the cause of this abrupt subsidence. The overall vertical movement can be understood best as the combination of two opposite effects, thermal uplift of around 3 km from heating and thinning of the continental lithosphere and subsidence of around 5 km resulting from the drastic crustal thinning. The net result at the north Biscay margin is a subsidence of about 2 km.

The mechanism by which the crustal thinning takes place is controversial. Three possibilities have been suggested. One suggestion is that the phase transition of gabbro to eclogite in the lower crust causes the Moho to migrate upwards, but this has not received much recent support because it is unlikely that the Moho is a phase transition. A second possibility is that extreme thinning of the continental crust when subjected to tension can occur by plastic necking (KINSMAN, 1975). The problem here is to understand how the brittle upper part of the crust can be stretched by a factor of two or more in a situation such as at the North Biscay margin where the estimated stretching at the top of the crust is about 10%. A third suggestion is that hot and ductile lower continental crustal material flows out into the newly forming oceanic lithosphere shortly after the time of splitting (BOTT, 1971c). The evidence decisively shows that thinning of the continental crust does occur at rifted margins but the cause of this thinning is still uncertain.

Sediment loading, thermal contraction of the cooling lithosphere and crustal thinning can thus together explain the subsidence characteristic of the drifting stage at passive margins. Although we should be wary in treating the north Biscay margin as typical, the evidence from there suggests that rapid subsidence of about 2 km may occur as a result of crustal thinning at the time of the split outweighing the thermal uplift. Later on, a further 3 km of subsidence at the oceanic contact, decreasing into the continent, would be expected to occur as the lithosphere cools. Thus about 5 km of subsidence can be caused by the combined effects of crustal thinning and cooling without sediment loading at starved margins. It is easy to see that over 10 km of subsidence can occur if sediments are loaded up to sea level. A model of the drifting stage evolution based on these three subsidence factors is shown in Fig. 5.6.

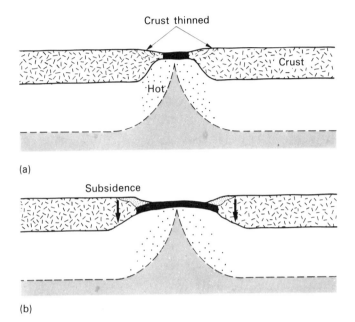

(a)

(b)

Fig. **5.6** The thermal hypothesis with major crustal thinning caused by processes other than uplift and erosion. (**a**) Crustal thinning and faulting at the time of continental splitting. (**b**) Subsidence of continental margins as underlying lithosphere cools.

Offset margins mark the juxtaposition of continental and oceanic crust resulting from transform faulting. They undergo an initial active stage as a transform fault, at first with continental crust on both sides but later bringing oceanic crust on one side against continental on the other. A point on such a margin becomes passive once the spreading centre has passed by. In general, offset margins show much less subsidence than rifted margins. This may be caused by absence of strong heating of the lithosphere during the early stages of evolution.

Sedimentary basins of the continental interior

Subsidence features akin to those at passive margins occur within the interior of the continents. The two main types of continental interior basin closely resemble the subsidence features formed at rifted margins during their two main stages of evolution. Narrow trough-like basins appear to form as a result of faulting in the underlying basement. Examples include the Carboniferous and Permo-Triassic basins of Great Britain and the Basin and Range grabens of the western United States. Such basins probably form by the wedge subsidence mechanism in response to crustal tension as described in Chapter 2 with reference to rift valleys. The other main type of quite distinct origin comprises the more nearly circular-shaped basins of wide areal extent formed by downsagging of the continental crust without conspicuous faulting, such as the Palaeozoic Michigan basin and the Tertiary North Sea basin.

Sedimentary basins of the continental interior, such as the Michigan and North Sea basins, are known to be in approximate isostatic equilibrium. The same regions were probably also in equilibrium prior to basin formation, so that a load must have been added to the continental lithosphere to cause the subsidence. The crust beneath the North Sea is known to thin by about the amount required to account for the load. Some recent ideas which have been put forward as possible explanations of interior basin subsidence are as follows. The subsidence of the Michigan basin was attributed by SLEEP and SNELL (1976) to thermal contraction of the cooling continental lithosphere following a thermal event which caused doming and crustal thinning by surficial erosion. HAXBY and others (1976) proposed an alternative thermal mechanism for the Michigan basin in which heating of the lower crust caused metastable gabbroic rocks to revert to the stable form eclogite, thus increasing the density of the lower crust. Subsidence would then occur as the lithosphere cooled back to normal without need for surficial erosion to thin the crust. According to McKENZIE (1978), subsidence such as that of the Michigan and North Sea basins can be explained by an initial sudden stretching of the continental lithosphere by a factor of two or thereabouts, causing thinning of the crust and upwelling of hot asthenospheric material. This would give rise to a rapid initial subsidence caused by the crustal thinning followed by a slower exponential subsidence as the lower part of the lithosphere cooled back to normal. All these hypotheses appeal to a thermal event but differ in mechanism of loading the lithosphere. They still remain controversial and each of them is confronted by some difficulties. For instance, some geologists are reluctant to accept that the Mesozoic sediments of the North Sea region have been stretched by a factor of 1·5.

Microcontinents

Certain isolated fragments of continental crust occur as shallow rises within the ocean basins. These are known as microcontinents. They need to be distinguished from other shallow regions such as the Iceland-Faeroe ridge which are underlain by anomalously thick oceanic crust. The best known example of a microcontinent in the Atlantic Ocean is the Rockall-Faeroe plateau (SCRUTTON, 1972; ROBERTS, 1975). This is separated from the British continental shelf by Rockall Trough. The continental reconstruction of BULLARD and others (1965) first suggested that this region might be continental as otherwise an unacceptable gap was left in the reconstruction of the pre-Atlantic continental mass. Subsequently SCRUTTON (1972) showed that the crust is about 30 km thick and of apparently continental type. Geological sampling has since conclusively demonstrated the continental nature by the discovery of Precambrian rocks. Another well known example of a proved microcontinent is the Seychelles Bank in the Indian Ocean, which is formed of Precambrian granite and underlain by crust about 30 km thick.

Microcontinents occur within the Atlantic and Indian Oceans as a result of one or more major jumps in the location of the spreading axis during the opening up of the ocean (Fig. 5.7). They are

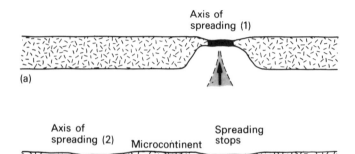

Fig. 5.7 The formation of a microcontinent. (a) Narrow ocean forms as a result of continental splitting and sea-floor spreading. (b) The spreading axis jumps causing a second continental split which isolates the microcontinent.

bordered on all sides by passive margins of rifted or offset type. The evolution of the Rockall-Faeroe microcontinent provides a good example of how this can come about. The initial break between North Europe and Greenland occurred along the Rockall Trough, the eventual microcontinent still being attached to Greenland at this stage. Active spreading in the Rockall Trough probably only occurred during the mid-Cretaceous, after which the spreading axis migrated to form the Labrador Sea. At this second stage, the southern part of the Rockall-Faeroe microcontinent split off Canada by rifting and transform faulting, producing a southern margin of the future microcontinent with short rifted and offset segments. At the third stage, the spreading axis again migrated to split off the microcontinent from Greenland, leaving it as an isolated continental mass surrounded by oceanic regions. The south-eastern and north-western margins of the microcontinent are of rifted type formed respectively about 100 and 54 My ago whereas the southern margin of mixed type formed about 85 My ago. The subsequent geological history of Rockall-Faeroe microcontinent has been dominated by subsidence of its passive margins.

5.3 Active continental margins and island arcs

Most of the present day active continental margins and island arcs occur around the Pacific Ocean along its eastern, northern and western boundaries. This belt is continuous except for a short section where the East Pacific rise intersects the North American continent, and the tectonic boundary is formed by the San Andreas transform fault. Some of the Earth's most vigorous tectonic and volcanic activity takes place within this circum-Pacific belt and about 85 % of the global release of earthquake strain energy occurs here. Island arcs are much less prominent in the other oceans. These include the Caribbean and Scotia arcs in the Atlantic, the Sunda and Banda arcs in the Indian Ocean and the Aegean arc in the Mediterranean.

A deep bathymetric trench occurs on the seaward side of active continental margins and island arcs, except where masked by thick sediments as opposite the Lesser Antilles (Caribbean) arc and off Alaska. Trenches are the largest linear subsidence features affecting the Earth's surface. They are remarkable for their length and continuity. The Peru-Chile trench is 4500 km long without serious interruption and the Tonga trench is continuous and straight at a depth of 9 km for a length of 700 km. They reach depths of 2 to 4 km beneath the adjacent ocean floor. The ocean's greatest depths of 10 to 11 km are found in the trenches of the deep west Pacific. The average width is less than 100 km and in cross section they have an asymmetrical V-shape with the steeper slope (about 8–20°) on the landwardside. The apex of the V may be truncated by a flat bottom a few kilometres across formed by infilling turbiditic sediments.

Each trench is bordered on its landward side either by an active continental margin or by an island arc. At active margins, the trench is bordered by the continental mainland. The continental shelf is narrow or absent and an Andean type mountain range rises straight from the coast. The mountains are penetrated by a line of andesitic volcanoes lying about 150 km from the trench axis. Island arcs differ from active margins by being separated from the mainland by a marginal sea. Sumatra and Java are separated from the mainland by shelf seas, but more typically marginal basins such as the Sea of Japan are underlain by oceanic crust. Island arcs are arcuate features which are convex towards the oceanic side. They may consist of fragments of continental crust such as Java which are penetrated by volcanoes, or they may be chains of volcanic islands such as the Lesser Antilles which are built up on top of pre-existing oceanic crust. A tectonic ridge or outer sedimentary arc may occur between a volcanic arc and trench. On the oceanward side of trenches, a broad uparching of the ocean floor of a few hundred metres height is commonly observed. This is called the outer rise.

The deep oceanic trenches are associated with some of the largest known negative isostatic gravity anomalies, first discovered in the East Indian region by the pioneering submarine gravity measurements of Vening Meinesz. The majority of the abundant earthquakes occur as shallow focus events on the landward side of trenches. However, the circum-Pacific belt and other arc-trench systems are characterized by a belt of earthquakes dipping away from the ocean at around 45° on average which extends down to several hundred kilometres depth. These belts, known as Benioff zones, intersect the surface near to, but landward of, the axis of the trench.

Both active continental margins and island arcs are tectonic features produced at convergent plate boundaries where oceanic lithosphere is being recycled into the mantle. The trench axis normally marks the surface position of the plate boundary and the Benioff zone maps the downward path of the sinking tongue of lithosphere. Thus in the circum-Pacific belt the trenches mark the junction between the outward spreading Pacific lithosphere and inward migrating continental regions which are progressively overriding the Pacific and reducing its areal extent. The relative motion of plates converging at a given trench can be computed from global analysis of plate motions (p. 134), this being estimated as about 90 mm y^{-1} across the American active margins and also about 90 mm y^{-1} across the western Pacific island arcs.

Structure of subduction zones from seismology

It has been known for some time that earthquake foci tend to cluster on a plane which reaches the surface beneath the trenches or their continentward flanks and dips on average at about 45° away from the main ocean (GUTENBERG and RICHTER, 1954; BENIOFF, 1955). The earthquakes of Benioff zones extend downwards for several hundred kilometres in most regions, the deepest detected ones being just over 700 km deep in the Tonga region. An important discovery of the late 1960s was that the deep earthquake belts map out the upper part of the sinking lithosphere being recycled into the mantle. The structure of the subducting lithosphere is discussed here as an essential preliminary to understanding the associated shallower crustal features.

The critical evidence which led to the identification of the Benioff zones with sinking tongues of oceanic lithosphere came from seismological studies made in the Tonga region of the south-western Pacific. The first step was a detailed study of the pattern of earthquake foci in the Tonga-Kermadec region made by SYKES (1966) using the greatly improved World-Wide Standardized Seismograph Network (WWSSN) which had been established by that time. His results confirmed the existence of the steeply dipping belt of earthquake foci (Fig. 5.8) and showed that the belt is less than 100 km wide and in some places probably not more than 20 km wide. He also showed the close geometrical relationship between the deep earthquake belt and the surface expressions of trench and arc, both belts bending westwards together at the north end of the Tonga trench (Fig. 5.8). The other important evidence came from the study of body-wave amplitudes (especially S) from deep events beneath the Tonga region reaching the surface by a series of different paths (OLIVER and ISACKS, 1967). The waves reaching a station on Tonga traverse the length of the earthquake belt and these show much larger amplitudes and higher frequency content than waves which reach stations on Fiji to the west or Rarotonga to the east (Fig. 5.9). The waves reaching Tonga also travel with higher average velocity. These observations suggested to Oliver and Isacks that the deep earthquake zone is also a belt of high Q of about 1000, in contrast to the more normal Q of 150 averaged over the paths to Fiji and Rarotonga. The seismic zone appears to lie near the upper surface of the high Q belt which is about 100 km or less in thickness. They interpreted the high Q belt as the Pacific lithosphere which is disappearing into the mantle as a result of sea-floor spreading and which is considerably cooler than the adjacent asthenosphere through which it sinks.

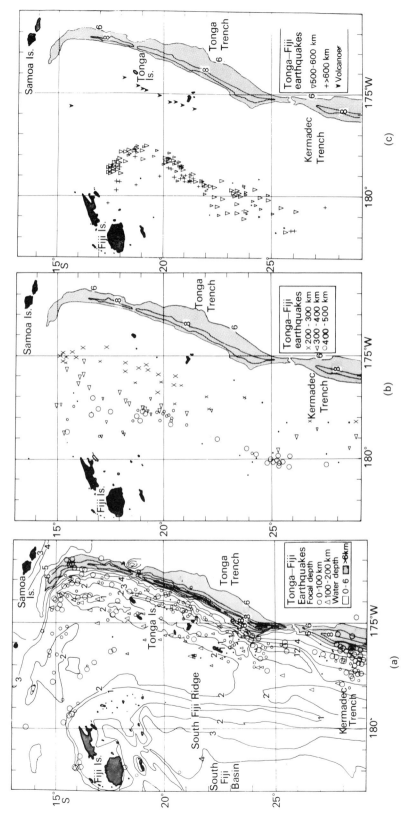

Fig. 5.8 Tonga–Fiji earthquake epicentres: (**a**) shallow, showing water depth in km; (**b**) intermediate; and (**c**) deep. Redrawn from SYKES (1966), *J. geophys. Res.*, **71**, 2984–2986.

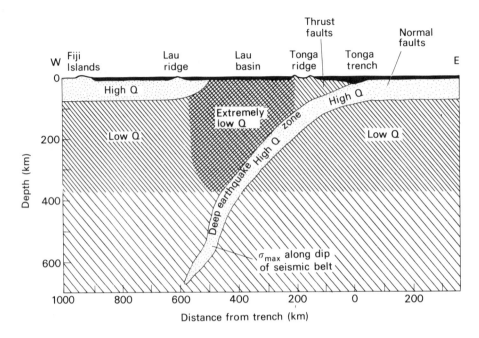

Fig. 5.9 Schematic section perpendicularly crossing the Tonga trench and arc, with marginal basin to the west, showing the high and low seismic attenuation (Q) regions and the high Q tongue of sinking lithosphere marked by the deep earthquake zone. Note the extremely low Q region beneath the Lau marginal basin. The prevalent types of earthquake mechanisms are shown. Rarotonga lies off the diagram about 1400 km east of the Tonga trench. Adapted from BARAZANGI and ISACKS (1971), *J. geophys. Res.*, **76**, 8511.

Subsequently a more detailed study of the pattern of seismic wave attenuation above the Tonga inclined earthquake zone made by BARAZANGI and ISACKS (1971) located a zone of extremely low Q of about 50 for P and less than 20 for S lying in the mantle below the marginal Lau basin to the west of the Tonga volcanic ridge (Fig. 5.9). This study made use of local earthquakes. A later investigation using teleseismic P arrivals originating from various deep earthquake zones showed that similar high attenuation occurs above most of them (BARAZANGI and others, 1975). The overall situation, which is probably generally applicable, is shown in Fig. 5.9.

The work of Oliver, Isacks and Sykes showed that the intermediate and deep focus earthquakes occur within the high Q slab interpreted as the sinking oceanic lithosphere, and that they occur nearer to the upper than the lower surface of the slab. It has been generally assumed that this situation applies elsewhere so that the deep earthquake zones can be used to map the shape of the subducting tongues of lithosphere. An important recent study by HASEGAWA and others (1978) has revealed the relationship between sinking lithosphere and the deep earthquake belt of the northern Honshu region in Japan in even greater detail. The focal depth distribution of small earthquakes here, projected onto a vertical east-west plane, shows that intermediate and deep foci are concentrated around two dipping planes (Fig. 5.10). The lower plane lies about 30 to 40 km below the upper one. How are these two planes related to the upper surface of the sinking slab? Hasegawa and his co-workers used the ScS phase from a nearby large earthquake to locate this surface. This S phase is reflected from the core-mantle boundary at nearly vertical incidence. As this ray passed upwards through the sinking lithospheric slab, they found that part of the energy was converted from S to P on passing through the relatively sharp boundary formed by the upper surface of the

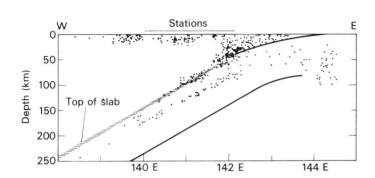

Fig. 5.10 Section showing the focal depth distribution of small earthquakes along an east-west line across the north-eastern part of the Japan arc, showing the double-planed deep seismic zone detected by the Tohoku University seismic network. The hatched zone shows the position of the boundary between the descending slab and the overlying mantle as inferred from ScS and ScSp arrivals, assuming that the descending slab has seismic velocities 6% above those of the adjacent mantle. Redrawn from HASEGAWA and others (1978), Geophys. J. R. astr. Soc., **54**, 288.

slab. The difference in arrival time at the surface between the main *ScS* arrival and the earlier *ScSp* converted arrival gave them an accurate estimate of the distance along the ray path from the receiving station down to the slab. Ray tracing gave the direction, so that positions on the upper surface of the sinking slab could be located accurately (Fig. 5.10). The figure shows that the narrow upper zone of earthquakes lies astride the upper surface of the slab and the lower zone lies about 30 to 40 km below it. Two other groups of earthquakes are shown on Fig. 5.10, notably the shallow events at the east of the section which may be related to the downbending of the lithosphere near the trench, and the shallow events on the overriding plate to the west where most of the energy is released.

A further important seismological tool for studying convergent plate margins and the sinking lithosphere is the determination of focal mechanism (p. 320). This yields the directions of maximum compression and extension resulting from an earthquake and thus the type of faulting which occurs. Reliable studies became possible with the establishment of the World-Wide Standardized Seismograph Network. Such studies (e.g. STAUDER, 1968; ISACKS and others, 1969) showed that the shallow focus events on the landward side of the trenches are characteristically caused by thrust faulting with the axis of horizontal compression orientated perpendicular to the trench axis. In contrast, normal faulting predominates beneath the axis of the trenches and on the oceanward side of them. This is consistent with underthrusting on the landward side and extension of the upper lithosphere by downward bending on the oceanward side.

A different situation occurs within the deep earthquake belts. The principal strain directions are not horizontal and vertical as in the shallow events but are aligned parallel and perpendicular to the dip of the earthquake belt. In the uncontorted parts of deep earthquake belts, ISACKS and MOLNAR (1969) found that either the maximum or the minimum compression is orientated parallel to the dip of the belt whereas the intermediate direction is horizontal and parallel to the strike of the belt. This suggested to them that the intermediate and deep focus earthquakes are predominantly caused by the release of stress within the sinking plate of lithosphere rather than by shearing at its upper boundary. To explain this, the sinking slab must be much stronger than the adjacent part of the upper mantle because of its lower temperature. Within the double-planed deep earthquake zone beneath north-eastern Japan (Fig. 5.10), the earthquakes of the upper belt show downdip compression or reverse faulting and those of the lower belt show predominantly downdip extension.

ISACKS and MOLNAR (1969) have suggested that the downward gravitational pull of the cool and consequently dense sinking plate would produce a tension orientated parallel to the dip of the plate, provided that the downward movement was unimpeded. On the other hand, if the plate abutted against the resistant lower boundary of the asthenosphere, then at least the lower part would be thrown into compression parallel to the dip. These two possible situations and variants on them are shown diagrammatically in Fig. 5.11. Isacks and Molnar have been able to find examples of the predicted situations; they found that downdip extension predominates, with gaps in the belt in some regions, in the Middle Americas, New Hebridean and Chilean parts of the circum-Pacific belt. In contrast, the Tonga and north Honshu parts of the belt are continuous in depth and indicate compression in the dip direction.

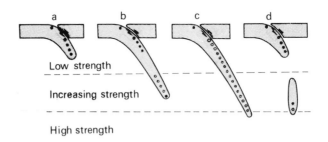

Fig. 5.11   Models to illustrate four possible types of stress distribution associated with sinking slabs of lithosphere. Open circles represent downdip compression and solid circles represent downdip extension.
(a) Slab of lithophere sinking into the asthenosphere without encountering resistance, producing extension at all depths.
(b) Slab starts to meet resistance to sinking near the bottom of the asthenosphere, producing extension near the top and compression near the base.
(c) Slab meets strong resistance to sinking at the base of the asthenosphere, producing compression at all depths.
(d) Bottom part of slab breaks off.
Horizontal extension affects the upper part of the slab, in all four models, where the slab bends downwards, and horizontal compression affects it where the downgoing slab underthrusts the adjacent plate of lithosphere. Redrawn from ISACKS and MOLNAR (1969), *Nature, Lond., 223*, 1123.

Although this explanation of the deep earthquake mechanism has been fairly widely accepted, not all observations can be explained by the simple model of axial compression or extension. Another possible mechanism is the elastic unbending of the sinking slab (ISACKS and BARAZANGI, 1977). If the slab suffers plastic deformation as it bends downwards near the trench, then elastic strains must occur as it is straightened out again. This involves downdip compression near the upper surface and downdip extension near the lower surface of the elastic slab. It is possible that the double earthquake belt beneath North-eastern Japan (Fig. 5.10) is caused by such unbending of the slab, although an alternative explanation is that the upper belt represents shear motion and the lower belt is caused by downdip tension related to the downpull of the slab. Another possible cause of the earthquakes is the occurrence of phase transitions within the sinking slab (WOODWARD, 1977). In addition to producing additional negative buoyancy, the phase transitions change the elastic moduli causing compression near the edges of the slab and tension in the middle.

In summary, the seismological study of Benioff zones has led to the recognition that these mark downsinking tongues of cool oceanic lithosphere characterized by high Q. Earthquake foci can now be used to map the downsinking slabs with some accuracy. Focal mechanism studies show that the majority of the earthquakes are probably caused by the downward pull of the dense slab and the resistance to its motion when it penetrates below the asthenosphere. Some earthquakes may also result from elastic unbending of the sinking slab and from passage through phase transitions. One unsolved problem is to understand how shear fracture can occur at such high pressure, where the types of fracture mechanism applicable nearer the surface may not apply. Despite such outstanding problems, seismological studies since the establishment of much improved world-wide and local networks in the 1960s have made a very important contribution to understanding of the deep structure at convergent plate boundaries.

Thermal structure of subduction zones

The high Q and high density of the subducting oceanic lithosphere and the possibility of sudden fracture occurring within it are all probably consequences of its relatively low temperature in relation to the adjacent mantle at the same depth. Evidently the cool oceanic lithosphere sinks to considerable depths before it has had time to warm up to the ambient temperature of the mantle. It is thus the anomalous thermal state of the sinking lithosphere which controls its mechanical properties and causes its recycling. The sinking slab, however, must progressively heat up as a result of thermal conduction from the adjacent mantle and frictional heating on the slip zone at its upper surface. There will also be some adiabatic rise in temperature as the pressure increases and as the slab passes down through regions of phase transition.

The temperature distribution within a model of the sinking lithosphere can be calculated from the equations of heat conduction provided that thermal properties and boundary conditions can be specified. It is also necessary to know the rate of subduction as the depth penetration of the cold tongue is roughly proportional to this. Some of the earlier attempts produced widely different results depending on different assumptions. For instance, much lower temperatures occur within the slab modelled by TURCOTTE and SCHUBERT (1973) than in those of TOKSOZ and others (1971), probably mainly because of the thicker lithosphere assumed in the Turcotte-Schubert model. All the models, however, show the same general structure with the low temperatures within the slab persisting to considerable depths. An example of a thermal model based on a relatively thick lithosphere is shown in Fig. 5.12. This takes into account frictional heating at the slip zone and latent heat at the two main phase transitions, the 'olivine-spinel' transformation being exothermic and the 'spinel-oxides' being assumed endothermic. In this model, the temperature of the slab is on average about 450 K below the adjacent mantle down to 700 km depth. Taking the volume coefficient of thermal expansion to be $3 \times 10^{-5} \, K^{-1}$ and the mantle density to be $3300 \, kg \, m^{-3}$, the slab is on average about $45 \, kg \, m^{-3}$ denser than the mantle at the same depth as a result of its thermal contraction.

A series of important phase transitions occur at anomalous depths within the sinking slab as a result of the changing temperature and pressure. The oceanic crust would be expected to transform to high density eclogite, with release of water, at a depth of less than 100 km. This would increase the negative buoyancy of the slab. A minor increase in density would also result from the transformation of plagioclase or pyroxene pyrolite (or lherzolite) to garnet pyrolite. Deeper down, the low temperature of the sinking slab allows the exothermic 'olivine-spinel' transition to occur at shallower than normal depth, producing additional negative buoyancy. The effect of the deeper transitions is unclear as it is not known whether these are endothermic (as assumed in Fig. 5.12) or exothermic. If endothermic, they would have the effect of opposing the sinking.

A further thermal consequence of the subducting lithosphere is the development of a hot (very low Q) upper mantle above it and the formation of andesitic and basaltic magma. The mechanism by which the magma which feeds the volcanoes of the overriding plate is formed is not yet well understood. It used to be assumed that the magma was produced as a result of shear heating within the slip zone raising the temperatures sufficiently high to cause partial fusion of the subducting

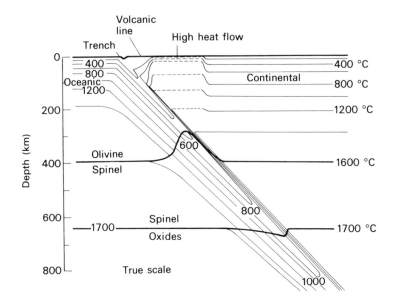

Fig. 5.12 Thermal structure of the descending lithosphere, including effects of frictional heating on the slip zone and latent heat of transitions. It is assumed that the oceanic plate approaches the trench at a velocity of $80 \, mm \, y^{-1}$. Note the relatively large uparching of the exothermic olivine-spinel transition and the relatively small depression of the spinel-oxides transition which is assumed to be endothermic. Redrawn from SCHUBERT and others (1975), *Geophys. J. R. astr. Soc.*, **42**, 728. For detailed assumptions see this paper.

oceanic crust. This interpretation has been placed in serious doubt by YUEN and others (1978) on the basis that the viscosity of the mantle is strongly temperature dependent, as seems likely (p. 298). Under such conditions shear heating is effective at low temperature when the viscosity is high, but not at high temperature when the viscosity is much lower. According to their calculations, shear heating cannot cause shear melting near the slip zone. RINGWOOD (1977) has therefore suggested that the magma is formed by release of water from oceanic crustal amphibolite transforming to eclogite at 80 to 100 km depth and from relict masses of serpentinite reverting to olivine rock at depths greater than 100 km. According to Ringwood, the water from the amphibolite rises above the Benioff zone to lower the melting temperature and produce tholeiitic basalt magma by partial fusion. This fractionates to produce andesitic magma at shallower depth (Fig. 5.13). The water released at greater depths causes partial fusion of the quartz eclogite (high-pressure oceanic crust) which indirectly produces andesitic to rhyolitic magmas.

Crustal structure

As noted earlier, some of the Earth's largest negative isostatic gravity anomalies occur over the ocean trenches, indicating strong deviation from local isostatic equilibrium. The negative anomalies were originally attributed by Vening Meinesz to a downbuckle of the crust, known as a tectogene, which is held down against gravity either by compression of the crust or by drag of converging convection currents. More recently, the combined use of gravity and seismic refraction surveys has given a much more reliable idea of the crustal structure across active margins and island arcs than is possible using gravity alone. Seismic refraction lines are used to give control points of crustal structure and gravity enables the modelling to be completed. The isostatic

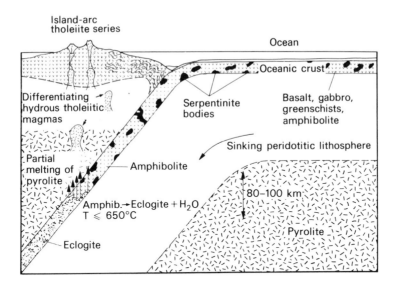

Fig. 5.13 Production of island arc tholeiitic series magmas by dehydration of amphibolite in subducted oceanic crust and subsequent partial melting of overlying mantle. Dehydration of the serpentinite bodies with subsequent partial melting of the quartz eclogite may form the calc-alkaline magmas at greater depth and at a later stage of development. Based on NICHOLS and RINGWOOD (1973) and redrawn from RINGWOOD (1977), *Island arcs deep sea trenches and back-arc basins*, p. 321, American Geophysical Union.

anomalies are now related to the influence of the downsinking slab and the downbending of the oceanic lithosphere rather than to the older tectogene hypothesis.

The western margin of South America provides a relatively simple example of the structure across an active continental margin. The Peru-Chile trench borders this margin on the seaward side and the Andean mountain range on the continental side. It has relatively uniform geophysical characteristics along most of its length although the trench is subdivisible into provinces depending on topography and gravity anomalies (HAYES, 1966). The main trench province, extending between about 8° and 32°S, has a narrow V-shaped topography reflecting absence of significant sediment thickness. A prominent belt of low free air gravity anomalies follows the axis of the trench along its length, the minimum free air anomaly reaching about −200 mgal in typical profiles across the main trench province.

The crustal structure across the active margin of South America at about 23°S, as interpreted by GROW and BOWIN (1975), is shown in Fig. 5.14. This is based on a gravity profile, partly after HAYES (1966), and seismic refraction and surface wave control of crustal structure. The refraction lines indicate that the oceanic crustal structure is uniform to the west of the trench and beneath it. To the east of the trench, the crust thickens to nearly 70 km beneath the Andes. If the upper mantle density is assumed to be uniform across the whole structure as in the earlier interpretation by Hayes, then a good fit to the observed gravity can only be obtained if the oceanic crust is considerably thinned beneath the trench, conflicting with the seismic refraction evidence. Grow and Bowin were able to show that the gravity profile could be interpreted without disagreement with the seismic evidence of uniform oceanic crustal thickness provided that the relatively high density of the subducting oceanic lithosphere is taken into account (Fig. 5.14). This includes the presence of high density eclogite formed from the subducting oceanic crust below about 30 km depth.

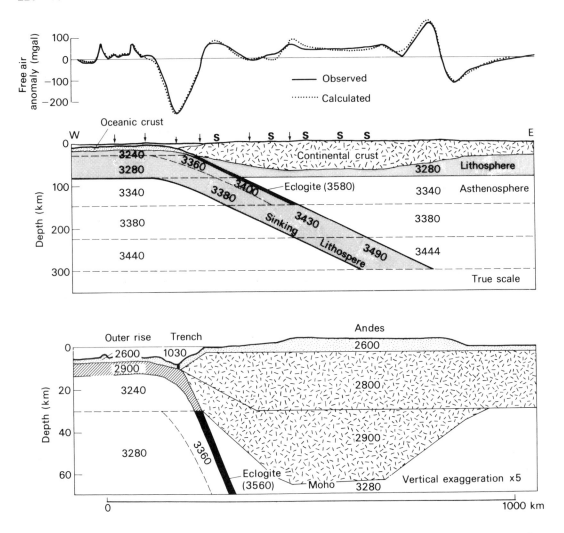

Fig. 5.14 Gravity model of the crustal and upper mantle structure across the Chile trench and the Andes near 23° S, based on a peridotitic mantle, and oceanic crust which transforms to eclogite at 30 km depth. The upper section is at natural scale and shows the whole gravity model (densities in kg m^{-3}). The lower section shows the crustal structure and the top of the subducting slab at a vertical exaggeration of ×5. The gravity model is controlled by crustal refraction lines at locations marked by arrows at the surface and surface wave determinations marked by S. Modified from GROW and BOWIN (1975), *J. geophys. Res.*, **80**, 1454.

The boundary between the Nazca and South American plates occurs along the axis of the Peru-Chile trench. To the west of the plate boundary, the oceanic crust and lithosphere of the Nazca plate has almost uniform structure apart from local seamounts. The main structural feature west of the trench is the slight shallowing of the seafloor to form the outer rise. The outer rise is formed by a slight uparching of the whole oceanic lithosphere bordering the trench and consequently it displays a characteristic positive free air gravity anomaly. The outer rise is not as well developed here as it generally is in the western Pacific. The trench itself is about 8 km deep and corresponds to a free air gravity minimum of −270 mgal. If allowance is made for the positive gravity anomaly caused by the downsinking slab, then the gravity anomaly profile across the outer rise and trench can be explained by the visible topography of the seabed which reflects the deformation of the oceanic lithosphere as it approaches the subduction zone. There is no indication of a tectogene as envisaged by Vening Meinesz. On the continental side of the plate boundary, the main feature is the abrupt thickening of the crust to about 65 or 70 km beneath the Andes. The gravity anomaly here is accounted for by the combined effect of surface topography, crustal root and dense subducting lithosphere. The thick crust is probably mainly caused by the addition of andesitic magma from the mantle below rather than by crustal shortening.

Figure 5.15 shows a generalized model of the shallow structure of an overriding plate in the forearc region between trench axis and the line of andesitic volcanoes, as revealed by seismic reflection surveys. This model is applicable both to active continental margins and to island arcs. The most characteristic feature is the subduction complex which underlies the inner (landward) trench slope. This consists of a series of landward dipping thrust slices which may consist of sediment scraped off the subducting oceanic crust or derived from the overriding plate and may contain slivers of remnant oceanic crust or continental basement. The thrust planes characteristically dip at about 20° at the surface and apparently flatten off at depth to merge into the plane of decollement formed by the contact between the plates. The subduction complex forms in response to the stresses caused by the relative motion between overriding and underthrusting plates. Several types of sedimentary basin may form within this environment (SEELY, 1979). The basin shown in Fig. 5.15 overlies a remnant of oceanic crust and is known as a residual forearc basin. A structural high occurs near the junction between the subduction complex and the basin and this may form a bathymetric ridge. The structure of the overriding plate in the forearc region varies greatly from region to region, and within each region it evolves with time. A full discussion is outside the scope of the book.

Island arc-trench systems differ from simple active continental margins in that the trench is bordered on its landward side by a normally arcuate island or chain of islands rather than by the

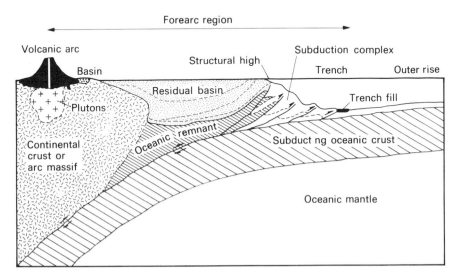

Fig. 5.15   Generalized forearc model. Note that the features shown are transient in space and time and may become superposed. The model is applicable to active margins and island arcs. Redrawn with some simplification from SEELY (1979), *Geological and geophysical investigations of continental margins*, p. 246, American Association of Petroleum Geologists.

mainland. The islands of the volcanic arc are formed from dominantly andesitic volcanoes. A submarine ridge of tectonic origin may occur between the volcanic arc and the trench (Fig. 5.15) and this may give rise to an outer chain of islands such as the Mentawai Islands off Sumatra. The island arc is normally separated from the continental mainland by a marginal basin as described in the last part of the chapter. Some island arcs, such as Sumatra and Java, may be underlain by continental crust and the structure of these may closely resemble that of active continental margins. Others, such as the Lesser Antilles and Scotia arcs, are built up by the volcanic activity on older oceanic crust.

As an example of an island arc probably built on old oceanic crust, an east-west crustal section across the Lesser Antilles island arc is shown in Fig. 5.16. This is based on gravity, magnetic, seismic reflection and crustal seismic refraction observations (WESTBROOK, 1975; BOYNTON and others, 1979). In general, the forearc and trench structures associated with island arcs are closely similar to those of active continental margins, but here the situation is anomalous in that the trench has been completely filled by a thick pile of sediments which have probably mainly come from the South American continent. The upstanding Barbados ridge now occupies the position of the trench. The oceanic lithosphere beneath Barbados and to the east of it has been depressed by this large sediment load which has developed by eastward migration of the large subduction complex over the Atlantic floor. The total sediment thickness reaches about 20 km beneath Barbados. This subduction complex is at an advanced stage of development. The western flank of the Barbados ridge is faulted but further west lies the residual forearc basin of the Tobago Trough above trapped oceanic crust.

The crust of the Lesser Antilles volcanic arc is about 35 km thick beneath St Vincent. At the surface there is a superficial layer 1 to 5 km thick formed of pyroclastic deposits, lavas and sediments. Beneath this there is a highly variable upper crustal layer, thickness estimated to vary from 2 to 20 km and seismic velocity varying between 5·3 and 7·0 km s^{-1} (average 6·2 km s^{-1}). This is probably formed of igneous rocks similar to those exposed on the islands, with plutonic rocks of intermediate composition being dominant. The lower crustal layer has an average velocity of about 6·9 km s^{-1} and is interpreted as the remnant layer 3 of the pre-existing oceanic crust which has been penetrated and thickened by basic to ultrabasic igneous rocks. To the west of the volcanic arc, the thick, flat-lying sediments of the Grenada Trough are underlain by oceanic crust. This is a marginal basin which may represent old trapped oceanic crust or may have formed by back-arc spreading as described later in the chapter. Even further west, the Aves Ridge (KEAREY, 1974), which is underlain by crust nearly 40 km thick, may represent an older island arc system which is now inactive and forms an aseismic ridge.

Downbending of the oceanic lithosphere

Deep sea trenches and their outer rises are a consequence of the downbending of the oceanic lithosphere at convergent plate boundaries. This downbending can be modelled by the bending of an elastic or elastic-plastic plate as shown in Fig. 5.17. The bending occurs in response to the bending moment M and the vertical force Q which are caused by the downpull of the dense sinking slab. The influence of the horizontal force S is relatively unimportant. If the elastic properties and thickness of the lithosphere and the densities of the material above and below it are known, then the bending profile can be evaluated using beam theory. Theoretical profiles based on various assumptions can then be compared with the observed bathymetric profiles.

Simple elastic bending theory produces theoretical profiles in good agreement with certain observed profiles such as that across the Mariana trench (Fig. 5.18), provided that the elastic lithosphere is assumed to be of the order of 20 to 30 km in thickness. The lithosphere as defined by

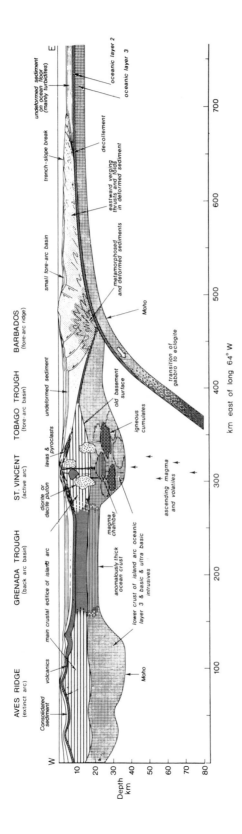

Fig. 5.16 Crustal and upper mantle structure across the Lesser Antilles island arc along an east-west profile at about 13° N, extending from the Atlantic Ocean floor to the Aves ridge, which is interpreted as an extinct arc. The section is based on seismic refraction and gravity interpretation (WESTBROOK, 1975; BOYNTON and others, 1979) and has been reproduced from WESTBROOK (1982). *Trench and fore-arc sedimentation and tectonics in modern and ancient subduction zones, Spec. Pub. geol. Soc. Lond.* (in the press).

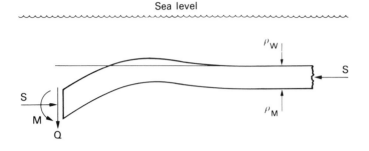

Fig. 5.17 Bending of an elastic or elastic–perfectly plastic plate in response to applied bending moment M, vertical force Q and horizontal force S, with hydrostatic restoring force applied by the seawater above and assumed fluid mantle below. Redrawn with simplification from TURCOTTE and others (1978), *Tectonophysics*, **47**, 194.

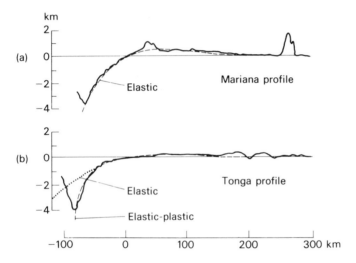

Fig. 5.18 Observed and theoretical profiles of bending at the outer rise and trench. (**a**) Mariana trench profile fitted by flexure of an elastic plate 29 km thick, with development of maximum deviatoric stress of 977 MPa. (**b**) Tonga trench profile only satisfactorily modelled by elastic–perfectly plastic plate 32 km thick with yield stress of 1000 MPa. Redrawn from TURCOTTE and others (1978), *Tectonophysics*, **47**, 202 and 203.

seismological observations is much thicker than this but its lower part is known to deform in response to large stresses of long duration by ductile creep. A 90 km thick elastic lithosphere could not possibly bend as sharply as the observed bathymetric profiles. Even with a 20 to 30 km thick elastic lithosphere, the maximum deviatoric bending stresses approach 1000 MPa (10 kbar). Certain other observed profiles, such as that across the Tonga trench (Fig. 5.18), bend too steeply to fit any reasonable elastic model. TURCOTTE and others (1978) have shown that such bending can be modelled by the flexure of an elastic – perfectly plastic beam which deforms elastically below a deviatoric stress of 1000 MPa and perfectly plastically above it. More complicated models of the downbending oceanic lithosphere can also be found to satisfy the observed bathymetric profiles. For instance, CHAPPLE and FORSYTH (1979) found that a two-layered elastic – perfectly plastic lithosphere 50 km thick, the yield strength of the upper 20 km being 100 MPa and that of the lower 30 km being 60 MPa, satisfies the observed profiles. This model has the further advantage of accounting for observed tensional earthquakes down to 25 km at the downbend and some small compressional events which occur locally down to 40 to 50 km depths. The main point of all these studies is that quite simple bending theory applied to the oceanic lithosphere can adequately explain the existence of outer rise and trench, and that very large stress differences occur in the vicinity of the bend.

Another geometrical consequence of the downbending of the oceanic lithosphere is that it may possibly explain the curvature of some island arcs. FRANK (1968) pointed out that if a flexible but inextensible spherical shell is bent inwards by an angle θ, then the indented part forms an intersecting spherical surface having the same radius as the shell (Fig. 5.19). The edge of the indented part is a circle whose diameter subtends an angle θ at the centre of the sphere. No other type of deformation can occur provided that the shell remains inextensible, as can be confirmed practically on an old table tennis ball. The theorem applies equally validly to a plate which forms part of a spherical shell. Trenches mark the position where oceanic lithosphere is bent downwards into the mantle at a dip averaging about 45°. The theorem suggests that each arc-trench system should form part of a circle which subtends an angle of about 45° at the Earth's centre. This would give a typical arc radius on the Earth's surface of about 2500 km. This is in agreement with the curvature of some island arcs but others do not agree well with Frank's hypothesis. Although Frank's argument may well be an oversimplification, no doubt the nearly universal convexity of island arcs towards the oceanic side is a consequence of this principle of spherical geometry.

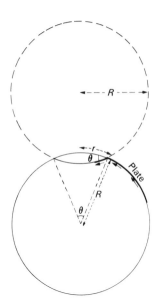

Fig. 5.19 Diagram to show the geometry of an inward-bent inextensible plate on the surface of a sphere. The radius of the resulting arc measured along the spherical surface is given by $r = \frac{1}{2}R\theta$ where R is the radius of the spherical surface and θ is the angle by which the plate is bent inwards, measured in radians.

The shape of the subducting oceanic lithosphere beneath the outer rise and trench is fairly similar from region to region, but there are considerable variations in the dip and shape of the subducting slab at greater depths. In this respect, UYEDA and KANAMORI (1979) pointed out that the subduction zones of the Pacific are divisible into two main types, dependent on whether or not back-arc basins are actively forming behind them. The Chilean margin is an extreme example of the type which lacks back-arc spreading and which is characteristic of the eastern Pacific border. The subduction at Chilean-type margins tends to occur at a relatively shallow angle which is less than 45° and may be as low as 10°. These margins are strongly seismic and are the location of all the large thrust earthquakes. Compression structures on the continental side and forced retreat of the trench on the oceanic side suggest that the plates are being forced against each other. In contrast, the Mariana subduction boundary is an extreme example of the type with active back-arc spreading and which is characteristic of the western edge of the Pacific The subducting oceanic lithosphere dips steeply or even vertically downwards and the release of seismic energy is much smaller than at Chilean-type margins. The oceanic trench appears to be spontaneously retreating, being decoupled from the back-arc region which is subjected to strong tension. Possible factors which may influence the mode of subduction include the horizontal velocity of the subducting plate relative to the overriding plate and the mantle beneath, the negative buoyancy which is a function

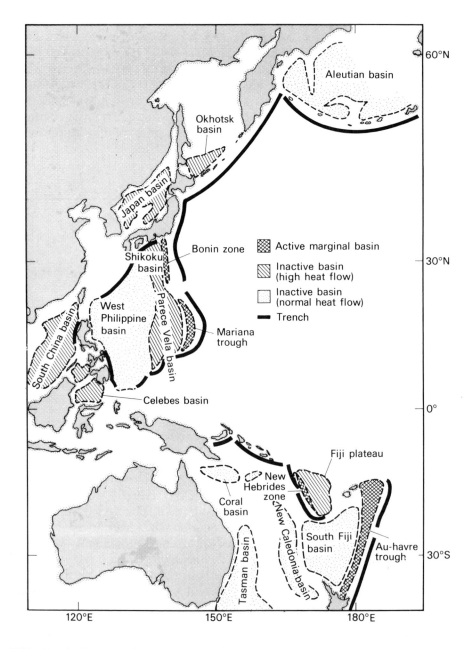

Fig. 5.20 Marginal basins of the western Pacific. Redrawn from KARIG (1971), *J. geophys. Res.*, **76**, 2543.

of lithospheric age, the evolutionary history of the zone, the depth extent of the slab and whether or not it reaches the mantle transition zone. It is notable that the shallow-dipping Chilean-type subduction affects the relatively young east Pacific ocean floor whereas the steeper Mariana-type subduction takes place where the ocean floor of the western Pacific is oldest, coolest and most dense.

5.4 Marginal basins

According to KARIG (1971), marginal basins may be defined as the semi-isolated basins of intermediate to normal ocean depths which lie behind island arc systems. Such basins predominate along the western and northern parts of the circum-Pacific belt (Fig. 5.20) and are absent along the eastern Pacific borderlands. Marginal basins also occur behind the Caribbean and Scotia arcs and in the Mediterranean Sea.

The crust beneath most marginal basins is closely similar to that of normal ocean basins (KARIG, 1971). Relatively thin sediments are characteristically underlain by oceanic layers 2 and 3 of fairly typical thickness. The crust of the basins has therefore probably formed by some form of sea-floor spreading rather than by subsidence and thinning of continental crust. Assuming an oceanic origin, a marginal basin might form by the entrapping of older oceanic crust between a newly formed island arc and the mainland. Alternatively, the basin could form by sea-floor spreading behind the arc. The latter process is called back-arc spreading and this is probably the mechanism of formation of most marginal basins. A few of them, such as some of the Caribbean basins and the Bering Sea, may have formed by entrappment of older oceanic crust.

The marginal basins of the western Pacific can be subdivided into those which are actively forming and those which are now inactive. The active basins and some of the inactive ones exhibit high heat flow. Other inactive basins have a normal oceanic geothermal regime. Some marginal basins, such as the Fiji and Shikoku basins, have formed by normal sea-floor spreading, producing dateable magnetic lineations. The more normal situation exemplified by the Japan Sea is that the magnetic lineations are irregular and undateable, suggesting that the magma emplacement forming the new crust is not restricted to a single spreading axis (Fig. 5.21).

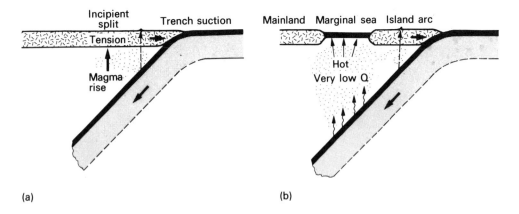

Fig. 5.21 Formation of a marginal basin by separation of an island from the mainland by back-arc spreading, attributable to tension caused by trench suction (Chapter 9) and to the hot mantle above the subducting oceanic lithosphere.

Why does back-arc spreading take place? Almost all the presently forming marginal basins are in the western Pacific. Here the rate of plate convergence is greater than elsewhere. Also the subducting lithosphere is oldest and therefore its negative buoyancy is greater than elsewhere. The actively forming basins overlie subduction zones and associated regions of very low Q in the upper mantle (p. 214). Two main hypotheses have been suggested. Firstly, the spreading may take place because of the abundant presence of basalt magma forcing its way up from the very low Q region below. The magma may come from processes associated with the subduction zone and convective circulation in the mantle above it. Secondly, it may occur because of tensile stress in the back-arc region caused either by the suctional drag on the overriding plate produced by the sinking lithosphere (p. 358) or by drag of local convection currents in the very low Q zone. Probably the basins form as a result of the combined effects of tension and magma source.

Between Japan and New Guinea, two island arc belts occur between the main Pacific Ocean and the mainland. The marginal basin crust is being subducted at the inner belt of trenches. In the absence of sufficiently rapid back-arc spreading here, the two arc-trench systems would inevitably collide. The geological structure of mountain belts suggests that this type of collision has occurred in the past. Evidently marginal basins can be formed and then be destroyed, possibly several times during the lifespan of an active margin.

6 The core

The core forms 16 % of the Earth by volume and 32 % by mass. It extends from a depth of about 2885 ± 5 km to the centre of the Earth, the core-mantle boundary having a radius of about 3485 km. The process of core formation has certainly been an important factor in the thermal and geochemical evolution of the Earth. Subsequently the continuing loss of heat from the core has probably had some influence on mantle convection. Furthermore, the main geomagnetic field originates within the core; it is the behaviour of the magnetic field that has made it possible to use palaeomagnetism to reveal past movements of the continents and magnetic surveys to show up the pattern of sea-floor spreading. Thus the core has direct relevance to the evolution of much shallower parts of the Earth and our knowledge of this.

This chapter aims at summarizing our present knowledge of the structure, evolution and processes of the core. For a more detailed account the reader is referred to JACOBS (1975).

6.1 Structure of the core

Until about 1970, knowledge of the seismological structure of the core was mainly based on travel-times of body waves traversing the core as P. Such was the basis for the well-known Bullen model of the core, which is still sometimes used for reference purposes. Zone E in Bullen's model comprised the fluid outer core (2900 to 4980 km depth range) with P velocity increasing steadily downwards from $8\cdot10$ to $10\cdot44\,\mathrm{km\,s^{-1}}$. Zone F represented a transition zone from 4980 to 5120 km with a complicated velocity-depth pattern between outer and inner cores. Zone G (5120 to 6371 km) comprised the inner core with P velocity rising downwards from $11\cdot16$ to $11\cdot31\,\mathrm{km\,s^{-1}}$, believed to be solid but without definite proof before the 1970s.

A much more accurate picture of the velocity-depth structure of the core and of its density distribution has come about during the 1970s, particularly as a result of the introduction of two new types of technique. Firstly, body wave phases can now be more convincingly recognized and their onsets more accurately timed by use of phased seismological array stations (p. 143). Secondly, improvements in the identification and measurement of overtones of the Earth's free oscillations has yielded important new evidence on core structure and density. Use of these techniques has demonstrated that the inner core transmits S waves and thus that it must be solid. The outer and inner subdivisions of the core have been found to be separated by a sharp boundary and the need for a complicated transition zone between them has been removed. The present model of the core is as follows:

	Depth extent	Radial extent
Outer core	2885–5144 km	1227–3486 km
Inner core	5144–6371 km	0–1227 km

The usual nomenclature for the many types of seismic ray which are associated with the core is built up from the following symbols:

P   compression wave in the mantle
S   shear wave in the mantle
K   compression wave in the outer core
I   compression wave in the inner core
J   shear wave in the inner core
c   incident mantle ray reflected at the core-mantle boundary
n   incident outer core ray internally reflected n times at the core-mantle boundary
i   incident outer core ray reflected at the inner core boundary

These letters are strung together in sequence to describe the complete path of a given ray through the Earth. For example, a ray which traverses the mantle as P both downwards and upwards and which penetrates the outer core is PKP; if it also penetrates the inner core as P it is PKIKP or as S it is PKJKP. Other possible rays which undergo conversion at the core-mantle boundary but are not reflected are SKS, PKS, SKP, SKIKS, SKJKP, etc. PcP, ScS, PcS, ScP are mantle rays reflected once at the core boundary and PKiKP is reflected at the inner core boundary. PnKP has been internally reflected n times at the core-mantle boundary.

The ray paths for the simplest direct body waves which traverse the core, PKP and PKIKP, are shown in Fig. 6.1 and the corresponding time-distance graphs are shown in Fig. 6.2. The estimated velocity-depth distributions in the core (Fig. 6.3) make use of other phases such as SKS, PKJKP and PKiKP, but PKP and PKIKP adequately illustrate the principles.

The PKP ray which has the shallowest penetration into the outer core emerges at an epicentral distance of about 188° at A, which is 8° beyond the anticentre (Figs 6.1a and 6.2). With increasing depth of penetration into the outer core, the epicentral distance of the emerging ray progressively decreases from about 188° to 143° (B in figures) and then it increases again to 170° (C in figures). The cusp in the time-distance curve at B is a geometrical effect of the total ray path and does not indicate a discontinuity of any sort. The time-distance graph for PKP is generally regarded as being approximately consistent with steady rise in velocity with depth in the outer core, although this interpretation cannot be unique because no PKP or even SKS rays have their lowest point of penetration in the topmost part of the outer core. The rays which penetrate the inner core form the branch DE of the time-distance graph (Figs 6.1b and 6.2). The ray with the shallowest penetration emerges at an epicentral distance of about 110° at D, and with increasing penetration the emerging ray passes from D to the anticentre E. It was LEHMANN (1936) who discovered the inner core by recognizing the significance of the branch DE, which represents sharp bending of the rays at the inner-outer core boundary. Here the velocity increases abruptly and continues to increase for some depth into the inner core.

The core-mantle boundary

The core-mantle boundary is marked by a discontinuous drop of P velocity from about $13.6 \, km \, s^{-1}$ at the base of the mantle to around $10.0 \, km \, s^{-1}$ at the top of the core. The S velocity correspondingly drops from $7.3 \, km \, s^{-1}$ at the base of the mantle to zero in the outer core. The density increases downwards across the boundary from 5500 to $10000 \, kg \, m^{-3}$. The boundary gives rise to strong reflected phases such as PcP, ScS and PnKP demonstrating that it is a sharp discontinuity over less than 2 km rather than a gradational boundary. It is normally interpreted as marking the junction between the solid silicon-magnesium-rich mantle and the fluid iron-rich outer core, and as such it is the most important and drastic compositional boundary within the Earth.

The estimate of the depth to the core-mantle interface must be treated as a mean value for the

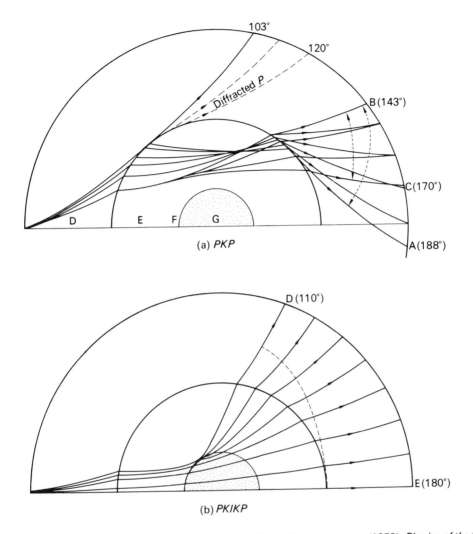

Fig. 6.1   Ray paths for *P* waves passing through the core. Adapted from GUTENBERG (1959), *Physics of the Earth's interior*, p. 104, Academic Press.

following reason. The Earth's rotation causes surfaces of equal density within the Earth, as well as the external surface, to be distorted very nearly into ellipsoids of revolution. The flattening at depths within the Earth can be calculated provided that the internal density distribution is known. BULLEN (1963) gave an estimate of about 1/390 for the flattening at the core-mantle boundary, which implies that the polar radius of the core is about 9 km less than the equatorial radius and that the corresponding difference of depth to the core-mantle boundary is about 12 km.

Estimation of the mean depth to the core-mantle boundary has mostly been based on reflected body waves and requires knowledge of the velocity-depth distribution of *P* or *S* throughout the mantle. The modern method based on body waves compares the observed travel times of *PcP* and *ScS* with their times computed for the assumed velocity-depth distributions in the mantle, adjusting the depth of the interface until consistency is obtained. Using this approach, JEFFREYS

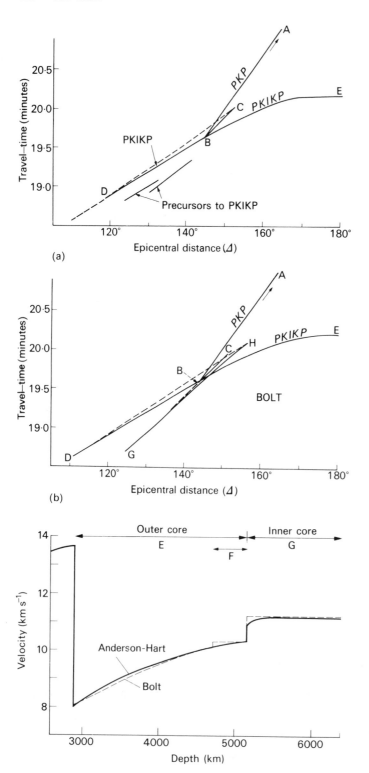

(a)

(b)

Fig. 6.2 Time-distance graphs for *PKP* and *PKIKP* (**a**) modified from GUTENBERG (1959) to show arrivals now interpreted as precursors, and (**b**) BOLT (1962), *Nature, Lond.,* **196**, 122, showing interpretation of the precursors in terms of an additional layer near the inner core boundary.

Fig. 6.3 Suggested velocity-depth distributions for *P* in the core, showing the older interpretation of BOLT (1962, 1964) involving a double-stepped transition zone, and the more recent distribution of HART and others (1977) which lacks a transition zone.

(1939a and b) obtained a depth of $2898 \pm 2\cdot5$ km. More recent estimates make use of the difference in travel times of P and PcP, or S and ScS, for given events, thereby removing the major systematic errors caused by uncertain focal data and inhomogeneities in crustal and upper mantle structure. These estimates show that the boundary is about 10–15 km shallower than the value obtained by Jeffreys. An alternative approach to body waves is to use the free oscillation data for the Earth. Prior to 1971, the available free oscillation data indicated that existing models were unsatisfactory, but it could not resolve whether this arose from too high velocity in the lower mantle or too great an estimated depth for the boundary. Subsequent accurate determination of the periods of a series of additional overtones has given the resolution necessary to show that the boundary is about 15 km shallower than Jeffreys' determination. Based on both body waves and free oscillations, DZIEWONSKI and HADDON (1974) thus concluded that the best available estimate of the mean core radius is 3485 ± 3 km corresponding to a mean depth of 2886 km.

A suggestion that there may be long wavelength undulations or 'bumps' on the core-mantle boundary originally stemmed from geomagnetic rather than seismological studies. In order to explain fluctuations in the length of day occurring over periods of a few years (decade fluctuations), some coupling mechanism for transfer of angular momentum between core and mantle is needed (p. 265). HIDE (1970) suggested the idea of topographic coupling between the fluid core and the solid mantle as a result of the presence of undulations in the depth of the core-mantle boundary. To be effective, the bumps need to be about 1 km or more in height and their horizontal extent would be of the order of a few hundred to a few thousand kilometres. Because of the large density contrast of about $4500\,\mathrm{kg\,m}^{-3}$ at the core-mantle interface, such undulations in depth should give rise to a significant contribution to the longer wavelength components of the Earth's gravity field; the gravity effect of undulations of wavelength less than about 3000 km at the boundary would be greatly attenuated at the Earth's surface so that they would not contribute significantly. If such bumps exist, they might also be expected to influence the main geomagnetic field. Perhaps the most convincing evidence for the presence of very long wavelength undulations on the boundary is the correlation of the gravity and geomagnetic harmonics of degree four or less found by HIDE and MALIN (1970), with the geomagnetic harmonics being displaced eastwards in longitude by about 160° relative to the corresponding gravity harmonics. Although this correlation appears to be statistically significant, it does not conclusively demonstrate the presence of the undulations. It might also arise from thermal anomalies in the lower mantle which directly produce the gravity anomalies because of the associated density anomalies. Such thermal anomalies may also influence the geomagnetic field because of their effect on the fluid convection patterns in the outer core (HIDE, 1970). If the undulations do exist, then the change in density at the interface would give rise to stress differences in the lowermost mantle which would be of magnitude proportional to the height of the bumps, reaching up to about 20 MPa for a topographical height of 1 km. Unless the viscosity of the lowermost mantle is much higher than suspected (Chapter 8), such bumps would be expected to dissipate by viscous flow in a relatively short time, suggesting that they must be in some sort of dynamic equilibrium if present.

Present seismological evidence is not sufficiently precise to reveal the presence of long wavelength undulations of the type suggested by Hide. Such undulations of ten kilometres or more in amplitude appear to be ruled out but those less than two kilometres in height cannot be excluded as this is below the resolution of the travel-time data. There is, however, some evidence from the scattering of seismic waves that much shorter wavelength topography does occur at the boundary. For instance, DOORNBOS (1978) suggested that some precursors of the PKP phase may be caused by scattering by topography of the order of a few hundred metres height and 10–20 km length at the boundary. CHANG and CLEARY (1978) interpreted consistent precursors to $PKKP$ from Novaya

Zemlya explosions arriving at LASA up to 65 seconds prior to the main phase as caused by scattering on underside reflection at the core-mantle boundary, attributable to short wavelength bumps on the boundary. For some other travel paths such precursors are absent, suggesting that short wavelength topography varies in character from region to region on the boundary. On the other hand, BUCHBINDER (1972) earlier suggested that phases of the type $PnKP$ which have been multiply reflected at the core-mantle boundary from below are consistent with a smooth boundary which does not cause significant scattering.

The fine structure in the vicinity of the core-mantle boundary is also subject to controversy. The main evidence comes from amplitude versus frequency studies of the reflected phases, especially PcP, and from mantle body waves diffracted at the boundary. Several detailed structures have been suggested, particularly involving either a significant rigidity for the outermost core or a series of layers of rather improbable properties imbedded at or near the boundary. However, MÜLLER and others (1977) found such models unsatisfactory and considered that a single sharp boundary can best explain the observations, if their wide scatter is taken into account. Furthermore, DOORNBOS and MONDT (1979a and b), on the basis of their studies of the diffracted phases of P and SH, find no evidence for finite rigidity of the outermost core. Their observations are best explained by a 70–100 km wide zone at the base of the mantle in which both P and S velocities decrease slightly with depth, without need for further complication.

Thus the depth to the core-mantle boundary has now been determined with considerable precision but there is still uncertainty about the detailed nature of the boundary and the possibility of bumps a few kilometres high on it. It is to be hoped that these problems concerning the Earth's most fundamental discontinuity will be to some extent resolved over the next ten years.

The outer core

The outer core extends between average depths of about 2885 and 5144 km, forming 95·6 % of the core by volume and about 95 % by mass. The presence of a transition zone at the base of the outer core has been discounted in most recent models of seismological structure for reasons which are explained later in this section. Thus the outer core is now regarded as extending down to the inner core boundary.

Our knowledge of the seismological structure of the outer core is mainly dependent on travel times of PKP, SKS, and associated phases which have been internally reflected at the core-mantle boundary. All of these phases travel as P in the outer core. An independent check on the P velocity distribution and a direct estimate of the broad level of density are given by inversion of gross earth data including free oscillations. For body wave data, it is possible to strip off the mantle parts of the travel path to obtain travel-time curves through the outer core alone, starting and finishing at the core-mantle boundary. This can then be inverted by the Herglotz-Wiechert method to yield a velocity-depth distribution in the outer core. The problem with the PKP phase is that the shallowest ray penetrating the core bottoms about 1000 km below the core-mantle boundary, with the velocity distribution in the overlying region being indeterminable. This unsatisfactory situation can be rectified by using the phases SKS and $SKKS$ (HALES and ROBERTS, 1971) which yield evidence of the velocity-depth distribution from about 40 km below the core-mantle boundary down to about 100 km above the inner core boundary. This works because the velocity of S at the base of the mantle is only slightly higher than P at the top of the core, so that the SKS rays only steepen very slightly on passing down from mantle to core.

Estimates of the P velocity just below the core-mantle boundary vary between 7·90 and 8·26 km s^{-1} but the free oscillation data supports a value of about 8·03 km s^{-1}. Most seismological models show a smooth increase in velocity with depth throughout the outer core rising to about $10 \cdot 2 - 10 \cdot 3 \, \text{km s}^{-1}$ just above the inner core boundary. Such distributions are approximately consistent with the Adams-Williamson equation. However, an alternative interpretation has been suggested by KIND and MÜLLER (1977) who found that the pattern of relative amplitudes of SKS and SKKS indicates the presence of an anomalous zone of slightly high velocity between depths of 3700 and 4000 km. Comparison with shock wave results has also suggested that there may be deviations from perfect homogeneity near the upper and lower boundaries of the outer core. Such deviations suggest that there could be some sort of chemical inhomogeneity within the outer core, perhaps stable stratification. This would have far-reaching repercussions on our understanding of the origin of the geomagnetic field if eventually proved to be correct, ruling out the possibility that the outer core is intimately mixed by convection currents as has been widely supposed.

One of the main controversies about core structure during the 1960s concerned the region about 150 km wide just above the inner core boundary, which used to be regarded as a transition zone. The problem arose because of the observations of short period P arrivals between epicentral distances of about 125° and 143° which precede the main PKIKP phase by about 15 seconds (Fig. 6.2). GUTENBERG (1959) interpreted these precursors as caused by dispersion in a transition zone in which velocity continuously increases with depth on passing from outer to inner core, the shorter period waves travelling with higher velocities. Subsequently BOLT (1962, 1964) suggested that these early arrivals represent an additional branch GH of the time-distance graph (Fig. 6.2b). He suggested that the precursors can best be explained by refraction at an abrupt increase in velocity about 400 km above the top of the inner core (Fig. 6.3); some other workers suggested two such steps. However, CLEARY and HADDON (1972) advanced a radically different interpretation of these puzzling precursors which avoids the need for velocity irregularities above the inner core boundary. They interpreted the early arrivals as PKP rays which had been scattered by irregularities in the lowermost 200 km of the mantle, or as later suggested for PKKP at the core-mantle boundary (CHANG and CLEARY, 1978). Other observations of phase velocity by array stations has given full support to this interpretation, which is becoming widely accepted. Thus the need to postulate a core transition zone has now probably been removed, and P is generally regarded as increasing smoothly down to the inner core boundary. Its exact value just above the boundary is still rather uncertain.

The straightforward interpretation of the lack of any observed shear waves passing through the outer core is that it is liquid and cannot therefore transmit S. This argument is not quite conclusive because shear waves might be cut out by strong attenuation in the core. However, other body wave phenomena corroborate the interpretation of the region just below the core-mantle boundary as being liquid. If it were solid, then the amplitude of the phase SKS would be much smaller. This is because a much larger fraction of the incident energy is converted from S to P, and then back to S on return, at a solid-liquid interface than if both media are solid. The observations of diffracted P and SH phases of DOORNBOS and MONDT (1979a and b) also indicate zero rigidity of the core just below the boundary. On a broader basis, the observed periods of the lower modes of the Earth's free oscillations suggest that the outer core as a whole has negligible rigidity modulus and therefore is liquid.

There are two other geophysical phenomena which when taken with the seismology of the core make the interpretation of the state of the outer core as liquid virtually conclusive. One of these is the Earth's magnetic field and its secular variation which are described later in the chapter and which could not be explained if the core were solid. The other is the response of the Earth to tidal forces and to impulses affecting the axis of instantaneous rotation which provides estimates of the rigidity of the core, as described in the next section.

Earth tides and the fluid outer core

The *earth tide* (MELCHIOR, 1966, 1978) arises as follows. The gravitational attractions between the Earth and the Sun and Moon only exactly balance the centrifugal forces at the centre of the Earth. Elsewhere within the Earth and at its surface there is a small residual gravity potential which causes both the ocean tide and a periodic deformation of the solid Earth which is known as the earth tide. The tide-raising potential at a point on or within the Earth depends on its position and also that of the Sun and Moon in relation to a frame of reference fixed within the Earth and rotating with it. Semi-diurnal tides are the most prominent effect but longer period tides such as diurnal, fortnightly and semi-annual tides do also occur both in the ocean and in the body of the Earth. The earth tides are effectively in equilibrium with the tide-raising potential and are therefore in phase with it, but this is not true for the ocean tides especially in shallow water. Earth tides can be observed by recording the tidal variation of gravity, and tilting of the Earth's surface, and variation in linear strain using a strain seismometer (Fig. 6.4). The amplitude of the earth tide at the equator is about 20 cm.

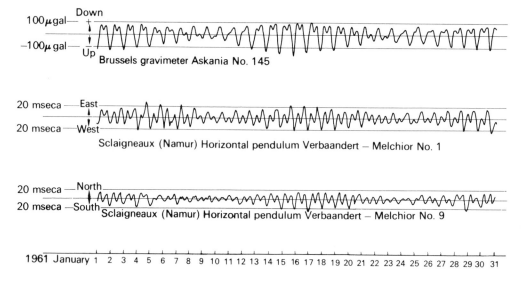

Fig. 6.4   The earth tides simultaneously recorded *above* by Askania gravimeter at Brussels, and *below* by tiltmeters at Sclaigneax (Namur). Reproduced from MELCHIOR (1966), *The earth tides*, p. 18, Pergamon Press.

The response of the Earth to the tide-raising potential depends on the rigidity distribution within it. It is therefore convenient to relate the amplitude of the actual tidal variations of gravity and tilting to the values which would be observed for a perfectly rigid Earth and which are readily calculable. The ratio of observed tidal gravity variation to the theoretical rigid Earth value is known as the *gravimetric factor* and the corresponding ratio for tilt is the *diminishing factor*. The deformation of the Earth can more generally be expressed in terms of two quantities h and k which are known as *Love's numbers** and which can be computed for specified distributions of density and elasticity within the Earth assuming radial symmetry; h is the ratio of the heights of the earth tide

* In this section, k denotes one of Love's numbers; elsewhere in this book it denotes bulk modulus unless otherwise specified.

to the corresponding theoretical ocean tide assuming equilibrium is reached; k is the ratio of the gravitational potential caused by the tidal deformation of the Earth to the tide-raising potential. Shida introduced a further number l which is the ratio of the horizontal displacement of the Earth tide to that of the corresponding equilibrium ocean tide. The numbers h, k and l can be determined from earth tide observations. The quantity $(1 + h - \frac{3}{2}k)$ is equal to the gravimetric factor determinable from the tidal variation of gravity and $(1 + k - h)$ is equal to the diminishing factor determinable from tilt observations. Love's number k can also be determined directly from the period of the Chandler wobble (p. 22).

The most satisfactory values of Love's numbers obtained from observations are $h = 0.58$, $k = 0.29$ and $l = 0.045$. These values may be in error by about 5 % as a result of scatter of observations. This scatter is mainly attributable to the following four factors:

(*i*) The ocean tides affect the observations because of the gravitational attraction of the varying water masses, the deformation of the surface by the water loading, and the influence of both these effects on the Earth's gravity potential. These are collectively known as the indirect effect. Correction can now be applied for the indirect effect, the error mainly depending on the accuracy with which the ocean tides can be predicted. Stations furthest from oceans are least affected.

(*ii*) Local effects arise from topography, geological structure, presence of aquifers and meteorological variations. Most of these can be reduced by good siting of stations.

(*iii*) Lateral variation in the structure of the crust and upper mantle may cause regional variations in the numbers, particularly near continental margins.

(*iv*) Resonance effects, due to free vibration of the fluid core, affect the tidal components of comparable period. This effect is particularly applicable to the K_1 lunisolar tide of 23.93 hour period, reducing the gravimetric factor and increasing the diminishing factor.

The observed values of Love's numbers can be compared with values theoretically computed for various earth models. TAKEUCHI (1950) did this by taking a distribution of density and rigidity similar to that of Bullen and assuming a range of possible values of rigidity modulus for the core. He found that the observed values of Love's numbers require the rigidity modulus of the core to be less than about 1000 MPa. A later analysis by ALSOP and KUO (1964) placed the upper limit at about 15 000 MPa. On the basis of the evidence from the static earth tides, one can be reasonably confident that the rigidity modulus of the outer core is significantly less than 10 000 MPa, suggesting a fluid rather than solid state.

Further important evidence comes from the resonance effects mentioned above, dependent on free natural oscillations of a fluid core at periods much longer than the free vibrations of the whole solid Earth. These have the effect of modifying the amplitudes of the earth tides and also the comparable wobbles (nutations) at nearly the same period. Study of this effect requires separation of the deformation of the various tidal constituents by appropriate methods of harmonic analysis. The results to date clearly indicate the presence of the effect on both the earth tides and the nutations, giving strong added support to the inference of a fluid outer core.

The inner core

The existence of an inner core was first postulated by LEHMANN (1936) on recognition of P arrivals of the branch DE (Fig. 6.2) within the P shadow zone. These were interpreted as rays strongly refracted at a steep rise in velocity with depth of about $1\,\mathrm{km\,s^{-1}}$ marking the boundary between outer and inner core. Subsequent body wave and free oscillation studies have amply confirmed the existence of the inner core and have added much more precise interpretations than were originally possible.

The phase *PKiKP* which travels through the outer core as *P* and is reflected at the inner core boundary has been particularly useful in determining the depth and sharpness of the boundary. This phase was originally observed as a wide-angle reflection and the smaller amplitude near-vertical incidence reflection was later recognized by ENGDAHL and others (1970) using recordings from seismological array stations including LASA. The difference between the travel times of *PKiKP* and *PcP* at nearly vertical incidence, taken in conjunction with an assumed velocity-depth structure for the outer core, yielded an estimate of 1227·4 ±0·4 km for the mean radius of the inner core (ENGDAHL and others, 1974). The high precision of this estimate marks the internal consistency. The absolute accuracy is probably more like 5–10 km. This estimate is more than 20 km lower than the value in the Jeffreys-Bullen models but it is in good agreement with other recent estimates based on body wave and free oscillation studies. The nature of the reflected phase *PKiKP* demonstrates that the inner-outer core boundary is sharp, being short in comparison with the wavelength of one second period *P* waves and therefore a discontinuity or perhaps gradational over 1–2 km at the most.

The earlier *P* velocity-depth distributions for the inner core showed an abrupt increase of about $1·0\,\mathrm{km\,s^{-1}}$ at the boundary and a constant value of about $11·2\,\mathrm{km\,s^{-1}}$ throughout the inner core. More recent studies based on the position of the cusp D on the *PKIKP* branch of the travel-time curve (Fig. 6.2), which is placed now at 120° rather than 110° (BUCHBINDER, 1971), on the amplitudes in the vicinity of this cusp (QAMAR, 1973), and on the amplitude characteristics of *PKiKP* in relation to other phases (MÜLLER, 1973), all indicate that the abrupt increase in *P* velocity at the inner core boundary is only about $0·6\,\mathrm{km\,s^{-1}}$. There must therefore be a pronounced increase of velocity with depth in the uppermost 200 km or thereabouts of the inner core, below which the velocity-depth curve flattens off towards the centre of the Earth where the value is about 11·2 to $11·3\,\mathrm{km\,s^{-1}}$.

BULLEN (1963) suggested that the inner core must probably be solid for the following reason. At that time it was believed that the compressional wave velocity increases by about 11 % across the inner core boundary. As it is inconceivable that the density could decrease downwards across the boundary, a zero rigidity modulus of the inner core would imply an increase of bulk modulus of at least 23 %. According to Bullen's compressibility-pressure hypothesis, one would not expect this modulus to rise more rapidly across the boundary than in the outer core. Thus Bullen inferred that the steep increase in *P* velocity between outer and inner core probably indicates a rigidity modulus of the order of 3×10^5 MPa in the inner core.

More definite evidence that the inner core is solid has come from three sources. Firstly, the only plausible interpretation of certain overtones of the spheroidal free oscillations first observed on recordings of the Alaskan earthquakes of March 1964 is that the inner core has a finite rigidity modulus corresponding to an average *S* velocity of about $3·5\,\mathrm{km\,s^{-1}}$ (DZIEWONSKI and GILBERT, 1971). Secondly, JULIAN and others (1972) claimed to have identified the phase *PKJKP*, which travels through the inner core as *S*, using the LASA array station; they estimated the average *S* velocity in the inner core to be $2·95\,\mathrm{km\,s^{-1}}$ which does not agree well with the other evidence. This has led to the suggestion that the phase they actually detected was *SKJKP* which would have an *S* velocity of $3·5\,\mathrm{km\,s^{-1}}$ in the inner core in good agreement with other estimates. Thirdly, MÜLLER (1973) used the amplitude pattern of *PKiKP* to show that the inner core just below the boundary has a finite rigidity corresponding to an *S* velocity of about $3·5\,\mathrm{km\,s^{-1}}$. Taken together, this evidence provides an almost conclusive argument for the solidity of the inner core, with an average

S velocity of about $3 \cdot 5 \, \text{km s}^{-1}$. Thus the inner core appears to be solid although the low *S* value and low *Q* (next section) indicates that it must also be soft.

Density, elastic properties and seismic attenuation

Early estimates of the density distribution of the core made by BULLEN (1963) assumed the validity of the Adams-Williamson equation in both outer and inner core and were further based on arbitrarily assumed density values at the Earth's centre. Much more reliable density values can now be obtained directly from the free oscillation overtones sensitive to the core. According to the model of HART and others (1977), density in the outer core increases from $9970 \, \text{kg m}^{-3}$ just below the core-mantle boundary to $12\,120 \, \text{kg m}^{-3}$ just above the inner core boundary. A small but sharp increase of $180 \, \text{kg m}^{-3}$ occurs at the inner core boundary and the density then rises to $12\,570 \, \text{kg m}^{-3}$ at the Earth's centre. The density of the inner core is not well resolved, the uncertainty being of the order of $500 \, \text{kg m}^{-3}$. However, the density at the Earth's centre is clearly much less than in most of the early models of Bullen. Once the density distribution is defined, the elastic moduli of the core can be determined directly from the seismic velocities (Fig. 6.5).

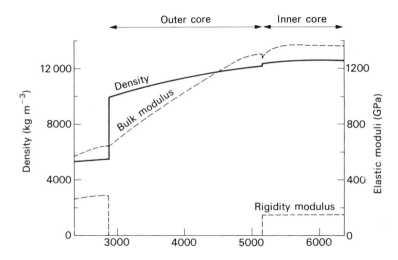

Fig. 6.5 The estimated distributions of density and elastic moduli in the core, according to the model of HART and others (1977).

The seismic attenuation in the core is mainly known from body wave studies. In the outer core, very high values of *Q* are indicated by the slight dissipation of the multiply reflected phases *PnKP* (BUCHBINDER, 1971; QAMAR and EISENBERG, 1974). Estimates of the value of Q_α in the outer core are of the order of 5000 or greater. Q_β is zero as *S* is not transmitted. In contrast, much lower values of Q_α of about 200 to 600 are obtained for the inner core, based on the amplitude of *PKIKP* relative to *PKP*. Q_β appears to be similar magnitude. It was indicated in Chapter 4 (p. 169) that all or nearly all of the seismic dissipation in the mantle could be attributed to attenuation in shear, the attenuation in bulk compression being negligible with Q_k approximating to infinity. Some of the free oscillation spheroidal overtones, however, do indicate that there must be some dissipation in pure compression somewhere within the Earth. According to ANDERSON and HART (1978), such dissipation probably occurs within the inner core where Q_k is probably of similar magnitude to $Q_\mu \, (= Q_\beta)$. Alternatively, it may occur in the upper mantle.

6.2 Composition of the core

Shortly after the discovery of the two main classes of meteorites in the mid-nineteenth century, it was suggested that the Earth has a core similar to the iron meteorites surrounded by a silicate shell analogous to the stony meteorites. Wiechert later adopted this hypothesis to explain the high mean density of the Earth but he over-estimated the radius of the core because he did not allow for self-compression. After the discovery of the core through the use of earthquake waves (OLDHAM, 1906), the concept of an iron core was taken over by seismologists and remained essentially unquestioned until 1941.

KUHN and RITTMAN (1941) suggested that the core consists of a condensed form of hydrogen. This hypothesis was based on the idea that the Sun and Earth have the same composition. It stimulated a lot of discussion but it is now known to be untenable because the pressure within the Earth is nowhere high enough to cause condensation of hydrogen. A few years later RAMSEY (1949) suggested that the core may be formed of a high pressure modification of mantle material. He postulated that the silicate material is stripped of an outer electron to produce a metallic material with high density, low melting point and very high electrical conductivity. Ramsey's hypothesis gained serious support for a few years but has since lost favour for reasons outlined below. A few geophysicists still hold to the hypothesis, notably some supporters of expanding or contracting earth hypotheses.

The point at issue during the 1950s and 1960s was whether the core-mantle boundary represents a change in composition from the silicon-magnesium rich mantle to a core of metallic iron with subordinate nickel and lighter impurities, or whether it is a radical phase boundary such as Ramsey postulated. The iron meteorites can no longer be taken as convincing evidence for an iron core as they are no longer regarded as originating by break-up of the core of a planetary-sized body, although they do indicate the abundant presence of an iron-nickel phase in the material which accreted to form the inner planets. The Earth without an iron core would be grossly depleted in iron relative to silicon in comparison with the general abundance in the solar system. Furthermore, the sharp boundary between mantle and core indicated by seismological studies is much better explained by a chemical discontinuity than by a phase transition in a multi-component system which would give rise to a gradational boundary. The most decisive evidence, however, for an iron core comes from comparison between the seismologically inferred hydrodynamic sound velocity and density of the core and the values yielded by shock wave experiments as described below. This evidence also indicates that a proportion of one or more lighter elements must be admixed with the iron and nickel.

Core pressures have recently been attained in static experiments, but inferences on core composition are mostly based on the momentary attainment of such pressures in shock waves produced by explosives (e.g. McQUEEN and others, 1964). A shock wave differs from a seismic wave in that the particle velocity U_p is comparable to the velocity of propagation of the wavefront U_w and the temperature rise associated with the passing wave is higher than would be caused by adiabatic compression. The velocities U_p and U_w can be observed in experiments. The pressure and the density within the wave can be deduced from U_p and U_w by the following equations which are based on the principles of conservation of mass and momentum (Birch; in CLARK, 1966, p. 99)

$$p = \rho_0 U_w U_p$$
$$\rho = \rho_0 U_w / (U_w - U_p)$$

where ρ_0 is the density at zero pressure. The resulting relationship between pressure and density is called the Hugoniot. By applying thermodynamic equations to correct for the temperature rise, the Hugoniot can be reduced to a more general relationship between pressure and density.

The shock wave results for metals can be conveniently compared with the seismic and density distribution within the Earth's interior by plotting $(dp/d\rho)^{\frac{1}{2}}$ against density for both (Fig. 6.6). The quantity $(dp/d\rho)^{\frac{1}{2}}$ for metals can be determined from the Hugoniot; the error introduced by using the Hugoniot instead of an adiabatic equation of state is not serious. Within the Earth, $(dp/d\rho)^{\frac{1}{2}}$ is the square root of the seismic parameter ϕ, and in the outer core it is the P velocity.

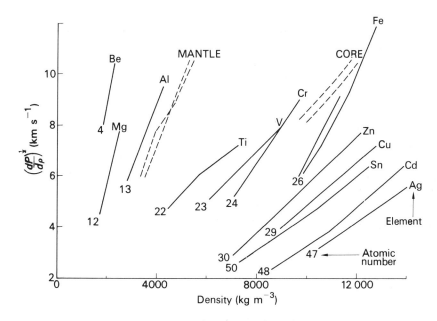

Fig. 6.6 Hydrodynamic sound velocity as a function of density shown for a selection of metals as obtained by shock wave experiments (solid lines) and for the mantle and core as obtained by seismic observations and density models (dashed lines). The numbers shown are atomic numbers. Redrawn from BIRCH (1961b), *Geophys. J. R. astr. Soc.,* **4,** 309.

Birch's diagram (Fig. 6.6) and associated shock wave evidence on the density and hydrodynamic sound velocity of elements as a function of pressure effectively rule out any element of atomic number less than 23 (vanadium) as the major constituent of the core. This strongly supports the concept of an iron-rich core in that the only heavy element present in sufficient abundance in the solar system to form the bulk of the core is iron. In fact the agreement between the shock wave data for pure iron and the observed seismologically-derived data for the outer core is not perfect. According to the shock wave experiments, the density of iron at 2270 K would be 11 000 kg m^{-3} at the core-mantle boundary and 12 500 kg m^{-3} at the inner core boundary, whereas the estimated values are about 9960 and 12 100 kg m^{-3} respectively. This discrepancy would be removed if a significant proportion of a lighter element is present in the outer core, as discussed in the next section. In contrast to the outer core, the seismic velocity and density inferred for the inner core are in good agreement with the shock wave data for pure iron. Thus the outer and inner parts of the core, while both being predominantly formed of iron, appear to have significantly differing compositions.

Iron and nickel are associated together in the metallic phase of meteorites. The abundance of Ni relative to Si in the crust and mantle is less than in the solar system and it therefore seems likely that the Earth's nickel supply has been preferentially concentrated in the core. It is thus possible that about 4 % by weight of nickel may be incorporated in the core. The presence of such a proportion of nickel, however, would slightly accentuate the discrepancy between observed and predicted density and velocity in the outer core, requiring a small additional proportion of a lighter element to satisfy the shock wave data.

Silicon, sulphur or oxygen in the core?

The question now arises as to the nature of the light element or elements which must be present in significant proportion in the core. Several candidates have been suggested but all of these apart from silicon, sulphur or oxygen can probably be ruled out on geochemical grounds or because of insufficient abundance. It is generally agreed that 5 % to 15 % by weight of one of these light elements would be required to lower the density and raise the seismic velocities of a predominantly iron core to the observed levels.

The idea that silicon is the light element alloyed with the iron of the core was suggested by RINGWOOD (e.g. 1966, 1975). Some difficulties associated with this hypothesis have been summarized by BRETT (1976). The incorporation of the silicon into the core depends on an unduly complicated hypothesis of Earth accretion involving the outgassing and escape of vast amounts of CO of which no trace is now left. The hypothesis implies that the mantle and core are grossly out of chemical equilibrium with each other. It also requires the Earth to have a rather high overall content of silicon. The hypothesis has not been promoted further in the more recent papers of Ringwood.

A more widely accepted hypothesis is that sulphur in quantity has been incorporated into the core as FeS. MURTHY and HALL (1970, 1972) put forward this hypothesis on the assumption that the Earth accreted homogeneously from grains of iron-nickel, iron sulphide and silicates. They suggested that the core would be able to form by outmelting of a eutectic mixture of Fe–FeS at a temperature of 500 K or more below the melting point of pure iron, thereby explaining how mantle silicates remained unmelted at the time of core formation. The sulphur hypothesis is not without difficulties as pointed out by RINGWOOD (e.g. 1979). The abundance of sulphur relative to silicon within the Earth would need to be at least half the solar system abundance to account for so much sulphur in the core. This conflicts with the inference that the Earth was heavily depleted in volatile elements at its time of formation, as indicated for instance by the low abundances of alkali metals in the mantle. Sulphur is more volatile than the alkali metals and should therefore be much less abundant within the Earth than implied by the hypothesis.

Another possibility is that large quantities of oxygen may be present in the core, probably as FeO. This was originally suggested by DUBROVSKIY and PAN'KOV (1972) in explanation of the density and was later adopted by RINGWOOD (1977, 1979) on geochemical grounds. The observed density of the outer core could be accounted for by an admixture of between 30 % and 60 % by weight of FeO, that is 7 % to 13 % by weight of oxygen. Extrapolation of experimental solubility studies to temperatures of around 2750 K and pressures appropriate to the lower mantle indicates that FeO is probably soluble enough in molten iron to be incorporated in sufficient abundance into the core when it formed. According to Ringwood, incorporation of FeO into the molten iron during core formation would lower the melting point similarly to FeS, thus aiding the process of core formation from a hot homogeneous Earth.

The abundant light element present in the outer core is thus probably either sulphur or oxygen or a mixture of both of them. As discussed below, there has also been some speculation that a substantial proportion of the Earth's store of potassium is located in the core. Other elements such as carbon may also be present in the core in significant proportion.

Potassium in the core?

The alkali metals potassium, rubidium and caesium are notably deficient in the crust and upper mantle relative to silicon. This is normally attributed to a general deficiency of the more volatile elements in the Earth. An alternative speculation is that these metals were originally present in about their solar abundances in the proto-Earth but that the bulk of them has been taken into an iron–iron sulphide core when it formed (LEWIS, 1971; GOETTEL, 1976). According to this idea, up to 75% of the Earth's potassium may possibly be locked up in the core. The hypothesis suggests that potassium and the other two alkali metals were chalcophyle (sulphur-loving) under the reducing conditions assumed to be prevalent when the core formed, so that they were preferentially concentrated in the Fe–FeS melt rather than in the solid silicates left to form the mantle. There has subsequently been an unresolved controversy initiated by OVERSBY and RINGWOOD (1972) and summarized by GOETTEL (1976) as to whether potassium would in reality be partitioned into the iron melt. Further doubts are raised by the possibility that oxygen rather than sulphur is the abundant light element present in the outer core. Thus the hypothesis of potassium in the core is interesting but highly speculative.

An important implication of the presence of potassium is that it would provide a significant heat source within the core which would remain active over the Earth's whole lifespan. MURTHY and HALL (1972) suggested that the initial abundance of potassium in the Earth was 0·05% and that about three-quarters of this was taken into the core when it formed. Based on these assumptions, the present rate at which heat is being evolved by decay of the long-lived radioactive isotope ^{40}K within the core would be about 10^{12} W amounting to about 2·5% of the present heat loss from the Earth. This might be adequate to power the geomagnetic dynamo (p. 263). However, the geomagnetic dynamo could alternatively be powered by cooling of the core and/or by release of low density material during progressive solidification of the almost pure-iron inner core, so that the requirement for potassium in the core is not compelling.

In conclusion, it needs to be emphasized that the core cannot be sampled directly by man and that our ideas on its composition are based on extrapolation and indirect arguments. Thus all the speculations should be treated with caution.

6.3 Formation of the core

The core probably formed at a very early stage in the evolution of the Earth. It almost certainly predates the earliest crustal rocks nearly 4000 My old and it must have been functioning as the source of the main geomagnetic field prior to the oldest rocks displaying remanent magnetization. According to OVERSBY and RINGWOOD (1971), an appreciable amount of lead would have been taken into the core on differentiation from the mantle, thus re-setting the initial date of the U–Pb clock in the mantle. This initial date for the mantle does not differ significantly from the age of formation of the meteorites and the Moon, indicating that the core formed at or shortly after the Earth's formation and certainly within about one hundred million years of it. Most modern hypotheses go even further, dating core formation within at most a few million years of the Earth's accretion, as adequate heat sources are probably available at this stage but would cease to be effective shortly afterwards. Within this framework, one group of hypotheses attributes core formation to the process of inhomogeneous accretion of the Earth, the Fe-rich core accreting first and the silicate mantle subsequently. The other group assumes homogeneous accretion with subsequent segregation of core and mantle.

Core formation by inhomogeneous accretion

If the Earth condensed from the solar nebula by inhomogeneous accretion, then iron particles may have agglomerated first to form the core with later accretion of silicate particles to form the mantle. This idea avoids the need for later segregation of core and mantle from an initially homogeneous Earth. Early segregation of the iron-rich grains could have occurred if they had been able to stick together more easily than the silicate grains. OROWAN (1969) suggested that the metallic particles in the solar nebula may have stuck together because they are likely to have been plastic-ductile, making it possible for them to lose kinetic energy on collision as a result of plastic deformation; alternatively they may have stuck together because of their magnetic properties. Yet another possibility suggested by GROSSMAN (1972) is that the iron and nickel grains agglomerated before the silicate grains because they were inferred to condense from the gaseous nebula at rather higher temperatures than the silicates.

A series of difficulties raised by the inhomogeneous accretion hypothesis has been summarized by BRETT (1976). Firstly, the temperature at which iron would condense out of the gaseous nebula is now believed to be not significantly higher than that at which the predominant silicate mineral forsterite would condense. Secondly, inhomogeneous accretion might be expected to produce a core of almost pure iron and nickel without a sufficient proportion of a lighter element. Thirdly, there is difficulty in understanding how the core could have melted early in the Earth's history without an adequate heat source.

Core formation after homogeneous accretion

The idea of homogeneous accretion of the Earth has been rather more widely accepted than that of inhomogeneous accretion. The associated process of core formation is believed to have occurred broadly along the following lines. The primitive Earth, by analogy with meteorites, probably consisted of an intimate admixture of silicate grains with high melting point and low density and metallic iron–nickel grains with low melting point and high density. A significant amount of iron sulphide or iron oxide may also have been present. The primitive Earth probably heated up rapidly during and shortly after accretion as a result of trapping of energy of collision and by decay of now extinct short-lived radionuclides. As the primitive Earth warmed up, eventually the melting point of the Fe–Ni phase, or more probably a eutectic mixture of it with FeS or FeO, would be reached at some depth within the outer shell deep enough not to have been affected by cooling by thermal conduction to the surface. A thin shell consisting of a mixture of liquid metallic phase and solid silicates would form at this depth (Fig. 6.7). The denser liquid phase would then drain towards the Earth's centre by processes akin to magma intrusion but in the opposite direction. Once core formation had started, a large amount of gravitational energy would be released as a result of the concentration of the high density phase towards the Earth's centre. Most of this would be evolved as heat by friction, with the result that the process of core formation, once started, would avalanche to completion over a relatively short period of time.

Assuming that the Earth did form by homogeneous accretion, then the temperature somewhere within it must have reached the melting point of the iron phase during the first 100 My of its life. Otherwise the core could not have formed. Our problem is to understand how the Earth could have heated up so rapidly. The main present-day source of heat within the Earth is the decay of the long-lived radioactive isotopes of uranium, thorium and potassium (p. 276), but these could only raise the average temperature by about 100 K during the first 100 My. Such a small contribution would be insignificant to the problem of core formation. Probably the main source of rapid early heating is the release of gravitational energy as the separate small bodies which coalesced to form the Earth were pulled together. As these bodies accelerated towards the growing proto-planet,

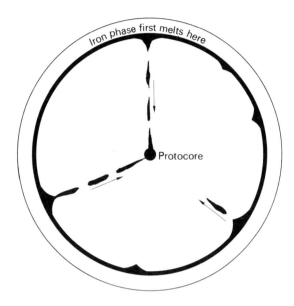

Fig. 6.7 The start of the process of core formation.

their gravitational energy would have been converted to kinetic energy which was mostly released as heat on impact. Some of this heat would have been re-radiated back into space but some of it would have been trapped within the growing Earth. The faster accretion occurs, the more of the heat gets trapped. Heating is also aided by the presence of an opaque atmosphere. In addition to the collisional heat, a further rise in internal temperature would have occurred as a result of adiabatic self-compression of the growing Earth. Yet another possible source of heat for up to 2–3 My after start of accretion is the decay of short-lived radionuclides such as ^{26}Al (p. 278). All these sources of heat would only be effective during the accretion period or at most within about 2 My of it. This is why it is widely believed that the core formed either during accretion or at most a few million years after it.

HANKS and ANDERSON (1969) modelled the rise in temperature in the newly accreting Earth caused by released energy of collision and adiabatic self-compression, but neglecting any contribution from short-lived radionuclides. Their calculations are based on an arbitrary specified history of radial growth of the Earth during accretion, starting slowly, then accelerating as the planet grows and finally slowing down as it approaches its final size. They assumed that the surface of the growing Earth remained at a temperature such as to maintain equilibrium between heat released by collisions and that re-radiated back into space. The level of radiation was assumed to be 1 % of the blackbody value. The resulting temperature distribution on completion of accretion is higher for rapid than for slow accretion (Fig. 6.8), the maximum computed temperature in the newly formed Earth being 2300 K if it accreted over 0·2 My but reaching only 1300 K if accretion dawdled over 1 My. The highest temperatures occur between a third and half way down to the centre, with relatively low temperatures in the central region because of the small amount of heat trapped there during the early stages of accretion. According to this model, high enough temperatures would be reached to melt an iron–iron sulphide eutectic mixture provided that accretion occurred over about 0·5 My or less (Fig. 6.8).

Other accretion models allowing a rather longer time scale of core formation have also been suggested (see RINGWOOD, 1979). If ^{26}Al was a significant heat source then core formation could occur up to 1 or 2 My after start of accretion. If certain other extinct radionuclides such as ^{236}U

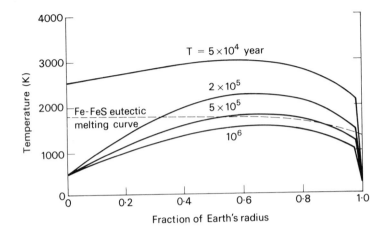

Fig. 6.8 Temperature-depth distributions within the Earth on completion of accretion according to the Hanks-Anderson model (see text), compared with the melting curve for the Fe-FeS eutectic mixture according to MURTHY and HALL (1972). T is the time period of accretion. Note that melting first occurs about one third way down to the centre provided that accretion occurs over a period of 5×10^5 years or less. Note also that short-lived radioactive heat sources are not taken into account.

were sufficiently abundant, an even longer time scale would be possible. The available evidence, however, favours a time scale of core formation much nearer 1 My than the upper allowable limit of 100 My.

6.4 Temperature distribution of the core and origin of the inner core

Knowledge of the thermal regime of the core is of great importance in understanding the relationship between the inner and outer core and the origin of the main geomagnetic field. It has been determined by the highly uncertain extrapolation of the melting point of iron to high pressure, complicated by the likely presence of an admixed lighter element. Consequently there remain some significant unresolved controversies.

The temperature at the core-mantle boundary must be below the melting temperature of the silicon-magnesium rich rocks of the lowermost mantle but above that of the iron of the outer core. The inner core boundary appears to mark the depth within the core where the temperature equals the melting point of the iron present, with the iron being solid at the higher pressure below and liquid at the lower pressure above. Such a situation can only occur if the melting point gradient is steeper than the actual temperature-depth gradient (Fig. 6.9). It implies that the temperature at the core-mantle boundary must be significantly above the melting point of the outermost part of the core. Within this general framework, JACOBS (1953) suggested that the inner core has progressively grown by solidification of the liquid iron as the core of the Earth has slowly cooled. This idea still remains the most viable explanation of the origin of the inner core, with the proviso that the nearly pure iron–nickel of the inner core is probably crystallizing from a liquid outer core containing iron sulphide or iron oxide or both. The timescale of formation of the inner core depends on the rate of cooling of the core, which is not known. At one extreme, there may have been a nucleus of it present when the core formed. At the other extreme, the inner core may not have started to form until quite late in the Earth's history.

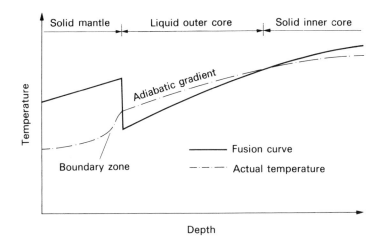

Fig. 6.9 Possible explanation of the inner core boundary as caused by solidification at high pressure, which continues to occur provided that the core is slowly cooling.

Three known cubic phases of solid iron are stable at low pressure. These are, with increasing temperature, alpha, gamma and delta iron. Shock wave experiments have shown that a further low temperature phase with hexagonal close packed structure called epsilon iron is stable at high pressures (BIRCH, 1972). The delta iron is the phase stable at melting point down to about 200 km depth where a triple point occurs between the delta and gamma phases and liquid iron, gamma iron being the melting phase through most of the lower mantle below this depth. LIU (1975) has extrapolated the epsilon-gamma phase boundary to very high pressure to infer that there is a further triple point between the liquid iron and the gamma and epsilon phases at a pressure slightly less than that at the core-mantle boundary. This suggests that the epsilon iron is probably the phase in equilibrium with liquid iron at core pressures.

The melting curve of iron has only been measured up to pressures which fall far short of those of the core. Most estimates of core temperature are therefore gross extrapolations of the low pressure melting curves using semi-empirical laws and they take no account of the solid phase transitions. An assessment by BOSCHI and others (1979) based on the Ross melting criterion (p. 171), using a variety of possible interatomic potential functions, yielded melting temperatures within the range 3500 to 4300 K at the core-mantle boundary and between 4500 and 7000 K at the inner core boundary. They estimated that the fusion gradient in the outer core lies between 0·6 and 1·1 K km^{-1}.

The presence of oxygen or sulphur in the core would lower the melting temperature below that of pure iron. The melting temperature at atmospheric temperature of the Fe–FeS eutectic mixture, which contains about 27 % by weight of sulphur, is about 550 K below that of pure iron. USSELMAN (1975a) has experimentally investigated the effect of increasing pressure on this melting point. He found that the temperature-pressure gradient steepens at about 5200 MPa, possibly corresponding to the delta-gamma transition. At 10 000 MPa, the highest pressure reached by the experiments, the melting temperature of the eutectic mixture (23·6 % sulphur) was found to be 1431 K, which is about 1000 K below that of pure iron. USSELMAN (1975b) extrapolated his results to core pressures assuming linear dependence on isothermal compressibility, obtaining a eutectic temperature of 2070 K (17·5 % sulphur) at the core-mantle boundary and 2370 K (15 % sulphur) at the inner core boundary. The composition of the core, with about 10 % sulphur, is thus on the iron side of the eutectic mixture and the extrapolated melting temperatures at the upper and lower boundaries of the outer core were estimated to be 3250 and 3500 K respectively. The inferred fusion gradient in the outer core is about 0·1 K km^{-1} which seems to be exceptionally small, possibly arising from neglect of the gamma-epsilon iron transition. According to Usselman, the effect of admixed nickel on the melting temperature is relatively minor. The potential influence of admixed iron oxide on the melting point of iron at high temperature and pressure has not yet been studied in a comparable way to that of sulphur, but it is anticipated that the effect would be similar, in significantly lowering the melting temperature. In conclusion, the temperature at the core-mantle boundary is likely to be between 3000 and 4000 K and that at the inner core boundary some 1000 K higher.

The core paradox

It was pointed out earlier in this section that the actual temperature gradient in the core must be less steep than the melting point gradient if the inner core boundary is at the melting point. It was also generally assumed, at least until about 1970, that the actual gradient must be slightly steeper than the adiabatic gradient, so that thermal convection can take place in the outer core to produce the main geomagnetic field. According to STACEY (1977) the estimated adiabatic gradient in the core decreases downwards from about 0.7 to $0.2 \, \text{K km}^{-1}$, with an uncertainty of the order of 50 % arising from estimation of the ratio of thermal expansion to specific heat capacity. Thus the adiabatic gradient in the core cannot be much shallower than the fusion gradient. KENNEDY and HIGGINS (1973) went even further by pointing out that the adiabatic gradient may even be steeper than the melting point gradient in the outer core. If so, the adiabatic gradient must also be steeper than the actual temperature gradient. Such a situation would mean that the outer core is stably stratified and cannot undergo thermal convection in generation of the main geomagnetic field. This problem has been referred to as the *core paradox*.

There are two possible ways around the core paradox. One of these is to accept it at face value, with the implication that the main geomagnetic field must be generated by processes other than thermal convection in the outer core (p. 263). In such a situation, there is no compelling requirement for a substantial heat loss from core to mantle, except if the inner core is still growing significantly. The temperature-depth gradient might even be negligible.

The alternative explanation is that the estimated temperature gradients in the Kennedy-Higgins model must be in error, the fusion gradient in reality being steeper than the adiabatic gradient. Later work on the melting point of iron (LIU, 1975; BOSCHI and others, 1979) favours this latter interpretation but suggests that the fusion gradient cannot be much steeper than the adiabatic gradient. Hence the temperature at the core-mantle boundary is probably not more than 100–300 K higher than the melting point there. Thermal convection would be able to take place in the outer core to generate the geomagnetic field and also to allow the inner core to cool significantly. However, this viewpoint also implies that there must be a substantial outflow of heat from the core to the mantle by thermal conduction. The reason for this is that the thermal conductivity of the core is estimated to be five to ten times greater than that of the mantle, so that a significant amount of heat will be conducted down an adiabatic gradient of the order of $0.25 \, \text{K km}^{-1}$, amounting to about 3 % of the geothermal heat loss from the Earth's surface. This must originate either by cooling of the core or from heat sources within it or both.

6.5 The Earth's magnetic field

Introduction—the geomagnetic field in historical times

Since about 1950 it has been recognized that the origin of the main geomagnetic field is related to fluid motions in the outer core, probably involving a dynamo process. A comprehensive theory has not yet been developed but what we do know adds considerably to our understanding of the structure of the core and processes which occur in it.

The geomagnetic field at a point is a vector quantity and therefore it needs three quantities to describe it completely (Fig. 6.10). It can be specified by the vertical component measured downwards (Z), the horizontal component (H) and the declination (D) which is the angle between true north and the direction of the horizontal component (taken as positive towards the east). Another common way of describing the field is to use the magnitude of the total field (F), the angle of its dip (I) and the declination. The variation with time of the elements of the geomagnetic field are studied on a world-wide basis by recording them at permanently established magnetic observatories, such as the one run by the Institute of Geological Sciences at Eskdalemuir in south Scotland. Short-period variations can be studied using temporary observatories and the areal distribution of the geomagnetic field can be mapped by making magnetic surveys on land and at sea. The standard work on the description and analysis of the geomagnetic field is by CHAPMAN and BARTELS (1940).

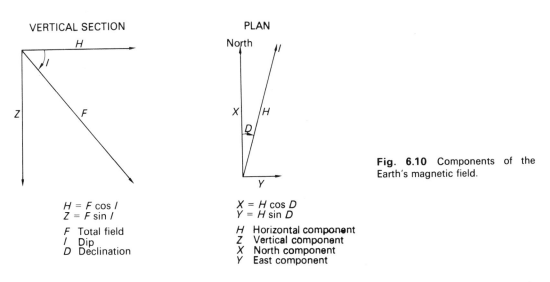

VERTICAL SECTION

$H = F \cos I$
$Z = F \sin I$

F Total field
I Dip
D Declination

PLAN

$X = H \cos D$
$Y = H \sin D$

H Horizontal component
Z Vertical component
X North component
Y East component

Fig. 6.10 Components of the Earth's magnetic field.

About 90% of the present-day geomagnetic field can be represented by the field of a magnetic dipole at the Earth's centre which makes an angle of about $11.5°$ with the Earth's axis of rotation. An appreciable *non-dipole field* remains after the best-fitting *dipole field* has been subtracted from the observed field of the present day.

The mean annual values of all the magnetic elements vary in a regular way from year to year. This long-period change in the geomagnetic field from one year to the next is known as the *secular variation*. It was discovered by Gellibrand in 1634 when he recognized that the declination at London changes with time (Fig. 6.11). The secular variation affects both the dipole and non-dipole parts of the field. Over the last century the dipole field has been decreasing by about 0.04% per year. The percentage annual change in the non-dipole field is on average somewhat larger but it varies from region to region and involves both increase and decrease of field strength.

It can be shown by spherical harmonic analysis that both the main field and its secular variation originate within the Earth. This contrasts with the short-period variations of the magnetic elements (p. 175) which are primarily caused by electric current systems above the Earth's surface. Spherical harmonic analysis has also been used to obtain a representation of the internal geomagnetic field for the year 1965.0 including harmonics up to and including degree eight (that is, wavelengths of 5000 km and longer). This representation is known as the International Geomagnetic Reference

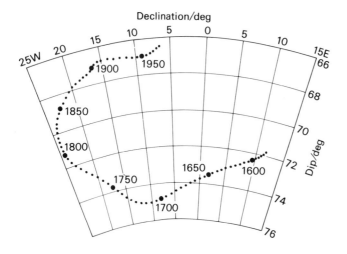

Fig. 6.11 Secular variation of the declination and inclination at London since 1570, plotted every fifth year. Redrawn from MALIN and BULLARD (1981), *Phil. Trans. R. Soc.*, **299A**, 367.

Field (IGRF) and it includes coefficients to allow for secular variation up to about 1972. Because of unpredictable changes in the rate of secular variation, the IGRF of 1965 is now becoming unsatisfactory but a generally accepted revised reference field has not yet been agreed upon.

A map showing the secular variation of a specific magnetic element for a given epoch is called an *isoporic chart*. Figure 6.12 is an example. This shows a series of centres where the secular variation is particularly large. These are known as *isoporic foci* and they are particularly associated with large changes in the non-dipole field.

Isoporic foci are conspicuously lacking from the Pacific Ocean, but whether this is an accident of chance or has some more fundamental significance is still not clear. When isoporic foci for successive epochs are compared (Fig. 6.12), a most interesting fact emerges. It is found that these foci are drifting westwards at a rate determined by BULLARD and others (1950) as $0.32° \pm 0.067°$ in longitude per year. The non-dipole field is also drifting westwards but at a slower rate of about $0.18°$ per year. The different rates of drift of the foci of secular variation and the non-dipole field can best be accounted for by assuming that there are stationary and drifting components of the non-dipole field. The secular variation, and particularly its westward drift, cannot be adequately accounted for by any process occurring entirely within solid subdivisions of the Earth. The relevant part of the Earth, notably the outer core, must therefore be fluid. The westward drift also implies that the outermost part of the core may possibly be rotating at about $0.3°$ per year more slowly than the crust and mantle.

History of the geomagnetic field in the geological past

The palaeomagnetic method (p. 80) and the analysis of oceanic magnetic anomalies (p. 104) have together made it possible for us to study in remarkable detail certain aspects of the geomagnetic field and its secular variation going back beyond the historical record into the geological past.

The secular variation can be traced back into prehistory by measuring the thermo-remanent magnetization of ancient pottery and bricks from archaeological sites and historically-dated lava flows. It can be taken further back to around 50 000 years ago by studying the depositional remanent magnetization of lake sediments, dated by counting varves or by use of the radiocarbon dating technique. Such studies show that the secular variation occurred then in much the same way as in historical times. Some periodicities of longer duration have additionally been revealed, particularly ones of about 2000 and 5000 year periods (OBERG and EVANS, 1977). Going further back into geological time, the palaeomagnetic method is not normally able to trace the actual course of the secular variation because most rock successions are deposited too slowly to record it in detail. However, its presence is clearly revealed by the angular dispersion of observations after correction for scatter arising from other sources. For instance, an analysis of data over the last 5 My shows an average angular dispersion attributable to secular variation which increases with latitude from 13° at the equator to 20° at the poles (McELHINNY and MERRILL, 1975). There is some indication that the amount of angular dispersion may vary with time, suggesting periods of more and less vigorous secular variation.

The secular variation is likely to include separate contributions from the time variation of (i) intensity and direction of the non-dipole field, (ii) intensity of the main dipole (dipole oscillations) and (iii) orientation of the main dipole relative to the rotation axis (dipole wobble). According to McELHINNY and MERRIL (1975), the secular variation over the past 5 My, as evidenced by angular dispersion of palaeomagnetic observations, can be modelled in terms of a dipole wobble of about 11° upon which non-dipole field variations similar to those of historical times are superimposed.

An important geomagnetic inference from the historical and archaeological record of the past field and from palaeomagnetism of deep-sea cores extending back a few million years is that the ancient field of the Earth as a whole, when averaged over about 10 000 years, is to the first order that of a dipole at the Earth's centre orientated along the axis of rotation. This is confirmed by the average pole position obtained for palaeomagnetic samples of lavas up to 20 My in age, which are too young to have been significantly displaced by continental drift; this average pole coincides with the pole of rotation. Thus the non-dipole field appears to average out to zero over a period of the order of 10^4 years and the average axis of the dipole field corresponds to the Earth's spin axis. The use of the palaeomagnetic method to determine the palaeolatitudes and palaeoazimuths from older rocks tacitly assumes that this situation applies further back in the past and to individual localities. A certain amount of caution, however, is needed. At certain specific localities, such as Hawaii, the secular variation appears not to average out even over periods longer than 10^4 years and a palaeomagnetic measurement based on such a region might show an error in azimuth after averaging of several degrees.

Turning to the intensity of the ancient field, this can be estimated by comparing the natural remanent magnetization of cooled igneous rocks or baked pottery with experimentally determined thermo-remanent magnetization of the same samples acquired by cooling in fields of varying intensity. Studies using archaeological material show that the typical secular variation of the field strength over the last few thousand years amounts to about 10–15% of the average value. Going further back in time, palaeointensity measurements indicate that the general level of the field has on average been much the same as it is at present. The oldest rocks displaying remanent magnetization are 3800 My old (MURTHY, 1976) so that we know that the main geomagnetic field was in existence at the beginning of the Precambrian.

A most important discovery of palaeomagnetism relating to the past history of the magnetic field is that some rocks have picked up a permanent magnetization in the opposite sense to the present field. This is known as *reverse magnetization*. For instance, HOSPERS (1951) showed that the late Tertiary and Pleistocene lavas of Iceland could be stratigraphically subdivided into groups of normally and reversely magnetized rocks. Later work showed that reverse magnetization is about as common as normal magnetization through the geological column and that reversals occur in sediments as well as in igneous rocks. Either the reversely magnetized rocks have somehow picked up a magnetization in the opposite direction to the ambient field when they were formed, or the Earth's field itself has suffered periodic reversal.

In this respect, it has been shown that a dacitic pitchstone from Mt. Asio in Japan picked up a

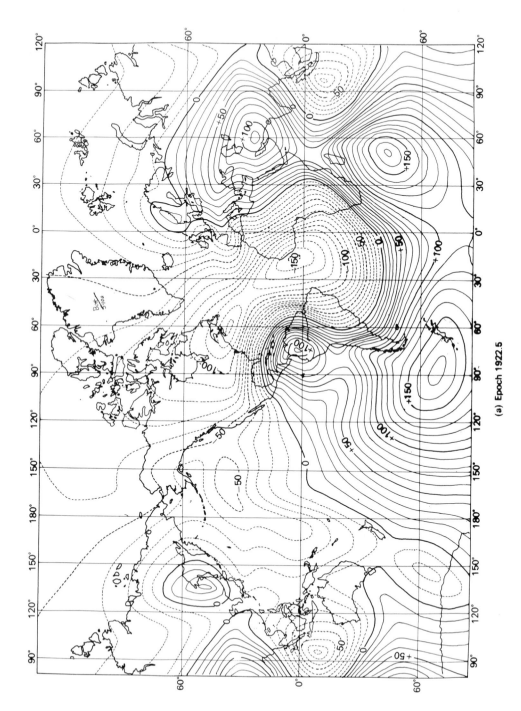

(a) Epoch 1922.5

Fig. 6.12(a) Isoporic chart showing secular change in gamma per year, vertical intensity, for epoch 1922·5.

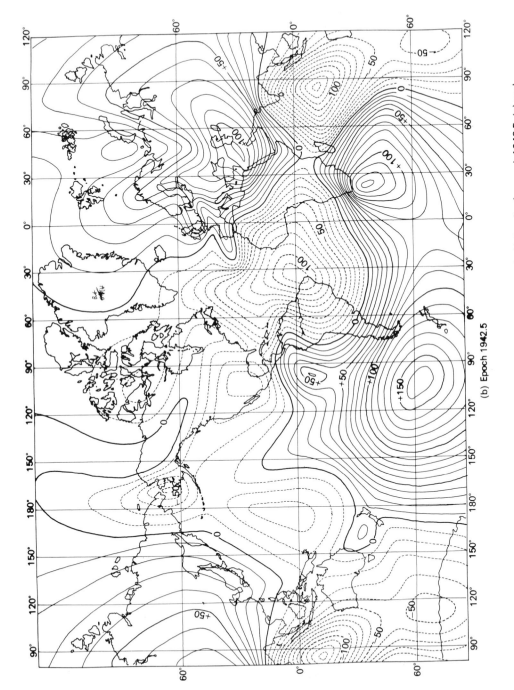

(b) Epoch 1942·5

Fig. 6.12(b) Isoporic chart showing secular change in gamma per year, vertical intensity, for epoch 1942·5. (a) and (b) redrawn from VESTINE and others (1947), *Description of the main magnetic field and its secular variation, 1905–1945*, p. 384, Carnegie Institution of Washington.

reverse magnetization in the present normal Earth's field. This rock and other dacites from Haruna in Japan also undergo self-reversal in laboratory experiments (NAGATA, 1953). This self-reversal can be explained theoretically by several processes (NÉEL, 1951). The simplest of these is as follows. Two magnetic phases with different Curie points may be intergrown in a rock. As the phase with the higher Curie point cools, it picks up a normal magnetization, but the resulting lines of force within the mineral intergrowths are in the opposite direction and may exceed the Earth's field strength. As the second phase cools through its Curie point, it therefore picks up a magnetization in the reverse direction. As further cooling occurs, the reverse magnetization of the phase with the lower Curie point may become dominant, imparting an overall reverse magnetization to the rock. This simple mechanism is probably unimportant in relation to the other processes suggested by Néel, but it illustrates the principle that self-reversal is possible.

However, the dacites of Asio and Haruna are exceptional. Most igneous rocks contain only a single important magnetic phase and they do not undergo self-reversal in laboratory experiments. There is circumstantial evidence from depositional magnetization, and also from the observation that normally magnetized lavas adjacent to a reversely magnetized dyke become reversely magnetized in the vicinity of the dyke as a result of thermal metamorphism (Fig. 6.13). The almost perfect agreement between the sequence of reversals over the last 5 My recorded in lava sequences and deep-sea sediments, and that displayed by oceanic magnetic anomalies at the ocean ridge crests (p. 103) adds further convincing evidence that the polarity of the main field changes. It is therefore now generally agreed that most observed reversals represent true switches in polarity of the geomagnetic field. The occurrence of such polarity reversals is a remarkable property of the ancient geomagnetic field which must be taken into account in any viable theory of its origin.

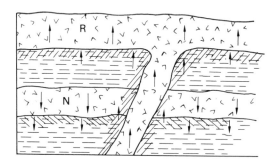

Zone baked by flow N

Zone baked by flow R

Fig. 6.13 Magnetic directions in the baked zone adjacent to a dyke, showing that the polarity of the normally magnetized lava has been reversed during thermal metamorphism. Redrawn from COX and DOELL (1960), *Bull. geol. Soc. Am.*, **71**, 736.

Palaeomagnetic measurements can be used to track in some detail how certain reversals have taken place. Rapidly erupted piles of lavas have been used but the possible irregularity of eruption presents problems over timing. It is better to use a rapidly deposited sequence of lake or oceanic sediments, or to study the palaeomagnetism of samples from an igneous intrusion across which a cooling front has migrated at a calculable rate. An example which uses the cooling of the Tatoosh intrusion in the Mount Rainier National Park in Washington is shown in Fig. 6.14. According to FULLER and others (1979), the main features of a typical transition are as follows. Where it has been possible to estimate the duration of the reversal, this has been found to take about 4000 years.

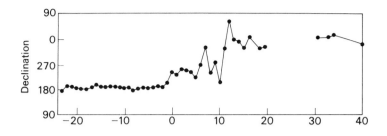

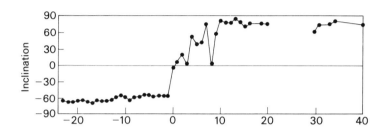

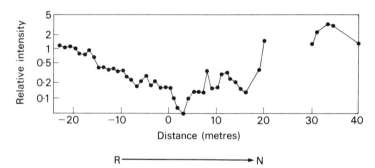

Fig. 6.14 A record of the reversal of the geomagnetic field based on progressive cooling across the contact of the Tatoosh intrusion, Mount Rainier National Park in Washington, with change of polarity from reverse to normal. Redrawn from FULLER and others (1979), *Rev. Geophys. space Phys.*, **17**, 194.

During the reversal, the apparent position of the pole relative to the locality migrates to its opposite polarity position by an irregular path, somewhat constrained in longitude, which makes several loops in transit. The reversal may typically be accompanied by a reduction in the field intensity to about 10 % of its normal value; the intensity starts to weaken well before the onset of the migration of the apparent pole and does not regain its full strength until well after it. The observations indicate that the reversals probably do not take place by rotation of the dipole field from one orientation to the opposite one. Rather, the initial dipole field appears to decay to zero and then build-up in the opposite direction, accompanied by the dominant presence of a nearly axially symmetrical non-dipole field. It is possible that the reversal is triggered by a weakening of the dipole field.

The pattern of geomagnetic reversals over the last 150 My is displayed in the geomagnetic polarity time scale based on the interpretation of oceanic magnetic anomalies (p. 104 and Fig. 3.15). The periodicity of reversals over the last 45 My, averaged over 10 My windows, has been approximately constant at 0·3 My. In detail, the reversals show an irregular pattern which according to cox (1975) may possibly be attributable to a randomly-occurring instability in the generation mechanism of the main field. Prior to 45 My ago, the frequency of reversals displayed in the Mesozoic polarity time scale appears to be significantly lower. The most striking feature of the pre-Tertiary record is the occurrence of two periods of dominantly normal polarity from 85 to 110 My ago in the mid-Cretaceous and prior to 150 My ago in the Upper Jurassic. Further back in time, land palaeomagnetic studies have shown an even longer period of dominantly reverse polarity spanning Upper Carboniferous and Permian time. The available evidence thus shows that the frequency of reversals, although remaining similar over periods of a few tens of million years, has varied considerably over longer periods. According to cox (1975), the most likely explanation of these small but significant changes in the geomagnetic generation process, which occur at about 50 My interval, relates to change in the physical properties at the core-mantle boundary. Such a phenomenon might well be associated with changes in the pattern of convective circulation in the mantle, particularly the location of the cold downsinking material which may reach the core-mantle boundary.

Origin of the main field and the secular variation

In 1600 William Gilbert published his famous treatise *De Magnete*, in which he pointed out that the dip of the Earth's field resembles that of a sphere of the magnetized material lodestone. Between then and the early 20th century it was widely held that the geomagnetic field is caused by a strongly magnetized region within the Earth. For instance, a dipole at the centre with a magnetic moment of $8{\cdot}05 \times 10^{22}$ A m^2 gives a reasonable approximation to the Earth's field as it was in 1955. Exactly the same effect would be caused by a uniform magnetization of 75 A m^{-1} affecting the whole Earth, or 490 A m^{-1} affecting the whole core, or about 10^4 A m^{-1} affecting the inner core only (Fig. 6.15). The dipole field could also be produced by a current system such as is shown in Fig. 6.15(d).

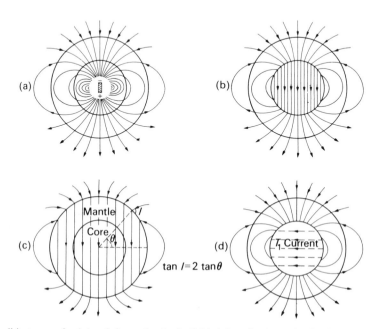

Fig. 6.15 Possible types of origin of the main dipole field: (**a**) a dipole at the Earth's centre; (**b**) a uniformly magnetized core; (**c**) a uniformly magnetized Earth; and (**d**) an east-west current system flowing along the core-mantle boundary with current density proportional to the cosine of latitude (toroidal type T_1).

About the most highly magnetized common rocks of the crust form layer 2 of the oceanic crust (p. 100). Their magnetization is about $10\,A\,m^{-1}$ and they cause local disturbances of the magnetic field amounting to about 1 %. A much stronger magnetization is needed to account for the main field. One of the obvious difficulties of the large magnet theory is that the Curie point of iron is about 1040 K and this temperature is probably exceeded at and below a depth of 200 km at most. The effect of increased pressure is to reduce the Curie temperature slightly, although it is just possible that the effect is reversed at exceptionally high pressures.

Several new theories have been suggested during this century. One of the more exotic of these was the idea of BLACKETT (1947) that massive rotating bodies have an inherent magnetic field associated with them. This theory was disproved when (i) it was shown that the magnetic field increases with depth in mines, contrary to Blackett's predictions (RUNCORN and others, 1951), and (ii) a laboratory experiment made by Blackett failed to reveal the effect. Nowadays, the most decisive evidence contrary to most of the early theories is the realization that the Earth's field has periodically reversed itself in the geological past.

The only modern theory which seems capable of explaining the past and present features of the main field is the dynamo theory. It was originally suggested by LARMOR (1920) as a mechanism to explain the Sun's magnetic field, but COWLING (1934) showed that Larmor's mechanism would not work. Cowling's objection applies to any dynamo based on axially symmetrical fluid motions. The theory was revived again in a modified form as an explanation of the geomagnetic field by W. M. Elsasser and E. C. Bullard.

The dynamo theory attributes the observed geomagnetic field to a system of electric currents in the core and lower mantle. The electric currents must be maintained because otherwise they would die out in less than one million years. This is accomplished by fluid motions in the outer core which cause it to act as a dynamo. It was usually assumed that the fluid motion is caused by thermal convection. The outer core needs to be a good electrical conductor.

To illustrate the basic principle, a disc dynamo is shown in Fig. 6.16. A rotating circular disc which is a conductor cuts a weak axial magnetic field. An e.m.f. is produced between the centre and edge of the disc and this is used to drive an electric current through the attached coil. Provided that the coil is wound in the correct sense, the current causes an axial magnetic field which reinforces the original field. If the disc is rotated fast enough, a very slight stray magnetic field at the start can be amplified so that the dynamo maintains its own magnetic field. This is the principle of the self-exciting dynamo, although in practice the disc is replaced by an armature with many windings and the coil also has many turns, both being wound on soft-iron cores. A self-exciting dynamo of this type differs fundamentally from the core in one respect. The ability of the disc dynamo to work at all depends on its asymmetry; but the core is geometrically homogeneous (singly-connected) and

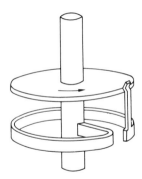

Fig. 6.16 A simple disc dynamo. After BULLARD and GELLMAN (1954), *Phil. Trans. R. Soc.*, **247A**, 214.

cannot have the same type of asymmetry as the disc. The crucial question is whether a homogeneous self-exciting dynamo can exist at all.

The branch of applied mathematics dealing with the interaction of fluid motions and electromagnetic fields is called *magnetohydrodynamics*. Thorough investigation of the dynamo theory would involve looking for relevant solutions to the following simultaneous set of partial differential equations subject to specified boundary conditions:

(*i*)   electromagnetic equations connecting the magnetic field with the velocity field;

(*ii*) the Navier-Stokes hydrodynamic equation connecting the velocity field in a viscous rotating fluid with boundary and body forces of electromagnetic, thermal and mechanical origin;

(*iii*) the equation of heat flow appropriate to a fluid in motion;

(*iv*) the equation of continuity for an incompressible fluid.

These equations are non-linear and the problem of their solution is more than formidable. Furthermore, the boundary conditions and the relevant physical properties are not properly known.

In order to simplify the solution of the dynamo equations, the most widely adopted approach has been to assume a pattern of fluid flow in the outer core, thereby reducing the problem to a much simpler one of solving the electromagnetic equations (*i*) above. This is known as the *kinematical approach*, in contrast to the *hydromagnetic approach* which is an attempt to solve the whole system together. The kinematical approach neglects the influence of the electromagnetic force (Lorentz force) acting on the fluid and altering its motion. It is rather like driving an ordinary dynamo at constant speed irrespective of load, implying that extra mechanical force is applied as and where needed to annul the Lorentz force at each point in the outer core. The kinematical approach is useful for demonstrating the feasibility of the homogeneous dynamo mechanism and gaining insight into significant facets of the process, but it cannot produce an actual model of the geodynamo.

The first attempt at modelling a kinematical dynamo was made by BULLARD and GELLMAN (1954) but it has been shown subsequently that the exact model they proposed will not actually work. Later HERZENBERG (1958) and BACKUS (1958) gave rigorous mathematical demonstrations of the possibility of homogeneous dynamo mechanisms, but the models they used are unlikely to represent true fluid motions in the core. A few years later, LOWES and WILKINSON (1963) constructed an experimental model of the Herzenberg dynamo which works and has the added interest of being capable of reversing its field polarity. Perhaps the most significant insight was that of PARKER (1955) who showed how the dynamo mechanism can work as a result of the Coriolis force produced by the Earth's rotation causing the upwelling convection currents to undergo spiralling motion rather like that undergone by wind in atmospheric depressions (cyclones). This spiralling motion removes the axial symmetry as it is in opposite senses in the two hemispheres – otherwise the dynamo would not work as Cowling showed.

Both the velocity of a fluid at a point within the outer core and the magnetic field there are vector quantities. Each of them can be represented in space by a vector field which specifies the magnitude and direction of the vector at each point. Such a vector field, when expressed in terms of spherical polar co-ordinates, can conveniently be divided into two complementary types of field as follows: (*i*) a *toroidal field* in which the vector at every point is perpendicular to the radial direction; and (*ii*) a *poloidal field* which contains radial and tangential components. These are represented by the symbols T and S respectively.

Here is a broad qualitative outline of the Parker dynamo (PARKER, 1955; LEVY, 1972a, 1976).

Two types of magnetic field are assumed to exist in the core. There is a dipole field orientated along the spin axis which occurs both within the core and outside it and which is of poloidal type symmetrical about the spin axis or nearly so, and an east-west orientated toroidal field which wraps itself around lines of latitude within the core. The toroidal field can only exist within the conducting region and for this reason it does not penetrate up to the Earth's surface. Three superimposed types of motion in the outer core are required by the Parker dynamo. These are (*i*) a pattern of thermal convection currents, (*ii*) a differential rotation of the core such as would be expected to occur in a convecting and rotating fluid shell, and (*iii*) spiralling motion in the upwelling convection currents produced by the Coriolis force and having opposite polarities in the two hemispheres. Each of the two main stages of the dynamo process involves the electromagnetic interaction of a magnetic field of T or S types with a component of the fluid motion to produce an electric current system which flows in the electrically conducting core. The electric current then produces a further magnetic field. To simplify the account, we omit reference to the electric current system. Then each stage of the process can be described as an interaction of magnetic field X with fluid motion Y to produce a new magnetic field Z. If X is a toroidal field then Z is poloidal and vice-versa. Looking at it in a slightly different way, the fluid motions in a highly conducting core of large dimension have the effect of dragging the magnetic lines of force with them, thereby increasing the energy of the magnetic field at the expense of the fluid motions and changing their pattern (Fig. 6.17). This effect is opposed by the decay of the field with time which causes the lines of force to slip backwards. The velocity of the motions in the outer core is sufficiently fast for some stretching of the lines of force to take place.

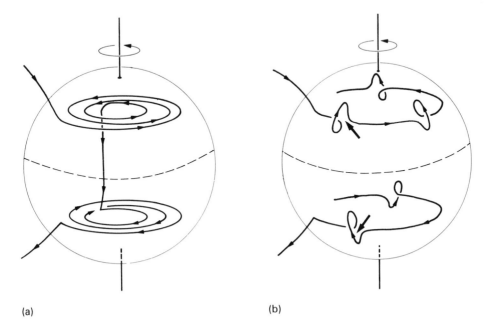

(a)   (b)

Fig. 6.17 The stretching and dragging of magnetic lines of force by fluid motions in the conducting core: (a) differential rotation (T_1 motion) producing a toroidal field of T_2 type out of an initial dipole field of type S_1; (b) spiralling cyclonic motions forming loops in the T_2 lines of force. Redrawn from LEVY (1976), *Ann. Rev. Earth & planet. Sci.*, **4**, 164 and 169. Reproduced, with permission, by Annual Reviews Inc.

The simplest Parker dynamo generating process involves two stages of interaction as follows (Fig. 6.18):

Stage 1: A weak initial dipole field (type S_1) interacts with the toroidal velocity field (type T_1) caused by differential rotation within the core to produce a toroidal magnetic field of type T_2, which wraps itself round the core along lines of latitude, in different directions in opposite hemispheres.

Stage 2: The T_2 magnetic field then interacts with the upwelling and spiralling fluid motions in the convection cells to produce magnetic field loops which coelesce with each other over the core as a whole to produce an axial dipole field of type S_1 of the same polarity as the original field; the opposite polarity of the cyclonic motions in the two hemispheres is essential to the process.

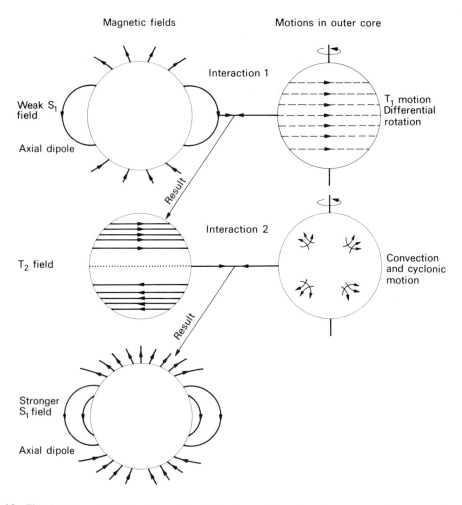

Fig. 6.18 The electromagnetic interactions of the Parker-Levy self-exciting homogeneous dynamo mechanism, as described in the text. The diagrams showing the S_1 magnetic field are diametrical sections passing through the poles. The other diagrams depict magnetic lines of force and flow lines at the surface of the core as they would be viewed from a point vertically above the equator.

An interesting feature of this and other dynamo models is that a toroidal field must be present in the outer core and may extend into the significantly conducting lower mantle. Earlier work tended to imply the existence of a very strong toroidal field but later models have removed the need for the toroidal field to be significantly stronger than the dipole field.

Subsequent developments in kinematical dynamo theory have shown that large and well-organized convection cells in the outer core are not necessarily needed to produce a feasible dynamo mechanism. A system of smaller cells of periodic type (e.g. ROBERTS, 1972), or even the presence of turbulent eddies possessing some coherence of pattern, can under suitable circumstances amplify an initially weak dipole field and thus act as a homogeneous dynamo. All that is basically required is (i) some differential rotation in the outer core to produce the toroidal magnetic field from the initial dipole field, and (ii) an organized system of fluid motions acted upon by the Coriolis force which under suitable conditions can regenerate the dipole field from the toroidal field. The kinematical investigations of dynamo theory have brought us to a state of confidence that such a process can account for the generation of the Earth's magnetic field, and also those of the planets Mercury and Jupiter, the Sun, other stars, and even that of the galaxy. The homogeneous dynamo process is thus not an exceptional phenomenon but rather an expected feature of any sufficiently large body of rotating fluid which is conducting.

It is a much more complicated matter to model a true hydromagnetic dynamo which takes into account the electromagnetic force acting on the fluid. This problem must be tackled if we are to progress to a proper understanding of the generation of the geomagnetic field. Initially the motions in the viscous fluid will result from the pressure gradient caused by the driving mechanism as modified by the Coriolis force, with the Lorentz force having a negligible effect. At this stage the kinematical dynamo theory is applicable. As the dynamo builds up towards its full power output, the Lorentz force will have an increasingly significant effect on the balance of forces driving the fluid motion. Eventually an equilibrium state is reached. One of the only attempts to model a hydromagnetic dynamo is that of BUSSE (e.g. 1975), who used a cylindrical rather than spherical model, but found good agreement with the known properties of the main geomagnetic field. In his model, the Lorentz force is small in comparison with the Coriolis force and the toroidal field is of similar magnitude to the poloidal dipole field.

The secular variation is probably caused by processes associated with the main dynamo mechanism. Variation in the magnetic moment and orientation of the main dipole field can be explained by fluctuations in the strength and geographical pattern of the centres of cyclonic motion which drive the dynamo. Turning to the non-dipole field, LOWES and RUNCORN (1951) showed that the centres of maximum secular variation (Fig. 6.12) could be attributed to horizontal current loops of varying strength at the surface of the core. These current loops are probably caused by interaction between local centres of fluid motion and the toroidal field there. One suggestion is that they result from eddies in the upper reaches of the core, these drifting westwards because of the differential rotation of core and mantle (BULLARD and others, 1950); such eddies might possibly mark the top of the main centres of cyclonic activity. Alternatively, HIDE (1966) has suggested that the motions may arise as hydromagnetic waves of a type which propagates slowly westwards. These waves may be associated with topographical irregularities at the core-mantle boundary. Such waves may interact with the poloidal field to produce the westward drifting isoporic foci.

More spectacular than the secular variation is the ability of the geomagnetic dynamo to reverse its polarity at apparently random intervals of time. In this respect, the Earth's dynamo differs from those of certain stars which reverse themselves periodically. The geomagnetic dynamo retains an approximately constant intensity and direction for periods of the order of 0·1 to 1 My and then abruptly reverses polarity over a period of a few thousand years. Within the framework of the Parker dynamo theory, such reversal of the field can be explained by variation in the spatial distribution of the fifteen to twenty centres of cyclonic upwelling in the core which are supposed to generate the main field (PARKER, 1969; LEVY, 1972b). If the main dipole field at a given time is maintained by cyclonic centres predominantly situated at high latitudes, then a sudden burst of cyclonic activity in low latitudes can cause the field to reverse, and vice-versa. During the generation of the geomagnetic field, regions of both normal and reverse toroidal field are probably present in the core, one of them predominating to define the polarity of the dipole field. A sudden burst of cyclonic activity at latitudes where their activity was previously subdued produces local loops of poloidal field there which interact with the differential rotation to flood the core with a strong toroidal field of opposite polarity to the one previously dominant. The interaction between the main cyclonic motions and the newly reversed toroidal field then causes the dipole field to reverse polarity; it then remains stable until a further burst of suitable new cyclonic activity takes place. As the occurrence of a reversal depends on the development of a suitable pattern of cyclonic centres, each one growing and migrating in a random way, the random time interval between reversals is readily accounted for. Although the Parker-Levy model of a field reversal is not the only possible mechanism of reversal, it does demonstrate that the apparently random timing of reversals is readily explicable within the framework of the dynamo theory.

It has only been possible to give here an introductory and qualitative account of a particularly complicated subject and one which is developing rapidly. Reviews which incorporate extensive reference lists have been written by GUBBINS (1974) and LEVY (1979).

Powering the geomagnetic dynamo

A significant source of power is needed to keep the geomagnetic dynamo running. Otherwise it would decay over a period of less than 100 000 years. Although energy is stored in the magnetic field, the continuing drain on an energy source is needed to drive the associated electric currents which suffer ohmic dissipation as heat. In a steady-state dynamo, the ohmic heat loss is equal to the power consumed by the dynamo. The electrical resistivity of the core is estimated to be about 2×10^{-6} ohm m, so that if the electric current system is known, then the energy dissipated as heat can be estimated. The pattern of electric currents depends on how the dynamo works, the energy dissipated being much greater if a strong toroidal field has to be maintained. Estimates of the energy consumed by the dynamo vary between 10^9 and 10^{12} W, a realistic estimate being about 10^{10} W. This energy contributes to the outflow of heat from the core into the base of the mantle. It is small in comparison with the rate of heat loss from the Earth's surface which is about 4×10^{13} W (Table 7.1).

Thermal energy used to be the most widely favoured power source for the geodynamo (e.g. GUBBINS, 1976). The causitive fluid motions in the outer core would then be driven by thermal convection. If the outer core is stably stratified as a result of a sub-adiabatic temperature gradient or compositional layering, then thermal convection cannot take place. But if it is homogeneous and the temperature gradient exceeds the adiabatic gradient by a small margin, then it probably does occur. Assuming that convection does take place, then the rate of release of thermal energy by the core must greatly exceed the mechanical energy needed to drive the dynamo because of thermodynamical considerations. Firstly, heat which must be lost by thermal conduction down the adiabatic gradient of about 0.25 K km^{-1} is not available for powering the dynamo. Taking the value of about 30 W m^{-1} K^{-1} (Table 7.3) for the thermal conductivity of the outer core, this heat loss by thermal conduction is 1.1×10^{12} W, with an error of around 50%. Secondly, the second law of thermodynamics requires that the 'useful' energy consumed by the dynamo can only be a small fraction of the thermal power input to the system. The efficiency must be less than the ratio of temperature drop across the outer core (in excess of the adiabatic gradient) to its absolute temperature. The actual efficiency cannot be much better than 1 %, suggesting that the thermal energy needed to drive the dynamo is probably about 10^{12} W. Adding this to the heat loss down the adiabatic gradient by thermal conduction, the total thermal energy flowing out of the core needed to make the dynamo work would be about $2 \cdot 1 \times 10^{12}$ W, which is about 5% of the Earth's whole heat loss.

Where can this heat come from? One possibility is that about three-quarters of the Earth's store of potassium is contained in the core (p. 243). The radioisotope ^{40}K would then generate heat at a rate of about 10^{12} W, possibly sufficient to drive the dynamo. Another source of heat consistent with modern ideas of the Earth's thermal evolution is that the core has cooled by say 100–300 K over the Earth's lifespan, thereby releasing heat which might be effectively transferred to the surface by convection in the mantle (p. 283). Taking c_p to be 7×10^2 J kg^{-1} K^{-1} and ignoring the contribution from solidification of the inner core, the average rate of release of heat for a 100 K drop in temperature over 4500 My would be about 10^{12} W. These computations show the feasibility of a thermal power source for the geodynamo provided that the actual temperature gradient is above the adiabatic gradient and that an excessively strong toroidal field is not part of the essential dynamo mechanism. If thermal energy does drive the dynamo, then at least 5% of the Earth's present day heat loss must come from the core.

Two other possible sources of power for the geodynamo have been suggested. Neither of these involves conversion of thermal to mechanical energy and thus they are not limited by thermodynamical considerations. The precession of the Earth (p. 22) has been suggested as one possible mechanism. The precession of the solid mantle above the fluid core would be expected to induce some fluid motions in the outermost core which may become turbulent. MALKUS (1963) suggested that these turbulent currents might power the dynamo, but later work has ruled out this mechanism as only about 10^8 W at most would be available (ROCHESTER and others, 1975).

Another possibility is progressive release of gravitational energy as the dense and pure material of the inner core has separated from the outer core. As almost pure iron and nickel crystallize out to form the inner core, the lighter component FeO or FeS is released to lower the density of the bottom of the outer core. This flows upwards because of its buoyancy causing fluid currents which drive the dynamo while maintaining the outer core in a well-mixed state. According to LOPER (1978), the average release of gravitational energy as the inner core has grown over 4500 My would be $1 \cdot 76 \times 10^{12}$ W. The gravitationally driven dynamo may be up to 50 % efficient, so this provides a viable alternative to thermal power. The inner core can only grow if the core is being cooled. The gravitational dynamo must therefore be associated with a significant loss of heat from the core although the temperature gradient in the outer core does not need to be as steep as the adiabatic gradient. Thus the gravitationally-powered dynamo provides a simple way of avoiding the core paradox.

In summary, precession appears to be inadequate but either thermally or gravitationally driven convection in the outer core (or a combination of both) are possible feasible mechanisms of driving the geodynamo. The gravitational mechanism is becoming increasingly favoured. With either mechanism, the main bulk of the outer core would be maintained in a well-stirred condition but with the gravitational mechanism it is possible that a low density layer may accumulate just below the core-mantle boundary.

Irregularities in the Earth's rate of rotation and core-mantle coupling

The records of ancient eclipses and banding in fossil corals described in Chapter 1 indicate that the Earth's rate of rotation is progressively slowing down mainly as a result of tidal friction. At the other extreme, there are small seasonal fluctuations in the rate of rotation associated with movements of air and water masses. Intermediate between the seasonal fluctuations and the secular increase in length of day are irregular variations which have been revealed by astronomical studies over the last 200 years. These occur over a period of a few years and are commonly known as *decade fluctuations*. They affect the Earth alone. A particularly sharp increase in the length of day, corresponding to a slowing up of the rate of rotation, occurred in 1899 and a corresponding decrease occurred between 1910 and 1925 (Fig. 6.19). Of all the possible explanations of the decade fluctuations, the only adequate one is that angular momentum is at times exchanged between core and mantle. If the crust and mantle speed up, then the core must slow down so as to maintain overall constancy of angular momentum, and vice-versa. To account for this, there must be some type of coupling between core and mantle by which angular momentum can be transferred to and fro. According to MORRISON (1979), the torque which operates on the Earth's mantle varies with a period of about 30 years and reached a maximum of 10^{18} N m in about 1900 A.D.

It is of great interest that the changes in length of day appear to correlate with certain changes in the geomagnetic secular variation. For instance, the rate at which the field has drifted westwards appears to have changed several times over the past hundred years. The effect is well shown by the apparent motion of the eccentric dipole which best fits the main field (Fig. 6.19). The rate of drift was fairly steady until about 1900 A.D. Just before 1900 the rate decreased and it increased again at about 1910, corresponding to points of inflexion in the length of day (or year) curve. The most

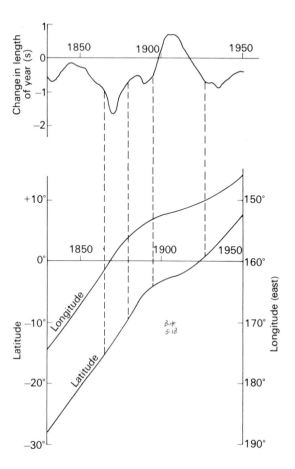

Fig. 6.19 Correlation between the irregular change in the length of the day with latitude and longitude of the eccentric dipole which best fits the Earth's observed field. Redrawn from VESTINE (1962), *Proc. Benedum Earth Sci. Symp.*, University of Pittsburgh Press.

obvious explanation of the correlation between change in length of day and the rate of westward drift is that the rotation rates of outer core and mantle have changed rather abruptly and that fluid motions causing the magnetic field have moved with the outer core.

How is the angular momentum exchanged between core and mantle? Viscous forces appear to be inadequate. One possibility is electromagnetic coupling between outermost core and mantle but computations suggest that sufficient torque could not be exerted unless there are much stronger short period features of the secular variation which are screened from observation at the Earth's surface by the high electrical conductivity of the lower mantle. The other possibility is topographical coupling between bumps on the core-mantle interface and the fluid motions of the outer core, as suggested by HIDE (1970). We need to know more about the detailed structure of the core-mantle boundary before a realistic choice can be made between these two postulated mechanisms of core-mantle coupling.

7 Terrestrial heat flow

7.1 Introduction

The study of thermal processes within the Earth is one of the more speculative branches of geophysics. This is because the available evidence on heat flow at the surface and on temperatures within the Earth can be interpreted in different ways. Over the last ten years, however, there has been a considerable advance in our knowledge of how heat escapes to the surface from the Earth's deep interior. The subject is a particularly important one because the process of heat escape from the Earth is probably the cause, directly or indirectly, of most tectonic, metamorphic and igneous activity.

The major energy transactions which affect the Earth are summarized in Table 7.1. The largest item is the heat received from the Sun, but this is mainly re-radiated back into space and only a minute fraction penetrates below a depth of a few hundred metres. It is the main source of energy for processes on and above the solid Earth's surface. It also controls the Earth's surface temperature, aided by the blanketing effect of the atmosphere. Nevertheless its influence on the interior of the Earth is negligible in comparison with that of heat generated within the Earth. Energy released by earthquakes and by the tidal slowing down of the Earth's rate of rotation is also small in comparison with the geothermal heat loss. The main source of the heat flowing out of the Earth at the present time is believed to be radioactive decay of long-lived isotopes supplemented by a slow cooling of the Earth as a whole. As heat escapes from the Earth, a small fraction of it is converted to other forms of energy which drive tectonic processes, cause igneous and metamorphic activity and possibly give rise to the main geomagnetic field.

Table 7.1   The major energy transactions which affect the Earth.

	W	J y^{-1}
Solar energy received by Earth (and re-radiated)	$1 \cdot 8 \times 10^{17}$	6×10^{24}
Geothermal loss of heat	4×10^{13}	$1 \cdot 3 \times 10^{21}$
Tidal dissipation of the Earth's rotational energy	3×10^{12}	1×10^{20}
Elastic wave energy released by earthquakes	3×10^{10}	1×10^{18}
World power production (1970)	6×10^{12}	2×10^{20}

7.2 The measurement of terrestrial heat flow

Thermally conducted heat flow per unit area is equal to the temperature gradient multiplied by the thermal conductivity. Applying this to the Earth, the heat escaping by conduction from the interior at a place on the surface can be determined by measuring (*i*) the temperature gradient just below the Earth's surface, and (*ii*) the thermal conductivity of the rocks.

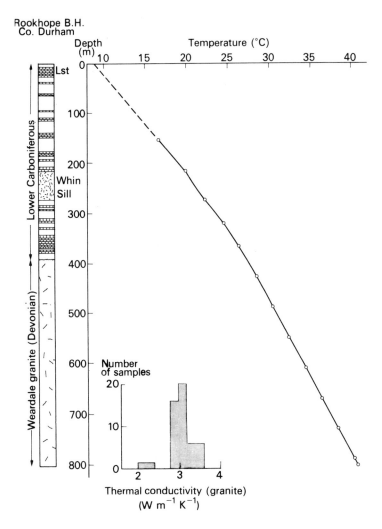

Fig. 7.1 Heat-flow determination in the Rookhope borehole, Stanhope, north England. The temperature-depth profile shown was obtained by BOTT and others (1972) using a thermistor probe three years after completion of drilling. Thermal conductivity measurements were made on 49 regularly spaced granite samples by ENGLAND and others (1980) and are plotted on a histogram shown as an inset. The results are:

observed temperature gradient (427–792 m) = (32.45 ± 0.10) K km^{-1}
correction for topography = $-(1.55 \pm 0.50)$ K km^{-1}
corrected temperature gradient = (30.90 ± 0.51) K km^{-1}
thermal conductivity = (3.10 ± 0.04) W m^{-1} K^{-1}
heat flow corrected for topography = (95.8 ± 2.1) mW m^{-2}.

If a glacial correction is applied, the estimated heat flow increases by about 5% to 10%.

On land, the temperature gradient is usually measured in a borehole by lowering a temperature measuring device, normally a thermistor probe (BECK, 1965). The measurement should if possible be made below a depth of about 200 m to avoid the transient effects of climatic changes such as were associated with the Pleistocene glaciations. In non-porous strata, the borehole needs to be left

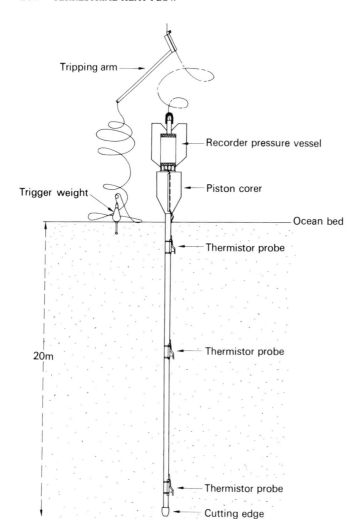

Tripping arm

Recorder pressure vessel

Trigger weight

Piston corer

Ocean bed

Thermistor probe

20m

Thermistor probe

Thermistor probe

Cutting edge

Fig. 7.2 An ocean-floor temperature gradient probe attached to a Ewing piston corer. When the trigger weight hits the seabed, the corer is released to fall freely into the soft sediments. The thermistor readings are recorded in the attached vessel. Redrawn from LANGSETH (1965), *Terrestrial heat flow*, p. 62, American Geophysical Union.

for several times as long as the drilling took in order to re-establish thermal equilibrium; alternatively temperature may be measured at the bottom of the borehole at intervals during drilling provided it can be left to settle for about a day before each measurement. The most reliable measurements are obtained from boreholes in non-porous basement rocks because the flow of water in porous sedimentary strata can carry away heat, thus disturbing the true geothermal gradient. A correction can be applied for disturbances to the geothermal gradient caused by irregular topography in the vicinity of the borehole. The thermal conductivities of samples from the borehole are measured in the laboratory, or occasionally the thermal conductivity is measured *in situ* by putting a heating coil in the borehole and measuring the temperature response. An example of a heat flow determination in the Rookhope borehole, Stanhope, northern England, is shown in Fig. 7.1. Here, the heat flow is estimated to be 95·8 mW m^{-2} with an uncertainty of about 2%.

The method used for measuring heat flow at sea was devised by Bullard in 1950. A probe two or more metres in length is dropped into the soft sediments of the ocean-floor and the temperature gradient is measured between two or more thermistors attached to the probe. A sample of the sediment is collected in the hollow barrel of the probe for later determination of thermal conductivity. One version of the probe is shown in Fig. 7.2. The probe is left in position for a few minutes to ensure that thermal equilibrium has been reached. The method works because the temperature of the ocean-floor remains almost constant and has probably done so in the past. Some small errors may be caused by heat exchange between the ocean-floor and the water near the bottom, by movement of interstitial water in the sediments and by sedimentation. The best confirmation of the reliability of the method comes from the close agreement between measurements made by probe and those made in DSDP boreholes.

Although measurements at sea have only been possible since about 1950, there are now a larger number of reliable values at sea than on land. The reason is the expense of drilling suitable holes on land; many of the measurements have been made in boreholes drilled for other purposes which are often in unsuitable regions.

7.3 The pattern of terrestrial heat flow

Over 5400 heat flow measurements are reported in a compilation by JESSOP and others (1976). About 30 % of these measurements are from continents and these are badly distributed, leaving large gaps in Antarctica and parts of Africa, South America and Asia. The oceanic observations are more evenly distributed but show serious gaps in the Arctic and Antarctic regions. The available data is still hardly ideal for statistical investigation.

Heat from below reaches the Earth's surface by two main processes, that is by thermal conduction and by discharge of hot fluids such as water and lava. Until the mid 1970s, the contribution from hot fluids was regarded as relatively small. It has now been recognized, however, that about a quarter of the global heat loss is transported to the surface by circulation of seawater in the upper oceanic crust beneath ocean ridges. The Earth's heat loss has therefore been underestimated in the past.

Estimates of the global heat loss which take into account the hydrothermal heat discharge at ridges have been made by SCLATER and others (1980) and by DAVIES (1980a). Most of the continental heat flow probably reaches the surface by thermal conduction but about a third of the oceanic heat flow is contributed by the hydrothermal activity at the ocean ridges. According to SCLATER and others (1980), the break-down of the global loss of heat from the Earth's surface is as follows:

	Average heat flow $(mW\ m^{-2})$	Area $(\times 10^6\ km)$	Rate of heat loss $(\times 10^{13}\ W)$
Continents and shelves, including lavas	57	202	1·15
Oceans			
conduction	66		2·03
hydrothermal etc.	33		1·01
total	99	309	3·04
Worldwide	82	510	4·2

DAVIES (1980a) obtained a closely similar estimate of the worldwide rate of heat loss of $4 \cdot 1 \times 10^{13}$ W. The uncertainty is probably about 10 % and this arises as follows. Firstly, the estimate of the oceanic hydrothermal heat loss is dependent on an assumed model of the cooling of the oceanic lithosphere as is explained later in the chapter (p. 289). Secondly, the continental heat flow is uncertain because of poor distribution of observations and because of the effects of past climate and erosion.

One of the central problems of terrestrial heat flow used to be to explain why the average heat flow values of continents and oceans are equal to within a few percent. This problem no longer exists. When the hydrothermal contribution is taken into account, the average continental heat flow is only about 60 % of the average oceanic value. The shapes of the continental and oceanic heat flow distributions also differ significantly (Fig. 7.3). The oceanic values are more scattered than the continental values and the continental distribution is more stumpy, with a smaller proportion of very high or very low values. About 75% of the global heat loss occurs through the oceans and about 25 % through the continents.

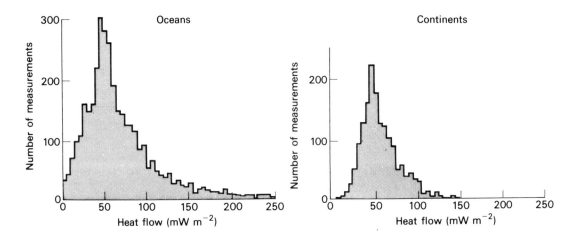

Fig. 7.3 Comparison of oceanic and continental heat-flow distributions shown as histograms of individual heat flow values from the tabulation of JESSOP and others (1976). Redrawn from POLLACK and CHAPMAN (1977b), *Scientific American*, **237**, No. 2, 66. © (1977) by Scientific American, Inc. All rights reserved.

Both oceanic and continental regions can be subdivided into heat flow provinces which broadly correlate with the major geological subdivisions depending on age in particular (Table 7.2). Within oceanic regions, the heat flow generally falls off with increasing age of the ocean floor. Three main oceanic provinces can be recognized: (*i*) ocean ridges, characterized by high average and very variable heat flow values, (*ii*) ocean basins, with low average and fairly uniform values, and (*iii*) certain marginal basins overlying subduction zones having anomalously high heat flow. Within continental regions, the heat flow generally falls off with increasing age of the last tectono-thermal event to affect a region. Thus the lowest values occur over the Precambrian shields and the highest values over Tertiary mountain belts and volcanic regions. As would be expected, active volcanic regions including the rift valley systems are associated with high and very variable heat flow patterns. Histograms of certain oceanic and continental heat flow provinces are shown in Fig. 7.4 and a map showing a selection of observations in Australia and the adjacent oceanic regions is shown in Fig. 7.5. The origins of the continental and oceanic heat flow are discussed later in the chapter.

A general way of studying the global pattern of heat flow is to carry out a spherical harmonic analysis on the observations. The main practical difficulty is that the observations are not evenly

Table 7.2 Mean heat flow values for continental (after POLYAK and SMIRNOV, 1968) and oceanic (after CHAPMAN and POLLACK, 1975) provinces as a function of tectonic age. It should be noted that the younger oceanic values are likely to underestimate the true heat loss because of hydrothermal circulation in the young oceanic crust.

Tectonic province Continents	Heat flow (mW m^{-2})	Tectonic province Oceans (crustal age)	Heat flow (mW m^{-2})
Precambrian shields	38	100–136 My	49
Precambrian platforms	44	76–100	58
Caledonian orogenic areas	46	63–76	59
Hercynian orogenic areas	52	38–63	56
Mesozoic orogenic areas	59	20–38	65
Tertiary orogenic areas	73	10–20	83
Tertiary volcanic areas	92	0–10	103

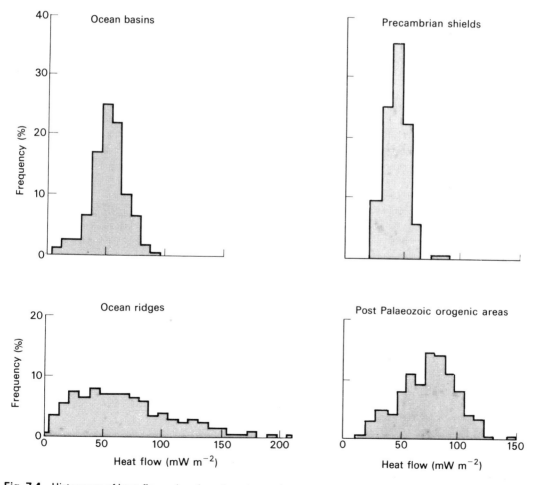

Fig. 7.4 Histograms of heat-flow values for selected oceanic and continental heat-flow provinces. Modified from LEE (1970), *Phys. Earth planet. Interiors*, **2**, 336.

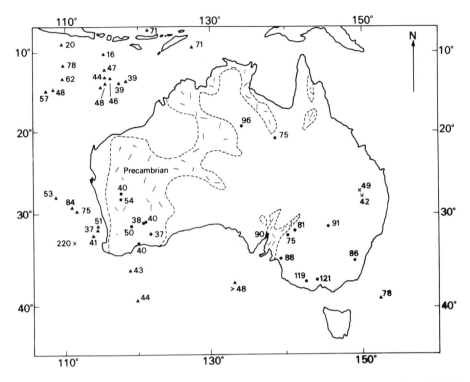

Fig. 7.5 Heat-flow values in and around Australia, showing the low values over the Precambrian shield of western Australia and near the oceanic trench, and high values in south-eastern Australia. The individual values are in units of mW m^{-2}. Redrawn with change of units from LEE and UYEDA (1965), *Terrestrial heat flow*, p. 106, American Geophysical Union.

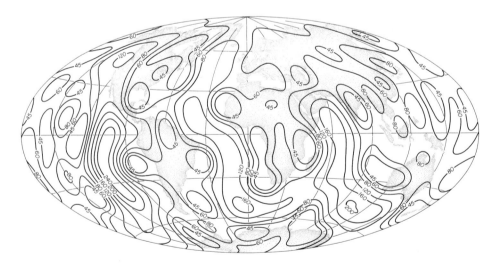

Fig. 7.6 The global heat flow pattern as represented by spherical harmonic coefficients up to degree eighteen, based on observations supplemented by predictions using tectonic and sea-floor age in unsurveyed regions. This is an updated version of a map published by CHAPMAN and POLLACK (1975) presented at the spring meeting of the American Geophysical Union in 1980, and is reproduced by kind permission of the authors. Units are mW m^{-2}.

distributed over the Earth's surface. This problem has been overcome by CHAPMAN and POLLACK (1975) by predicting the heat flow in regions without observations, by using the observed correlation between continental heat flow and age of the last tectono-thermal event and between oceanic heat flow and the age of the ocean floor. They assigned mean heat flow values to each of the $5° \times 5°$ grid areas on the globe using the observed values where available and the predicted values elsewhere. They then carried out a spherical harmonic analysis of the grid means to degree twelve (Fig. 7.6). The advantage of this map over earlier analyses of the global heat flow is that it should be free from distortions introduced by regions lacking data.

Figure 7.6 is a smoothed version of the true global heat flow pattern, with wavelengths less than about 3300 km being absent. Some distortion has probably been introduced by the theoretical prediction of heat flow values in the regions lacking data. The map shows high values over the ocean ridge system and over the marginal basins of the western Pacific Ocean. Low heat flow values visibly correlate with the Precambrian shields and with the oldest ocean floor of the western Pacific Ocean. Otherwise there is no obvious correlation with the major subdivision of the Earth's surface into continents and oceans. Nor is there any obvious correlation with most of the features on the global map of geoid heights (Fig. 1.2).

7.4 Thermal properties of rocks

At low pressure and low to moderate temperature, experimental measurements of most thermal properties of common rocks have been made. At high temperature and pressure, it is necessary to rely on the theoretical predictions of solid state physics and thermodynamics.

The *thermal conductivity* (K) of a rock or mineral is the sum of conductivities caused by lattice vibrations (K_L) and by transfer of heat by radiation (K_r). Below about 750 K the thermal conductivity is due almost entirely to lattice vibrations. In general, the lattice thermal conductivity of rocks and minerals decreases with increasing temperature according to the formula

$$K_L = (a + bT)^{-1}$$

where a and b are constants determinable by experiment (SCHATZ and SIMMONS, 1972). Above temperatures of around 1500 K, K_L becomes temperature independent and can be related to the sound velocity V, density ρ and mean atomic weight m by the theoretical equation

$$K_L = kV(\rho/m)^{\frac{2}{3}}$$

where k = Boltzmann's constant.

The effect of increasing pressure is to cause a slight increase in lattice conductivity with depth (KIEFFER, 1976).

Above about 750 K a significant amount of heat may be transferred through rocks by radiation, thereby increasing the thermal conductivity above the value for lattice vibrations alone. The effectiveness of radiative heat transfer depends on the transparency of the appropriate silicate minerals to radiation at the red end of the visible spectrum. An approximate expression for the radiative contribution to the thermal conductivity at temperature T is given by

$$K_r = 16n^2 sT^3/3e$$

where n is the refractive index, s is the Stefan–Bolzmann constant and e is the opacity (CLARK, 1957). The quantities n and e are understood to be weighted means over the appropriate band of wavelengths. At first sight it would appear that K_r should increase very strongly with increasing temperature. However, the opacity, which depends on the scattering and absorption of radiation, also increases strongly with temperature. Thus SCHATZ and SIMMONS (1972) found that the radiative thermal conductivities of materials relevant to the Earth increased more nearly linearly with rising temperature than according to a T^3 law within the temperature range of 500 to 1800 K.

The thermal conductivity of most types of non-porous rock measured at room temperature lies between 1·7 and 5·9 W m^{-1} K^{-1} (BIRCH and CLARK, 1940). The range narrows to 1·7–3·8 at 470 K, mainly because the value for poorly conducting feldspar rises with increasing temperature and the values for the other minerals, which are better conducting, fall with increasing temperature. A realistic bulk estimate of 2·5 W m^{-1} K^{-1} applies to both continental and oceanic crust with an accuracy of about 10%. Estimates for the upper mantle are based on experimental results on olivines and pyroxenes at high temperature, corrected for pressure, yielding a value falling from about 3·4 W m^{-1} K^{-1} just below the oceanic Moho to a minimum of about 2·8 at 40 km depth and then rising because of increasing K_r to about 7·3 at 400 km depth (SCHATZ and SIMMONS, 1972). The value is probably marginally higher beneath oceans than continents because of the higher sub-oceanic temperatures. The contribution from K_r is probably less below the mantle transition zone than above it because of the strong increase in opacity expected to accompany the increase in electrical conductivity between 500 and 1000 km depths (p. 179), both phenomena depending directly on the density of free electrons. Using the estimate of K_L obtained by KIEFFER (1976), the thermal conductivity of the lower mantle probably lies between about 4 and 10 W m^{-1} K^{-1}. The theoretical value for the core increases downwards from 27 W m^{-1} K^{-1} at the core-mantle boundary to 36 W m^{-1} K^{-1} at the Earth's centre (STACEY, 1977).

The *specific heat capacity at constant pressure* (c_p) is best computed from the observed values of the constituent minerals. The value of c_p for most crystalline rocks rises from 630–840 J kg^{-1} K^{-1} at 270 K to 1130–1340 at 1050 K. At higher temperatures c_p can be fairly accurately predicted by theory. Assuming that the mean atomic weight of the mantle is 21 and that the specific heat capacity is principally due to lattice vibrations, solid state theory predicts that the specific heat capacity at constant volume (c_v) approaches a maximum value at high temperature of 1250 J kg^{-1} K^{-1}. Thermodynamic relationships show that c_p should be about 2·5% higher than c_v throughout the mantle. The specific heat capacities are the most accurately predictable thermal property of the mantle. A good estimate for c_p throughout the mantle is 1250 J kg^{-1} K^{-1}, which is probably accurate to better than 20%. The estimated value for the core, with higher mean atomic weight, is about 680 J kg^{-1} K^{-1}.

The *thermal diffusivity* (κ) is needed for solving problems in heat conduction in which the temperature changes with time and for assessment of the feasibility of convection. It is defined as $\kappa = K/\rho c_v$.

The *volume coefficient of thermal expansion* (α) of common types of crystalline rock is calculated from observed linear coefficients of expansion and lies in the range $1·5 - 3·3 \times 10^{-5}$ K^{-1}. The value of α for olivine and pyroxenes between 300 K and 1250 K is about 3×10^{-5} K^{-1} which is a good estimate for the upper mantle. ANDERSON (1965) has shown that estimates of α at all depths in the mantle can be obtained indirectly from knowledge of the seismic velocities, using the theory of solid state physics. He first showed that the Grüneisen ratio γ can be estimated from the increase in bulk modulus with pressure (dk/dp) which can be obtained from the body wave velocities (p.162). The estimate of α can then be obtained from the thermodynamic relationship $\alpha = \gamma c_v/k\rho$. Using an improved method for estimating the Grüneisen ratio from the body wave velocities, STACEY (1977) obtained values of α which decrease from about $3·0 \times 10^{-5}$ just below the Moho to $1·0 \times 10^{-5}$ K^{-1} at the base of the mantle. The uncertainty is probably less than 50%.

The pressure dependence of the temperature at which a phase transition occurs is related to the change in specific volume ΔV and the enthalpy change ΔH (e.g. heat of fusion or of solid-solid transition) by the Clapeyron equation which states that

$$\Delta H \, dT/dp = T\Delta V.$$

The experimentally observed heat of fusion of basalt is about 400 kJ kg^{-1} in good agreement with the predicted value using the equation. The Clapeyron equation is useful in discussing the melting point gradient and the occurrence of solid-solid transitions within the Earth. The adiabatic gradient can be computed from the thermal properties using the relation

$$dT/dp = \alpha T/c_p \rho.$$

Table 7.3 shows estimates of the thermal properties of the main subdivisions of the Earth, including the adiabatic gradient.

Table 7.3 Thermal properties of the main subdivisions of the Earth, based mainly on the thermal model of STACEY (1977). Where a range of values is given, the first figure refers to the shallower depth.

Subdivision		Temperature (K)	Thermal conductivity (W m^{-1} K^{-1})	Specific heat (c_p) (kJ kg^{-1} K^{-1})	Volume thermal expansion (10^{-5} K^{-1})	Adiabatic gradient (K km^{-1})
Continental crust		280– 650	2·5	1·17	(3·0)	
Oceanic crust		280– 550	2·5	1·17	(3·0)	
Upper mantle continental	35–120 km	750–1270	3·1	1·25	3·0	0·15–0·34
	120–370 km	1270–2025	3·3–6·5	1·26	2·9–2·0	0·32
oceanic	10–120 km	550–1670	3·3	1·25	3·0	0·15–0·34
	120–370 km	1670–2035	3·7–6·9	1·26	2·9–2·0	0·32
Mantle transition zone		2030–2250	6·9–7·3(?)	1·26	1·9–1·5	
Lower mantle		2250–3160	7·3–9·6(?)	1·27	1·9–1·0	0·34–0·25
Outer core		3160–4170	27–35	0·71–0·66	1·6–0·8	0·76–0·22
Inner core		4170–4290	36	0·64	0·7	0·20–0·00

7.5 The Earth's internal sources of heat

Before radioactivity was discovered, the flow of heat out of the Earth was believed to be the result of the cooling by conduction of an initially hot body. On this supposition Lord Kelvin deduced that the Earth could not be older than about 80 My (p. 10). After the discovery of radioactivity, it was recognized that radioactive decay of long-lived isotopes within the Earth may provide a source of heat adequate to explain the observed heat flow without recourse to the cooling hypothesis. The most recent idea is that the Earth has actually cooled slightly over its lifespan as a result of vigorous mantle convection. A major part of the heat now escaping is regarded as coming from the decay of long-lived radioactive isotopes but a significant fraction also comes from the slight cooling.

Two types of heat source have contributed to the thermal evolution of the Earth. Once the Earth had formed and the core had separated, the slow evolution of heat over the Earth's lifespan has mainly been by decay of long-lived radioactive isotopes. Other, more short-lived sources of heat must have been present at the time of the Earth's formation to account for the high temperatures established on completion of core formation, which were probably slightly higher than the present internal temperatures. The various long-lived and short-lived sources of the Earth's internal heat are reviewed in this section.

At the outset, it is instructive to compare estimates of the heat loss over the Earth's lifespan and heat required to raise the internal temperatures to the present level. Taking the present rate of heat loss to be 4×10^{13} W and assuming this to apply over the Earth's lifespan, the estimated heat loss is 5.8×10^{30} J. More realistically, the average rate of heat loss may have been about twice the present value, so that the overall heat loss has been about 10^{31} J. Assuming the thermal model of the Earth shown in Table 7.3 and allowing for heat of fusion of the outer core, the heat required to raise the Earth's temperature from the accretion temperature of 400 K to the present values is about 1.6×10^{31} J, which is rather higher than the overall heat loss as estimated here. Thus the total heat produced within the Earth during formation and over its lifespan is probably between 2 and 3×10^{31} J. Probably over two-thirds of this was contributed by short-lived sources during accretion and core formation and less than one third has subsequently been contributed by decay of long-lived radioactive isotopes.

Long-lived radioactive isotopes

There is a loss of molecular binding energy when a radioactive isotope decays. This provides the energy of the γ-rays and imparts kinetic energy to α- and β-particles. This energy is dissipated as heat in the immediate vicinity of the decaying isotope.

The radioactive isotopes which contribute significantly to the present heat production within the Earth are ^{238}U, ^{235}U, ^{232}Th and ^{40}K. These have half-lives comparable to the age of the Earth and hence they are still sufficiently abundant to be important heat sources. Uranium consists essentially just of these two isotopes, the present-day proportion of ^{235}U being 0.71%. ^{40}K forms 0.0118% of present-day potassium.

The rate of heat production for each of these four isotopes has been determined experimentally, and is shown together with the half-life in Table 7.4. An isotope with decay constant λ was more abundant in the Earth by a factor of $e^{\lambda t}$ at time t before the present. This means that the radioactive heat production from these four isotopes was larger in the past and has progressively decreased since the Earth's formation, simply because they were more abundant in the past. The abundances of the isotopes at various times in the past, relative to the present values, are shown in Table 7.4. The rate at which heat has been produced for each kilogramme now present of U, Th and K can then be computed from the known isotopic composition of each element, and is shown in Table 7.5. The past and present heat production in rocks can be estimated using this table, provided that the average content of the radioactive elements can be estimated. Estimates for some materials relevant to the Earth's geothermal history are shown in Table 7.6.

It is of interest to investigate to what extent the estimated radioactive content of the Earth can account for the present heat loss of about 4×10^{13} W. It used to be suggested that the Earth has a composition equivalent to that of ordinary chondrites. Assuming the abundances of the long-lived radioactive isotopes measured in chondrites (Table 7.6) and taking the Earth's mass to be

Table 7.4 The half-lives, rates of heat production and abundances in the past (relative to present) for long-lived radioactive isotopes.

	Half-life (1000 My)	Heat production (μW kg^{-1})	Abundances in past relative to present				Average over 4570 My
			0 My	2000 My	4000 My	4570 My	
^{238}U	4.47	94	1.00	1.36	1.86	2.03	1.45
^{235}U	0.70	570	1.00	7.17	51.4	90.1	19.8
^{232}Th	14.01	26.6	1.00	1.10	1.22	1.25	1.12
^{40}K	1.25	27.9	1.00	3.03	9.18	12.6	4.58

Table 7.5 Rate of heat production per kilogramme of U, Th and K now present (in $J\,y^{-1}$). Computed from the data in Table 7.4 and the isotopic ratios of uranium and potassium as stated in text.

	0 My	2000 My ago	4000 My ago	4570 My ago	Average over 4570 My
U	3077	4901	11667	16769	6679
Th	840	924	1025	1050	932
K	0·104	0·306	0·905	1·236	0·457

Table 7.6 Average content of long-lived radioactive isotopes in granite and basalt (after MACDONALD, 1965), average upper continental crust (after HEIER and ROGERS, 1963), chondrites (after LARIMER, 1971) and the Earth as estimated by ANDERS (1977) and the resulting past and present estimates of heat production in these rocks.

	Composition (ppm)			Heat production ($J\,kg^{-1}\,My^{-1}$)				Average over 4570 My
	U	Th	K	0 My	2000 My	4000 My	4570 My	
Granite	4·75	18·5	37900	34100	52000	108700	–	66300
Basalt	0·60	2·7	8400	4990	8000	17400	–	10400
Average upper continental crust	1·7	6·0	19000	12250	19700	43200	–	25600
Ordinary chondrite	0·012	0·043	850	160	360	950	1300	509
Average Earth	0·018	0·065	170	130	200	430	580	258
K-depleted chondrite	0·012	0·043	170	90	150	340	460	200

$5\cdot974 \times 10^{24}$ kg, then the rate of heat production in a chondritic Earth would be $3\cdot0 \times 10^{13}$ W. Prior to the recognition of the large hydrothermal heat loss from ocean ridges, this was almost exactly equal to the computed heat loss from the Earth's surface. This led to the inference, which is now known to be fallacious, that the present-day heat loss can be attributed entirely to the on-going radioactive decay of the long-lived isotopes in the Earth. This has been referred to as the 'chondritic coincidence'. If the Earth is chondritic, then most of the potassium must be concentrated in the core as suggested by GOETTEL (1976). More probably it is heavily depleted in potassium and other volatile elements (p. 21) and the average rate of heat production within the Earth is below that of the chondritic model. According to the potassium depleted model of Anders (Table 7.6), the rate of present heat production would be $2\cdot5 \times 10^{13}$ W. Alternatively, if the Earth is chondritic apart from depletion in the volatile elements, with potassium being depleted by a factor of five, the present heat production would be $1\cdot7 \times 10^{13}$ W. Taking these models as an indication of the possible range of compositions, the present radiogenic heat production within the Earth would amount to somewhere between 43 % and 75 % of the present rate of heat loss. This suggests that somewhere between about 25 % and 57 % of the heat loss comes from slow cooling of the Earth, assuming there are no major undiscovered sources of heat. A present rate of temperature drop of between 5 and 12 K per 100 My would provide this heat. At this rate, the Earth would have cooled by 230 to 500 K since the core formed, although cooling may have been more rapid during the earlier periods.

Another important factor in the Earth's thermal history is the strong concentration of the heat-producing isotopes into the rocks which form the upper continental crust (Table 7.6). In contrast, the granulites believed to form the lower crust appear to be depleted in the radioactive elements relative to the upper crust, so that the main heat sources of the continental crust probably occur within the uppermost 10 to 20 km. This is borne out by the heat flow patterns associated with continental plutonic intrusions as described later in the chapter (p. 291). Assuming the model of continental heat production of Table 7.6 extending down to 15 km depth with an average density of 2800 kg m^{-3}, then the contribution from radioactive decay in the crust to the observed heat flow would be 17 mW m^{-2}. More exacting analyses by BLACKWELL (1971) and by SMITHSON and DECKER (1974) estimate the contribution from crustal radioactivity to be between 18 and 38 mW m^{-2}. Recently, POLLACK and CHAPMAN (1977a) estimated that 40% of the continental heat flow comes from radioactive decay within the crust. In contrast, the thin basaltic oceanic crust would be unlikely to contribute more than about 4 mW m^{-2}, so that over 93% of the oceanic heat flow must come from beneath the crust. Other aspects of the highly significant difference between continental and oceanic heat flow were touched on earlier (p. 270) and the problem will be discussed further later in the Chapter.

The long-lived radioactive isotopes produced heat at too slow a rate for them to be a significant factor in the initial heating up of the Earth after accretion. If all the heat produced by them remained trapped inside the Earth, then the consequent rise in average temperature would only be 600 K over the first 1000 My or about 1200 K over the 4570 My since the Earth's formation. Other types of heat source must have caused most of the initial rise in temperature. On the other hand, the long-lived isotopes have probably been the main source of internal heat production within the Earth after the initial period involving accretion and core formation.

Short-lived radioactive isotopes

As UREY (1955) first suggested, the decay of the short-lived radionuclide ^{26}Al, of half-life 0·74 My, may have contributed to the initial heating of accreting bodies of asteroidal or planetary size for a period of about 5 My after nucleosynthesis by carbon burning in a supernova explosion. The detection of excess ^{26}Mg correlating with the Al/Mg ratio in the Allende meteorite shows that it was present in sufficient quantity in some meteorite bodies to melt them within about 0·3 My of accretion (LEE and others, 1977). The two other radionuclides ^{36}Cl and ^{60}Fe may also have contributed to this heating, but with shorter half-lives of about 0·3 My they would loose their effectiveness as heat sources after about 2 My. Thus ^{26}Al was probably the main source of energy available for heating up the parent meteorite bodies. It also provides the most plausible mechanism of rapidly heating the Moon after accretion (SINGER, 1978).

According to JACOBS (1975), the radionuclides ^{236}U, ^{146}Sm, ^{244}Pu and ^{247}Cm, which have half-lives between 16 and 80 My, may have contributed to the heating up of the planetary bodies over 100 to 200 My after their formation. Their initial total heat production may have been about twenty times as great as that produced by decay of ^{40}K over the same time interval, thus raising the temperature within the primitive Earth by up to about 400 K. A more speculative suggestion by RUNCORN and others (1977) is that decay of the supposed relatively stable transuranic elements with atomic numbers between 114 and 126 and suggested half-lives of 100 to 1000 My may have been an important source of heat in the young planetary bodies.

The extent to which extinct radionuclides were effective in heating up the primitive Earth's interior depends on how long it took for the Earth to form by accretion and on how abundant these isotopes were then. If accretion took only about 5 My after termination of nucleosynthesis of ^{26}Al, then heat evolved by its decay may have been a dominant cause of the Earth's high internal temperature. On the other hand, if accretion took more than 10 My then ^{26}Al, ^{36}Cl and ^{60}Fe would only raise the temperature of the innermost part, but longer-lived isotopes such as ^{244}Pu may have contributed modestly but significantly to the initial increase of the Earth's internal temperature.

Heat generated during accretion and core formation

Two physical processes other than radioactivity would raise the internal temperature of the Earth as it grew by accretion. These are: (*i*) kinetic energy released by the accreting bodies as they were gravitationally accelerated towards and eventually collided with the growing Earth, and (*ii*) rise in temperature caused by progressive adiabatic compression. These processes were discussed in Chapter 6 (p. 244), where it was shown that even in absence of other heat sources they would cause the temperature to rise sufficiently to trigger core formation provided that accretion occurred over less than about 0·3 to 0·5 My.

A further major release of gravitational energy would occur as a result of core formation, assuming that this occurred by differentiation of an originally homogeneous Earth. Part of the energy released would be converted to elastic strain energy, because concentration of the dense material towards the centre would increase gravity and thus the pressure. Most of the remaining energy would be dissipated as heat through friction and viscous flow. TOZER (1965) has estimated the total change of gravitational and strain energy during the process of core formation, and he found that the release of heat amounted to $1·18 \times 10^{31}$ J with an uncertainty of about 20 %; this is equivalent to an average of about 2×10^6 J kg^{-1} which would raise the Earth's average temperature by about 1500 K. A more recent estimate by FLASAR and BIRCH (1973) gives a total release of heat energy of $1·47 \times 10^{31}$ J.

The significance of core formation on the thermal history of the Earth is emphasized by comparing the minimum likely temperatures within the Earth just prior to core formation with the present inferred temperature distribution (Fig. 7.7). The energy needed to raise the internal temperature distribution from the initial to the present distributions of Fig. 7.7, allowing for fusion, is estimated to be about 8×10^{30} J which is only about two-thirds of the energy released by core formation. The released heat energy would thus be sufficient to raise the internal temperatures to an average of at least 500 K above the present ones. Evidently the process of core formation from a homogeneous primitive Earth would have been a major factor in raising the temperatures within the Earth, resulting in the immediate establishment of a thermal regime somewhat hotter than the present one.

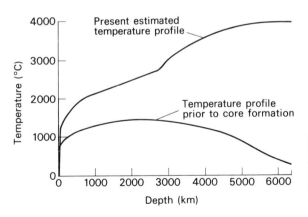

Fig. 7.7 The minimum likely temperature distribution of the Earth prior to core formation, based on the 0·5 My accretion model of HANKS and ANDERSON (1969), compared with the present sub-oceanic temperature distribution according to STACEY (1977).

Conversion of the Earth's rotational energy to heat

Another possible source of internal heat is the dissipation of the Earth's rotational energy as it slows down through tidal interaction with the Moon and less significantly with the Sun (p. 24). A relatively small fraction of this energy is taken up in increasing the Moon's orbital energy as it recedes from the Earth. Most of it is dissipated by tidal friction in the oceans but up to about 10% of it may possibly be dissipated in the interior by the solid earth tides, amounting to about 1% of the present global heat flow. MACDONALD (1964) has shown that a total of 1.5×10^{31}J would be released overall on the Earth's slowing down from a 3 hour period of rotation to the present 24 hour period. If as much as 10% of this was dissipated by earth tides and then trapped within the Earth, it would only raise the average temperature by about 200 K. This source of energy is thus unlikely to have been of much significance in the thermal history of the Earth.

7.6 Transfer of heat within the Earth

Heat escaping from the deep interior through the lithosphere is mostly transferred by lattice thermal conduction. The temperature is too low for radiation and the rocks are too stiff for convection, except by hydrothermal activity such as in the shallow crust at ocean ridges. The steep geothermal gradient implied by thermal conduction in the lithosphere raises the following problem. Taking the heat flow to be $60 \, \text{mW m}^{-2}$ and the thermal conductivity to be $2.5 \, \text{W m}^{-1} \, \text{K}^{-1}$, the temperature gradient is $24 \, \text{K km}^{-1}$. If this gradient is extrapolated to a depth of 100 km then the temperature would be over 2700 K which would cause wholesale melting. Therefore there must be a substantial decrease in the temperature-depth gradient before a depth of 100 km is reached, and in general the gradient must be shallower than the fusion gradient of about $3 \, \text{K km}^{-1}$ throughout the mantle below the lithosphere. The flattening of the temperature-depth curve could be caused either by a concentration of the Earth's internal heat sources near the surface, or by a more effective mechanism of heat transfer than thermal conduction at depth, or by a combination of these.

If the transfer of heat below 100 km depth occurs mainly by thermal conduction in rocks possessing thermal conductivities of about $3–6 \, \text{W m}^{-1} \, \text{K}^{-1}$, then 80–90% of the radioactive heat sources within the Earth must be concentrated above a depth of 100 km. Otherwise there would be wholesale melting of the mantle. It is believed that about 40–50% of the radioactive heat sources in continental regions are concentrated in the crust. But the problem is more acute beneath the oceans, where radioactive heat sources are insignificant in the underlying lithosphere. We are forced to conclude that some other mechanism of heat transfer than thermal conduction occurs in the mantle below the lithosphere.

Thermal conduction would also be a relatively ineffective mechanism for removal of heat to the surface from the Earth's deep interior. The time constant for the cooling of a uniform layer by conduction is proportional to the square of its thickness. For a 100 km thick lithosphere, the time constant of cooling is about 50 My. Assuming the same thermal properties, the time constant for cooling by conduction of the uppermost 1000 km of the Earth would be 5000 My, which is more than the age of the Earth. Thus lattice thermal conduction could not remove much heat from the deep interior of the Earth even over its lifespan, except through any pre-existing temperature gradient established when the Earth formed. The main geomagnetic field probably could not be generated under such conditions. Again, other mechanisms of heat transfer in the mantle below the lithosphere are indicated.

The three regions of the Earth where thermal conduction is probably the main mechanism of heat transfer are (i) the lithosphere, (ii) the lowermost 100 km of the mantle where heat flowing out of the core must be transferred to the mantle above by conduction, and (iii) the inner core where the heat transfer is relatively small. Elsewhere lattice thermal conduction must occur wherever there is a temperature gradient but it is probably subordinate to other heat transfer processes which are now to be discussed.

Transfer of heat by radiation

During the 1960s it was widely suggested that at temperatures above about 1000–1700 K substantial amounts of heat could be transferred through rocks by radiation, providing a possible alternative to mantle convection for removal of heat from the deep interior. As shown earlier in the chapter (p. 273), the result of radiative heat transfer is to increase the thermal conductivity by an additional amount K_r. It is known that the thermal conductivity of glasses and ceramic materials is increased at high temperature by radiation, and KANAMORI and others (1968) found that the thermal diffusivity of some common minerals, including quartz and olivine, started to increase at about 700 K due to this effect. The theoretical dependence of K_r on T^3 suggested that radiative conductivity might be at least an order of magnitude greater than lattice conductivity through most of the mantle.

The more recent experimental results of SCHATZ and SIMMONS (1972) do not encourage this view. These show that K_r does not increase rapidly with temperature because the increase in opacity counteracts the T^3 effect. They found that the predicted thermal conductivity of olivine at 400 km depth, including K_r, is less than twice the surface value, and that radiative conductivity was even less effective in enstatite. Going to greater depths, it is significant that both absorption of radiation and electrical conductivity depend on the density of free electrons. The electrical conductivity increases by two orders of magnitude between depths of 500 and 1000 km in the mantle (p. 179). Thus a comparable increase in opacity should occur which would swamp the T^3 effect and cause radiation to be less significant in the lower mantle than at 400 km depth. Radiation is thus unlikely to dominate the heat transfer processes in the mantle.

Transfer of heat by thermal convection

The process of thermal convection can transfer substantial quantities of heat upwards through fluids in the presence of a relatively small temperature gradient. The mechanism of thermal convection is described in more detail in Chapter 9 with particular reference to its role in global tectonics. As explained there, the onset of convection in a Newtonian fluid layer of the Earth of thickness d depends on whether the non-dimensional Rayleigh number $R = \alpha\beta g d^4/\kappa v$ exceeds about 10^3, where α is the coefficient of thermal expansion, β is the temperature gradient in excess of the adiabatic gradient, g is gravity, κ is the thermal diffusivity and v is the kinematic viscosity. It is shown in Chapter 9 (p. 346) that the Rayleigh number in the mantle for a temperature gradient only 0.1 K km^{-1} in excess of the adiabatic gradient is of the order of 10^5 to 10^7, indicating that fairly vigorous convection would be expected to take place in the mantle. Thermal convection would also be expected to take place in the outer core provided that the temperature gradient is marginally above the adiabatic gradient.

The efficiency of convection as a mechanism of heat transfer is measured by the Nusselt number, which is the ratio of total heat transferred to heat transferred by thermal conduction. In general, the higher the Rayleigh number, the greater the proportion of heat transferred by convection. The experimentally determined relationship between Nusselt number and Rayleigh number for Newtonian fluid convection is shown in Fig. 7.8. At marginal stability when the actual Rayleigh number R equals the critical value R_c, the Nusselt number is unity which means that a negligible amount of heat is transferred by convection. At higher Rayleigh number, the Nusselt number is approximately equal to $(R/R_c)^{0.3}$. Thus the Nusselt number appropriate to mantle convection is probably about ten, corresponding to a Rayleigh number of about 10^5 to 10^7.

An important geothermal implication of mantle convection is that heat produced at depth can be transported towards the Earth's surface much more rapidly than by conduction. The temperature gradient within the convection cell would not be expected to be more than a few times the adiabatic

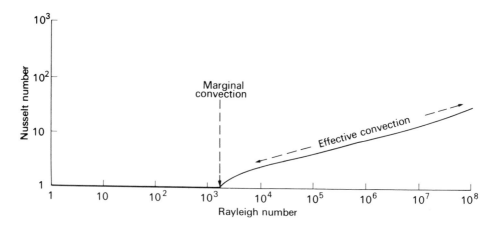

Fig. 7.8 Nusselt number as a function of Rayleigh number for free convection in a viscous fluid. Adapted from ELDER (1965), *Terrestrial heat flow*, p. 234, American Geophysical Union.

gradient which is about $0.3\,\mathrm{K\,km^{-1}}$. Thus the mantle convection hypothesis can explain why the temperature gradient decreases substantially with increasing depth at about 50–100 km below the surface. This is illustrated in Fig. 7.9 which shows the contrast in the steepness of the temperature gradient between layers of the Earth affected by conduction and convection.

Penetrative convection (ELDER, 1965) is another mechanism which can rapidly transfer heat upwards to the Earth's surface. It is the name given to the upward flow of hot, low density fluids, such as magma and hydrothermal solutions, through a denser porous or fractured region. It is an irreversible process in that the magma or hot water is deposited at or near the Earth's surface where it remains. This is probably the principal method of upward transfer of heat in geothermal regions and at ocean ridge crests.

Mantle convection and the Earth's thermal evolution

The preceding discussion suggests that heat escapes from the deep interior of the Earth by thermal conduction through the lithosphere but probably by thermal convection in the mantle below the

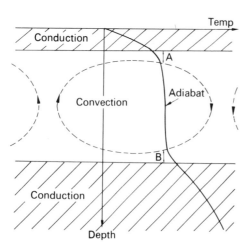

Fig. 7.9 Schematic diagram showing the temperature-depth distribution associated with a convecting zone in the Earth, overlain and underlain by layers in which heat is transported by conduction. A is the boundary zone affected by cooling of the upper part of the convection cell, and B is the zone affected by heating near the lower surface. Based partly on ELDER (1965), *Terrestrial heat flow*, p. 235, American Geophysical Union

lithosphere. This accounts for the steep rise of temperature with depth across the lithosphere and the much shallower temperature-depth gradient below it. The recognition that thermal convection is probably the dominant heat transfer process in the mantle has led to a radical revision in the understanding of the Earth's thermal history.

The vigour of mantle convection depends most critically on the viscosity of the mantle. TOZER (e.g. 1972) has pointed out some very significant consequences to the thermal evolution of the Earth and other inner planets of the probable rapid decrease of mantle viscosity with increasing temperature. A planet containing sufficient heat resources must inevitably warm up until convection is established. Most importantly, if a local temperature disturbance occurs within or below the convecting mantle, then the changed viscosity will alter the vigour of the convection in such a way as to re-establish the stable temperature regime. The time scale for removal of a thermal anomaly in this way is very much shorter than by thermal conduction. The convecting planet thus settles down into a quasi-equilibrium temperature distribution with a heat loss dependent on the rate of internal heat production. Such a state was probably first established inside the Earth within 500 My after formation of the core.

Tozer inferred that the convective equilibrium would imply an almost exact balance between heat production within the Earth and the heat loss, thus explaining the now discredited 'chondritic coincidence'. Numerical experiments on mantle convection made by McKENZIE and WEISS (1975) also indicated that equality would be maintained between internal heat production and heat loss. However, the exponential decay of the Earth's long-lived radiogenic heat sources over its lifespan had not been taken into account in these studies. This factor indicates that the Earth must probably be cooling slightly, so that the heat now escaping is partly of radiogenic origin and partly represents cooling.

Numerical experiments on mantle convection which take into account the progressive decay of the Earth's radiogenic heat sources are important in showing that the Earth must be cooling slightly. As the complete convection problem is far too complex to solve, these studies are based on approximate empirical relationships between the rate of convective heat transport and the temperature difference across the convecting layer. The approach adopted by DAVIES (1980b) was to start with the present heat loss and a representative internal temperature. For a variety of values of assumed present radiogenic heat production as a fraction of the heat loss, he then used the empirical equations to extrapolate backwards in time towards the early history of the Earth. He used the results to determine the range of present radiogenic heat production which would be consistent with two main constraints: (i) that the heat flow 2500 My ago was probably not more than five times the present value, and (ii) that the mantle temperatures in the early Archean were about 200 K higher than the present values, as indicated by the presence of the high temperature igneous rock known as komatiite. Using preferred convection parameters, he inferred that the radiogenic heat production now amounts to between 45 % and 65 % of the present heat loss. Using a similar approach, SCHUBERT and others (1980) independently estimated that radiogenic heat production supplies 65 % to 85 % of the present heat loss and that the Earth is cooling at between 5 and 10 K per 100 My, that is by about 230 to 460 K over its lifespan. These numerical estimates can be regarded as preliminary values which will be modified and improved in the future. They are, however, in excellent agreement with the modern geochemical assessment that radiogenic heat sources are at present producing heat within the Earth equal to 50 % to 75 % of the heat loss.

The vigour of mantle convection must have decreased as the Earth has cooled slightly since core formation. Because of the strong temperature dependence of mantle viscosity, quite a small rise in the temperature of the deep interior would substantially increase the convective heat transport. Thus DAVIES (1980b) suggested that the heat flow in Archean time (3000 My ago) was probably between two and five times the present value, even though the internal temperature was in general only about 200 K above the present values. An important implication would be that the lithosphere was then much thinner than it is now, possibly by a factor of two or more. This, in turn, would have an important influence on the contemporary style of tectonics.

7.7 Oceanic heat flow

Ocean ridges and ocean basins

The heat flow pattern in oceanic regions is most obviously related to the age of the underlying oceanic crust (Fig. 7.10). The highest average values occur above newly formed crust at the ridge crests and the lowest average values occur at the ocean trenches. In general, the heat flow falls off at a decreasing rate from the ridge crest towards the oldest oceanic crust near the margins of the ocean, but this pattern is interrupted by the occurrence of low values on the flanks of the ridges. Individual heat flow values over the ocean ridges are strongly scattered whereas those over the ocean basins are much more uniform (Fig. 7.11). In this way, the oceans can be subdivided into two main heat flow provinces, ocean ridges characterized by high average and highly variable values and ocean basins characterized by below average and less scattered values averaging about 40 mW m^{-2}. This general pattern applies to all the oceans, but the age of the boundaries between these provinces varies, being about 20 My on the flanks of the fast-spreading East Pacific rise, 40 My in the Indian Ocean (Fig. 7.11) and 70 My in the slow-spreading Atlantic Ocean. Certain marginal basins, mainly found in the north-western Pacific, comprise a further oceanic heat flow province characterized by high average heat flow.

The decrease of oceanic heat flow with increasing sea-floor age is broadly what would be expected in terms of the spreading oceanic lithosphere. High heat flow occurs above the hot newly-formed oceanic lithosphere at the ridge crests and the heat flow then falls off as the lithosphere

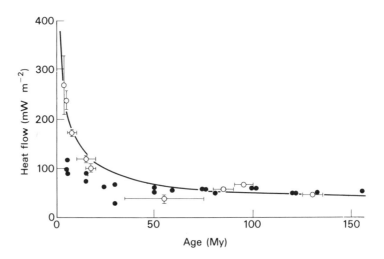

Fig. 7.10   Reliable regional heat-flow mean values for the northern Pacific (open circles) plotted as a function of the age of the ocean floor. The reliable means were selected by SCLATER and others (1976) from regions of thick and uniform sediment cover where crustal hydrothermal circulation is probably minimal. The solid circles denote other average values from the Pacific, Atlantic and Indian Oceans. The theoretical curve has been computed for the cooling of a lithosphere 125 km thick. Redrawn from PARSONS and SCLATER (1977), *J. geophys. Res.*, **82**, 819.

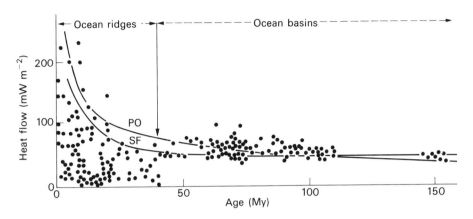

Fig. 7.11 Heat-flow values of the Indian Ocean plotted as a function of age of ocean floor. The more unreliable or uncertain values have been omitted. The solid curves are the theoretical values for the Sclater-Francheteau (SF) and Parker-Oldenburg (PO) models. Modified from ANDERSON and others (1977), *J. geophys. Res.*, **82**, 3399.

progressively cools as it spreads away from the ridge. In detail, the heat flow values over the ocean ridges are significantly lower than predicted by this hypothesis but this can readily be accounted for by seawater circulation in the crust as explained later. This concept of the spreading and cooling oceanic lithosphere explains with some accuracy the oceanic bathymetry, the increase in depth with age reflecting thermal contraction of the lithosphere as it cools. SCLATER and others (1980) have estimated that the cooling of the lithosphere accounts for about 85 % of the total oceanic heat flow. Much of the remaining 15 % is explained by heat flowing into the base of the lithosphere below the ocean basins from the underlying mantle and there is a small contribution from radiogenic heat sources within the lithosphere.

The cooling oceanic lithosphere is amenable to simple thermal modelling once the boundary conditions are fixed. Two basic types of model have been adopted. In the original models of McKENZIE (1967) and SCLATER and FRANCHETEAU (1970), the oceanic lithosphere was taken to be of constant thickness (Figs 7.12 and 7.13). The base of the lithosphere and the vertical boundary beneath the ridge crest were both assumed to be at a constant temperature. Assuming a value for the thermal diffusivity and neglecting internal heat sources, the temperature distribution in a lithosphere spreading at a given rate can be deduced by analytical or numerical methods. The theoretical heat flow distribution at the surface can then be obtained. Knowing the coefficient of thermal expansion and assuming isostatic equilibrium, the oceanic topography can be obtained from the temperature distribution. A more sophisticated approach is to use the temperature-pressure field of the spreading lithosphere to determine the boundaries between the different mineral assemblages characteristic of the upper mantle and to allow for their differing densities in the modelling of the topography.

A second type of model of the cooling lithosphere, introduced by PARKER and OLDENBURG (1973), involves the progressive thickening of the lithosphere as it spreads away from the ridge crest as a result of the release of latent heat at its lower boundary. The asthenosphere is regarded as wholly or partly molten and as it solidifies it becomes part of the overlying lithosphere. Except in the immediate vicinity of the ridge crest, the predicted thickness of the lithosphere is proportional to $t^{\frac{1}{2}}$ where t is the age. The sea depth also increases from the ridge crest as $t^{\frac{1}{2}}$ and the heat flow falls off as $t^{-\frac{1}{2}}$. This model gives a good fit to the observed bathymetry over the age range 0–70 My, except that rather too great an elevation is predicted at the ridge crest. This is possibly attributable to the

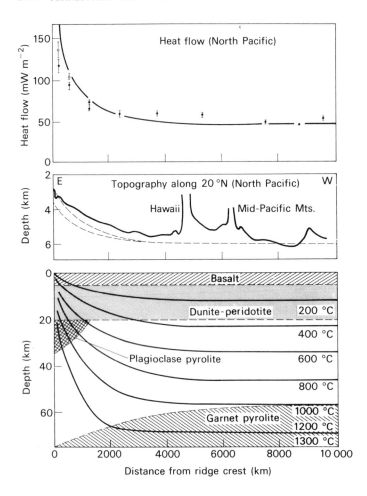

Fig. 7.12 The Sclater-Francheteau model of the cooling and spreading oceanic lithosphere 75 km thick spreading at 50 mm y⁻¹, with initial temperature 1300° C (1570 K), thermal conductivity 2·57 W m⁻¹ K⁻¹ heat production 0·042 μW m⁻³ adiabatic gradient 0·3 K km⁻¹, specific heat 1050 J kg⁻¹ K⁻¹, and volume coefficient of expansion 4×10^{-5} K⁻¹. The densities at 0° C for plagioclase, pyroxene and garnet pyrolites are 3260, 3330 and 3380 kg m⁻³ respectively. *Above*, observed heat-flow averages in the North Pacific are compared with the theoretical heat-flow profile; *centre*, observed and theoretically predicted topographical profiles are compared for thermal expansion (upper dashed curve) and for the phase transition model (lower dashed curve); *below*, lithospheric model, showing isotherms and rock types. Redrawn from SCLATER and FRANCHETEAU (1970), *Geophys. J. R. astr. Soc.*, **20**, 531.

assumption made by Parker and Oldenburg that the asthenosphere is fully melted, implying too great a release of latent heat at the boundary. More realistically, the asthenosphere may be between 1 % and 10 % partially fused at most but this introduces a further difficulty in that the lithosphere would thicken too rapidly – for a 5 % partial melt in the asthenosphere it would be 178 km thick at 100 My age (OLDENBURG, 1975).

How do these two basic models of the cooling lithosphere stand when tested against the observations? According to the constant thickness model the heat flow should asymptotically approach a uniform level of about 32 mW m⁻² in the old parts of the ocean basins, but according to the thickening lithosphere model it should continue to fall-off as $t^{-\frac{1}{2}}$. Unfortunately the heat flow pattern does not provide a useful test, as the observed values are depressed well below the theoretical predictions of both models at the ocean ridges as a result of hydrothermal circulation in the crust, and the observations over the ocean basins are not sufficiently accurate to discriminate between the two models.

A better test is to compare the theoretical and observed bathymetric profiles (Fig. 7.14). The oceanic topography is more sensitive than heat flow to the overall temperature-depth distribution in the underlying lithosphere and is less dependent on extraneous disturbing factors such as

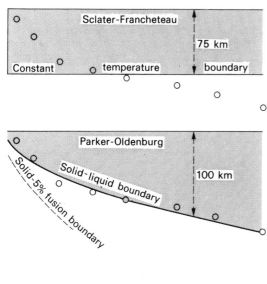

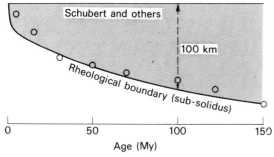

Fig. 7.13 Comparison of the cooling oceanic lithosphere models of SCLATER and FRANCHETEAU (1970), PARKER and OLDENBURG (1973) and SCHUBERT and others (1976). The estimates of the thickness of the lithosphere as a function of age based on surface wave dispersion obtained by LEEDS and others (1974) and LEEDS (1975) are shown for comparison with the models.

seawater circulation in the crust. Both models give an acceptable fit to the present oceanic topography out to ages of about 80 My. Within the age range 0–80 My, the depths increase fairly accurately as $t^{\frac{1}{2}}$ (DAVIS and LISTER, 1974). However, the $t^{\frac{1}{2}}$ relationship is a property of all thermal cooling models within this limited time range (PARSONS and SCLATER, 1977) and thus the bathymetry out to 80 My fails to provide a distinctive test. On the other hand, evidence from surface wave dispersion studies (p. 159) does indicate that the oceanic lithosphere thickens away from the ridge crests approximately as predicted by the Parker-Oldenburg model. Beyond 80 My, the depths appear to decay asymptotically towards a constant value rather than continuing to increase as $t^{\frac{1}{2}}$, thus favouring the Sclater-Francheteau model.

Neither of the two basic models satisfies all the observations and some sort of compromise is needed. Out to at least 80 My age, it is clear that the lithosphere does thicken at the expense of the asthenosphere, although it is unlikely that more than a small fraction of the latter is molten. However, beyond about 40 My age or less, the thickening of the lithosphere must be much less than predicted by the Parker-Oldenburg or OLDENBURG (1975) models. Such a suppression of the thickening of the lithosphere can occur if an alternative source of heat in the upper asthenosphere is available to keep the temperature of the lithosphere-asthenosphere boundary more nearly constant as it spreads beyond 40 My. Possible heat sources include heat conducted through a temperature gradient in the asthenosphere, thermal convection in the asthenosphere with increasing vigour at

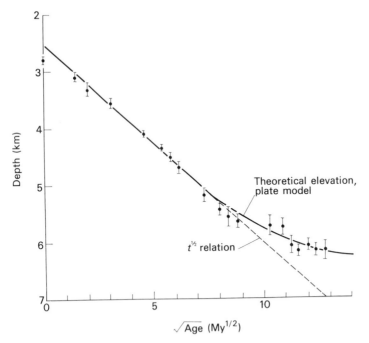

Fig. 7.14 The mean regional depths of the north Pacific Ocean plotted as a function of the square root of ocean-floor age. Note that the 125 km thick plate model fits better than the $t^{\frac{1}{2}}$ model beyond an ocean-floor age of about 50 My. Redrawn from PARSONS and SCLATER (1977), *J. geophys. Res.*, **82**, 818.

greater lithospheric ages (RICHTER and PARSONS, 1975), or shear heating resulting from differential horizontal motions of lithosphere and asthenosphere.

A model of the cooling lithosphere in which viscous dissipation of heat takes place in a boundary layer between a rigid, moving lithosphere and a viscous, stationary asthenosphere was suggested by SCHUBERT and others (1976). In this model, the lithosphere-asthenosphere boundary is below fusion temperature and is defined arbitrarily but realistically in terms of the depth at which the velocity is reduced by 10%. The viscous heat production has the effect of causing the lithospheric thickening with age to be rather less rapid than in the Parker-Oldenburg model based on a liquid-solid boundary. This model agrees better with the bathymetric observations and independent seismological estimates of lithospheric thickness than the earlier models. It is, however, dependent on the rather uncertain estimates of the rheology of olivine in the upper mantle. An equally good fit to the observations can probably be obtained by asthenospheric convection which increases in vigour away from the ridges (p. 352), or by a combination of convection and viscous dissipation.

Hydrothermal seawater circulation in the young oceanic crust

The anomalously low heat flow values of ocean ridges (Fig. 7.15) and their wide scatter are best explained by the circulation of thermally driven seawater through the oceanic crust. Only a small part of this scatter can be explained by other effects such as seabed topography. The cold seawater, which sinks through cracks into the crust, is heated up and carries the heat back into the sea by convection. Direct evidence for the occurrence of such convection comes from detailed suites of heat flow observations observed near ridge crests, such as a survey over the Galapagos spreading centre (WILLIAMS and others, 1974) and one over the Juan de Fuca ridge (DAVIS and LISTER, 1977). The Galapagos survey shows a regular variation in the heat flow pattern with a wavelength of about 6 km, attributable to convection cells of this wavelength in the underlying crust. The vents where

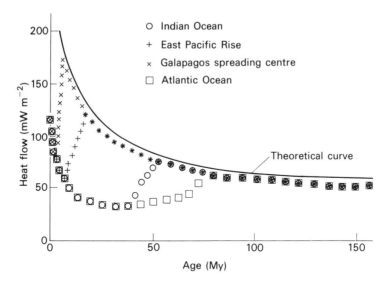

Fig. 7.15 Composite heat flow versus ocean-floor age curves for the main oceans, showing the variable widths of the ocean ridge provinces of depressed heat flow caused by hydrothermal circulation in the crust. Redrawn from ANDERSON and others (1977), *J. geophys. Res.*, **82**, 3402.

the hot water is discharged have been observed from manned submersibles. Evidently the water may penetrate to about 5 km depth and carry away about 80% of the overall heat loss here. The Juan de Fuca results suggest that the circulation is most active where sediments are thin or absent. Most heat flow observations over ocean ridges are too widely spaced to record the convection pattern and they merely record the resultant scatter.

The convecting seawater evidently flows through an interconnected series of cracks and pores in layer 2, possibly penetrating through much of layer 3 except near the ridge crest. This convective heat transfer dies out beyond a crustal age of 20–70 My, possibly partly as a result of deposition of an impermeable lid of sediments over the porous layer 2 and possibly partly by the plugging of the pores by the deposition of hydrothermal minerals. Some support for the idea of 'plugging' comes from the observation that convective heat transfer stops at about the location where the porous layer 2A (p. 95) ceases to exist.

The heat flow transported to the surface by hydrothermal activity at ridges can be estimated in the following way. The rate of heat loss at ridges from the cooling of the oceanic lithosphere can be calculated on the assumption that it falls off as $t^{-\frac{1}{2}}$ from the ridge crest towards the ocean basins. The contribution from the hydrothermal activity is then obtained by subtracting the average measured heat flow at ridges from this value. Using this method, DAVIES (1980a) and SCLATER and others (1980) determined the hydrothermal contribution to be about one third of the total oceanic heat flow and therefore about a quarter of the total global heat flow. Previous neglect of this factor has led to a significant underestimate of the total global heat loss. The associated water circulation has other important implications, notably in the exchange of elements between oceanic crust and seawater.

Marginal basins

The marginal basins of the western Pacific Ocean (p. 227), lying between the island arc systems and the continental regions, can be subdivided into two main groups based on their heat flow patterns.

Most of those occurring in the north-western Pacific have high heat flow values averaging slightly in excess of 80 mW m^{-2} and much shallower bathymetry than the deep west Pacific (SCLATER, 1972). In contrast, most of the marginal basins of the south-western Pacific have a more normal heat flow related to their age of formation by sea-floor spreading.

The marginal basins of the north-western Pacific probably formed by back-arc spreading with formation of normal oceanic lithosphere similar to that produced at ridge crests. Their high heat flow and shallow bathymetry might at first sight be explained in terms of the cooling of relatively recently formed lithosphere. However, SCLATER (1972) showed that the plot of heat flow versus depth for these basins falls outside the range of values consistent with the models of the cooling oceanic lithosphere; the average heat flow is significantly higher than that to be expected for the observed depths. On the other hand, the basins of the south-western Pacific mostly fall within the expected range of the spreading lithosphere models.

The best explanation of the high heat flow and bathymetric elevation of the marginal basins of the north-western Pacific is that raised temperatures cause the underlying lithosphere to be thinned. These raised temperatures are the indirect product of the underlying subduction zones. This may occur by frictional heating, release of water and melting in the vicinity of the upper surface of the subducting plate or above it, with subsequent penetrative convection by the magma. Alternatively, the heating may result from mechanically driven convection in the asthenosphere above the sinking tongue of lithosphere.

7.8 Continental heat flow

Continental heat flow is less well understood than that of oceanic regions. The more complicated and uncertain origin means that simple models such as that of the cooling oceanic lithosphere cannot satisfactorily be fitted to the observations. A substantial proportion of the heat flow comes from radiogenic heat sources in the upper crust and the contributions from deeper sources are not easy to categorise. Nevertheless, continental heat flow has recently become much better understood as a result of (i) the recognition that regional heat flow falls off with increasing tectonic age, and (ii) the discovery that local variability in heat flow can often be substantially attributed to the variable radioactivity of the upper crustal rocks.

Continental heat flow shows a general decrease with increasing age of the last tectono-thermal event to affect a region, as shown in Table 7.2 and Figs 7.4 and 7.16. The Precambrian shields show

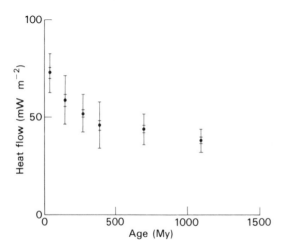

Fig. 7.16 Average regional heat-flow values for continents plotted against age of the last major tectonic event that affected the region, based on the analysis of POLYAK and SMIRNOV (1968). The outer error bars show the standard deviation and the inner bars show the standard error of the estimated mean value. Redrawn from POLLACK and CHAPMAN (1977b), *Scientific American*, **237**, No. 2, 73. © (1977) by Scientific American, Inc. All rights reserved.

the lowest and least scattered values averaging about $40 \, \mathrm{mW \, m^{-2}}$ whereas the Tertiary fold mountains and volcanic belts show the highest and most scattered values. The broad pattern, however, is not the same as that of the oceans. The overall scatter of values is less and the decrease in heat flow with increasing tectonic age is at least a factor of five less rapid than the decay of oceanic heat flow with sea-floor age. The full significance of the decay of continental heat flow with age is at first sight masked by the contribution arising from the decay of the long-lived radioactive isotopes in the crust, which is of similar magnitude to the escape of heat into the crust from the underlying mantle. An approximate method of separating these contributions is next described.

A new understanding of the contribution of crustal radioactivity to continental heat flow and its local variability has come from the discovery of an approximately linear relationship between heat flow and the surface radioactive heat production in plutonic rocks within specific provinces such as New England and the Basin and Range province of North America (BIRCH and others, 1968; ROY and others, 1968; LACHENBRUCH, 1970; BLACKWELL, 1971). Similar relationships have subsequently been found to apply in other continental regions and also to metamorphic terrains. They all take the form

$$Q = Q_r + bA$$

where Q is the observed heat flow, A is the measured radioactive heat production per unit volume of the surface metamorphic or plutonic rocks, b is the slope of the line fitted to the observations when plotted as in Fig. 7.17, and Q_r is its intercept on the Q axis. Q_r is known as the reduced heat flow and is interpreted as representing the contribution to the observed heat flow coming mainly from sources below the upper crust. The product bA is interpreted as the contribution to the heat flow from radioactive sources within the upper crust, that is within the pluton or the upper crustal metamorphic rocks. Two extreme types of distribution of the radiogenic heat sources with depth have been suggested, one type assuming uniform radioactivity down to a depth equal to the characteristic depth b, and the other assuming an exponential decrease with depth z taking the form $Ae^{-z/b}$. The observations show that Q_r and b are approximately constant within each continental heat flow province. This suggests that much of the local scatter of continental heat flow observations can be attributed to the long-lived radioactive isotopes concentrated in the igneous

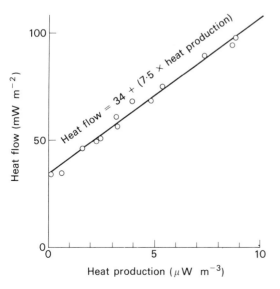

Fig. 7.17 Heat flow for plutons in the eastern United States plotted against heat production in the surface rocks, showing a typical linear relationship characteristic of individual continental heat flow provinces. Redrawn from ROY and others (1968), *Earth & planet. Sci. Lett. (Neth.)*, **5**, 4.

and metamorphic rocks of the upper crust. This is consistent with the inferred widespread occurrence of granulites in the lower crust. Granulites are known to be deficient relative to the upper crust in the radioactive heat producing elements.

Diagrams such as Fig. 7.17 make it possible to separate approximately the contributions to regional heat flow arising from upper crustal radioactivity and those arising from deeper sources. The results from various provinces have been summarized by VITORELLO and POLLACK (1980) and the determinations based on nine or more observation points are as follows:

Province	Mean heat flow Q (mW m^{-2})	Reduced heat flow Q_r (mW m^{-2})	Characteristic depth b (km)
Basin and Range	92	69	10·0
Sierra Nevada	37	18	10·1
Eastern United States	57	33	7·5
Superior, Canada	34	21	14·4
Ukraine	37	25	7·1
England and Wales	59	23	16·0
Western Australia	39	26	4·5
Central Australia	83	27	11·1

These show that the characteristic depth b is normally within the range of 7 to 10 km. An exceptionally high value of 16 km was obtained by RICHARDSON and OXBURGH (1978) for England and Wales, which they mainly attribute to the occurrence of upper crustal slaty rocks in which the radiogenic heat sources have not been concentrated upwards to the extent of other regions. Once a crustal thermal model has been set up for a province, then the temperature-depth profile in the crust can be estimated, albeit with considerable uncertainty. For instance, BLACKWELL (1971) estimated the temperature at the Moho at 30 km depth beneath the Basin and Range province to be between 1000 and 1270 K, and that at 35 km depth beneath the eastern United States to be between 700 and 800 K. According to RICHARDSON and OXBURGH (1978), the temperature at 30 km depth beneath England and Wales, which is about the depth of the Moho here, is estimated to be about 700 K.

Taking the results from such analyses of the heat flow of different provinces at their face value, it is estimated that about 40 % of the average continental heat flow is contributed by the radioactivity of the upper crustal rocks and about 60 % comes from apparently deeper sources. VITORELLO and POLLACK (1980) have further shown that the fall-off of heat flow with tectonic age affects the two contributions about equally, so that the 40 % to 60 % partition of the heat flow between shallow radiogenic and deeper contributions appears to apply to provinces of all tectonic ages, at least approximately. They also showed that the average measured heat production in the surface igneous or metamorphic rocks falls off similarly with tectonic age. The enrichment of the upper crust of young mountain ranges in radiogenic heat producing elements can be explained in terms of crustal thickening caused by tectonic or igneous processes, followed by upward concentration of the radioactive elements by metamorphic or hydrothermal activity. The subsequent fall-off with age since the orogeny can be accounted for by progressive erosion removing the surface rocks and cutting down into the upper crust where radiogenic heat producing elements decrease in abundance with depth.

The reduced heat flow attributable to sources below the upper crust is estimated by subtraction of the inferred upper crustal radiogenic contribution from the observed heat flow. In absence of firm evidence, the reduced heat flow of a region is generally interpreted in terms of a transient contribution depending on tectonic age superimposed on a uniform background heat flux. It should be emphasized that this is only a tentative model. The transient contribution may originate in several ways and the background heat flux may vary from region to region.

The transient contribution has generally been attributed to progressive cooling of the continental lithosphere following the latest tectono-thermal event (CROUGH and THOMPSON, 1976; VITORELLO and POLLACK, 1980). The initial heating can variously be attributed to upward transfer of heat by granitic or basaltic magma or hydrothermal activity, or to thinning of the continental lithosphere by stretching or otherwise. According to Vitorello and Pollack, the transient contribution falls off exponentially with increasing tectonic age with a time constant of about 300 to 400 My. It yields about 27 mW m^{-2} in Tertiary tectonic regions and effectively falls off to zero for terrains older than about 1000 My. The main problem is to understand the long time-constant of decay which requires the continental lithosphere to grow more than 300 km thick beneath the shields by cooling. This implies that fixed lithospheric keels must extend down almost to the mantle transition zone beneath the shields. The hypothesis also appears to imply that the continental lithosphere subsides by more than 5 km as it cools over 1000 My, which is hard to recognize in reality.

Alternatively, the transient heat flow in orogenic mountain belts can be partly or wholly attributed to the effects of uplift and erosion (ENGLAND and RICHARDSON, 1980). It was earlier pointed out that erosion probably removes progressively some of the upper crustal radiogenic heat sources. Another important effect of erosion is that hotter rocks from depth are progressively brought nearer to the surface by uplift and erosion. This has the effect of increasing the surface heat flow during the period of uplift and erosion with subsequent exponential decay towards its equilibrium value. According to England and Richardson, the transient contribution to the reduced heat flow of orogenic belts could be completely accounted for by the erosion of about 30 km of continental crust over a period of around 50 to 200 My. In reality, both lithospheric cooling and erosion probably contribute to the transient heat flow in orogenic belts. On geological grounds, cooling is likely to be the dominant cause in plateau uplift regions such as the Basin and Range province, where the lithosphere is known to have been thinned and erosion may be more limited than in orogenic belts.

According to VITORELLO and POLLACK (1980), the background heat flux is estimated to be about 27 mW m^{-2}. They attribute somewhat over half of this to radiogenic heat sources in the lower crust and the mantle part of the lithosphere. The remainder is assumed to flow upwards into the continental lithosphere from the underlying mantle. These estimates are very tentative.

In summary, the continental heat flow can be interpreted in terms of three main contributions as follows (Fig. 7.18): (i) about 40% is contributed by radiogenic heat sources in the upper crust; (ii) about 20% or more comes from cooling of heated continental lithosphere or erosion effects or both; and (iii) up to about 40% is interpreted as a uniform heat flux contributed by radiogenic heat sources in the continental lithosphere below the enriched upper crust and by upward flow from the mantle below the lithosphere. As emphasized earlier, there are considerable uncertainties and the model should be treated with caution.

Rift valley systems

A special type of continental heat flow province falling outside the scope of the above analysis is the rift valley system. As would be expected, active volcanic regions including the rift valley systems tend to be associated with high and very variable heat flow patterns. These regions are associated with plateau uplift structures which may be attributed to the presence of hot, thinned lithosphere beneath.

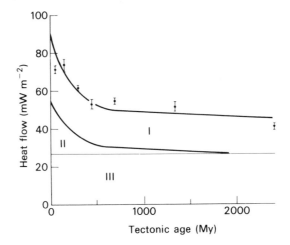

Fig. 7.18 Decrease in continental heat flow with tectonic age interpreted in terms of three principal components. Component I is radiogenic heat from the upper crust which decreases slowly with age as a result of erosion. Component II is the transient thermal perturbation and component III is the uniform background heat flux from radiogenic sources below the upper crust and from the mantle below the lithosphere. The bars indicate the standard error of the observed mean values. Redrawn with modification from VITORELLO and POLLACK (1980), *J. geophys. Res.*, **85**, 984.

No substantial heat flow study of the East African rift system has as yet been published but heat flow data on the Baikal rift system of south-central Siberia is available (e.g. LYSAK, 1978). In the Baikal rifts, the heat flow values vary locally and range from 25 to 145 mW m^{-2} but surprisingly low and uniform values of about 45 mW m^{-2} occur over the adjacent Siberian platform. Evidently the thermal activity here has not affected the heat flow of the uplifted regions adjacent to the rift zone and only occurs within the rift zone in certain localized regions.

7.9 Escape of heat through the lithosphere

The main sources of heat escaping to the Earth's surface through the lithosphere are approximately summarized in Table 7.7. The table emphasizes the differing origins of the continental and oceanic heat flows. The oceanic heat flow is dominated by the contribution from the cooling of the spreading oceanic lithosphere. In contrast, the main contribution to continental heat flow probably comes from radiogenic heat sources in the lithosphere with only about 20 % arising from cooling of the lithosphere after thermal events, or uplift and erosion.

Table 7.7 also shows the proportion of the global heat loss contributed by each individual source. Long-lived radioactive isotopes within the lithosphere contribute 19 % and heat from the region beneath the lithosphere contributes about 81 %, the error being about 5 %. It is particularly striking that over 75% of the heat escaping from the mantle below the lithosphere is at present being brought to the surface by formation and cooling of the oceanic lithosphere. The contributions from cooling of the continental lithosphere and by heat flow from the mantle into the base of the lithosphere are relatively small in comparison. It is suggested in Chapter 9 that the escape of the underlying heat through the lithosphere is responsible for global tectonic activity, the contribution escaping through the oceanic lithosphere by its cooling causing the plate motions and the local heating of the continental lithosphere being an important factor in continental splitting.

Table 7.7 Approximate estimates of the proportions of the heat escaping from the Earth's surface from the various sources within and below the lithosphere in oceanic and continental regions.

	Proportion of oceanic or continental flux (%)	Proportion of global flux (%)
Oceans		
Cooling lithosphere	85	62
Flux from mantle into base of lithosphere	10	7
Radiogenic heat sources within lithosphere	5	4
Continents Radiogenic heat sources:		
upper crust	40	11
rest of lithosphere	15(?)	4(?)
Flux from mantle into base of lithosphere	25(?)	7(?)
Cooling lithosphere, etc.	20	5

The heat generated within the Earth by the long-lived radioactive isotopes was probably about three times the present value in the early Archean (Table 7.6). The heat loss from the Earth must have been about two or three times the present loss. This could reach the surface by conduction through the lithosphere if it was then about a third of its present thickness and the geothermal gradient about three times as steep as the present value. BURKE and KIDD (1978) have pointed out that the Archean continental geothermal gradient cannot have been as steep as this because of the lack of evidence of widespread crustal melting implied. They suggested that the excess heat generated at that time was mostly removed by cooling of a much more rapidly spreading oceanic lithosphere. This emphasizes the probable importance of the rate of sea-floor spreading in controlling the loss of heat from the Earth's deep interior.

7.10 Thermal history of the Earth

This chapter is concluded by a somewhat speculative summary of the present model of the thermal history of the Earth. This differs radically from the accepted opinion about twenty five years ago when the Earth's mantle was treated as a simple thermal conductor rather than a convecting region. This older model involved the slow escape of heat from the deep interior by thermal conduction on a timescale much longer than the age of the Earth, so that thermal equilibrium would be far from established. The modern concept depends on the escape of heat out of the deep interior by mantle convection controlled by a heavily temperature-dependent viscosity, with the consequent establishment of a thermal equilibrium between internal radiogenic heat production, slight cooling and loss of heat from the surface. The slight cooling is a consequence of the exponential decrease in the radiogenic heat sources over the Earth's lifespan but it probably contributes between 25 % and 50 % of the present heat loss. The four stages into which the thermal history of the Earth can be subdivided are as follows.

The *first stage* was the initial heating of the Earth during accretion resulting from release of gravitational energy of the colliding bodies, adiabatic compression and possibly the heat released by the decay of short-lived radioactive isotopes, notably ^{26}Al. This stage probably lasted less than about one million years as otherwise the heat could not be retained within the growing Earth, unless medium-lived radioactive isotopes were significantly contributing to the heat. This stage was terminated when the temperature at some depth within the outer half became high enough to melt the iron-nickel phase with admixed FeS or FeO.

The *second stage* involved the substantial release of gravitational energy as heat during the process of core formation from an initially homogeneous Earth. Once core formation had started it would be expected to avalanche to completion over a short period of time. The result would have been that the Earth's internal temperatures were raised to levels significantly above the present day values, possibly causing extensive melting in the upper (but not lower) mantle with formation of a proto-crust. This stage was probably complete within a million years of the Earth's formation and certainly within 100 My. The result would be to establish a vigorous thermal regime within the Earth, with a molten convecting core and a mainly solid convecting mantle. The long-lived radioactive isotopes were of negligible significance over the relatively short period of the initial heating up of the Earth but subsequently they became the main source of internal heat generation.

The *third stage* lasted for up to a few hundred million years during which thermal equilibrium was established between heat production by long-lived radioactive isotopes, steady cooling and heat loss from the surface. Exceptionally vigorous mantle convection would remove heat at five to ten times the present rate, the proto-lithosphere probably being less than 20 km thick. There is no direct record of this stage as it was probably over before the formation of the oldest known Precambrian rocks.

The *fourth stage* represents the establishment and maintenance of a stable thermal balance between heat production, slow and steady cooling, and heat loss. This stage probably started about 4000 My ago and persists to the present day. The Earth has probably cooled by a few hundred degrees over this stage but the heat flow has fallen off at a decreasing rate by a factor of about three as the radiometric heat sources have progressively decayed. The lithosphere has probably thickened but the main control on the heat loss through the lithosphere may have been by the rate of sea-floor spreading. Under the present thermal regime, about 75 % of the heat loss from below the lithosphere occurs by cooling of newly formed oceanic lithosphere as it spreads laterally from the ocean ridges. In the early Precambrian, a more chaotic mechanism of heat escape through the lithosphere may have occurred to produce the distinctive but ill-understood style of tectonics characterized by the Archean greenstone and gneiss complexes.

What of the future? The radioactive heat sources will continue to decay in their abundance and the Earth's tectonic processes will become more sluggish and eventually die. In the words of POLLACK and CHAPMAN (1977b), ' . . . for the diminishing band of earth scientists who still adhere to a nonmobile view there may be some small solace in the fact that the Earth will eventually conform to their concept of it. They must be patient, however, since that time is probably some two billion years hence'. In yet another two billion years after that, according to the predictions of the astrophysicists, the Sun itself will die.

8 Rheology of the crust and mantle

This chapter deals with the non-elastic response of the crust and mantle to applied stress systems. The applied stresses which affect the Earth vary in duration from less than a second to over 1000 My. A serious difficulty is that our knowledge of the mechanisms of deformation which occur over long time periods is particularly uncertain.

8.1 Fracture and flow in solids

The type of deformation which a solid material undergoes is expressed in terms of the relationship between stress and strain. Stress is defined as force per unit area. The stress acting on any plane can be resolved into a normal stress or pressure acting at right angles to the surface and two components of shear stress acting within the plane. In any stressed medium there are three planes mutually perpendicular to each other in which the shear stress is zero. The three *principal pressures* (σ_{max}, σ_{int}, σ_{min}) act perpendicular to these planes. The state of stress at a point is completely specified by the magnitudes and directions of the three principal pressures. The shear stress is maximum in the two planes which bisect the directions of maximum and minimum principal pressure and is numerically equal to half the stress difference, i.e. $\tau_{max} = \frac{1}{2}(\sigma_{max} - \sigma_{min})$. In dealing with stresses within the Earth it is convenient to treat compression as positive and tension as negative (i.e. opposite to the usual convention).

Strain is defined as change in length per unit length. There are two types, linear strain in which the change in length is in the same direction as the initial length, and tangential strain in which they are at right angles. The state of strain can be completely specified by the directions and magnitudes of the three principal extensions.

Below the elastic limit solids deform according to Hooke's law with stress almost proportional to strain. Solids may be classified as *brittle* or *ductile* substances depending on how they deform above the elastic limit. Brittle substances fracture without appreciable deformation; ductile substances deform appreciably by flow. Substances which are brittle under atmospheric conditions become ductile at high temperature and/or pressure. The brittle-ductile transition for the substance is a function of temperature and pressure.

The two main types of brittle fracture are *extension fracture* and *shear fracture*. Extension fracture occurs when one or more of the principal pressures is a tension; the plane of fracture occurs perpendicular to the direction of maximum tension. Shear fracture occurs under compression. In a homogeneous and isotropic substance the two complementary planes of fracture each subtend an angle of less than 45° to the maximum principal pressure (characteristically about 30°) and contain the axis of intermediate principal pressure (Fig. 8.1.). According to the Coulomb-Navier hypothesis of shear fracture, the reason for this angle being less than 45° is the influence of internal friction on the attitude of fracture planes.

The Griffith theory of fracture (GRIFFITH, 1921) and subsequent developments of this theory (e.g. McCLINTOCK and WALSH, 1962; MURRELL and DIGBY, 1970) attribute brittle fracture to the presence of

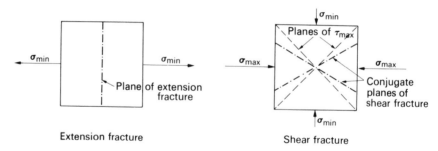

Fig. 8.1   Relationship between fracture planes and principal pressures for extension fracture and shear fracture.

microscopic (or possibly larger) cracks within the material. The stress field is locally modified by the presence of the cracks, causing the stress difference to be greatly amplified at the tips of suitably oriented cracks. If the applied stress field is large enough, fracture will propagate from the tips of the most favourably oriented cracks. This theory explains why the strength of most solid materials is much less than the intrinsic strength of the lattices of individual crystals. Under tension, or a small compressive confining pressure, the cracks remain open. If an applied tension $T (= -\sigma_{min})$ exceeds the tensile strength of the material and $(\sigma_{min} + 3\sigma_{max}) < 0$, then the cracks oriented perpendicular to the applied tension are the first to yield. Consequently extension fracture is predicted to occur. At the other extreme, all the cracks become closed at sufficiently high confining pressure so that the normal stress across each crack and the coefficient of friction of its surface come into play. This causes shear fracture of the type predicted by the Coulomb-Navier hypothesis to occur when the compressive strength is exceeded. The theory implies that the compressive strength at low confining pressure is about ten times greater than the tensile strength and that it increases with increasing confining pressure. Intermediate types of fracture occur between these two regimes, such as extension shear which occurs while the cracks all remain open with $(\sigma_{min} + 3\sigma_{max}) > 0$.

Brittle substances may undergo a certain type of flow which results from repeated shear fractures which progressively reduce the grain size, or by the rolling and sliding of fragments over one another. This is known as *cataclastic flow* (Fig. 8.2a).

The flow properties of crystalline solids are mainly known from experiments on metals. Metals have the advantage that most of them deform by ductile flow at room temperature and low confining pressure whereas silicate minerals do not. Enough is known about the deformation of rocks and minerals, however, to be reasonably confident that they undergo the same mechanisms of flow as metals but generally at higher temperature and pressure. Useful accounts relevant to the crust and mantle have been given by NICOLAS and POIRIER (1976) and KELLY and others (1978).

At low temperature and stress, most materials including rocks suffer a small amount of deformation of up to about 1 % by transient creep (creep being the name given to very slow flow caused by a constant load). Otherwise, ductile materials deform by flow when their strength is exceeded. There are three common types of ductile flow which may cause significant deformation of metals, these being low temperature plastic flow, power-law flow or creep, and diffusion creep. Each of these flow mechanisms can be described by a constitutive law relating strain rate to stress under specified conditions. All three main flow mechanisms are thermally activated processes, the temperature dependence of strain rate being of the form $e^{-Q/kT}$, where Q is the activation energy, k is Boltzmann's constant and T is the absolute temperature. The strain rate increases very rapidly with rising temperature when $T = Q/k$, being negligible for much lower

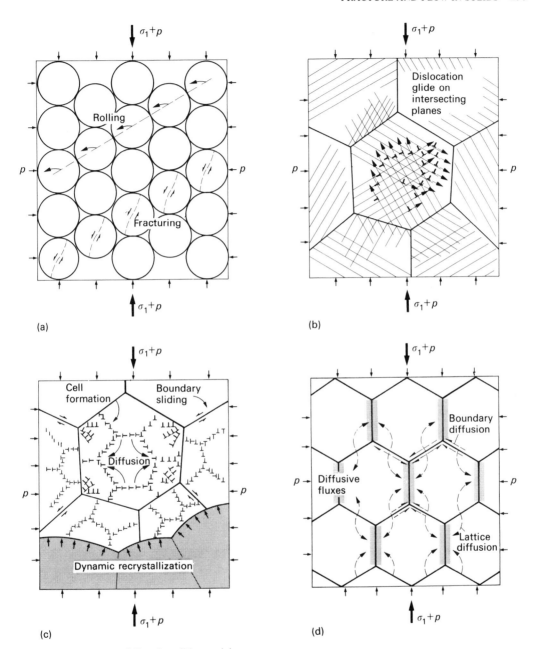

Fig. 8.2 Mechanisms of flow in solid materials.

(a) Cataclastic flow, by which granular or previously fractured material can deform by rolling or sliding of granules or fragments over each other.

(b) Low temperature plasticity, in which gliding motions on several sets of dislocations permit compatible deformation of the grains of a crystalline solid.

(c) Power-law creep, caused by motion on dislocations which take up a polygonal cellular pattern within the grains. This mechanism may be complicated by grain-boundary sliding and by recrystallization.

(d) Diffusion creep, caused by migration of atoms in a stress gradient, occurring within the crystal lattices (Nabarro-Herring creep) or along grain boundaries (Coble creep).

Redrawn from ASHBY and VERRALL (1978), *Phil. Trans. R. Soc.*, **288A**, 67–79.

temperatures and attaining a constant plateau for much higher temperatures. An increase in confining pressure makes creep or flow more sluggish, but the effect is much less marked than that of temperature except on cataclastic flow in the brittle regime. The pressure dependence of strain rate commonly takes the form $e^{-pV_a/kT}$ where V_a is the activation volume. The pressure dependence is sometimes approximately incorporated in the flow laws by assuming that the activation energy is proportional to the melting temperature T_m, which increases with rising pressure in the same way. Thus the ratio T/T_m, known as the *homologous temperature*, is often used as a guide to the type of flow likely to prevail.

In transient creep occurring at temperatures below about $0.2T_m$, the strain is proportional to the logarithm of time; this is known as *logarithmic creep* or *α-creep*. It cannot produce large strains. It is caused by movement of dislocations (atomic mismatches in lattice planes) and the progressive slowing down of the creep rate is attributed to work hardening produced by the pile up of dislocations against obstacles, including dislocation tangles, more rapidly than these can be removed by diffusion. Between about $0.2T_m$ and $0.5T_m$ *transitional creep* or *β-creep* occurs and at higher temperatures this merges into power-law creep described below. Transient creep is unlikely to produce significant deformation within the Earth but it is probably of considerable importance in the attenuation of seismic energy (p. 166) and in the release of certain stresses in the lithosphere (membrane stress, thermal stress, bending stress).

Metals at low temperature deform by *plastic flow* when their yield stress is exceeded. In a perfectly plastic substance the shear stress at yield cannot be exceeded whatever the strain rate, but metals deviate from such ideal behaviour in that the strain rate increases exponentially with rising stress above the yield point (Fig. 8.3). Several yield criteria have been suggested, such as the widely accepted Von Mises criterion which depends on the elastic strain energy reaching a critical value. Plastic deformation in metals is produced by the movement of edge or screw dislocations on glide planes accompanied by some movement at crystal boundaries (Fig. 8.2b). The finite yield strength is a measure of the stress needed to overcome the resistance of the lattice to dislocation glide or the presence of obstacles. Most plastic substances show an increase in yield stress with progressive plastic strain (work hardening). Plastic flow in a single crystal requires the presence of two independent glide planes. In a polycrystalline solid, plastic deformation without unequal distortion of the crystals requires the presence of five independent glide planes within each crystal; commonly the crystals are unequally distorted and flow can proceed with only three or four sets of glide planes being active. Plastic flow as observed in the tensile testing of metals is not strictly a steady state

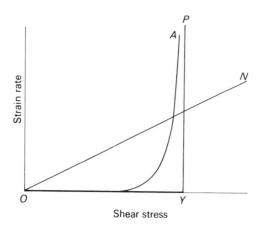

Fig. 8.3 Strain rate plotted against stress for Newtonian viscosity (*N*), non-Newtonian (Andradean) viscosity (*A*) and ideal plasticity (*OYP*). Redrawn from OROWAN (1965), *Phil. Trans. R. Soc.*, **258A**, 285.

process as the flow accelerates during necking and rupture eventually takes place. However, in contrast to transient creep, large strains can occur as a result of plastic flow. It has been suggested that plastic flow is a significant factor in the bending of the oceanic lithosphere at trenches (p. 224).

Power-law creep occurs in metals at temperatures above about $0.55T_m$. It is variously referred to as hot creep, recovery creep or Weertman creep. This type of flow also occurs by the movement of dislocations on glide planes (Fig. 8.2c). The main distinction from plastic flow is that thermally activated diffusion of atoms in the lattice at the higher temperatures prevailing enables the obstacles such as dislocation tangles to be removed as fast as they are produced. Thus work hardening is not experienced and steady state flow occurs in response to a given applied stress field. The dislocations tend to orientate themselves into polygonal cells within each crystal, their dimension decreasing as the applied stress increases. Adjacent crystals accommodate to the deformation by grain boundary sliding. As the yield strength is much less than at lower temperatures, power-law creep occurs at much lower strain rates and in response to much smaller stress differences than low temperature plastic flow. The main characteristic is that the strain rate is proportional to a power n of the stress, where $n \geqslant 3$. The constitutive equation relating strain rate $\dot{\varepsilon}$ to stress σ takes the form

$$\dot{\varepsilon} = c\sigma^n e^{-Q/kT}$$

where Q is the pressure dependent activation energy of creep and c is a constant. Thus power-law creep gives rise to a stress-dependent viscosity. By carrying out experiments at different strain rates and temperatures, the constants c and Q can be determined for a given material. Unlike plastic flow, power-law creep is not limited by a yield stress, although the power-law dependence implies that the strain rate at low stress is negligible. Power-law creep is probably the most important mechanism of deformation in the lower part of the lithosphere and in the mantle below it.

Diffusion creep is observed experimentally in metals at temperatures above about $0.85T_m$. Theory suggests that it also occurs at lower temperatures but is either swamped by power-law creep or occurs too slowly to be detected in experiments. It is caused by migration of atoms in a stress gradient (Fig. 8.2d). If the migration occurs through the crystal lattices it is called *Nabarro-Herring creep* but if it occurs along crystal boundaries it is called *Coble creep*. In strong contrast to plastic flow and power-law creep, the strain rate in diffusion creep is proportional to the applied stress. Substances showing this sort of creep behave as if they possessed Newtonian viscosity such that

$$\sigma = \eta\dot{\varepsilon}$$

where η is the coefficient of dynamic viscosity. Solid state theory can be used to relate it to the properties of the grains, giving

$$\eta = kTR^2/10DV_a$$

where R is the grain radius, D is the diffusion coefficient and V_a is the activation volume. D depends on whether Nabarro-Herring or Coble creep is occurring and it takes the form $e^{-Q/kT}$ where Q is the pressure-dependent free energy of activation. Diffusion creep is thus enhanced relative to the dislocation mechanisms of flow by high temperature and low strain rate. It may be of some importance in the mantle below the lithosphere.

Two other known mechanisms of creep may be of much greater importance at the slow strain rates within the Earth than in metals. One type is *fluid-phase transport creep*, resulting from solution and redistribution of the solid phase from a water film between the grains in response to a stress field. This is closely akin to Coble creep except that the diffusion can probably take place much more rapidly in a fluid than along solid grain boundaries. The other type is *recrystallization creep* which is observed to occur in metals and rocks above about $0.6T_m$ as a result of thermally activated nucleation and growth of new crystals in response to stress. Recrystallization creep is known to occur in metals deforming by power-law creep at low strain rates and its repeated occurrence can give rise to episodes of accelerated creep rate (SELLARS, 1978). These two mechanisms are probably of some importance in development of metamorphic fabrics. A further mechanism of flow which may be locally relevant within the Earth is *superplasticity*. This may occur when a second mineral phase is present in significant proportion provided that the grain size is small. Deformation proceeds by sliding at the interphase grain boundaries. This mechanism may be relevant to the movement at or near fault planes where mylonitization occurs.

It is often convenient to model the response of regions of the Earth to applied stress fields by using simple mathematical laws. The simplest of these is linear elastic deformation which can be used to model the response of the Earth to short period stressing such as the tidal forces or for investigating stress in the outermost brittle layer. Deformation in the mantle has often been modelled by Newtonian viscous flow. A visco-elastic substance is basically a Newtonian viscous substance showing an initial elastic strain, and this is useful in modelling the response of the lower part of the lithosphere to applied stress systems. In recent years the finite element method of stress analysis, using computers, has made it possible to model more realistic rheological behaviour, such as visco-elastic deformation associated with power-law creep.

8.2 Rheology of the crust and mantle – inferences from experiments and theory

Knowledge of the mechanisms by which rocks deform at different depths within the Earth comes partly from experimental evidence and partly from the study of the Earth's response to natural phenomena such as glacial loading. The most striking feature is the rheological stratification of the outermost 400 km, with the strong and brittle rocks near the surface giving way downwards to the weak and ductile rocks of the asthenosphere. There are also large lateral variations in rheology of the crust and upper mantle mainly related to lateral temperature variations. The experimental and theoretical evidence bearing on the rheological stratification of the crust and mantle is discussed in this section, starting at the surface and working downwards.

Brittle deformation

Unlike metals, silicate minerals are brittle at low temperature and pressure. The uppermost rocks of the crust, excluding the sedimentary layers where these exist, are therefore brittle and also relatively strong. At atmospheric temperature and pressure, the compressive strength of granite is 140 MPa and the tensile strength is about 4 MPa, this being typical of the topmost crust. The strength of brittle rocks increases with confining pressure but decreases with rise of temperature. Experiments indicate that the strength would be expected to increase with depth down to about ten kilometres, with the influence of pressure outweighing that of temperature. At greater depths, depending on the local temperature gradient, the rocks would be expected to weaken with increasing depth. Useful reviews of brittle deformation of rocks have been given by MURRELL (1977) and PATERSON (1978).

Beneath a depth of a few hundred metres, all three principal pressures are normally compressions. The measured closure pressure for Griffith cracks in gabbro specimens from the Michigan borehole is 145 MPa (WANG and SIMMONS, 1978). Probably most cracks are closed by 5 km depth. Closed-crack shear fracture would thus be expected to be the dominant failure mechanism in the brittle outer zone of the Earth. This is borne out by the prevalence of faulting. There are three main classes of faults observed by geologists. These were interpreted according to the Coulomb-Navier criterion by ANDERSON (1951) to depend on which of the three principal pressures is the vertical one (Fig. 8.4). *Normal faults* have fracture planes dipping at about 60–70° and the downthrown side overlies the plane. According to Anderson, normal faults are formed by shear fracture when σ_{max} is vertical. The fault trace is perpendicular to σ_{min} and faulting results in local extension of the crust in this direction. Normal faulting, in common with the two other types, may occur on either of two conjugate planes but local inhomogeneities may cause one of them to be dominant. *Thrust faults* are caused by shear fracture when σ_{min} is vertical and in theory the fault plane should dip at 15–30°. *Strike-slip faults* show horizontal movement on vertical fault planes and they occur when σ_{int} is vertical.

Anderson's theory of faulting was one of the early successes of the application of mechanics to geological problems. There are, however, observed features of faulting which do not fit into the theory without some modification of it. For instance, the theory is unable to explain the common occurrence of oblique slip faulting. This may be explained without recourse to obliquely orientated stress systems by the presence of pre-existing planes of weakness (BOTT, 1959; JAEGER, 1960). A more

(a)

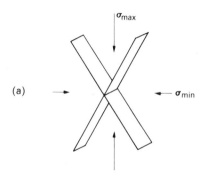

(b)

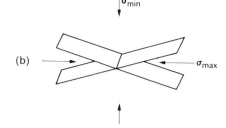

(c)

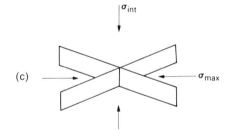

Fig. 8.4 The subdivision of faults into three major classes by Anderson depends on (*i*) the Coulomb-Navier hypothesis of shear fraction, and (*ii*) the three principal pressures near the Earth's surface being orientated horizontally and vertically. The three classes shown are:
  (**a**) normal fault—σ_{max} vertical;
  (**b**) thrust fault—σ_{min} vertical;
  (**c**) strike-slip fault—σ_{int} vertical.
Redrawn from ANDERSON (1951), *The dynamics of faulting*, pp. 14–16, Oliver and Boyd.

serious difficulty recognised by HUBBERT and RUBEY (1959) is that the theory requires unrealistically high shearing stress for initiation of faulting at high confining pressure. This problem arises because the Coulomb-Navier criterion for faulting requires the shearing stress to reach $(\tau_0 + \mu\sigma_n)$, where μ is the coefficient of internal friction and σ_n is the normal stress. Dry rocks are observed to have values of μ ranging between 0·5 and 1·0. The effect of such a high level of friction would be to inhibit faulting. The problem is avoided if either internal friction or normal stress is reduced. In the crust both of them are probably reduced for the following reasons. Faults are usually lined by a gouge of saturated clay and WANG and MAO (1979) have shown that the coefficient of internal friction of such material is about 0·1. Furthermore, if the cracks are filled by pore fluid at pressure p, then it can be shown by extending the Griffith theory that the effective normal stress is reduced to $(\sigma_n - p)$ which can be quite small if the fluid pressure is high enough. The importance of pore fluid in reducing the shear fracture strength at high confining pressure is borne out by experiments on rocks containing hydrous minerals (RALEIGH and PATERSON, 1965; MURRELL and ISMAIL, 1976).

Dyke intrusion and the formation of certain types of joint are caused by extension fracture. Dyke intrusion is believed to occur by the wedging effect of a sheet of magma under higher hydrostatic pressure than the mean pressure in the intruded rocks (ANDERSON, 1951). This causes an actual tension to develop ahead of the advancing wedge of magma and this is relieved by extension fracture as shown in Fig. 8.5. Both dykes and faults should form at right angles to the minimum principal pressure and can therefore indicate past directions of tension in the continental crust.

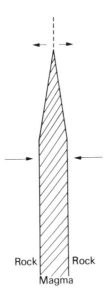

Rock   Rock
Magma

Fig. 8.5 The mechanism of dyke intrusion: a vertical extension fracture develops ahead of the upward rising wedge of magma which has a hydrostatic pressure exceeding the confining pressure of the intruded solid rocks. Redrawn from ANDERSON (1951), *The dynamics of faulting*, p. 24, Oliver and Boyd.

The occurrence of quite large local gravity anomalies caused by lateral density variation at shallow level in Precambrian shields, such as large igneous intrusions, indicates that the brittle rocks of the upper continental crust are able to withstand stresses of around 10 MPa for periods greater than 1000 My. These rocks, however, are probably able to suffer insignificant amounts of deformation by low temperature transient creep and this may be of considerable importance in relieving certain non-renewable types of strain, such as that produced by bending.

The rocks of the brittle zone are known to undergo local cataclastic flow in the vicinity of large faults and possibly also in downbending of the oceanic lithosphere at trenches. They are also affected by *dilatancy*, which is the increase in volume suffered by brittle rocks just prior to faulting as a result of the growth and opening up of cracks. As a result, water drains out of the pores into the newly-opened cracks and returns to the pores again when the cracks close on completion of faulting. This phenomenon is not yet well understood although it is probably of considerable importance in the rheological behaviour of the rocks of the brittle zone. It may give rise to detectable changes in *P* and *S* velocities of the affected rocks, this being relevant to the problem of earthquake prediction.

The brittle-ductile transition

Experiments suggest that a very important change in the mechanical properties of rocks occurs within the depth range of about 10 to 50 km depending mainly on geothermal gradient. This is the downward transition from brittle to ductile behaviour. GRIGGS and others (1960) showed that no sudden fracture was experimentally observed in any rock examined apart from quartzite above 500 MPa confining pressure and 770 K, which would represent a depth of about 20 km in a region where the geothermal gradient is somewhat above average. The brittle-ductile transition within the Earth thus marks off an overlying brittle layer about 10 to 50 km thick from the underlying region where ductile flow occurs and strength decreases with depth. A problematical exception, discussed later in the chapter, is the apparent brittle behaviour indicated by earthquakes in the subducting oceanic lithosphere down to 700 km depth.

Creep mechanisms in olivine

Below the brittle zone and outside the deep earthquake belts, creep is likely to be by far the most important deformation mechanism in the solid part of the Earth. The fabrics of naturally deformed rocks which have been brought up from depth by tectonic or igneous processes indicate that the same types of creep process occur in them as in experimentally deformed metals, despite the much slower strain rates involved. The best available approach to determining the creep mechanisms of the lower crust and mantle is therefore to assume the constitutive creep laws derived for metals and to determine the appropriate constants from experiments on rocks and minerals. Most of the relevant deformation experiments have been carried out on the mineral olivine or on olivine-rich rocks such as peridotite, the results being particularly relevant to flow in the upper mantle. This approach has several limitations. The most serious of these is that the creep rates within the Earth are of the order of 10^{-15} s^{-1} whereas the slowest rate attainable in experiments is about 10^{-8}. Further difficulties arise because the effect of pressure on creep rates is poorly known, because profound weakening of rocks can be caused by the presence of small quantities of water and because the presence of two or more crystal phases in a rock may greatly modify its rheology. Consequently the inferred strain rates under given conditions may be in error by several orders of magnitude and should be treated with appropriate caution.

Experiments suggest that low temperature plastic flow may be the dominant mechanism of deformation in olivine and peridotite at depths within the Earth just below the brittle-ductile transition (e.g. CARTER and AVE' LALLEMANT, 1970; CARTER, 1976). According to GOETZE (1978), the experimental results above a stress difference of 100–200 MPa fit a simple law of the form

$$\dot{\varepsilon} = \dot{\varepsilon}_0 e^{-Q(1 - \Delta\sigma/\sigma)^2/RT}$$

This type of equation makes it possible to estimate the low temperature plastic flow properties of an olivine upper mantle from the experimental data. The evidence therefore indicates that there may be a zone of plastic flow just below the brittle-ductile transition where temperatures are insufficiently high for the occurrence of power-law creep. Because of the significant yield strength, stress differences of the order of 100 MPa or greater may occur within this zone.

By analogy with the behaviour of metals, power-law creep might be expected to become the dominant flow mechanism in the mantle at depths where the temperature reaches about $0.55 T_m$. The value of T_m appropriate to the rheology is the melting point of the most abundant mineral phase rather than of the eutectic mixture (WEERTMAN, 1978). Thus power-law creep probably starts to occur at about 1100 K in the mantle but at a substantially lower temperature of about 700 K in the continental crust. Several experimental investigations have detected power-law creep in olivine-rich rocks at temperatures between 0.55 and $0.9 T_m$ and at strain rates of between 10^{-7} and 10^{-4} s^{-1}. The experiments yield a value of $n = 3$ for the power-law exponent at low stress and a value of about $n = 5$ at high stress. The activation energy and activation volume have also been determined (KOHLSTEDT and GOETZE, 1974; KOHLSTEDT and others, 1980). A strain rate equation can thus be determined governing power-law creep in the upper mantle, this being subject to considerable uncertainty because of the slower natural strain rate and other factors such as presence of water. One of the best pieces of evidence that power-law creep does actually occur within the mantle down to at least 200 km depth is the dislocation structures observed in peridotite nodules from the mantle (NICOLAS and POIRIER, 1976, p. 406). The stress-dependent viscosity associated with power-law creep would be expected to decrease by several orders of magnitude on going downwards towards the low velocity zone of the mantle where temperatures are closest to the melting temperature.

Because of the linear stress-strain relationship, diffusion creep becomes increasingly prominent relative to power-law creep at lower applied stress differences. It is also favoured by high temperature and the strain rate is inversely proportional to the square of grain size. Diffusion creep has not been observed in olivine but it can be modelled fairly accurately for specified grain size by solid state theory. It is normally assumed that the oxygen ion is the slowest moving species in the olivine lattice controlling the overall rate of lattice diffusion. There have been some measurements of the oxygen transport rate in olivine and the activation volume controlling the pressure dependence has also been estimated. Evidence from olivine nodules from the mantle down to 200 km depth suggest the grain size is about 1 to 10 mm. Consequently the viscosity associated with diffusion creep in an olivine mantle can be estimated for both Nabarro-Herring and Coble mechanisms. These estimates can then be compared with the power-law creep behaviour at comparable temperatures and pressures and the dominant mechanism can be identified. Conflicting results have been obtained (NICOLAS and POIRIER, 1976, p. 400) but the more recently published comparisons tend to favour power-law creep throughout the mantle except possibly at the slowest strain rates of about 10^{-16} s^{-1}. For instance, WEERTMAN (1978) calculated that power-law creep must be dominant throughout the mantle if the grain size exceeds 1 mm and the stress exceeds 0.1 MPa.

One method of visually displaying inferred rheological properties within the Earth is by use of deformation maps, a concept introduced by STOCKER and ASHBY (1973). This shows the fields where the different types of creep or flow are dominant as a function of shear stress and either temperature or pressure or depth (for an assumed geothermal gradient). Strain rates are plotted as contours on the deformation map. Two examples of deformation maps constructed for a dry olivine upper mantle of grain size 0.1 mm are shown in Fig. 8.6, showing the behaviour at zero pressure as a function of homologous temperature and as a function of depth for an assumed geothermal gradient of a type probably occurring beneath shield areas. These maps should be treated with caution for the reasons emphasized previously but they clearly indicate the general rheological zoning of the crust and upper mantle. According to the maps of ASHBY and VERRALL (1978), the brittle zone where flow can only occur by cataclastic processes extends under normal temperature

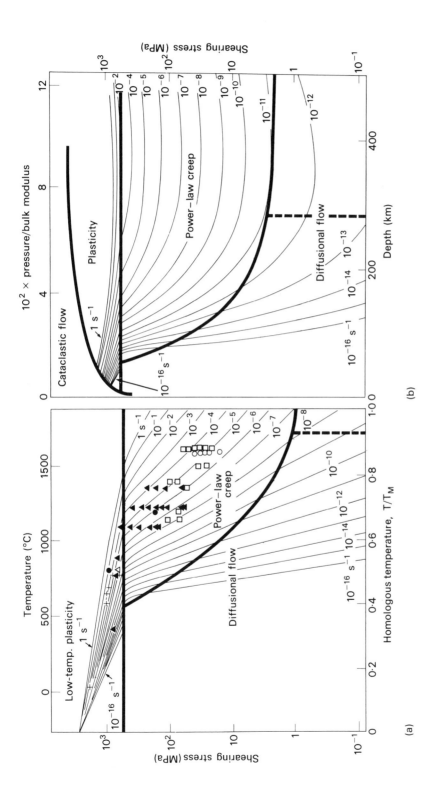

Fig. 8.6 Deformation maps for olivine of 0·1 mm grain size, showing the fields where the different flow mechanisms are dominant.

(a) Map showing shearing stress versus temperature at contoured strain rates for deformation at zero pressure. The symbols identify experimentally determined points.

(b) Map showing shearing stress versus depth at contoured strain rates for deformation under upper mantle conditions. This diagram assumes that the activation volume controlling the pressure dependence is equal to the oxygen ion volume at atmospheric pressure, and that temperature increases with depth according to a shield type geotherm.

Redrawn from ASHBY and VERRALL (1978), *Phil. Trans. R. Soc.*, **288A**, 84 and 88.

conditions to about 20 to 40 km depth. According to them, a low temperature plastic flow regime occurs below this down to about 100 km depth with power-law creep occurring at greater depths. Diffusion creep can only occur if the grain size is less than about 1 mm. In general, it appears that these maps probably overestimate the width of the plastic zone and that power-law creep starts rather shallower than they infer.

The lithosphere and the asthenosphere

The idea that the outer part of the Earth has a relatively strong lithosphere underlain by a weak asthenosphere was first postulated to explain how isostatic equilibrium is attained and has subsequently formed an essential basis for the plate tectonic theory. Such zoning is now readily understood in terms of the variation of the rheological behaviour of rocks with temperature in any planet with a cold surface and a hot, convecting interior.

The lithosphere itself is probably generally subdivisible into three main rheological layers of contrasting properties. At the top is a brittle layer, about 20 to 40 km thick, which is relatively strong even in response to long period stressing. It responds to stresses below the yield strength by elastic deformation and transient creep. Above the yield strength it deforms by fracture and locally by cataclastic flow. Below the brittle layer, there is probably a plastic layer which has a finite yield strength of the order of 100 MPa above which deformation occurs by plastic flow. This layer is not well understood and is not included in most rheological modelling of the lithosphere except in downbending at trenches. The lowermost zone, transitional between lithosphere and asthenosphere, is the region where power-law creep becomes the dominant process of deformation. This layer probably spans the region where the temperature increases downwards from about $0.55T_m$ to $0.8T_m$ or thereabouts, and it can be treated as visco-elastic with stress-dependent viscosity. The apparent viscosity probably increases by several orders of magnitude going down across this zone, as the temperature rises from $0.55T_m$ towards the melting point.

The boundary between the lithosphere and the asthenosphere occurs within the zone of power-law creep and must be regarded as gradational. It is dependent on the time duration of the load. For short-period loading of up to a few thousand years, most of the power-law creep zone between lithosphere and asthenosphere will be unable to deform significantly and will thus form part of the elastic lithosphere. On the other hand, for long-period loading over a few million years or more, significant flow can occur throughout the zone of power-law creep so that it will form part of the asthenosphere. The gradational and time-dependent nature of the boundary makes it impossible to define the lithosphere without ambiguity.

The boundary between lithosphere and asthenosphere is often taken at the top of the upper mantle low-velocity channel for S waves where the temperature may approach the melting point. For long duration loads it probably occurs at a depth where the temperature reaches about $0.55T_m$ which is probably normally about half way down to the top of the low velocity channel. The long-term elastic lithosphere may thus be only about half the thickness of the seismological lithosphere.

The thickness of the lithosphere and its constituent rheological layers varies laterally because of varying temperature-depth distributions on which it is critically dependent. In hot regions such as ocean ridge crests where temperatures exceed 1000 K at shallow depth, the elastic lithosphere may be only a few kilometres thick whereas beneath continental shields it probably exceeds 100 km. A further complication in relatively hot continental regions is that the lower crustal material may be weaker than the underlying topmost mantle because of its lower melting point.

The asthenosphere is the weak zone of the upper mantle where isostatic adjustments are accommodated. In the absence of satisfactory long-term rheological evidence, it is often regarded as being coincident with the seismological low velocity and low Q channel in the upper mantle although there is no proof that these anomalous seismological properties are necessarily identifiable with weak rheology. The main justification is the belief that both low velocity channel and asthenosphere mark the region in the mantle where the melting point is most closely reached. If this assumption is correct, then the asthenosphere is a channel extending on average between about 80 and 180 km depths and is deeper and less well developed beneath continents than oceans.

The apparent viscosities appropriate to both power-law and diffusion creep would be expected to reach a minimum in the zone within the Earth where the melting point is most nearly reached. The transition from asthenosphere to lithosphere above is a consequence of the steep decrease in temperature upwards caused by thermal conduction, which greatly outweighs the pressure effect. Below the asthenosphere, the viscosity would be expected to rise downwards slightly because the adiabatic temperature gradient appropriate to a convecting mantle is lower than the fusion gradient. It should be emphasized that the existence of the weak zone does not necessarily depend on the presence of a partially fused fraction. The theory of diffusion-controlled creep mechanisms clearly predicts the weak zone even if the temperature is considerably below the melting point. If power-law creep applies, then the stress-dependent viscosity would be further reduced as a result of the high strain rate caused by the differential motion between lithospheric plates and the deeper mantle occurring across the asthenosphere. In regions where partial fusion does occur, fluid phase transport creep is probably dominant and the viscosity is anomalously low.

According to WEERTMAN (1970), the viscosity of the mantle at a strain rate of $10^{-16} \, \text{s}^{-1}$ is calculated to increase by a factor of about a hundred between the asthenosphere and the base of the mantle for power-law creep, but by six or seven orders of magnitude for Nabarro-Herring creep. In general, rheologists consider that power-law creep is probably the dominant flow mechanism throughout the mantle unless the grain size is unexpectedly small and the strain rates very low, when diffusion creep might become significant. These inferences do not take into account the radical phase changes occurring in the mantle transition zone, and they appear to conflict with inferences based on post-glacial isostatic recovery as discussed later in the chapter.

8.3 Flexure of the lithosphere

The lithosphere is regarded as being relatively rigid in the sense that it does not undergo much horizontal deformation of significance except near plate boundaries. Nevertheless, substantial differential vertical movements occur as a result of bending of the lithosphere. This is the response, partly elastic and partly non-elastic, to loading or unloading such as may be caused by water, icecaps, sediments or erosion. For loads of limited areal extent the bending is insignificant but for loads substantially wider than the lithospheric thickness the bending can produce vertical displacement closely approximating that of simple Airy isostasy. As discussed in Chapter 5, acute bending of the oceanic lithosphere occurs at trenches.

The response of the lithosphere to a given load has usually been modelled by treating it as a thin elastic or visco-elastic plate underlain by a viscous fluid substratum. The rate of attainment of equilibrium depends on both the mechanical properties of the lithospheric plate and the viscosity distribution within the underlying substratum. For wide loads such as icecaps the underlying viscosity is the controlling factor in the recovery of equilibrium and for this reason glacial rebound phenomena are primarily studied to determine this viscosity, as described later in the chapter. On the other hand, the equilibrium configuration in response to a given load is independent of the viscosity of the substratum and depends only on the thickness and rheological properties of the lithosphere.

Consider first the lithosphere modelled as an elastic sheet. The resistance of an elastic sheet or beam to bending is measured by its flexural rigidity given by

$$D = ET^3/12(1 - \sigma^2)$$

where E is Young's modulus, σ is Poisson's ratio and T is the thickness of the sheet. As the elastic constants of the lithosphere are reasonably well known, determination of the flexural rigidity enables the thickness T to be estimated. The response of the sheet to loading is affected by the presence of underlying and overlying fluid layers of densities ρ_m and ρ_w respectively, representing the mantle below and seawater above the lithosphere. These exert a resultant upward pressure of $(\rho_m - \rho_w)gw$, where w is the vertical displacement of the sheet downwards. Treating the lithosphere as a two-dimensional horizontal sheet, w then satisfies the fourth order differential equation

$$D\frac{d^4w}{dx^4} + (\rho_m - \rho_w)wg = P(x)$$

where $P(x)$ is the load as a function of horizontal distance x and g is gravity. This equation can be solved for a given load distribution. It can also be extended to the general three-dimensional case. The solution can usefully be expressed in terms of the *flexural parameter a*, which has the dimension of distance and is a function of D and the fluid densities above and below, where

$$a = \sqrt[4]{4D/(\rho_m - \rho_w)g}$$

Insight into the response of the lithosphere to loading can be obtained by studying its elastic response to harmonic loads of the form $P(x) = P\cos 2\pi x/L$, where L is the wavelength. The resultant deformation obtained by solving the fourth order differential equation above is also harmonic, having the same wavelength L. The deformation for sufficiently large wavelengths closely approximates local isostatic equilibrium whereas sufficiently short wavelengths are borne without significant deformation. The degree of isostatic equilibrium attained can be expressed in terms of the flexural parameter. It can be shown that if $L = 2\pi a$ then isostatic equilibrium is 80% attained and if $L = 4\pi a$ it is 98·5% attained. On the other hand, for a shorter wavelength of $L = \pi a$ it is only 20% attained and for $L = \frac{1}{2}\pi a$ it is only 1·5% attained. Estimates of the flexural parameter of the lithosphere vary between about 55 and 200 km in different regions, so that loads narrower than about 100 km are effectively borne by the strength of the lithosphere whereas those of 1000 km or greater width reach approximate isostatic equilibrium. If other factors were equal, the oceanic lithosphere would have a slightly larger flexural parameter than the continental lithosphere because of the presence of the fluid layer above formed by the sea, but this effect is counteracted by the oceanic lithosphere being normally thinner than the continental.

Flexure of the continental lithosphere

An instructive example of flexure of the lithosphere is provided by the isostatic uplift following the last drying up of Lake Bonneville in western Utah (CRITTENDEN, 1963a and b). Lake Bonneville was a Pleistocene lake which covered an elliptical area of about $400 \times 250 \ km^2$ on the salt flats west of Salt Lake City. It was filled periodically during the Pleistocene, the last period extending from 25 000 to 10 000 years ago. The ground was depressed by the water load when the lake was filled and recovered towards its original level when the load was removed. The ancient shoreline caused by the last filling can be traced round the margins of the lake and on the hills which formed islands. This allows the depth of the lake and the deformation of the once level shoreline to be accurately worked out. The average depth of the last filling was about 145 m and the centre of the lake has subsequently risen 64 m relative to the margins (Fig. 8.7). Thus the ancient shoreline is now domed as a result of the isostatic readjustment after removal of the water load. Crittenden showed that the

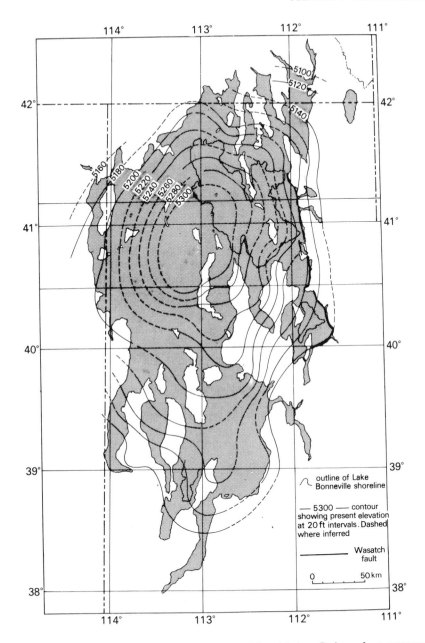

Fig. 8.7 Deformation of the shoreline of Lake Bonneville since it last dried up. Redrawn from CRITTENDEN (1963a), *Prof. Pap. U.S. geol. Surv.*, **454**-E, p. E9.

degree of isostatic equilibrium reached during the last period of loading and unloading is at least 75% and probably 90%. Thus the doming nearly mirror-images the depression caused by the last loading.

If the degree of isostatic equilibrium attained is known, the observed depression at the centre caused by the known water load can be used to determine the flexural parameter of the lithosphere below Lake Bonneville. WALCOTT (1970a) used this and other features of the deformation to show that the flexural parameter here lies between 48 and 60 km, the preferred value being 55 km. The corresponding value of the flexural rigidity is $(5 \pm 2) \times 10^{22}$ Nm. Putting $E = 8 \times 10^{10}$ Nm^{-2} appropriate to the continental crust and $\sigma = 0.25$, the estimated thickness of the elastic lithosphere beneath this part of the Basin and Range province is 19 km. This is exceptionally thin, reflecting the active extensional tectonism and high heat flow of the province. If the lithosphere was of normal continental thickness, then a load of such small areal extent would result in relatively minor vertical movement.

Other examples of the response of the continental lithosphere to flexural loading are not as well defined as Lake Bonneville where load and response are accurately determinable. They fall into two main categories depending on the time scale of loading. On a short time scale of a few thousand years, the best examples for estimating the flexural parameter are the drying up of large glacial lakes. On a much longer time scale, other loads can be recognized which have been in existence for periods of the order of 100 My or more. WALCOTT (1970a) studied both types of loading, and he found that the apparent flexural rigidity of the continental lithosphere appears to be time-dependent. For the short-period loading of around 10^4 year duration the flexural rigidity was found to be about 10^{25} N m but for loads of much longer duration of around 100 My the value is around 10^{23} to 10^{24} N m or even less. This indicates that the continental lithosphere is apparently generally just over 100 km thick for short-period loading but that it is typically 50 km or less for long-term loading. Walcott suggested that this time dependence of the apparent flexural rigidity of the continental lithosphere could be explained by treating the whole lithosphere as visco-elastic rather than elastic, the viscosity being about 10^{23} Pa s. An alternative suggestion can be made on the basis of the rheological model of the lithosphere described in section 8.2. On this model, the transitional power-law creep zone can be regarded as part of the lithosphere for short-term loading but as part of the asthenosphere for long-term loading.

Flexure of the oceanic lithosphere

Unlike the continents, oceanic regions do not provide recognizable examples of Pleistocene loading suitable for studying short-term loading. On the other hand, excellent examples of long-term loading are provided by individual volcanic seamounts or chains of seamounts which have erupted on top of the oceanic crust, causing the underlying lithosphere to sag under their weight. A good example of such loading is provided by the almost circular Great Meteor seamount situated west of the Canary Islands in the North Atlantic Ocean (Fig. 8.8) which was studied by WATTS and others (1975). The excess surface load was determined from the visible bathymetry along a profile across the seamount. The expected downbending of the lithosphere could then be computed for a range of possible values of the flexural rigidity and the corresponding theoretical gravity anomaly profiles could be calculated for these models and compared with the observed gravity profile (Fig. 8.8). The flexural rigidity was then estimated by picking out the value which gives the best agreement between calculated and observed gravity profiles, this value being 6×10^{22} N m corresponding to an elastic layer thickness of about 19 km. The seamount itself causes a large local positive free air gravity anomaly but this is superimposed on a broader negative anomaly of smaller amplitude caused by the downsagging. The negative anomaly is caused by the depression of the low density crust as there is unlikely to be a significant density contrast at the base of the elastic lithosphere.

One of the most comprehensive examples of loading of the oceanic lithosphere is provided by the Hawaiian-Emperor seamount chain which forms a ridge stretching for more than 4000 km in the central Pacific Ocean. The volcanoes along the ridge vary systematically in age and they were

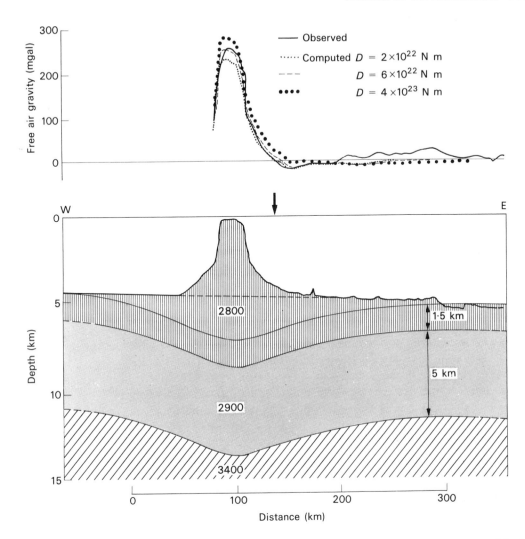

Fig. 8.8 Comparison of the observed free air anomaly across the Great Meteor seamount, north-eastern Atlantic Ocean, with theoretical profiles based on downbending of the lithosphere under the load. Note that the observed and calculated profiles agree best for a flexural rigidity D of 6×10^{22} N m. Densities are shown in kg m^{-3}. The arrow marks the position latitude 30° N, longitude 28° W. Redrawn from WATTS and others (1975), *J. geophys. Res.*, **80**, 1397.

erupted onto oceanic crust of varying age. The Hawaiian ridge has formed during the last 20 My on oceanic crust 80 to 105 My old, the age of the crust at the time of loading being about 77 to 87 My. The Emperor seamounts are 55 to 60 My old and they formed on top of oceanic crust 80 to 115 My old, the age of the crust at the time of formation varying from 20 My at the northern end to 70 My at the southern end near the junction with the Hawaiian ridge. WATTS (1978) studied the response of the oceanic lithosphere to the loading of this seamount chain along fourteen profiles across the ridge spanning its length. He carried out Fourier analysis on the gravity and bathymetric profiles before comparing them to obtain the flexural rigidity, thus using a more refined technique than the simple approach adopted to study the Great Meteor seamount.

Treating the lithosphere beneath the Hawaiian-Emperor ridge as elastic, Watts found that its estimated thickness apparently depends on the age of the lithosphere at the time of formation of the volcanic load and not on its present age. At the northern end of the Emperor chain where the volcanoes formed on crust then 20 My old, the analysis yields an estimated lithospheric thickness of 10 to 20 km. At the southern end of the Emperor chain and along the Hawaiian ridge, the lithosphere was about 80 My old when the volcanoes formed and its palaeo-thickness yielded by the analysis is 20 to 35 km. Watts also tried a visco-elastic model of the lithosphere but he rejected this as it gave inconsistent results between the two parts of the ridge. Watts thus suggested that the oceanic lithosphere behaves as an elastic plate of varying thickness. The thickness yielded by the loading studies is not the present value but that appropriate to the time of formation of the volcanic load. The deformation occurring at the time of loading and shortly afterwards remains frozen in, retaining its original shape as the underlying lithosphere subsequently thickened as it cooled further.

The results obtained from the Hawaiian-Emperor ridge can be combined with other flexural studies from the Pacific, including downbending at trenches, to estimate the apparent thickness of the long-term elastic lithosphere beneath the Pacific Ocean as a function of its age. Figure 8.9 shows such results compared with the temperature distribution and the thickness of the seismological lithosphere determined from surface wave studies. It is of considerable interest to note that the long-term elastic lithosphere appears to thicken in conformity with the isotherms, its base approximately corresponding to the 450°C (720 K) isotherm which is about half the absolute melting temperature. Taking into account the uncertainties in the flexural studies, this suggests that the long-term elastic lithosphere comprises the brittle and plastic zones and that its base is marked by the significant onset of power-law creep. On the other hand, the seismological lithosphere appears to be over twice the thickness of the long-term elastic lithosphere, including a substantial region where power-law creep is the dominant deformation mechanism. Thus short-term seismological and long-term flexural studies of the Pacific lithosphere appear to be in excellent agreement with the rheological model of the lithosphere described in section 8.2.

Can the lithosphere buckle?

If a thin elastic plate is compressed along its length it may buckle before the compressive strength is reached. If the plate is viscous or plastic, resting on a fluid substratum of lower viscosity, then buckling may still occur. The problem arises as to whether the lithosphere can deform by buckling

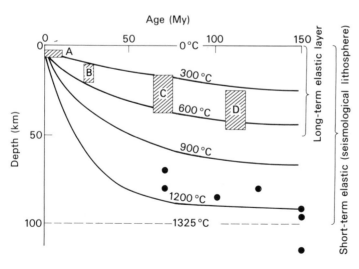

Fig. 8.9 Comparison of the estimated long-term elastic layer thickness of the Pacific lithosphere as determined by flexural studies with the short-term thickness as defined by surface wave studies. These are shown on an age-depth plot of the oceanic lithosphere displaying isotherms. Redrawn with modification from WATTS (1978), *J. geophys. Res.*, **83**, 6002.

in response to horizontal compression. Such a mechanism has been suggested for the formation of geosynclines and trenches.

The problem of buckling of the lithosphere has been investigated from both theoretical and experimental standpoints by RAMBERG and STEPHANSSON (1964). They assumed that the lithosphere is an elastic or viscous plate overlying a fluid substratum. They found that the compressive stress needed to overcome the gravitational body forces caused by buckling greatly exceeds realistic estimates of the strength unless the lithosphere is unrealistically thin – 250 m thick or less in the elastic model and 1·5 km or less in the viscous model. They point out that their conclusion would also apply to more complicated rheological models of the lithosphere and substratum.

Thus the buckling mechanism cannot cause deformation of the lithosphere or the crust as a whole. Failure must take place by fracture or flow long before the compressive stress is large enough to buckle such a thick layer.

8.4 Earthquakes

Earthquakes are produced by sudden release of strain energy in relatively localized regions of the lithosphere, including the deep tongues of sinking lithosphere at some convergent plate boundaries. They are classified according to depth of focus as follows:

0–70 km   shallow focus
70–300 km   intermediate focus
below 300 km   deep focus

No earthquakes have been detected below 720 km depth. Over 75 % of the energy is released in shallow focus events and only about 3 % in deep focus events.

The world-wide distribution of all epicentres located by the U.S. Coast and Geodetic Survey during the period 1961–1967 is shown in Fig. 3.29, and that of large earthquakes occurring between 1904 and 1952 is shown in Figs 8.10 and 8.11. These maps show that the great majority of earthquakes lie along belts which coincide with the plate boundaries. This suggests that most earthquakes occur as a result of release of strain associated with the plate driving mechanisms. These earthquake belts can be subdivided into two types. A significant belt of shallow focus earthquakes follows the crest of the ocean ridge system and extends along the East African rift belt, marking the divergent plate boundaries. These events are seen on Fig. 3.29 but are not large enough to show up on Fig. 8.10. In contrast, nearly all the large earthquakes and all the intermediate and deep focus events occur along the circum-Pacific and Alpine-Himalayan belts where plates are colliding or sliding horizontally past each other or being recycled into the mantle. About 75% of shallow earthquakes, 90% of intermediate and nearly all deep events occur beneath the circum-Pacific belt, where the foci tend to cluster near a plane which dips at about 45° beneath the adjacent continents and underlies the belt of active volcanoes.

Small earthquakes do occur within plate interiors but the total energy release from them is insignificant in comparison with that from plate boundary earthquakes. Such earthquakes, however, are of considerable interest as indicators of the state of stress within plates.

The most widely accepted explanation of shallow earthquakes is based on Reid's elastic rebound theory which was formulated after the San Francisco earthquake of 1906 when a right-lateral movement of up to 7 m occurred on the San Andreas fault. The theory attributes earthquakes to the progressive accumulation of strain energy in tectonic regions and the sudden release of this energy by faulting when the fracture strength is exceeded (Fig. 8.12). The common observation that the

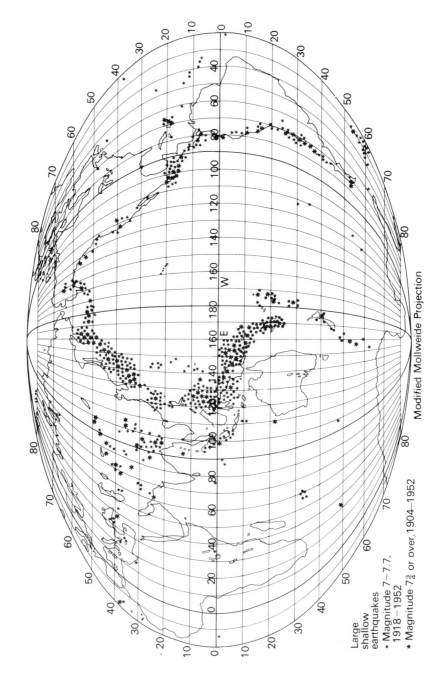

Fig. 8.10 World map of large shallow earthquakes, 1904–1952. Redrawn from GUTENBERG and RICHTER (1954), *Seismicity of the Earth and associated phenomena*, p. 14, Princeton University Press.

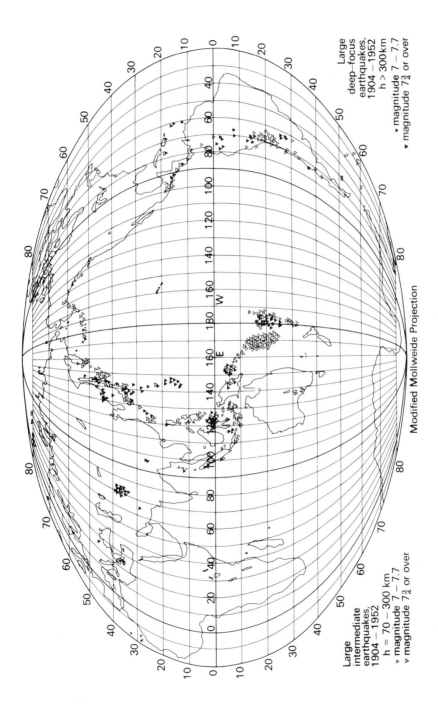

Fig. 8.11 World map of large intermediate and deep earthquakes, 1904–1952. Redrawn from GUTENBERG and RICHTER (1954), *Seismicity of the Earth and associated phenomena*, p. 15, Princeton University Press.

Large
intermediate
earthquakes,
1904 – 1952
h = 70 – 300 km
▽ magnitude 7 – 7.7
▼ magnitude 7$\frac{3}{4}$ or over

Large
deep–focus
earthquakes,
1904 – 1952
h > 300 km
▶ magnitude 7 – 7.7
▶ magnitude 7$\frac{3}{4}$ or over

Modified Mollweide Projection

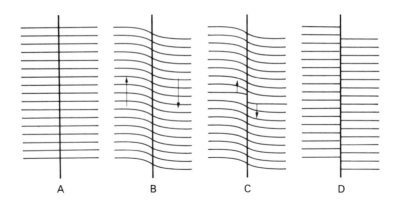

Fig. 8.12   Sketch illustrating Reid's elastic-rebound earthquake source mechanism. Redrawn from BENIOFF (1964), *Science, N. Y.*, **143**, 1400.

largest earthquakes are accompanied by visible slip on pre-existing faults adds support to the theory.

Strain release by slip along a fault can occur either by occasional sudden displacement or by steady creep or by a combination of both. In this respect, recent detailed study of the strain release along the San Andreas fault in California has yielded results of great interest (e.g. WALLACE, 1978; LINDH and others, 1979). The San Andreas fault is a transform fault which marks the boundary between the Pacific and North American plates. World-wide analyses of plate motions yield a right-lateral movement between the adjacent plates across the San Andreas fault of about 50 mm y^{-1} averaged over the last few million years but geodetic measurements indicate that the current rate is about 30 mm y^{-1}. The type of strain release associated with the relative motion of the plates varies from one segment to another along the visible length of the fault. Along most segments, the majority of the strain is released by the occasional occurrence of great earthquakes, with a relatively minor release by smaller earthquakes and slow creep. In between the earthquakes, the strain accumulates over a belt about 50 km wide along the line of the fault. At the other extreme, along a segment of the fault about 200 km long in central California, the strain is almost entirely taken up by slow creep of about 28 mm y^{-1}. This segment is characterized by a great deal of micro-earthquake activity but there are no large earthquakes. Segments of intermediate and gradational character also occur along the fault, as would be expected. The most plausible explanation of the difference in strain release mechanism between the segments of the San Andreas fault is variation in the strength of the rocks or the geometry of the fault plane from one segment to another.

The earthquake foci beneath different regions are in general not distributed over the whole thickness of the lithosphere but are restricted to an upper layer extending down to between about 10 and 40 km depending on the geothermal gradient (VETTER and MEISSNER, 1979). Beneath the San Andreas fault where the geothermal gradient is relatively high, the deepest foci are generally above 15 km depth whereas in the relatively cool shield regions the weak seismicity extends down to 30 or 40 km depth. This is in excellent agreement with the rheological model of the lithosphere developed earlier in the chapter, with the earthquakes occurring in the brittle upper zone of the lithosphere and ductile release of strain occurring at greater depths. According to Vetter and Meissner, comparison of the greatest depth of earthquakes in various regions with the local geothermal gradient indicates that the transition occurs at a temperature of about 0·6 to 0·65 T_m.

Focal mechanism of shallow earthquakes

The modern concept of shallow earthquake mechanism is an elaboration of Reid's elastic rebound theory. Some developments have come from experimental rock mechanics and some from intense seismological and geodetic study of specific regions such as the San Andreas fault. Perhaps, the most important contribution has come from the mathematical modelling of sudden fault motion and the prediction of the resulting distant (far-field) elastic-wave radiation pattern in terms of a few relatively simple parameters describing the orientation and nature of the source. Interest is now turning to the more complicated near-field seismic radiation patterns detectable using strong motion seismographs near the source. Such studies should in the future enable a much better idea of the complexity of the source mechanism to be obtained.

An indication of how the shearing stress on a slip plane may vary before, during and after a simple earthquake is shown in Fig. 8.13. The earthquake occurs as a result of the slow build-up of tectonic stress over a period of years resulting, for instance, from the relative motion of two adjacent plates. Laboratory experiments suggest that a small amount of slow stable sliding on the fault plane may precede an earthquake, during which fluids migrate out of the pores into opening cracks and the fault is made more homogeneous along its length. During this premonitory phase there is a small drop in shearing stress from its maximum tectonic value to a lower value shown as σ_1 in Fig. 8.13. Instability then occurs and sudden shear fracture is initiated on the fault plane. The fracture propagates at a rate somewhat lower than the shear wave velocity as the slip plane extends to a finite length. During the rapid fault motion, the shearing stress supported by the fault plane drops to its dynamic frictional value σ_F which is much less than σ_1. The consequent sudden release of strain energy causes the rocks on either side of the fault plane to accelerate in opposite directions. The rapid impulsive motion causes part of the released energy to be radiated as seismic waves. Energy is also dissipated by friction on the fault and, as a result of deformation, the gravitational energy associated with the affected region may change. The rapid fault motion decelerates and stops when the driving stress ceases to exceed the frictional stress or other constraints. The shearing stress after faulting is σ_2 which in general probably exceeds σ_F. The stress drop $\Delta\sigma$ is therefore $(\sigma_1 - \sigma_2)$. After the main event, a period of adjustment occurs during which small accumulated local strains on the fault plane are relieved by aftershocks and fluid migrates back into the pores as the

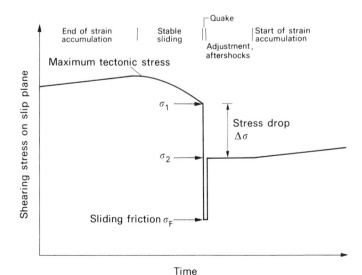

Fig. 8.13 Hypothetical stress history of an earthquake. The scales are probably non-linear. Redrawn with modification from FITCH (1979), *The Earth: its origin, structure and evolution*, p. 525, © Academic Press Inc. (London) Ltd.

cracks close again. Subsequently, strain starts to accumulate again and the earthquake cycle is repeated.

The length of the fault plane affected by slip in a large earthquake is typically about 50 km but in a few great earthquakes it exceeds 100 km. The dislocation u varies along the fault plane and its mean value is typically about 1 to 5 m. The maximum rate of slip reaches about 1 m s^{-1} in a large earthquake and the maximum acceleration at frequencies between 5 and 10 Hz may locally reach the acceleration due to gravity but probably does not exceed 2g. In practice, earthquakes undergo a more complicated pattern of slip than outlined above, varying in space and time and commonly lasting over a period of about one minute.

The modern method of studying strain release at the focus is to observe the pattern of ground motion at distance from the earthquake. This method was pioneered by BYERLY (1926). A great improvement in the quality of focal mechanism studies has been made by Sykes and his co-workers (e.g. SYKES, 1967) by using records from the World-Wide Standardized Seismograph Network which was established by the U.S. Coast and Geodetic Survey in 1962. This considerably improved the quality, consistency and geographical distribution of records. It enables the polarity of the first motion S arrivals to be picked reliably, using the long-period seismograph records. Sykes also found that the long-period records give the most consistent results for P and PKS arrivals. The outcome is that reliable focal mechanism determinations can now be made for earthquakes of magnitude 5·5 or greater, thereby considerably increasing the scope and accuracy of the method.

The usual method for studying the focal mechanism of an earthquake is to observe the polarity of the first motion of P and S waves at seismological stations spread over the Earth's surface. For each station, it is noted whether the first P pulse is a compression or dilatation, and the sense of the first S arrival is determined. The initial direction at source of the ray reaching the station is computed from a knowledge of the travel-path. The results from all stations are then plotted on a projection which shows the initial direction of the rays and their polarity. Figure 8.14 is an example for an Icelandic earthquake of 1963; this shows that the first pulse is compressional in the ENE and WSW quadrants and dilatational in the other quadrants.

The observed pattern for an earthquake needs to be interpreted in terms of a simple model of source mechanism. The two main models which have been suggested are a Type I source which is a single couple, and a Type II source which consists of two couples at right angles in the same plane (Fig. 8.15). The initial P motion is identical for both types of source. In two diagonally opposite segments, the initial pulse is compressional and the initial ground motion is away from the source; in the other two sectors the initial pulse is dilatational and the ground motion is towards the source. The two types of source can be distinguished by using the S radiation pattern. A Type I source gives rise to two shear wave lobes while a Type II source shows four lobes. More complicated source models, such as a moving source representing a fracture spreading along a fault plane, have also been studied.

It was once believed that the mechanism of elastic rebound could be represented best by a Type I source. However, the observed initial motions of S waves showed that the four-lobe pattern of a Type II source is characteristic of most earthquakes. At first this cast some doubt on the hypothesis of elastic rebound; but it was subsequently realized that faulting is better represented by a double couple, because fracture results in the relief of the maximum shearing stress on two conjugate planes at right angles to each other and not on just one of them.

The main value of focal mechanism studies is to determine the attitude of fault planes and the direction of movement associated with earthquakes. The directions of the three principal pressures at the source can be estimated, indicating qualitatively the state of stress in the lithosphere. The P wave radiation pattern (e.g. Fig. 8.14) shows two nodal planes at right angles to each other which

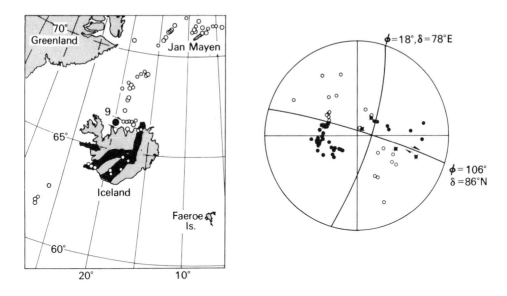

Fig. 8.14   Focal mechanism solution for the Icelandic earthquake of March 28, 1963, shown as event 9 on the map. The initial directions of the observed *P* arrivals are plotted on the lower hemisphere of the projection, compressions as solid circles and dilatations as open circles. The two great circles shown on the projection are the nodal surfaces, either of which could be interpreted as the fault plane. The movement must be perpendicular to their intersection and must therefore be strike slip; the movement would be in the right-lateral sense if the $\phi = 106°$ plane were the fault plane or left-lateral if the $\phi = 18°$ plane were the fault plane. Local geology suggests that the $\phi = 106°$ plane is the actual fault. Redrawn from SYKES (1967), *J. geophys. Res.*, **72**, 2143.

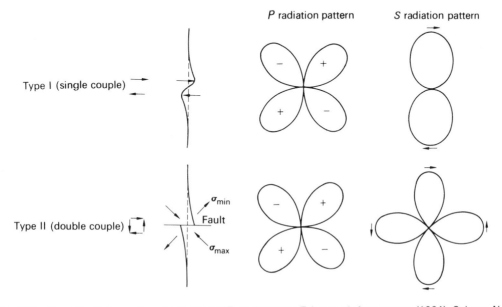

Fig. 8.15   *P* and *S* radiation patterns for Type I and Type II sources. Taken partly from BENIOFF (1964), *Science, N. Y.*, **143**, 1401 and 1402.

separate the compressional and dilatational quadrants. The fault plane is interpreted as being parallel to one of these nodal planes. The S wave radiation pattern for a Type I source would show which of the two planes should be chosen, but for a Type II source the ambiguity remains (Fig. 8.15) and it becomes necessary to use knowledge of the local geology to select the more likely fault plane. Once the fault plane is known, the P radiation pattern determines the sense of the fault movement uniquely. In Fig. 8.14 the two nodal planes are (i) strike 18°, dip 78° E; (ii) strike 106°, dip 86° N. Local geology suggests that (ii) is the more likely fault plane; it can now be deduced that the movement is almost horizontal and right-lateral. Even without knowing which is the actual fault plane, the directions of the principal pressures can be estimated from the polarity diagram. The maximum principal compression bisects the compressional quadrants, the intermediate direction lies parallel to the intersection of the two nodal planes, and the minimum principal compression bisects the dilatational quadrants.

Focal mechanism of intermediate and deep earthquakes

The seismic radiation patterns from intermediate and deep focus events suggest that the great majority of these occur as a result of sudden slip on shear fracture planes. The focal process, however, is less easy to understand in terms of our experimental knowledge of fracture mechanics. The problem is that brittle fracture would not normally be expected to occur at the temperatures and pressures predominant below about 50 km depth. The temperatures are anomalously low within sinking lithospheric slabs but the problem over pressure remains. The shearing stress required to overcome friction would seem to become excessive at such great depths. The most widely accepted explanation is that pore fluids are effective in overcoming the friction as suggested in shallower context by HUBBERT and RUBEY (1959). This might occur by the breakdown of hydrous minerals. RALEIGH and PATERSON (1965) showed that serpentine at 770 K and 350 to 500 MPa confining pressure can suffer shear fracture of the brittle type when decomposing to olivine, talc and water. The Raleigh-Paterson mechanism is probably restricted to a depth range of 20–60 km but other hydrous minerals may break down at greater depth.

Various other mechanism have been suggested in explanation of intermediate and deep earthquakes. These include creep instability, shear-melting, implosions, and rapidly running phase transitions. None of these has been widely accepted. It seems probable that the earthquakes are mostly caused by brittle shear fracture but that Coulomb friction is overcome in a way not yet fully understood.

Quantification of earthquakes

The old method of measuring the size of an earthquake was by use of an intensity scale based on the surface effects of the shock. By making field investigations of the area affected, an isoseismal map showing regions of equal intensity could be produced. Unfortunately the maximum intensity is not a good measure of the energy released because it is strongly affected by the depth of the focus and by local conditions.

A much better approach is to use the amplitudes of seismic waves at distance from the earthquake to define its size. The conventional method has been to determine a quantity related to strain release at the focus known as *magnitude* but more recently the newer concept of *seismic moment*, which has a more precise physical meaning, is becoming increasingly used. The far-field body wave spectra can also be used to determine a quantity known as *corner frequency* which yields an estimate of the dimension of the source. Once the seismic moment and the source dimension of an earthquake have been estimated, certain other source parameters such as the stress drop $\Delta\sigma$ can be determined from them.

RICHTER (1935) introduced the concept of magnitude as a measure of the size of shallow focus earthquakes of southern California. He defined magnitude in terms of the maximum trace amplitude observed at an epicentral distance of 100 km on a standard type of short-period torsion seismometer; this quantity is now known as local magnitude (M_L). GUTENBERG and RICHTER (1936) extended the concept to determination of magnitude (M) of shallow focus events at greater distances by measurement of the maximum ground amplitude in microns of Rayleigh waves at 20-second period. Subsequently there has been considerable discussion about the best relationship to use, although they are all of the form

$$M = \log(A/T) + af(\Delta, h) + b$$

where A is the maximum amplitude, T is the period, Δ is the epicentral distance, h is the focal depth, and a and b are constants, empirically determined. The formula suggested by BÅTH (1966) for shallow focus earthquakes is

$$M = \log(A/T) + 1\cdot66 \log \Delta + 3\cdot3$$

Because of the less effective generation of surface waves in deep earthquakes, GUTENBERG (1945) introduced another magnitude scale (m) based on the maximum amplitude of body waves. The empirical relationship of body wave magnitude to amplitude is also of the form

$$m = \log(A/T) + af(\Delta, h) + b$$

where f is as before an empirically determined function of epicentral distance and focal depth which allows for geometrical spreading and damping. For shallow focus earthquakes m and M are related with a very broad scatter by an empirical formula such as that given by BÅTH (1966):

$$m = 0\cdot56M + 2\cdot9$$

However, this relationship does not apply to surface explosions for which the ratio of surface wave to body wave amplitude is smaller than would be predicted, or for mid-ocean ridge earthquakes for which the ratio is larger than predicted.

The seismic moment of an earthquake is defined as

$$M_0 = \mu u S$$

where μ is the rigidity modulus, u is the mean slip and S is the area of the slip plane. Seismic moment is now regarded as a better measure of the source strength of an earthquake than magnitude. It has a more obvious physical meaning as it is the moment of either of the two equal couples at the source. It is less affected by the local source conditions or by the complexity of an earthquake, being an integrated measure of source strength. Furthermore the magnitude scale based on 20-second period Rayleigh waves saturates for the largest earthquakes for which the length of the slip plane exceeds the wavelength of about 60 km, whereas seismic moment as estimated from lower frequency arrivals does not saturate (KANAMORI, 1977). The seismic moment can sometimes be determined by direct field measurement of u and S as μ is known to reasonable accuracy; S can also be determined from the region affected by aftershocks, although this tends to yield a slight overestimate. The more general method of determining seismic moment is to measure the amplitude of the longer period seismic arrivals at teleseismic distance from the earthquake. The seismic moment is proportional to the amplitude, which is approximately independent of period for the longer periods. Corrections need to be applied for geometrical spreading and attenuation and for the radiation pattern which can be obtained from focal mechanism studies. For large, shallow focus earthquakes the preference has been to use dispersed surface waves for moment determination but for small or deep events the amplitudes of longer period P and S arrivals are commonly used. The accuracy of determining seismic moments from teleseismic arrivals or from field work is to within a factor of two or three. Seismic moments can be estimated for large past events, albeit with much less certainty, using an empirical relationship to magnitude.

A recent improvement in describing the seismic source is by use of the *seismic moment tensor*. This takes into account the orientation of the source as well as its strength. It has nine components, six of them representing the three double couples and the remaining three representing volume change. The seismic moment tensor is determined from the far-field seismic observations at several stations by linear inversion. This new concept has not yet been widely used but is likely to become increasingly prominent.

Another important earthquake parameter determinable from field observations or from the far-field seismic arrivals is the source dimension. This is usually measured as the radius r of the equivalent circular slip plane, although in reality fault planes are normally rectangular. Using a simple physical model of the source, BRUNE (1970) showed that its dimension could be obtained from a plot of amplitude versus frequency of teleseismic S arrivals. The method was later extended by HANKS and WYSS (1972) to use of P arrivals also. Both theory and observations show that the amplitude of the ground displacement for higher frequency P and S arrivals from a distant pulse-like source falls off as the square of frequency (Fig. 8.16); at lower frequencies, however, interference effects resulting from the finite dimension of the source cause the amplitude to be approximately independent of frequency. The frequency at which the two main segments of the amplitude-frequency curve intersect is known as the corner frequency and it can be shown theoretically that it is inversely proportional to the source dimension r. This actually gives quite a good approximation to the half-length of a rectangular slip plane. The horizontal, low frequency segment of the amplitude-frequency plot, after correction for radiation pattern and transmission losses, can be used to determine the seismic moment. Thus by examination of the far-field body wave spectra of an earthquake, both the seismic moment and the source dimension can be estimated.

Knowing the seismic moment and the source dimension, certain other dependent parameters

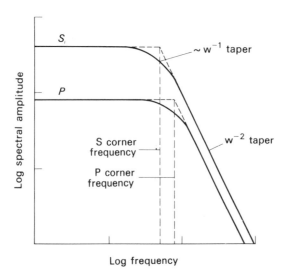

Fig. 8.16 Idealized form of the body wave spectra of an earthquake, corrected for all propagation effects. Redrawn with modification from HANKS and WYSS (1972), *Bull. seism. Soc. Am.*, **62**, 563.

such as the mean slip and the stress drop can be determined. The mean slip is given by

$$u = M_0/\mu S$$

and as μ is fairly accurately known, the slip can be determined once the seismic moment and the area of the fault plane are known. The area can be estimated either by use of the corner frequency or from field observations. BRUNE (1968) used this method to estimate the rate of relative motion across transform faults and convergent plate boundaries. The seismic moments of earthquakes occurring along a specific length of plate boundary are summed over a specified length of time. If the thickness of the brittle faulted part of the lithosphere can be assumed, then the area of the slip plane can be estimated and u can be determined. Results from this method agree with other estimates of relative plate motion to within a factor of about two. Alternatively, the above relationship can be used to determine the thickness of the faulted brittle layer if seismic moment, mean slip and the length of the slip plane are known. Application of this method suggests that the brittle zone of the San Andreas fault extends to about 20 km depth and that associated with oceanic transform faults extends to about 5 km depth.

BRUNE (1970) showed that the stress drop at the focus is proportional to the seismic moment and inversely proportional to the source volume. It can be estimated by use of the formula

$$\Delta\sigma = \tfrac{7}{16}M_0/r^3$$

The calculated stress drops for major plate boundary earthquakes are typically about 3 MPa and seldom exceed 6 MPa while those of the large plate interior earthquakes are about 10 MPa. Values down to 0·1 MPa are recorded for small earthquakes. It is not yet known whether the stress drop in an earthquake is in general almost complete or only partial. The evidence is conflicting. The lack of a major geothermal anomaly associated with fault lines such as the San Andreas suggests that friction on the fault must be quite small and thus that the stress drop may be almost complete. On the other hand, other evidence outlined in section 8.5 suggests that stresses in the brittle upper part of the lithosphere are generally much greater than indicated by the estimated stress drops in earthquakes.

Energy release in earthquakes

As pointed out by KANAMORI (1977), the energy release in earthquakes is a subject of considerable geophysical importance. It places a lower limit on the energy used up in driving the lithospheric plate motions and as such is relevant to the discussion of the mechanism of plate tectonics in Chapter 9.

The elastic energy in a seismic wave of given period is proportional to the square of its amplitude. Seismic arrivals at teleseismic distance can therefore be used to estimate the elastic-wave energy radiated by an earthquake. The normal procedure has been to estimate the energy release in an earthquake from its magnitude by use of an empirical relationship. The relationship between magnitude and energy release in joules has been subject to discussion and change. The normally used formula, due to Gutenberg and Richter, is

$$\log E = 1\!\cdot\!4\,M + 4\!\cdot\!8$$

KANAMORI (1977) showed that this formula considerably underestimates the energy release in the really great earthquakes such as the Chile earthquake of 1960 for which the magnitude scale saturates. Using the Gutenberg-Richter formula, he estimated the average annual energy release over the period 1900 to 1976 to be about 7×10^{17} J. Using a relationship between energy release and seismic moment, he estimated the average annual release in the great earthquakes alone during the period 1920 to 1976 to be 4.5×10^{17} J at a minimum. Furthermore, the rate of energy release in the great earthquakes was a factor of ten higher than normal during the period 1950 to 1965 when many great earthquakes occurred. An earlier maximum period probably also occurred at the turn of the century. During these periods of great earthquakes, the annual numbers of moderate to large earthquakes and the energy release according to the Gutenberg-Richter formula were both at a minimum.

Putting the two complementary estimates of energy release together, the average annual release during the present century is estimated to be about 1.1×10^{18} J which is equivalent to about 3×10^{10} W. Most of this energy comes from the few large shocks of magnitude greater than 7·0. It should be remembered that only part of the total energy released is converted into elastic strain energy, much of it probably being dissipated as heat by friction. WYSS (1970) estimated the seismic efficiency of South American intermediate and deep earthquakes to be 10 % at the maximum. If this is a realistic estimate of efficiency in general, then the total rate of energy release may be over 10^{11} W. This energy release is probably not constant from year to year but varies in cycles over a few tens of years in a way which is not yet properly understood.

8.5 Stress in the lithosphere

The lithosphere is much stronger than the deeper parts of the Earth. The occurrence of faulting and earthquakes indicates that it is locally stressed to breaking strength. The lithosphere must therefore be subjected to much larger stress differences than can occur at greater depths. The most significant stress system is that which arises from the boundary forces which drive the plate motions but lithospheric stress also originates in several other ways. The state of stress in the lithosphere is therefore the result of superposition of several stress systems of differing origin.

Two main categories of stress system affect the lithosphere. Stress systems of the first category are those which persist, even when their strain energy is progressively released, as a result of the continuing presence of the causitive boundary or body forces. The lithosphere therefore acts as a reservoir of strain energy which is fed into it by the activity of the forces at about the same rate as it is released by tectonic activity such as earthquakes and faulting. Such stress systems are capable of causing continuing tectonic activity. They will be referred to as *renewable stress systems*. The most important example is the stressing of the lithosphere by the boundary forces which cause the plate motions. Isostatically compensated crustal thickness variations can also give rise to this type of stress system.

Stress systems of the second category are those which can be dissipated by release of the strain energy which was initially present. These will be referred to as *non-renewable stress systems*. Examples include bending stress, membrane stress and stress of thermal origin. The associated stress differences are commonly quite large, perhaps exceeding 100 MPa, but the total release of strain energy over geological time must be negligible in comparison with that released by renewable stress systems. The initial strains may exceed 1 % but because they are non-renewable it seems likely that the stress can be dissipated by transient creep on a relatively short geological time scale, say less than 1 My. This may explain why they are not much in evidence in modern tectonic activity.

Stress caused by plate boundary forces

Plates move relative to each other because they are driven by renewable forces applied to their boundaries. Other forces act to resist the plate motions. As the inertial forces are negligible, the driving forces and resistive forces acting on any plate must be in dynamic equilibrium. The combined effect of these forces is to cause the plates to be stressed. If the distribution of the forces is known, the associated state of stress within the plates can be determined.

The plates may possibly move in response to horizontal forces applied to their vertical edges at the plate boundaries. A 'push' may be exerted by upwelling material at ocean ridges and a 'pull' may be exerted on each of the adjoining plates at a convergent boundary as a result of the downward pull of the sinking slab. Opposition to plate motion results from resistive forces near the convergent plate boundary and from drag on the base of the plate exerted by the underlying viscous asthenosphere. With this system of boundary forces, the plates would be expected to be in a state of horizontal compression near the ridges and tension near the convergent plate boundaries, with gradation between.

Alternatively, the plates may be stressed by the drag of underlying mantle convection currents acting on their undersides, with resistance to motion being at the plate boundaries. This would produce tension above the uprising convection currents and compression above the downsinking currents, with gradation between. The mechanism of plate motions is discussed more fully in Chapter 9, where it is shown that the plates are probably driven by 'push' and 'pull' at plate boundaries rather than by the drag of underlying convection currents.

Stresses caused by loading

The lithosphere is locally stressed by loading. The load may be a topographical feature on the surface or may arise internally from lateral variation in density or may be a combination of both. Loading can be studied approximately by using two simple models:

(*i*)   a load acting on the surface of a uniform half-space;
(*ii*) loading of an elastic layer underlain by a fluid substratum.

In this section those situations which are not associated with lithospheric flexure are described; bending stresses are discussed in the following section.

For loads which are relatively narrow compared to the thickness of the lithosphere, the resulting stress distribution can be approximated by a surface load acting on a uniform elastic half-space. Consider a uniform elastic half-space subjected to a uniform two-dimensional surface load σ extending from $x = +a$ to $x = -a$. The load causes a stress distribution as shown in Fig. 8.17(a). The maximum stress difference is equal to $2\sigma/\pi$ which occurs along the semi-circular arc containing the ends of the load. Stress differences beneath the load exceed half this value down to a depth of 1·5 times the width of the load. Applying this to a topographical load 40 km wide, 500 m high with a density of 2700 kg m^{-3}, the maximum stress difference of 8 MPa extends to 20 km depth. This is the sort of effect a topographical uplift such as the northern Pennines would cause; the negative load caused by a low density granite batholith associated with a -45 mgal gravity anomaly would cause stress differences of the same magnitude. JEFFREYS (1976) has shown that the stress difference below a large mountain range such as the Himalaya must locally exceed 100 MPa.

The stress differences shown in Fig. 8.17(a) would not be large enough to cause failure of the crust. However, an interesting situation arises if a local stress system such as this one is superimposed on a large regional stress field. The result is that the local stress system modifies the regional stresses in the vicinity of the load. Figure 8.17(b) shows the above local stress system superimposed on a regional horizontal tension of 100 MPa. The resulting stress differences are slightly increased beneath the load but are reduced elsewhere. Such interaction of local and regional stress fields may be an important factor in locating the position where failure such as faulting first occurs.

Another important situation occurs where a load acting on the surface of the lithosphere is

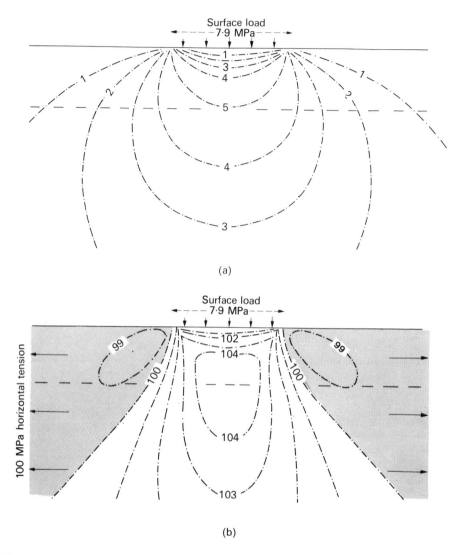

(a)

(b)

Fig. 8.17

(a) The stress differences in MPa produced by a uniform two-dimensional load of 7·9 MPa acting on the surface of a homogeneous elastic half space, assuming plane strain. This load would be equivalent to a topographic elevation of about 300 m, or a positive gravity anomaly of 25–30 mgal caused by half density rocks in the uppermost part of the crust.

(b) The stress differences in MPa caused by the superimposition of the same surface load on a horizontal tension of 100 MPa.

Both distributions are independent of scale, but if the load is assumed to be 50 km wide, then the dashed line gives the approximate depth of the continental Moho. Redrawn from BOTT (1965), *Submarine geology and geophysics*, p. 197, Butterworths.

counterbalanced by an equal and opposite upthrust acting on the base of the crust or at some other depth (BOTT, 1971a; ARTYUSHKOV, 1973). This is the appropriate model for local isostatic equilibrium and is particularly important in understanding the stresses associated with mountain ranges, passive continental margins and plateau uplifts. It can be approximated by studying the

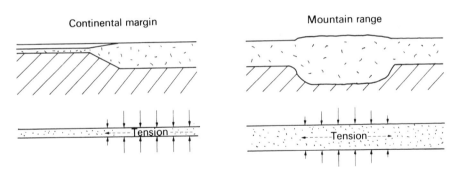

Fig. 8.18 Production of regional tensile stress in the crust by the combined effect of surface load and the upthrust of a root in isostatic equilibrium with the load, applied to continental margins and mountain ranges.

stress system caused by equal and opposite pressures applied to both sides of a two-dimensional beam (Fig. 8.18). The result is that the vertical pressure is increased between the load and the upthrust by an amount approximately equal to the load pressure whereas the horizontal pressure is not much affected. This means that the region between the load and the upthrust is thrown into a state of horizontal deviatoric tension relative to the adjacent regions. The resulting stress difference is approximately equal to the load pressure, being larger by a factor of $\pi/2$ than the maximum stress difference associated with a narrow uncompensated surface load (Fig. 8.17). For a compensated mountain range of 4 km elevation and 2700 kg m^{-3} density the resulting stress difference is about 100 MPa and for a plateau uplift of 2 km elevation it is 50 MPa. This phenomenon causes the continental crust adjacent to a passive continental margin to be thrown into a state of deviatoric tension relative to the oceanic side.

Bending of the lithosphere

Flexure of the lithosphere in response to loading was discussed earlier in the chapter. Flexure is associated with bending, which gives rise to horizontal compression on the concave side and tension on the convex side of each bend. The horizontal stress is given by

$$\sigma_x = Ez\frac{d^2w}{dx^2}$$

where E is Young's modulus, z is the vertical distance from the mid-plane of the lithosphere and w is the vertical displacement. Thus the maximum stress developed by elastic bending is proportional to the curvature and thickness of the lithosphere. Below a surface load the bending is concave upwards so that the upper part of the lithosphere is subjected to compression and the lower part to tension. The opposite situation applies beyond the flanks of the load where the bending is convex upwards.

The bending stresses in the lithosphere are greatest for loads of about 4.4 times the flexural parameter, that is for a wavelength of about 500 km. WALCOTT (1970b) showed that the maximum horizontal stress caused by such a load is about six times the load pressure. The bending stresses become negligible in comparison with the load pressure when the load wavelength is longer than about 50 times the flexural parameter (that is over about 5000 km) or shorter than about half the flexural parameter (about 50 km).

Substantial horizontal stresses must therefore develop in the lithosphere as it bends in response to surface loads a few hundred kilometres across. For instance, below a wide pile of deltaic sediments at a continental margin which are 5 km thick, the maximum bending stress is about 450 MPa which is near the breaking strength. It is a puzzle to understand why there is so little evidence of earthquakes or faulting in such regions. One possible explanation is that the bending stress, being non-renewable, is released shortly after the bending takes place by transient creep.

Even sharper bending occurs where the oceanic lithosphere turns downwards at subduction zones, causing extreme tension in the upper part and compression in the lower part. This is borne out by the nature of the earthquakes which are tensile above 25 km depth and compressive below down to 40 to 50 km depth (CHAPPLE and FORSYTH, 1979) as discussed in Chapter 5. The extreme stress developed at the downbend is probably partly relieved by plastic flow at some trenches.

Membrane stress

Because of the spheroidal shape of the Earth, the radii of curvature of the surface vary between the poles and equator. The two principal radii of curvature at the pole are equal, their value being 6400 km. At the equator, the radius of curvature along a parallel of latitude is 6378 km and that along a meridien of longitude is 6335 km. Consequently, if a lithospheric plate changes its latitude, it must undergo horizontal deformation to enable it to fit onto a part of the Earth's surface having different curvature. The horizontal stresses caused by this process are referred to as membrane stresses (TURCOTTE, 1974).

The origin of membrane stress can be understood as follows. Suppose a small spherical cap of circular outline fits without strain onto the surface of a sphere of radius R, subtending an angle of $2\phi_0$ at the centre (Fig. 8.19). The spherical cap is then forced to fit onto the surface of a sphere of slightly larger (or smaller) radius $(R + \Delta R)$. This can only occur if the outer part of the spherical cap is stretched and the inner part is compressed, or vice-versa if the radius is decreased. The deformed cap is supported by excess pressure on the sphere where it is stretched and deficient pressure where it is compressed. The maximum tension occurs on the edge of the cap in the direction of the circumference. Quite a good approximation to the value of the maximum tension, provided that ϕ_0 does not exceed 45°, is given by

$$\sigma = \tfrac{1}{4}E\Delta R \phi_0^2 / R$$

The maximum compression occurs at the centre and is half this value. It should be emphasized that these are not bending stresses as discussed in the previous section, but result from horizontal extension and shortening of the surface to enable it to fit. The main stresses are horizontal tensions

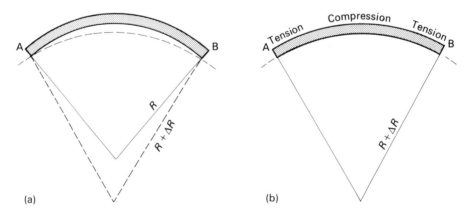

(a)   (b)

Fig. 8.19   The origin of membrane stress. The unstrained part of a spherical shell AB of radius R shown in (a) is deformed to fit onto a spherical surface of radius $R + \Delta R$ shown in (b), causing compression in the middle and tension near the edges.

and compressions which are independent of the vertical thickness of the cap. The stresses developed are of opposite sign if the cap is forced to fit onto a sphere of smaller radius.

Suppose we have a circular lithospheric plate which subtends $\phi_0 = 45°$ at the centre of the Earth and which is unstressed at the equator and then moves to the pole (or vice-versa). TURCOTTE (1974) showed that the maximum stress developed would be about 180 MPa. In practice, lithospheric plates are not circular in outline and their areal extent may change progressively as a result of accretion or subduction. Nevertheless, membrane stresses must develop when the latitude of plates changes significantly as a result of plate motions and polar wandering. Horizontal stresses of about 100 MPa may possibly develop in some plates from this cause. Membrane stresses, however, are of the non-renewable type and it is possible that they may be dissipated by transient creep without causing any significant effects. On the other hand, TURCOTTE (1974) suggested that the tensile stress may cause propagating fractures which give rise to rift valley systems and intra-plate volcanism such as along the Hawaiian Island chain.

Thermal stress

Stress in the elastic lithosphere may result from change of temperature. For instance, the new elastic part of the lithosphere forms near ocean ridges at a temperature of about $0.5T_m$. The upper part of it cools more rapidly than the lower part, causing tension above and compression below. Under such circumstances, thermal stress may exceed 100 MPa. There is a lack of evidence for tectonic activity resulting from such stress and it is possible that it is relieved by transient creep over a relatively short period of time.

Tidal stress

The Earth is subject to tidal deformation which gives rise to a small oscillatory straining of dominantly semi-diurnal period. The resulting stress in the lithosphere has an amplitude less than 10^{-3} MPa which is very much smaller than the stress from other causes. Tidal stress is unlikely to be a primary cause of tectonic activity but its peak values may sometimes supplement existing stress to the extent of triggering earthquakes.

Stress concentration in the upper elastic lithosphere

The subdivision of the lithosphere into an upper elastic layer overlying a lower ductile layer of transitional properties has important implications on the state of stress caused by systems of boundary forces acting on lithospheric plates. Suppose, for instance, that a plate is subjected to uniform horizontal normal pressure acting on two opposite vertical edges, as might be caused by upwelling at ocean ridges. KUSZNIR and BOTT (1977) modelled this situation by finite element analysis, approximating the lithosphere by a uniform visco-elastic layer 60 km thick underlying an elastic layer 20 km thick (Fig. 8.20). Initially a uniform horizontal compression equal to the boundary pressures affects the whole plate, but the stress within the visco-elastic layer decays with time except at the extreme ends of the plate. The time constant of decay is proportional to the viscosity, being about 0.3 My for a 2000 km long plate with viscosity of 10^{23} Pa s. As the stress in the visco-elastic layer decays, that in the upper elastic layer increases, eventually exceeding its initial value by a factor equal to the ratio of lithospheric thickness to elastic layer thickness. This phenomenon of stress concentration explains why moderate boundary pressures of around 20 MPa such as might be applied at ocean ridges or subduction zones can give rise to stress differences of the order of 100 MPa in the uppermost elastic part of the lithosphere.

Local variations in the thickness of the elastic layer within the interior of the plate should be accompanied by corresponding variations in the stress in the elastic layer, as stress is approximately

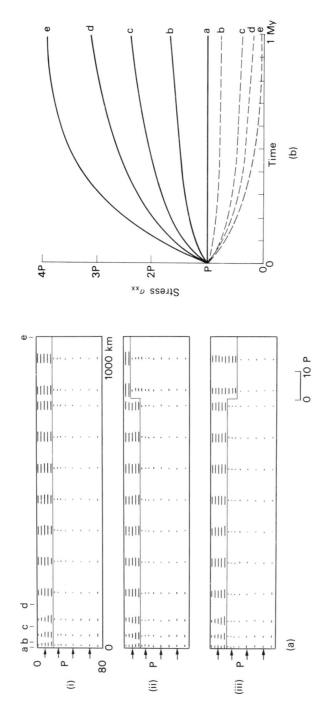

Fig. 8.20 Stress concentration in the upper elastic part of the lithosphere when a uniform pressure is applied at the two ends of a lithospheric plate 2000 km long, when the lithosphere is modelled by an elastic layer 20 km thick above a visco-elastic layer 60 km thick with uniform viscosity of 10^{23} Pa s.

(**a**) Principal stresses in the lithosphere at 0·5 My after the initial application of the pressure P (only one half of the plate is shown, the right end being the central plane of symmetry): (*i*) is for uniform thickness of the elastic layer, (*ii*) shows a localized thinning of the elastic layer with corresponding increase in stress, and (*iii*) shows a localized thickening of the elastic layer with decrease in stress.

(**b**) The horizontal lithospheric stress in the elastic layer (solid line) and visco-elastic layer (dashed line) at positions **a** to **e** of model (*i*), plotted as a function of time after the initial application of the pressure P at the ends of the plate.

Redrawn from KUSZNIR AND BOTT (1977), *Tectonophysics*, **43**, 249 and 250.

inversely proportional to its thickness. The thickness mainly depends on the local geothermal gradient. In Precambrian shield regions the geothermal gradient is anomalously low; the elastic layer must be substantially thicker than average so that stress caused by plate boundary forces should become reduced in such regions. The lowered stress combined with increased strength accounts for the tectonic stability of the shield regions. In contrast, regions where the geothermal gradient is anomalously high, such as western U.S.A., have thin elastic layers and the stress must be correspondingly greater than the average in the plate, helping to account for the tectonic activity characteristic of such regions.

The phenomenon of stress concentration in the upper elastic part of the lithosphere does not only apply to boundary forces but also to stress systems arising from isostatically compensated lateral variations of density. As an example, Fig. 8.21 shows a plateau uplift structure such as western U.S.A. or East Africa with a surface elevation of 2 km isostatically compensated by a low density region in the underlying upper mantle. If the whole structure is treated as elastic, the resulting stress differences in the crust reach about 40 MPa. If the lithosphere is more realistically modelled by an upper elastic layer overlying a lower visco-elastic layer, the resulting stress differences in an elastic layer 10 km thick exceed 200 MPa. This model also gives some insight into the process by which the stress concentration occurs. There is slow upward and outward flow of the buoyant mantle material which causes an outward drag on the base of the overlying elastic layer which produces the additional tension. This model explains the occurrence of high tensile stress in the upper crust in plateau uplift regions. Similar models account for tensile stress on the continental side of passive margins.

Thus the stress concentration phenomenon applies to stresses arising from plate boundary forces and loading. It does not affect the non-renewable systems such as membrane stress.

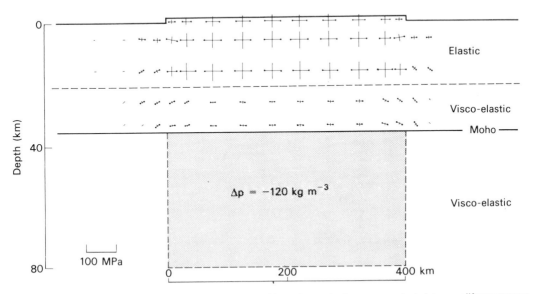

Fig. 8.21 Deviatoric stresses in the lithosphere produced by an isostatically compensated plateau uplift structure as a result of the loading of surface topography of 2 km elevation and the upthrust caused by the low density compensating region in the mantle beneath. The model has been computed assuming that the lithosphere consists of an upper elastic layer about 20 km thick above a visco-elastic layer. Even larger stresses than shown are produced in the elastic layer if this thins in the plateau uplift region. Modified from BOTT and KUSZNIR (1979), *Geophys. J. R. astr. Soc.*, **56**, 455.

Observed state of stress in the lithosphere

As explained earlier in the chapter, the orientation of the three principal pressures in the lithosphere can be estimated locally using earthquake focal mechanism studies. The stress drop in earthquakes can also be estimated but this does not necessarily represent the actual stress difference which may be much larger.

The orientation and magnitude of the principal stresses can be estimated experimentally at the Earth's surface by *in situ* methods. Two types of method reviewed by McGARR and GAY (1978) are commonly used. In the *stress relief* methods, strain measuring devices are fixed in a shallow borehole or on the free surface of rock. The strain in the rock is then released by overcoring and the resulting change in strain is recorded. In general, a number of measurements in different orientations are required to determine the complete state of stress. These methods are restricted to the vicinity of a free surface and the results may be badly affected by local inhomogeneities. The alternative *hydraulic fracturing* methods are used in boreholes and are effective to greater depths. A short length of a borehole is isolated and the pressure in this section is progressively increased by pumping fluid into it until the wall rock fails by tensile fracture. The pumping is stopped immediately this happens and the fluid pressure drops to an equilibrium value which is a measure of the minimum pressure in the plane, normally horizontal, perpendicular to the borehole length. Its direction can be determined by examining the borehole walls to locate the fracture orientation. Knowing the tensile strength and the pore-fluid pressure in the wall rock, the maximum pressure in the plane can be determined from the maximum fluid pressure just prior to fracture. The vertical pressure is obtained from the weight of the overburden. Thus the directions and magnitudes of the three principal pressures can be estimated, assuming that they are horizontal and vertical in orientation.

The measurement of the state of stress in the lithosphere is still at an experimental and exploratory stage. Results obtained to date have been reviewed by RICHARDSON and others (1979). Figure 8.22 is a summary of the results, which shows some consistency between the earthquake and in situ determinations of orientation. The observed stresses are probably mainly caused by plate boundary forces and by loading. RICHARDSON and others (1979) show that the observations are more consistent with plate motions being caused by edge forces than by underlying drag of convection currents.

8.6 Viscosity of the mantle

Theoretical considerations discussed in section 8.2 indicate that the mantle below the lithosphere probably deforms by power-law creep or possibly by diffusion creep. Further evidence on the rheology of the mantle can be obtained by studying the isostatic recovery of the Earth's surface after application or removal of a load (Fig. 8.23). The time constant of recovery is of the order of a few thousand years and therefore only recent loads can be used for this purpose. Such loads are conveniently provided at the present stage of the Earth by the melting of the late Pleistocene icecaps of North America, Fennoscandia and Greenland, and by associated phenomena such as the return of the meltwater to the oceans. These studies yield estimates of the apparent viscosity distribution in the mantle below the lithosphere.

The recovery towards equilibrium after application or removal of a load is affected by the flexural rigidity of the lithosphere as well as by the underlying viscosity distribution. Both factors are significant for narrow loads but the effect of the viscosity distribution is dominant for wide loads. If the load and the resulting pattern of vertical movements at the surface are known, then the viscosity can be estimated. The forward calculation involves the simultaneous solution of the

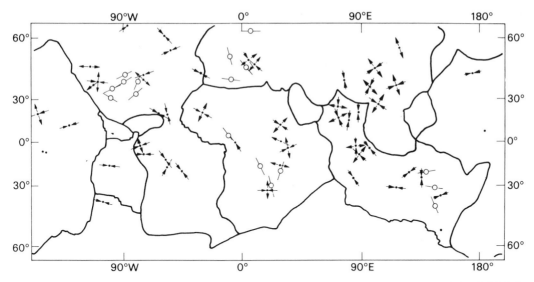

Fig. 8.22 Measured intraplate stress orientations shown with reference to the global plate boundaries. Solid circles denote earthquake focal mechanism determinations, with compression and tension represented by inward and outward pointing arrows respectively. A single tension denotes normal faulting and a single compression denotes thrust faulting, both being shown for strike-slip faulting. Solid circles without arrows represent thrust faulting with poorly constrained direction. Open circles represent *in situ* stress determinations, with the single line marking the direction of maximum compression. Redrawn with simplification from RICHARDSON and others (1979), *Rev. Geophys. space Phys.*, **17**, 996.

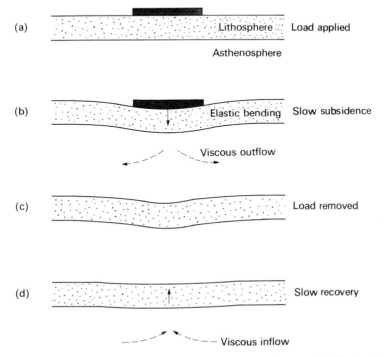

Fig. 8.23 The viscous response of the asthenosphere to the application and removal of a load such as an icecap on the surface of the elastic lithosphere.

Stokes-Navier equation of viscous flow for the substratum and the elastic equations for the lithosphere for a specified initial distribution of surface pressure. This was first done for a circular load on a flat Earth's surface assuming an infinite substratum of uniform viscosity by HASKELL (1935). As computational techniques have improved, the theoretical modelling has been extended to realistic load distributions acting on the spherical Earth's surface with radially variable viscosity distributions in the mantle. Analyses of this type, taking into account postglacial loading and unloading on a worldwide basis, were carried out using slightly different approaches by CATHLES (1975) and by PELTIER and ANDREWS (1976). The viscosity-depth distribution in the mantle, obtained from these and other investigations, is reviewed below, starting at the top and working downwards.

The viscosity of the asthenosphere is best determined using loads of small areal extent, such as Lake Bonneville. There is evidence from a series of narrow loading phenomena of a zone of minimum viscosity of 75 to 100 km thickness just beneath the lithosphere. The Lake Bonneville uplift is consistent with an estimated viscosity of 2×10^{20} Pa s down to 250 km depth or alternatively with a somewhat lower viscosity extending over a smaller depth range. Other estimates have been obtained from the uplift of the Mesiter Vig area of north-eastern Greenland following partial deglaciation about 9000 years ago and from the postglacial uplift of certain islands in the Arctic archipelago. Both yield viscosity values of 4×10^{19} Pa s for a channel 75 km or more in thickness just below the lithosphere. Further supporting evidence for such a low-viscosity channel is provided by the shorelines off the east coast of the United States. CATHLES (1975) has taken these observations to suggest that a universal 75 km thick low-viscosity zone of about 4×10^{19} Pa s is probably generally present below the lithosphere. This approximately corresponds in depth range with the seismological low velocity zone of the upper mantle. The assumption of a universal low viscosity channel in the upper mantle should be taken with some caution as lateral variation must certainly occur in upper mantle rheology and the Lake Bonneville and north-eastern Greenland examples occur in regions where the upper mantle is probably hotter than average.

The uplift of Fennoscandia after melting of the Pleistocene icesheet provides the most reliable information on the viscosity down to a depth of about 1000 km. The icesheet was about 2·5 km thick and covered an area of 2500×1400 km^2 with its centre near the Gulf of Bothnia. It began to melt 20 000 years ago and melting was practically complete about 10 000 years ago. The rate of uplift during the first half of the twentieth century is accurately known because precise levelling was carried out in Finland first between 1892 and 1910 and again between 1937 and 1953. The relaxation time is about 4400 years and the maximum rate of uplift occurs in the Gulf of Bothnia (Fig. 8.24). HASKELL (1937) used the rate of uplift to estimate the underlying viscosity to be 0·95 $\times 10^{21}$ Pa s. He also concluded that the absence of geological evidence of peripheral bulges at the end of the glaciation indicated that there is no sharp increase in viscosity between the upper and lower mantle. Although there has been controversy on this point, the work of CATHLES (1975) supports this conclusion and yields an average value of viscosity of 10^{21} Pa s, with an accuracy of about 10% claimed, down to about 1000 km depth. A low viscosity channel only 75 km wide just below the lithosphere is not inconsistent with this result as the Fennoscandian glacial load was rather too wide to be sensitive to such a channel, although a significantly wider channel is ruled out. CATHLES (1975) showed that the occurrence of phase transitions in the mantle transition zone should not affect the response of the mantle to loading provided that the temperature gradient is adiabatic and that phase equilibrium is rapidly attained.

During the 1960s it was widely held that the lower mantle has a viscosity of about 10^{25} Pa s, which is about four orders of magnitude higher than that of the upper mantle. The main basis for this opinion was the belief that a high lower mantle viscosity is needed to support an inferred non-

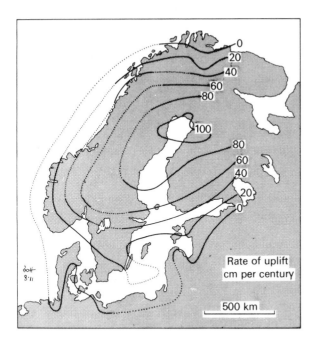

Fig. 8.24 Contemporary rate of uplift in Fennoscandia. Redrawn from GUTENBERG (1959), *Physics of the Earth's interior*, p. 194, Academic Press.

hydrostatic bulge of the Earth's equatorial regions. MUNK and MACDONALD (1960) had suggested that the non-hydrostatic bulge may represent a lag in attainment of hydrostatic equilibrium as the Earth's rate of rotation slows down as a result of tidal friction (p. 24). The existence of this fossil non-hydrostatic bulge was later disproved by GOLDREICH and TOOMRE (1969). They showed that the non-hydrostatic equatorial bulge is not in fact excessively large in comparison with other low degree harmonic components of the non-equilibrium shape of the Earth. In particular, they argued that previous workers had overestimated the gravitational energy associated with the non-equilibrium equatorial bulge in relation to other harmonics. They suggested that the pole of rotation follows the maximum principal non-hydrostatic moment of inertia, rather than vice-versa. Thus the non-equilibrium bulge loses its unique significance and the above argument for a high viscosity of the lower mantle loses its force. Goldreich and Toomre considered that the pole of rotation of the Earth has wandered relative to the mantle during geological time, following the axis of maximum principal moment of inertia. They used this hypothesis to put an upper limit of about 5×10^{23} Pa s on the viscosity of the lower mantle.

More precise estimates of the viscosity of the lower mantle have subsequently been obtained using the long wavelength postglacial recovery phenomenon (CATHLES, 1975; PELTIER and ANDREWS, 1976). This depends on worldwide modelling of postglacial recovery and the associated oceanic loading. Some of the most convincing evidence comes from the regions peripheral to the melting of the large North American icesheet. The central problem has been to determine whether the viscosity is relatively uniform throughout the mantle so that deep flow is associated with wide loads, or whether the lower mantle is much stiffer than the upper mantle so that recovery is effectively restricted to a channel of up to 1000 km thickness. The channel model should reveal itself by the occurrence of peripheral bulges as the load is applied or peripheral troughs as it is removed. Just the opposite effect is predicted for the deep adjustment model, with peripheral bulges developing as the load is removed. The sea level changes along the east coast of the United States show that the region underwent late glacial and early post-glacial uplift followed by subsidence;

this is in disagreement with the channel model and suggests a fairly uniform viscosity of about 10^{21} Pa s throughout the whole mantle. The pattern of uplift in Canada yields a similar estimate. A further test between the channel and deep adjustment models is the response of the Earth to the return of meltwater to the oceans. The sea level rises eustatically but the ocean bed also subsides in response to the increased load. The observed late glacial and post glacial sea level at distance from the icecaps can be compared with the values theoretically predicted for the models. Such a comparison shows that a viscosity of 10^{23} Pa s is too high for the lower mantle and that a value of 10^{21}Pa s throughout the mantle agrees well with the observations.

CATHLES (1975) proposed the following overall viscosity-depth model for the mantle. A 75 km thick channel with a viscosity of 4×10^{19} Pa s directly underlies the lithosphere, forming the asthenosphere. Otherwise the estimated viscosity of the mantle down to 1000 km depth is $(1.0 \pm 0.1) \times 10^{21}$ Pa s and that of the lower mantle below 1000 km depth is (0.9 ± 0.2) Pa s. This model implies that the mantle has an almost uniform viscosity except for the asthenospheric channel whose existence is still controversial. The model does not take into account the lateral variation which is likely to be prominent in the upper mantle particularly between the continents and oceans and elsewhere where there are lateral temperature variations. The actual uncertainty is probably much greater than the quoted errors above but the values should be correct to better than a factor of ten.

According to CATHLES (1975), the pattern of post glacial recovery is consistent with diffusion creep but not with power-law creep. The influence of power-law creep would be to concentrate the flow into the upper mantle causing surface deformation more akin to the channel model than to the deep flow model. The observations are against this. On the other hand, it was shown earlier in the chapter that theoretical and experimental evidence on rock deformation strongly favours power-law creep as the dominant flow mechanism in the mantle below the lithosphere. WEERTMAN (1978) has attempted to reconcile these opposing views by suggesting that glacial recovery occurs by Newtonian transient creep whereas mantle convection occurs by power-law creep. Another possibility suggested by RANALLI (1980) is that the small stresses involved in postglacial recovery in the mantle are superimposed on larger stresses of different origin such as convection, and that this gives rise to an apparent Newtonian viscous flow in a mantle with power-law rheology.

As a warning against too rigid acceptance of the present-day rheological models, the words of WALCOTT (1980) are timely: 'It is worth reminding ourselves of the unusual history of earth rheology where very persuasive arguments have been made for some particular model only to find, within a few years, the argument inverted to favour the opposing model.'

Can the poles wander?

This question is of interest in connection with the interpretation of palaeomagnetic and palaeoclimatological data. It has been pointed out (e.g. GOLDREICH and TOOMRE, 1969) that the apparent motion of continents relative to the poles over the last 300–500 My, as deduced from palaeomagnetic studies, is generally much larger than is needed to account for relative movement of the continents. This is taken to suggest that polar wandering has occurred as well as continental drift.

There are two basic types of polar wandering: (i) the pole of rotation may have wandered in relation to the Earth as a whole; or (ii) the pole of rotation may have remained fixed relative to the lower mantle, but the lithosphere has bodily slipped over the underlying part of the Earth causing apparent polar wandering. The possibility that the pole of rotation can wander relative to the Earth as a whole was discussed and rejected by Sir George Darwin during the last century. The question has been re-opened by GOLD (1955).

A rotating body is in stable equilibrium when the maximum principal moment of inertia coincides with the axis of rotation. This is the condition for minimum kinetic energy of rotation. Because of the equatorial bulge, the Earth rotates about its maximum principal axis of inertia and therefore the axis of rotation would be expected to remain stable both in space and relative to the fixed surface features. However, tectonic and meteorological activity cause small changes in the pattern of surface loading such as icecaps and newly formed mountain belts and these produce small changes in the directions of the principal axes of inertia amounting to a small fraction of a degree. The result would be that the pole would wander to a new position. For example, Gold showed that if the whole of South America were raised by 3 m, the change in moment of inertia would cause the pole to migrate by about $(1/1000)°$.

If the Earth behaves as a rigid body, the polar wandering caused by tectonic and glacial events would be negligible. Gold argued that if the Earth is capable of yielding by flow, the equatorial bulge (which is caused by rotation) will migrate towards the new equator $(1/1000)°$ away, and the deviation between the axis of rotation and the principal moment of inertia will be re-established. Thus the load will cause a progressive migration of the pole until the load itself has settled on the equator. The rate of migration depends entirely on how fast the equilibrium figure of the Earth can be re-established. Gold suggested that the decay time of about 10 years appropriate to the Chandler wobble would also apply to the re-adjustment of the equatorial bulge. Taking the 3 m uplift of South America as an example, the pole would take about 10 years to migrate $(1/1000)°$ or 10^5 years to migrate $10°$ of latitude. This argument suggested to Gold that quite large polar wandering would be expected to occur over periods of geological time.

GOLDREICH and TOOMRE (1969) have extended Gold's argument by showing that the redistribution of mass within the Earth associated with the movement of continents (and presumably associated plates) would itself cause the pole of rotation to follow the axis of maximum non-hydrostatic moment of inertia provided that the viscosity of the mantle is less than about 5×10^{23} Pa s. They showed that the movement of n continents would result in migration of the pole of rotation at a faster rate than that of the individual continents by a factor of $n^{\frac{1}{2}}$. The more recent inference that the viscosity of the mantle is probably about 10^{21} Pa s implies that wandering of the pole relative to the Earth as a whole is possible and has probably occurred.

9 The mechanism of global tectonics

9.1 Introduction

Large horizontal movements of the lithosphere, associated with sea-floor spreading and continental drift, affect the Earth's surface and give rise to tectonic activity at plate boundaries. One of the major outstanding problems of the solid Earth is to understand the causative mechanism underlying such global tectonic activity. This problem concerns both the normal driving mechanism of plate tectonics and the more spasmodic mechanism by which new continental splits are initiated.

The lithospheric plates move relative to each other in response to renewable forces acting on their boundaries. These boundary forces cause the plates to be repeatedly stressed while the strain energy is progressively dissipated as heat, elastic wave energy and gravitational energy of root formation predominantly in the mobile belts. The release of elastic wave energy by earthquakes alone is about 3×10^{10} W (p. 326). Thus the boundary forces must feed strain energy into the lithosphere at an average rate exceeding 3×10^{10} W by a process which almost certainly originates in the mantle below the lithosphere. This chapter discusses the nature of this underlying process.

The main known source of available energy within the Earth is the internal heat supply which escapes at a rate of about 4×10^{13} W. The simplest explanation of the fundamental tectonic process is that the lithosphere is strained as a by-product of heat escaping from the Earth. The Earth acts as a heat engine, converting a small fraction of the escaping heat into strain energy concentrated into the lithosphere. Most heat engines are highly inefficient, partly because of fundamental limitations on efficiency described by the second law of thermodynamics, and partly because of the practical difficulties in obtaining maximum efficiency. If the terrestrial heat engine were only 1 % efficient, with 99 % of the input energy escaping directly as heat flow, then about 4×10^{11} W would still be available for straining the lithosphere. This is over ten times more energy than is released annually by earthquakes. The idea is clearly feasible, and the remaining problems are (*i*) to find whether this is indeed the process by which tectonic energy is produced, and if so (ii) to determine how the terrestrial heat engine works.

9.2 The contraction hypothesis

The contraction hypothesis was suggested as a mechanism of mountain-building by DE BEAUMONT (1852) and by DANA (1847). The mathematical treatment of the contraction of a cooling Earth was developed by DAVISON (1887) and by DARWIN (1887). Until the 1950s the hypothesis had widespread support from many eminent geologists and geophysicists (e.g. JEFFREYS, 1959) but as a result of discoveries about continental drift and sea-floor spreading the contraction hypothesis is now generally regarded as superseded.

The basic idea is that the Earth contracts because it is believed to be cooling. This may occur through (*i*) normal thermal contraction, (*ii*) outgassing of volatiles from the Earth, or (*iii*) outward migration of phase boundaries as the temperature drops. Jeffreys has suggested that the available shortening of the Earth's circumference over geological time may be of the order of 200–600 km. In the region where cooling is most rapid, contraction causes tangential tension which may be relieved by flow if the strength is exceeded. The overlying shell is thrown into a state of tangential compression as gravity causes it to collapse inwards on top of the shrinking sphere beneath. DAVISON (1887) showed that there would be a 'level of no strain' between the regions of compression above and tension below. This used to be identified with the steepening of the circum-Pacific deep earthquake zone at about 300 km depth.

A fundamentally different version of the contraction hypothesis was subsequently proposed by LYTTLETON (1965), based on the assumption that the Earth has heated up from a cold initial state over its lifespan and that the core consists of a high pressure modification of mantle material as suggested by Ramsey (p. 240). After allowing for a slight expansion of the mantle as the transition zone deepened, Lyttleton calculated that the Earth's radius would have decreased by at least 370 km as the dense core has grown to its present size by outward migration of the core-mantle boundary in response to rising temperature. This requires that the melting point at the core-mantle boundary must *decrease* with pressure and that the heat transfer through the mantle must be dominantly by thermal conduction. Most of the basic assumptions conflict with the generally accepted interpretation of evidence on the Earth's interior, so that Lyttleton's hypothesis has not won much support.

According to the contraction hypotheses, tectonic activity is caused by the release of compressive stress in the outer shell, which contains the lithosphere. The lithosphere is too thick to buckle but it can yield by fracture if the compressive strength is exceeded. The strain can be most effectively relieved by two arcuate fracture systems of global dimension, roughly at right angles to each other. These form the two belts of Tertiary mountain building.

It has been increasingly recognized that the contraction hypothesis is inadequate as a fundamental theory of tectonic activity. Some of the main arguments against it are as follows:

(*i*) It is doubtful whether the theory can account for all the apparent crustal shortening in mountain ranges over the past 200 My, let alone through the whole of geological time; thus the contraction hypothesis is scarcely adequate to explain the one type of major surface feature for which it was originally postulated.

(*ii*) Contraction implies that the lithosphere is under continual and general compression, which conflicts with the widespread occurrence of tension features such as normal faults.

(*iii*) The hypothesis has no explanation to offer for other surface features such as ocean ridges, rift systems, plateau uplifts; it is quite inadequate to have any bearing at all on sea-floor spreading and continental drift, and if mountain building is related to plate movements there is no longer any need for a contracting Earth. It has become irrelevant to tectonic problems.

It was shown in Chapter 7 that the Earth has probably cooled by a few hundred degrees at most over its lifespan as a result of convection in the mantle removing heat from the deep interior. This suggests that the Earth may indeed have suffered a small thermal contraction amounting to a few tens of kilometres, assuming that the cooling effect has not been overprinted by expansion caused by decrease of the gravitational constant G. Such a small contraction is unlikely to have relevance to global tectonics.

9.3 The expanding Earth hypothesis

The expanding Earth hypothesis of LINDEMANN (1927) and HILGENBERG (1933) was originally proposed in explanation of the disruption of the continental masses and of the Earth's extension features such as rifting. The idea is that the sialic shell was once continuous and has been fragmented into the present continents as the Earth's surface area has increased about three-fold and the radius increased from an initial value of 3500–4000 km; the oceans have been formed by this process. The hypothesis has subsequently been adopted by several geologists and geophysicists as an explanation of continental drift, ocean ridge formation and the progressive withdrawal of the seas from the continental regions (Fig. 9.1).

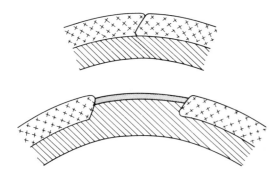

Fig. 9.1 The formation of ocean basins and the mechanism of continental drift according to the expanding Earth hypothesis. Redrawn from EGYED (1957), *Geol. Rdsch.*, **46**, 110.

More extreme versions of the expanding Earth hypothesis have been suggested in recent years by CAREY (1958, 1976), HILGENBERG (1962), OWEN (1976) and some others who believe that most of the expansion has taken place over the last 200–300 My during the disruption of Pangaea. They have adopted the expanding Earth hypothesis essentially because they find that the continents fit together much better on a globe of smaller size. Estimates of the rate of increase of the Earth's radius with time vary between the proponents of the hypothesis (Fig. 9.2). Carey, for instance, attributed the post-Palaeozoic separation of the continents to expansion, with the concurrent growth of all the major ocean basins including the Pacific; this implies that the Carboniferous radius was about 0.7 times the present value which is an average expansion rate of about 6 mm y^{-1}. These extreme forms of the expanding Earth hypothesis cannot be explained by known processes such as phase transitions or possible secular decrease of the gravitational constant G, so that a cause at present unknown to science must be postulated. Enormous membrane stresses (p. 330) must result from such gross expansion of the Earth's surface, but these do not leave any obvious trace of their present or former existence.

From quite a different standpoint, some physicists following DIRAC (1938) have suggested that the universal gravitational constant G may be decreasing with time as the universe expands and matter becomes more widely dispersed. DIRAC (1974) himself has also suggested that there may be complementary creation of mass, either uniformly throughout space or in proportion to the amount of matter already present. According to the Brans-Dicke formulation (DICKE, 1962), G is decreasing fractionally at a rate of about $10^{-11} y^{-1}$ to within a factor of three to five either way, corresponding to an Earth expansion rate of about 0.005 mm y^{-1}. According to DIRAC (1974) and HOYLE and NARLIKAR (1971) the fractional decrease of G is $5 \times 10^{-11} y^{-1}$, corresponding to an Earth expansion rate of 0·025 mm y^{-1} if mass remains constant, or less if mass is created. These rates are more than two hundred times slower than those required to explain continental disruption by Earth expansion.

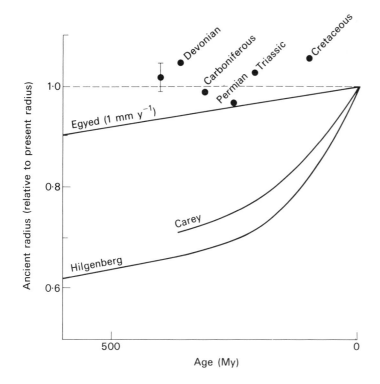

Fig. 9.2 Changes in the Earth's radius since the beginning of the Cambrian àccording to three versions of the expanding Earth hypothesis (after IRVING, 1964), compared with palaeoradius estimates obtained by palaeomagnetic measurements (McELHINNY and others, 1978). The palaeoradius estimated (with confidence limits) at 400 My age is the value obtained by fitting a least squares regression line to the other five determinations.

How can the expanding Earth hypothesis provide tectonic energy? As the Earth expands (whatever the cause), the soft interior deforms by flow and the strong lithospheric shell is stretched. The strain energy of a spherical shell, such as a football, increases as it is blown up. However, the reverse is true for the Earth because it is initially highly compressed. Stretching of the outer shell of the Earth causes the horizontal compression to be reduced slightly. This *reduces* the strain energy as it is blown up from inside. The energy which is released by normal faulting is gravitational energy; the appropriate mechanism depends on the wedge subsidence idea, which has been applied to rift valley formation (p. 65), and modifications of it. Thus the role of the tension in the lithosphere is to initiate tensile faulting, and the source of tectonic energy according to the expansion hypothesis must be gravitational energy rather than strain energy. It is not at all obvious how the energy dissipated by earthquakes and other tectonic activity could originate in the expanding Earth hypothesis.

The best available method of testing the expanding Earth hypothesis is to determine the ancient radius by palaeomagnetic measurements. The basic principle is as follows. The palaeolatitudes at a given time in the past are determined at two points as far apart in ancient latitude as possible on a continental mass which has kept its original dimensions. The results give an estimate of the ancient angle subtended by the two points at the Earth's centre; combining this with the known distance apart gives us an estimate of the radius at that time. The method devised by WARD (1963) makes use of all the available palaeomagnetic data of given age on the chosen continental mass and estimates the palaeoradius as the value which gives the smallest scatter of calculated pole positions. Much more precise estimates of ancient radius can now be made because the palaeomagnetic data has greatly improved in quality and quantity, and it has also been recognized that the method can be applied to reconstructed land masses such as Europe and North America in so far as these have not suffered internal deformation. A recent statistical assesment of palaeomagnetic estimates of the ancient radius made by McELHINNY and others (1978) yields a value of 1.02 ± 0.028 at the 95% confidence level for the Earth's radius 400 My ago relative to the present value, indicating a slight contraction or at most an expansion of only 0.8%. This estimate disagrees with the rapid expansion rates of Carey, Hilgenberg and Owen but it is not precise enough to discount the slower rates of expansion associated with hypotheses of decreasing G.

Rather more stringent limits can be placed on the possible decrease of G with time by considering the lack of major deformation seen on the surface of the Moon, Mars and Mercury (McELHINNY and others, 1978). Observations appear to suggest that the Moon has neither expanded nor contracted over the past 3200 My within limits of about 0.06%. Mars may possibly have suffered a slight expansion of around 0.5% early in its history, whereas a slight radial contraction of less than 2 km may have affected Mercury at an early stage. According to McELHINNY and others (1978), these place an upper limit of about 10^{-11} y^{-1} on the fractional rate of decrease of G, or about 3×10^{-11} y^{-1} if mass is created in proportion to the amount of matter already present as Dirac suggested.

To conclude, the rapid Earth expansion hypotheses of Carey, Hilgenberg and Owen are inconsistent with convincing evidence from palaeomagnetism and from the surface features of some inner planets. A slower rate of expansion amounting to a few tens of kilometres over the Earth's lifespan, as advocated by Dicke, cannot be ruled out. As the Earth is probably cooling slightly, the most viable mechanism for expansion is a possible slow secular decrease of G. Any such expansion would be of the same order of magnitude as the contraction produced by the slight cooling, so the two effects would tend to annul each other. Turning to the Earth's surface features, it seems clear that any slight expansion the Earth may have undergone cannot be the driving mechanism for sea-floor spreading, continental drift and associated tectonic activity. A slow expansion may be occurring but its tectonic effects would be swamped by sea-floor spreading and associated processes. Thus the expansion hypothesis appears to have no obvious relevance to the origin of the Earth's major surface features.

9.4 The convection hypothesis

Introduction

The hypothesis that sub-crustal convection currents cause mountain-building was originally suggested in 1839 by Hopkins and was later used by FISHER (1881). The idea that convection currents in the solid mantle cause both continental drift and mountain belts was introduced by HOLMES (1928, 1931, 1933) and others at about the same time. Holmes envisaged a convection current upwelling beneath a supercontinent such as Gondwanaland, causing it to split and dragging the fragments apart as shown in Fig. 9.3. The circum-Pacific and the Alpine-Himalayan belts were interpreted as compression features formed near the sinking currents. Holmes recognized that mantle convection would produce tension near the upwelling currents and compression near the sinking currents and between them. Since then, the ocean ridge system has been interpreted as the tension feature associated with the upwelling currents. This early version of the convection hypothesis is a remarkably accurate forecast of the modern concept. The main difference is that Holmes thought that sialic continents had floated over the simatic oceanic crust whereas we now believe that new oceanic crust is formed during drift.

There has been renewed interest in the convection hypothesis ever since palaeomagnetic results started to support the continental drift hypothesis. This is because some form of convection in the

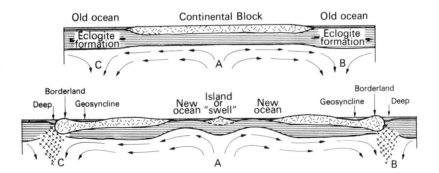

Fig. 9.3 The convection hypothesis according to Holmes. The upper diagram shows 'sub-continental circulation in the substratum with eclogite formation from the basaltic rocks of the intermediate layer above B and C where the sub-continental currents meet sub-oceanic currents and turn down'.

The lower diagram shows 'distention of the continental block on each side of A with formation of new ocean floors from rising basaltic magma. The front parts of the advancing continental blocks are thickened into mountainous borderlands with oceanic deeps in the adjoining ocean floor due to the accumulation of eclogite at B and C'. Redrawn from HOLMES (1933), *J. Wash. Acad. Sic.*, **23**, 188.

mantle is the only plausible mechanism yet suggested for drift. The demonstration that the ocean-floors are spreading away from the ridges at up to $60 \, mm \, y^{-1}$ adds further support to the convection hypothesis, but raises the new problem of how the convection currents are coupled to the moving plates of lithosphere.

Free convection occurs in fluids when the density distribution deviates from stable equilibrium. The resulting buoyancy forces cause flow to occur until equilibrium has been established. Irregularities in the density distribution can be produced both by thermal and by chemical disequilibrium. If the density anomalies are produced as rapidly as they are dissipated, then some form of regular convection pattern is produced; this is characteristic of thermal convection in a Newtonian viscous fluid. It is possible that episodic thermal convection may occur in fluids possessing more complicated rheological properties. The dominant type of convection believed to occur in the mantle is thermal, although penetrative convection following the production of a fluid phase is undoubtedly the mechanism which causes magma to rise to the surface.

In discussing the mantle convection hypothesis, our object is to find out as much as possible about the pattern of mantle convection and how it can be coupled to the rigid plates which form the lithosphere. Two threads run through the argument. Firstly, there is the theoretical knowledge about convection which places limitations on the allowable patterns; secondly, there is the search for a pattern of convection consistent with the observed facts of plate tectonics and other relevant aspects of geology and geophysics. We also need to bear in mind that convection is primarily the mechanism for upward transfer of heat through the mantle.

Feasibility of thermal convection in the mantle

As a first step, it is convenient to assume that the mantle behaves as a Newtonian viscous fluid. Although this may not necessarily be true (p. 309), the approach does give some indication of the feasibility of mantle convection and of the influence of sphericity and rotation on the convection pattern.

Free thermal convection may occur in a fluid when it is heated from below. The raised temperature at the bottom causes a reduction in density through thermal expansion; buoyancy forces cause the hot light fluid to rise to the surface where it loses heat and consequently increases in density and sinks to the bottom again. BÉNARD (1900) showed experimentally that convection in a fluid layer starts when the upward heat flow by conduction exceeds a critical limit. A regular pattern of convection cells then became established in his experiments, hexagonal in plan view and rectangular in cross-section. Hot fluid rises at the centre of each cell and cold fluid sinks near the margin. On substantially increasing the amount of heat transferred by convection, the regular pattern disappears and the flow eventually becomes turbulent.

The condition for the onset of convection was derived by RAYLEIGH (1916) in explanation of Bénard's observations. Rayleigh assumed a homogeneous and incompressible Newtonian fluid possessing a uniform kinematic viscosity v*. The upper and lower boundaries were assumed to be perfect conductors of heat and to be free in the sense that they exert no shear stress on the fluid. Rayleigh showed that convection starts when the dimensionless number $R = \alpha\beta gd^4/\kappa v$ exceeds $27\pi^4/4$ which equals 658, where α is the coefficient of thermal expansion, β is the temperature gradient, g is gravity, d is the thickness of the layer and κ is the thermal diffusivity. R is now called the Rayleigh number. The horizontal dimension of a cell is about $2\sqrt{2}d$.

JEFFREYS (1930) and KNOPOFF (1964) have shown that Rayleigh's result applies to a compressible fluid provided that β is regarded as the difference between the actual temperature gradient and the adiabatic gradient. This is called the *superadiabatic gradient*. The onset of convection requires a somewhat higher Rayleigh number if one or both boundaries are rigid or if the convection occurs in a spherical shell. For instance, Knopoff showed that the critical Rayleigh number for a spherical shell with outer radius twice the inner one and with both boundaries rigid is 2380.

The extent to which rotation affects convection depends on the magnitude of the dimensionless Taylor number $T = (2wd^2/v)^2$ where w is the angular velocity of rotation. The rotational influence becomes significant when T is unity or exceeds it. Applying the formula to the mantle by putting $\omega = 7\cdot27 \times 10^{-5}$ rad s^{-1}, $d = 3 \times 10^6$ m and $v = 10^{17}$ m^{2}s^{-1}, we find that T is about 10^{-16} which indicates that the Earth's rotation is unlikely to affect the pattern of Newtonian convection in the mantle. However, rotation would be expected to affect strongly the convection pattern in the much less viscous outer core.

Table 9.1 shows calculated values of the Rayleigh number for the mantle and its upper and lower divisions. The calculations are based on an assumed superadiabatic temperature gradient of $0\cdot1$ K km^{-1}; this is probably correct to within a factor of five in the lower mantle, but the true value in the upper mantle may be around 1 K km^{-1} with a corresponding increase in the calculated Rayleigh number by a factor of ten. Until a few years ago, the value of the viscosity in the lower mantle was the main uncertainty in such calculations, with estimated values of up to 10^{25} Pa s based on the now discredited fossil bulge of the Earth. Now the more recent calculations based on postglacial recovery yield a lower mantle estimate of about 10^{21} Pa s with accuracy of the order of 10% (p. 338). All the other relevant properties required for the computation of the Rayleigh number are known to an accuracy of better than 50%, so that the results shown in Table 9.1 are unlikely to be in error by more than a factor of five or ten, assuming uniform Newtonian viscosity.

Table 9.1 shows that fairly vigorous thermal convection with significant heat transfer probably occurs either in the lower mantle or in the mantle as a whole, depending on the effect of the transition zone (see below). The effectiveness of the convection may be enhanced by non-Newtonian viscosity if this is prevalent. On the other hand, the Rayleigh number computed for the upper mantle suggests much less vigorous convection here simply because it has a much smaller depth extent. If the temperature gradient is 1 K km^{-1} and the viscosity is significantly lower than

---

* Kinematic viscosity is the ratio of dynamic viscosity to density. The S.I. unit is m^2 s^{-1}.

Table 9.1 Rayleigh number $R = \alpha\beta gd^4/\kappa\nu$ computed for the mantle and its subdivisions assuming a superadiabatic temperature gradient of $\beta = 0.1$ K km^{-1} (10^{-4} K m^{-1}) and taking $g = 10$ m s^{-2} throughout.

	Thermal expansion α (K^{-1})	Layer thickness d (m)	Thermal diffusivity κ (m^2 s^{-1})	Kinematic viscosity ν (m^2 s^{-1})	Rayleigh number R
Upper mantle	2.5×10^{-5}	3×10^5	1.1×10^{-6}	3×10^{16}	6×10^3
Lower mantle	1.5×10^{-5}	2×10^6	1.3×10^{-6}	2×10^{17}	9×10^5
Whole mantle	1.9×10^{-5}	2.7×10^6	1.3×10^{-6}	2×10^{17}	4×10^6

our estimate, then the Rayleigh number may be as high as 10^5 which suggests that some upper mantle convection might certainly be expected to occur locally.

It can be tested whether convection currents in the mantle occur by laminar or turbulent flow by computing Reynolds number $Re = Vd/\nu$, where V is the velocity of flow. Putting $V = 0.5$ my$^{-1} = 1.5 \times 10^{-8}$ ms^{-1}, $d = 2700$ km $= 2.7 \times 10^6$ m and $\nu = 2 \times 10^{17}$ m^2 s^{-1} in the expression, we find that $Re \doteq 2 \times 10^{-19}$. A similar value is appropriate to the upper mantle. The very small value of Re indicates that the flow is laminar and that turbulence is highly unlikely to occur in any form of mantle convection.

To summarize, the results of applying the theory of convection in Newtonian fluids to the mantle are as follows:

(*i*) Convection might be expected to occur either through the whole mantle or in the lower mantle, depending on whether or not the transition zone inhibits mantle wide convection.

(*ii*) Independent convection may also occur within the upper mantle in regions where the viscosity is low and/or the temperature gradient is 1 K km^{-1} or steeper.

(*iii*) The Earth's rotation appears unlikely to influence the pattern of convection at low Rayleigh number.

(*iv*) Turbulent flow is unlikely to occur.

(*v*) Thermal convection is probably the most significant mechanism of heat transfer within the mantle as a whole (p. 281).

It must be emphasized that convection in the mantle, or part of it, is likely to be a very complex process and quite unlike the ideal Rayleigh-Bénard type described above. Factors may include: (*i*) a complicated rheological structure, not necessarily of Newtonian viscous type; (*ii*) a distribution of heat sources within and above the convection cell as well as below it; (*iii*) influence of the strong lithosphere above on the convection pattern; (*iv*) the presence of the mantle transition zone involving phase changes and possible chemical inhomogeneity in the mantle; and (*v*) the fact that convection is probably far from marginal. These and other factors make the problem quite inaccessible to analytical solution, even if we knew the relevant physical properties. However, we can get some way further by using approximate theory and by searching for relevant geological and geophysical manifestations of convection, as described in the following sections.

Convection at high Rayleigh number

As the Rayleigh number appropriate to the mantle is inferred to be of the order of 10^5 to 10^6, a more vigorous type of convection than that at marginal stability can be expected. Exact theoretical treatment is no longer possible but the flow pattern can be studied by approximate theory (TURCOTTE and OXBURGH, 1967) or by numerical analysis (McKENZIE and others, 1974). Such studies suggest that the flow pattern at moderately high Rayleigh number in uniform Newtonian viscous fluid resembles that shown in Fig. 9.4(a) and (b). Heat transfer essentially takes place by the fluid

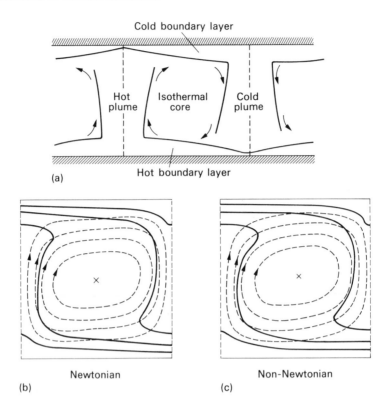

Fig. 9.4

(**a**) The boundary layer model of moderately vigorous two-dimensional thermal convection, with heating from below and cooling from above. Redrawn from TURCOTTE and OXBURGH (1967), *J. Fluid Mech.*, **28**, 33.

(**b**) and (**c**) Streamlines (dashed lines) and isotherms (solid lines) for Newtonian thermal convection at Rayleigh number of about 5×10^4 and for non-Newtonian convection under similar conditions. Redrawn from PARMENTIER and others (1976), *J. geophys. Res.*, **81**, 1842 and 1843.

flowing round the margin of each cell, with the isothermal core rotating in sympathy but not contributing to the heat transfer. The fluid is heated as it flows along the hot boundary layer at the bottom of the cell, becoming buoyant and rising to the surface in the hot plume. Cooling takes place by thermal conduction to the upper surface as the fluid flows along the cold boundary layer. The cold, dense fluid then sinks back to the base of the cell in the cold plume. An adiabatic gradient applies within the core of the cell and steep temperature gradients are restricted to the two boundary layers where heating or cooling occur by thermal conduction.

At significantly higher Rayleigh number of 10^8 or 10^9, a chaotic pattern of convection made up of time-varying eddies of varying size would be expected to occur. ELDER (1976) has suggested that such a pattern may apply to the Earth's mantle, but he appears to have overestimated the Rayleigh number by using the radius of the whole Earth rather than the thickness of the mantle in computing it and by adopting a much higher superadiabatic gradient than seems realistic. Here it is assumed that the Rayleigh number may reach up to about 10^7 but is not as high as 10^8. The convection cells in the mantle are thus likely to have some regularity, although they may not exactly conform to the simple pattern of Fig. 9.4 and may show some variation with time.

The pattern of convection is modified if part or all of the heat sources are distributed throughout the convecting material rather than being entirely at the base as shown in Fig. 9.4. If all the heat sources occur within the cell, then the lower boundary layer would not exist and the fluid flow which actively transports the heat would be more widely spread out over the lower part of the cell. Probably over 5% of the terrestrial heat loss originates from the core, as evidenced by the need to power the geomagnetic dynamo (p. 263). Further heat comes from the decay of long-lived radioactive isotopes within the mantle and from slight cooling of it. Thus part of the heat driving mantle convection is applied at the base of it and part is generated within it.

A hot boundary layer must occur at the base of the mantle if a significant proportion of the heat which causes mantle convection originates in the core. The thickness of this boundary layer and the temperature gradient across it can be estimated approximately by simple boundary layer analysis based on estimated thermal and mechanical properties of the layer. In this way JEANLOZ and RICHTER (1979) determined the boundary layer thickness to be about 100 km, this value not being very sensitive to the exact value of the heat flux across it. The average temperature gradient across it, which is directly proportional to the heat flux, was estimated to be 3 K km^{-1} on the assumption that a quarter of the Earth's heat loss comes from the core; the overall temperature drop across the layer is thus about 300 K. Supporting evidence for the existence of a layer of this type occupying the lowermost 100–200 km of the mantle comes from the seismological velocity-depth pattern (p. 156). Furthermore such a layer, whose thermal and mechanical properties vary laterally and change with time, provides the only satisfactory known framework to account for anomalous mechanical and electromagnetic interactions at the core-mantle boundary and the associated geomagnetic phenomena (p. 263).

The question now arises as to whether the cool boundary layer at the top of the mantle convecting system occurs beneath the lithosphere, or whether it can be identified with the lithosphere itself as TURCOTTE and OXBURGH (1967) suppose. If the cold boundary layer underlies the lithosphere, then the plate motions must be secondary consequences of the convection, with the recycling of the lithosphere representing only a small fraction of the overall mantle flow pattern. On the other hand, if the oceanic lithospheric plates form the main cold boundary layer, then the ocean ridges can be identified with the top of the hot plumes and the subduction zones with the cold plumes. This latter suggestion is highly plausible simply because the main loss of heat from the mantle does occur by formation and cooling of the oceanic lithosphere as it spreads (p. 294) and the main recycling of cool material occurs at the subduction zones. The relatively strong lithospheric plates are themselves likely to be one of the most important influences on the pattern of mantle convection.

The rheology of the mantle may not be Newtonian (p. 309). PARMENTIER and others (1976) carried out numerical investigations of convection at moderately high Rayleigh numbers in the mantle, assuming strain rate proportional to the cube of stress. They found that the pattern of non-Newtonian flow differed only slightly from that of simple Newtonian flow (Fig. 9.4c), implying that the type of creep law is not a particularly significant factor in defining the pattern of convection. Yet another possibility, albeit a rather improbable one, is that the mantle below the lithosphere possesses a small finite strength; however, convection would probably be able to take place even in the presence of a small permanent strength.

A more significant influence on the long term convection pattern is the temperature dependence of viscosity. As viscosity is a thermally activated process, an increase of about 100 K in the lower mantle would decrease the viscosity by an order of magnitude. This will have the effect of stabilizing the convecting system. If temperature rises significantly within the system, then viscosity will decrease and convection will be speeded up, carrying away the excess heat and tending to restore the initial situation. This will give rise to long period stability in the temperatures and heat transport as explained in Chapter 7 (p. 283).

Role of the mantle transition zone

Until quite recently it was widely believed that the mantle transition zone approximately coincides with a major rheological boundary between a low viscosity upper mantle able to convect and a lower mantle with too high a viscosity to convect under any realistic circumstances. This view now appears to be erroneous as the lower mantle viscosity is probably around 10^{21} Pa s (p. 338) rather than 10^{25} Pa s as then supposed. However, the mantle transition zone is likely to have an important influence on the overall pattern of mantle convection for other reasons. The main uncertainty now is whether convection currents can or cannot cross the transition zone. If they can, then mantle wide convection would predominate. If they cannot, then separate cells must exist in the upper and lower mantles.

The viscosity of the upper mantle may locally be two to three orders of magnitude less than that of the lower mantle. The possibility that this, by itself, might favour independent convection cells in the upper and lower mantles, with a thermal boundary layer between them, has been discounted by STEVENSON and TURNER (1979). This does not, however, rule out the possibility of secondary upper mantle cells superimposed on large scale whole mantle convection.

A much more significant influence of the transition zone on convection arises from the change in mineralogy across the zone. This depends on whether the zone merely consists of a series of solid-solid phase reactions affecting rock of uniform chemical composition or whether there is a difference of chemical composition above and below it. Opinions are still divided on this fundamental issue (p. 195). If the upper and lower parts of the mantle differ significantly in their chemical compositions, then the stable density stratification across the zone would prevent whole mantle convection. Separate sympathetic convection systems would be expected to occur in the upper and lower parts of the mantle, with the transition zone marking a thermal boundary layer of steep conduction gradient between them. The convection cells above and below would probably be coupled together so that cells of exaggerated horizontal dimension would occur in the upper mantle (Fig. 9.5a).

On the other hand, if the transition zone simply consists of a series of exothermic phase transitions, such as that of olivine to spinel, then whole-mantle convection can probably take place. This assumes that phase equilibrium is established almost instantaneously as material flows up or down across the transition zone, because any significant delay in attaining equilibrium would produce density anomalies which would seriously impede convection. There are two important effects associated with convection currents flowing across the transition. Firstly, because of the temperature dependence of the depth of an exothermic phase reaction, the transition must occur at greater than average depth in the hot upwelling plume and at shallower than average depth in the cold sinking plume (Fig. 9.5b). The associated density anomalies will help to drive the convection currents and may also give rise to significant long-wavelength gravity anomalies. Secondly, latent heat is absorbed in the upwelling current and released in the downsinking current, causing a local steepening of the temperature-depth gradient in both plumes. This has the effect of steepening the adiabatic gradient across the transition zone, so that the temperature in the lower mantle required for whole-mantle convection needs to be about 100–150 K higher in the presence of the transition zone than without it. To summarize, exothermic phase transitions are compatible with convection provided that equilibrium is maintained and the temperature field below is high enough to overcome the steepening adiabatic gradient. On the other hand, endothermic phase reactions would produce density anomalies that would tend to oppose convection.

At our present stage of knowledge, it is not possible to know with certainty whether or not the convection currents cross the mantle transition zone. Either way, the transition zone must influence the convection, particularly in steepening the temperature gradient across it either by

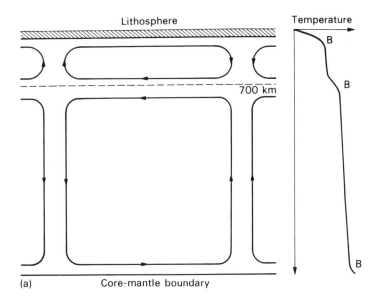

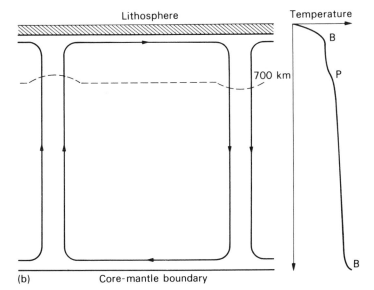

Fig. 9.5 Possible influence of the transition zone on mantle convection:
(a) Separate upper and lower mantle convection, with coupling across a hypothetical compositional boundary at 700 km depth (after STEVENSON and TURNER, 1979, © Academic Press Inc. (London) Ltd.). The temperature-depth gradient steepens across the three thermal boundary layers marked B.
(b) Mantle-wide convection crossing an exothermic phase transition. The temperature-depth gradient steepens at the two thermal boundary layers marked B and across the phase transition at P.

steepening of the adiabatic gradient or by acting as a boundary zone between separate upper and lower mantle systems of convection.

Evidence from observations

The theoretical treatment described above demonstrates that thermal convection must probably occur in the mantle but leaves us in the dark about the actual pattern of motions. Some slight indication of the fluid motions can be gleaned from certain observations, notably plate motions, heat flow, and long wavelength gravity anomalies.

As suggested previously, the oceanic parts of the lithospheric plates may be interpreted as the cold boundary layers of large scale mantle convection cells. If this is correct, the plate motions themselves must display the motions at the top of the convecting system, and a return flow broadly in the opposite direction must take place at some depth in the mantle beneath. The ocean ridges mark the top of the upwelling parts of the system and the subduction zones mark the return flow down to a depth of a few hundred kilometres. Whether the return flow is mainly restricted to the upper mantle or penetrates deeper is not clear from plate tectonics. But the occurrence of deep focus earthquakes down to 700 km depth in the Tonga region suggests that the circulation involves at least the transition zone and possibly also the lower mantle.

The pattern of surface heat flow might be expected to show up the locations of the hot upwelling and cool downsinking currents. The thermal structure of the oceanic lithosphere clearly supports the pattern deduced from the plate motions, with the ocean ridges above upwelling currents and the subduction zones marking the return of the cooled material to the interior. The situation for the sub-continental lithosphere is rather different. Here, the heat loss from the deeper parts of the mantle probably occurs partly by the spasmodic heating of the lithosphere resulting from the development of localized hot regions in the underlying upper mantle, with the subsequent cooling of the lithosphere on a long time-scale. The best explanation of such upper mantle hot spots is the convective upwelling of hot material from greater depths – possibly from the lower mantle. Continental hot spot activity is thus seen as a safety valve for the escape of heat from the deep interior in regions where new oceanic lithosphere is not formed to do this job. This probably represents a complementary but subordinate type of convective upwelling to that seen in oceanic regions. It is suggested that these two processes together form the main convective circulation of the mantle.

More detailed analysis of oceanic heat flow suggested to RICHTER and PARSONS (1975) that small scale secondary convection may be taking place in the sub-oceanic upper mantle at distance from the ridges. The relative uniformity of heat flow over oceanic crust older than about 50 My implies that heat is flowing into the base of the lithosphere (p. 287). According to Richter and Parsons, the best explanation is that small scale convection occurs between about 650 km depth and the base of the lithosphere, this being superimposed on the larger scale convection implied by the plate motions. In regions where the overlying lithospheric plates are moving rapidly such as in the Pacific region, they suggest that the cells may take the form of longitudinal rolls with their axes parallel to the spreading direction, as shown in Fig. 9.6. These postulated small scale convection cells would not be expected to exert any significant drag on the overlying plates and would not contribute materially to the plate driving mechanism.

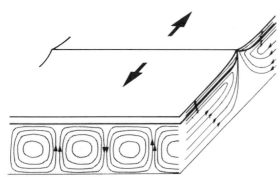

Fig. 9.6 Schematic illustration showing the two scales of convection which may occur in the sub-oceanic upper mantle. The larger scale convection involves the cooling oceanic lithosphere and the return flow (which may extend below the upper mantle). The small scale convection takes the form of superimposed horizontal rolls. Redrawn from RICHTER and PARSONS (1975), *J. geophys. Res.*, **80**, 2539.

The long wavelength gravity anomalies revealed by satellite orbital studies may give some further indication of the pattern of mantle convection. These anomalies are mainly caused by lateral density variation within the mantle below the lithosphere, with the longest wavelengths possibly partly representing fluctuations in the depth of the core-mantle boundary (p. 199). As the main body of the mantle appears to be devoid of strength on a long time-scale, lateral density variations would need to be supported by slow flow of material such as would be produced by convection or post-glacial recovery. Hot upwelling convection currents might be expected to show up as negative gravity anomalies and cold sinking currents as positive anomalies, but this may be an oversimplification as the associated vertical deformation of the surface would tend to produce opposite effects. The most marked lateral density variations associated with mantle-wide convection would be expected to occur at the transition zone, where a small change in temperature produces a significant change in the depth of the density step. How do the observed features of the global gravity field tie in with the other evidence on the convection pattern? The best agreement occurs around the circum-Pacific belt where the dominantly positive anomalies may reflect the subduction process. Elsewhere the correlation with the plate motions is less obvious, but it is possible that the large negative anomalies such as that in the north-eastern Indian Ocean may mark the locations where hot material upwells from the lower mantle into the upper mantle. A full understanding of the global gravity anomalies and their relation to convection and other processes has not yet been reached.

The pattern of mantle convection

The actual pattern of mantle convection is likely to be much more complicated than simple ideas suggest, with several complementary types of motion contributing to the transfer of heat from the deep interior of the Earth to the surface. Gathering together the various shreds of available evidence, at least three types of convective motion can be recognized.

(*i*) The oceanic lithosphere is interpreted as forming the cold surface boundary layer of the main convective system, with the hot plumes discharging over 60% of the Earth's total heat loss in forming new lithosphere beneath ocean ridge crests (Table 7.7). The cold plumes descend at the subduction zones which are mostly located around the Pacific Ocean. The resulting circulation pattern is more complicated than the simple closed cells of Fig. 9.4, in that the upwelling occurs beneath all the major oceans but recycling is mainly restricted to the margins of the Pacific. It is suggested later in the chapter that the normal plate driving forces arise from this convecting system. There is still controversy as to whether the circulation pattern is restricted to the upper mantle or penetrates deeper. Here the viewpoint is taken that most of the subducted oceanic lithosphere returns to the lower mantle, possibly by piecemeal sinking of large blocks which break off the bottom end of the sinking tongues of lithosphere. The complementary upwelling of lower mantle material into the sub-oceanic asthenosphere is less easy to detect although it may possibly be located beneath some of the negative global gravity anomalies caused by penetration of the transition zone by the hot plume.

(*ii*) A subordinate type of upwelling from the deeper parts of the mantle may occur locally beneath continents to produce upper mantle hot spots, the heat eventually being lost by cooling of the overlying continental lithosphere. It will be suggested later in the chapter (p. 363) that this type of upwelling may have an important role in causing new continental splitting.

(*iii*) It has also been suggested that some small scale convection of lesser global significance may occur in the upper mantle in regions where the temperature-depth gradient across the asthenosphere is steep, such as below the older parts of the ocean basins.

9.5 Mechanism of present plate motions

In absence of other satisfactory mechanisms, we assume that global plate motions are driven by a terrestrial heat engine, powered by heat escaping upwards from the core and mantle by convection. This section treats the problem as to how this heat engine can drive the present day steady plate motions. The more obscure problem of explaining the initiation of new continental splits is discussed in section 9.6.

The convection currents in the mantle are causing the lithosphere to be continually strained by application of renewable boundary stresses to the plates. This causes the plates to move relative to each other, with consequent release of strain energy by earthquake activity, friction and viscous dissipation, and by producing local crustal thickening. The relatively strong lithosphere thus acts as a reservoir for strain energy, which is applied by the boundary forces and released by the associated global tectonic activity. The fundamental problem in this section is to determine how the boundary stresses are applied to the lithospheric plates.

The two basic ways by which lithospheric plates can be stressed and consequently moved relative to each other are as follows:

(i) By drag exerted on the base of the lithosphere by the upper boundary layer of an underlying convection current; in the past this has been the most widely held hypothesis.

(ii) By application of boundary forces at the edges of lithospheric plates, notably by push exerted at ocean ridges and pull exerted at subduction zones; this developed from original ideas of OROWAN (1965) and ELSASSER (1969) and was initially associated with upper mantle convection, although its scope can now be widened.

Thermodynamics of mantle convection

A mantle convection system can be regarded as a heat engine. Heat is absorbed at the base of the mantle or at some depth within it and is transported to the Earth's surface where it is discharged. Part of the heat is converted into gravitational energy which drives the convective motion and strains the lithospheric plates near the surface. Within the context of plate tectonics, we can regard as useful work that which is done on the lithospheric plates by application of boundary stresses – even this is eventually mostly dissipated as heat in overcoming resistance to plate motions. Thermodynamics is important here because it places limitations on the proportion of heat transported through the mantle which can be converted into mechanical energy. In an ideal reversible mantle heat engine, the convective motion would occur at infinitessimal rate and the full amount of energy predicted by the second law of thermodynamics would be available for straining and moving the plates. Within the real mantle, motion occurs at finite rate and is irreversible. As a result, a significant proportion of the theoretically available energy is 'wasted' as heat evolved by viscous dissipation in the mantle.

Following STACEY (1977), the basic mechanism of the mantle heat engine can be understood by following a mass m of the working substance, which is the solid rock of the mantle, round a complete cycle (Fig. 9.7). Starting at point A on Fig. 9.7(a), heat absorbed by m as it travels at constant pressure along the hot boundary AB causes thermal expansion, which does external work on the system by increasing the gravitational energy. Further external work is done by m as it expands adiabatically from B to C in the hot plume, but work is done on m as it cools and contracts at constant pressure along CD and as it contracts adiabatically on sinking from D back to the starting point at A. There is a net output of useful work during the cycle because expansion of m occurs at higher pressure than its contraction. This energy is released during the convection cycle at the same rate as it is produced, by the uprise of the buoyant working substance in the hot plume and

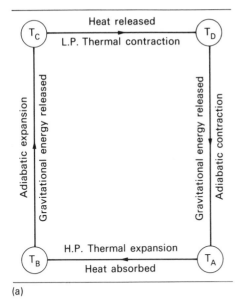

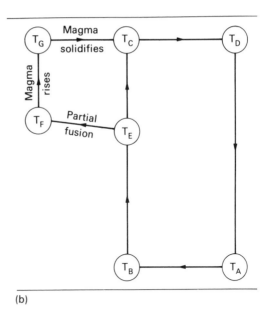

(a)   (b)

Fig. 9.7 Mantle convection as a heat engine. T_A, T_B, etc. are temperatures.
(a) Simplified cycle with the solid mantle material as working substance, heating at bottom and cooling at top, based on STACEY (1977).
(b) The supplementary magmatic heat engine, resulting from partial fusion at high pressure, upwelling of low density magma at the ridge crest and solidification at low pressure.

by sinking of the dense substance in the cold plume. From our point of view, part of this energy is wasted by viscous dissipation in the mantle, but part of it is available to stress the lithospheric plates and cause their motion.

A second, subordinate, type of heat engine may operate in parallel to the above solid state engine during mantle convection. This uses the basaltic fraction which melts at depth beneath ocean ridges as the working substance (Fig. 9.7b). The reduction in pressure in the upwelling hot plume causes partial fusion to occur at high pressure along EF in Fig. 9.7(b). The melting involves expansion, which does external work on the system in increasing its gravitational energy. The complementary solidification and contraction occurs at much lower pressure near the surface. The available energy is released as the low density magma fraction rises towards the surface.

The efficiency of a reversible heat engine measures the ratio of potentially useful work to heat transported. The mantle heat engine differs from the classical heat engine of the Carnot cycle in that heat is taken in at depth and given out at the surface at constant pressure rather than constant temperature.

STACEY (1977) has shown that the theoretical efficiency of the mantle convection heat engine shown in Fig. 9.7(a) is given by $(T_B - T_C)/T_B$. This is equal to the ratio of the adiabatic temperature drop between the base and top of the hot plume to the temperature at its base. This is a particularly important result as it emphasizes the essential part played by the adiabatic gradient in enabling mantle convection to do 'useful' work in driving the lithospheric plates, for without the adiabatic gradient there would be no power available. The result shows that the steepening of the adiabatic gradient by exothermic phase reactions across the transition zone and by partial fusion at a higher level must increase the efficiency, whereas the presence of endothermic phase reactions would have the opposite effect.

A rough estimate of theoretical efficiency of whole mantle convection can now be obtained. It is assumed that the average depth of the heat sources is 1500 km and the average source temperature is 2500 K. Taking the adiabatic gradient to be $0.33\,K\,km^{-1}$, the adiabatic temperature drop is 500 K, to which should be added 150 K for the steepening of the gradient across the transition zone. This yields a theoretical efficiency of about 25%. If convection is restricted to the upper mantle above 600 km depth, then the adiabatic temperature drop is correspondingly less and the theoretical efficiency is just below 10%. Whole mantle convection thus provides a more effective mechanism for stressing and driving the plates than separate upper and lower mantle convecting systems.

Looking at the same problem from an alternative point of view, the potentially useful energy which is released as a mass m of working substance completes a cycle in Fig. 9.7(a) can be estimated from the release of gravitational energy as the buoyant mass rises in the hot plume and the dense mass sinks in the cold plume. The difference in density between the hot and cold plumes is given by $\alpha\Delta T$ where α is the average volume coefficient of thermal expansion and ΔT is the temperature difference between the plumes at the same depth. The energy released is given by

$$W = mgh\alpha\Delta T$$

where g is gravity (assumed constant) and h is the vertical dimension of the convection cell. If m is treated as the mass flowing in unit time, then W becomes the useful power available. In practice, much of the available energy will be wasted by viscous flow in the mantle so that only a fraction will be available to drive the plates.

This method can be used to estimate the potentially useful power available as a result of the circulation of the oceanic lithosphere. The value of m can be estimated with some confidence from the rate of formation of new lithosphere at the ocean ridges. This is the product of the length of the ocean ridge system (60 000 km), the average rate of plate separation at ridges (56 mm y^{-1} = 18 $\times 10^{-7}$ mm s^{-1}), the thickness of the lithosphere (80 km) and its density (3200 kg m^{-3}), yielding an estimated value of 2.8×10^7 kg s^{-1} for m. Taking

$$\Delta T = 500\,K,$$
$$g = 10\,m\,s^{-2}$$
$$\alpha = 3 \times 10^{-5}\,K^{-1}$$
$$\text{and} \quad h = 1000\,km$$

the potentially useful power available is 4.2×10^{12} W. This estimate is halved if circulation is restricted to the upper mantle and top part of the transition zone. The heat loss through the oceanic lithosphere is about 2×10^{13} W, indicating a potential efficiency of 10% to 20%, in excellent agreement with the estimates previously obtained using Stacey's formula.

The potentially useful power available from the formation and rise of magma beneath ocean ridges can also be estimated. This is given by $m'gh'\Delta V/V$, where $\Delta V/V$ is the fractional increase in volume on melting. Taking the average depth of magma formation to be 50 km, this yields a potential power source of 10^{11} W, which is significantly smaller than the power of the main circulation but possibly of some importance in that it is entirely available at the ocean ridges.

Turning from the ideal situation to reality, an estimate of the actual efficiency of the plate driving heat engine can be determined from observations. The rate of heat transport through the mantle is about 2.5×10^{13} W. The rate at which the plates are being strained can be estimated from the rate of release of energy by earthquakes and associated processes. The average rate of release of energy in seismic waves by earthquakes is about 3×10^{10} W, but according to wyss (1970) the efficiency of conversion of elastic strain energy into seismic energy is only about 10%. Further energy will be released by friction and by forming mountain roots. The overall rate of release of energy in the earthquake belts is somewhat above 3×10^{11} W, say 5×10^{11} W. This suggests that the actual efficiency of the plate driving mechanism is about 2%, which is well within the theoretical limits placed by thermodynamics. This gives conviction to the hypothesis that the plates are in reality driven by thermal convection of some type in the mantle, and suggests that 5 to 10 times more energy is lost by viscous dissipation in the mantle than is used to drive the plates. This seems realistic.

Forces acting on plates

Lithospheric plates move relative to each other in response to pressure or shearing stress acting on their boundaries. These stresses can be regarded as forces acting on a specified length of the edge of a plate or on a given area of the bottom surface. The plate motion is opposed by another set of forces acting at or near the boundaries. When the plates are moving at constant velocity then the forces acting on each plate must be in equilibrium. Some of the more significant types of force which may cause or impede plate motions are listed below and shown in Fig. 9.8, in development of the system described by FORSYTH and UYEDA (1975).

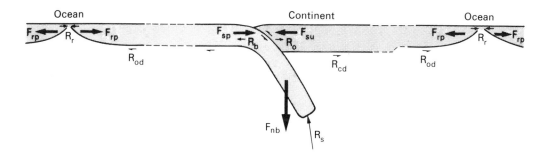

Fig. 9.8 Some of the forces which may act at plate boundaries or near them:
F_{rp} – ridge push
F_{nb} – negative buoyancy
F_{sp} – slab pull
F_{su} – trench suction
R_r – ridge resistance
R_s – slab resistance
R_b – bending resistance
R_o – overriding plate resistance
R_{od} – mantle drag beneath ocean, shown here as resistance
R_{cd} – mantle drag beneath continent, shown here as resistance
Transform fault and collision resistance are not shown. The diagram is not to scale.

A *mantle drag* force acts on the bottom surface of plates if their velocity differs from that of the underlying mantle. The shearing stress is proportional to the viscosity of the underlying asthenosphere and its vertical velocity gradient, assuming horizontal flow. The mantle drag force can act in two opposite ways. If the convecting mantle is moving faster than the overlying plate and in a similar direction, then the viscous drag exerted by the fast flowing mantle will act to drive the plate motion. On the other hand, if the plate is moving faster than the mantle, then the viscous drag caused by the differential motion will act to resist the plate motion. The resistance to plate motion is probably greater beneath continents than oceans because of the higher upper mantle viscosity beneath continents.

A *ridge push* force is inferred to act on the edges of the diverging plates at ocean ridges, helping to force them apart and thus contributing to the plate driving mechanism. There has been a widespread misconception that this force results from gravitational sliding of the newly formed oceanic lithosphere away from the elevated ocean ridge as it cools and contracts. Such a mechanism cannot work as the oceanic lithosphere outside subduction zones is in isostatic equilibrium, acting as if it is floating on the underlying asthenosphere. If gravity gliding of this type occurred from ocean ridges, one would expect even more impressive gliding to occur in the opposite direction from the continental margins. In reality, the ridge push force is caused by the progressive upwelling of low density asthenospheric material beneath the ridge crests. Gravitational energy is released as the low density solid material and magma rise to form new oceanic lithosphere. Some of the released energy may be dissipated by viscous flow but most of it is probably available to wedge the separating plates apart.

Some local resistance to the separation of plates at ocean ridge crests is indicated by shallow earthquake activity and faulting in the brittle uppermost part of the lithosphere. Such a *ridge resistance* force is probably a relatively minor influence causing a slight decrease in the effective magnitude of the ridge push force.

A large *negative buoyancy* force acts on the tongue of cold, sinking lithosphere at a subduction zone. The rate at which gravitational energy is released by the sinking slab can be calculated from its geometry and temperature-dependent density distribution (see later in chapter). The power available is potentially much greater than that associated with ridge push. This force will be enhanced if the slab penetrates part or all of the transition zone, as exothermic reactions such as olivine to spinel take place at shallower depth in the cold sinking slab than in the adjacent mantle. The sinking tongue of lithosphere also encounters viscous resistance to motion as it penetrates the upper mantle and transition zone. Such *slab resistance* forces act mainly on the bottom end of the tongue, being approximately proportional to slab velocity and becoming much greater as the tip of the slab passes into the more viscous transition zone. Thus the effectiveness of the negative buoyancy force in driving the surface plates may be greatly reduced by the slab resistance force, depending on the local circumstances at the subduction zone.

Part of the negative buoyancy force can be transmitted to the attached oceanic plate, causing it to be pulled towards the trench. This will be referred to as the *slab pull* force. Another less obvious driving force, however, which also acts at a subduction plate boundary is the *trench suction* force which pulls the overriding plate towards the trench. This force was originally recognized by ELSASSER (1971) but its physical nature has subsequently not been well understood. The force exists because the subducting slab must be sinking more steeply than the angle of the subduction zone, which has the effect of reducing the support for the overriding plate. Tensile boundary forces are effectively applied on the edge of the overriding plate, drawing it towards the trench. WOODWARD (1976) has convincingly demonstrated the existence of the suction force using finite element analysis. One of the difficulties often expressed in accepting the suction force is that thrusting and other compression features are observed to affect the overriding plate near the plate boundary, but these can probably be explained in terms of local stresses in the vicinity of the underthrust boundary. The negative buoyancy force also causes the trench to be held down strongly out of isostatic equilibrium and applies the couple which bends the lithosphere in the vicinity of the trench.

Resistance to plate motions, in addition to that caused by the deep-seated slab resistance force, occurs on both sides of subduction plate boundaries. The attached oceanic plate encounters considerable resistance in the region where plastic flow enables it to bend sharply downwards. This will be called *bending resistance*. The intense earthquake and tectonic activity characteristically affecting the overriding plate near the boundary shows that substantial resistance to motion of a different type occurs here. This will be referred to as *overriding plate resistance*.

Collision resistance to plate motions occurs in the vicinity of converging plate boundaries which mark the collision between continents, such as the Alpine-Himalayan belt. This is caused by the release of strain energy by earthquakes and deformation, and by the increase of gravitational energy associated with crustal thickening. Plate motion at conservative plate boundaries may be impeded by *transform fault resistance*.

To summarize, the potential plate driving forces include mantle drag on the bottom of the lithosphere, ridge push, slab pull and trench suction. Resistance to plate motion may include mantle drag (oceanic or continental), ridge resistance, slab resistance, bending resistance, overriding plate resistance, collision resistance and transform fault resistance. Our main problem in the following pages is to determine whether mantle drag or edge forces are the dominant influence in driving the present day plate motions.

Are plates driven by mantle drag?

According to the classical convection mechanism (Fig. 9.9a), the asthenosphere forms the upper cold boundary layer of the main mantle convecting system. The velocity gradient across the asthenosphere exerts a viscous drag on the bottom of the lithosphere, causing it to be stressed. Maximum tension in the lithosphere occurs above the upwelling currents and maximum compression above the downsinking currents. Lithospheric plates will respond to the underlying convection currents by moving away from the upwelling currents towards the downsinking ones.

The resulting stresses in the lithosphere can be approximately estimated as follows. Consider a two-dimensional system of convection cells of horizontal dimension L overlain by a lithosphere of thickness T (Fig. 9.10). The shearing stress acting on the base of the lithosphere is $\eta \dot{\varepsilon}$ where η is the

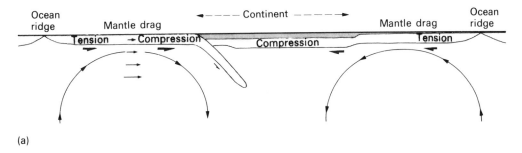

(a)

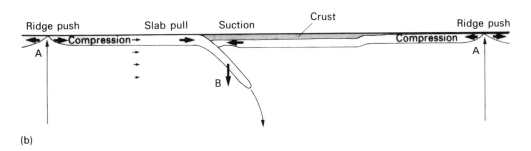

(b)

Fig. 9.9 Two concepts of the plate driving mechanism:
   (a) Cellular convection cells in the mantle, either whole mantle or upper mantle, exerting a drag on the overlying lithosphere.
   (b) The Orowan-Elsasser type of convection driven by the wedging effect of the upwelling low density material at A and the negative buoyancy of the cool, dense lithosphere at B.
Note that the velocities in the mantle implied by (a) are much higher than those implied by (b).

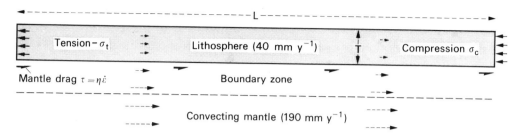

Fig. 9.10 The viscous drag of a horizontally flowing convection current on the base of the lithosphere. For equilibrium, $\tau l. = (\sigma_t + \sigma_c)\,T$. A simple boundary zone 75 km thick with uniform velocity-depth gradient across it has been assumed, the velocity above being the spreading rate of 40 mm y^{-1} and that below it being the maximum velocity within the convection cell of 190 mm y^{-1}.

coefficient of viscosity and $\dot{\varepsilon}$ is the strain rate. This causes a horizontal tension of $\eta\dot{\varepsilon}\,L/2T$ above the upwelling current and an equivalent tension above the downsinking one. According to CATHLESS (1975), a low viscosity channel 75 km thick with $\eta = 4 \times 10^{19}$ Pa s underlies the lithosphere. Suppose that the plate velocity just above the channel is 40 mm y^{-1} and the velocity in the convecting mantle just below it is 190 mm y^{-1}, then the strain rate across the zone averages $6 \times 10^{-14}\,\mathrm{s}^{-1}$ and the shearing stress is 2·4 MPa. Putting $L = 2500$ km and $T = 80$ km, the maximum values of tension and compression in the lithosphere are calculated to be 37 MPa assuming a linear velocity increase across the asthenosphere. Stresses of this order are probably adequate to explain the plate motions, but it is clear that quite large velocities must occur in the convecting mantle if this mechanism is to be effective.

The classical convection hypothesis is difficult to reconcile with the observed pattern of present and past plate motions. This is because cellular mantle convection cells would be expected to form a relatively simple pattern of motions and would need to have large horizontal dimensions of around 2500 km if they are to be effective in stressing the plates. On the other hand, the upwelling currents must occur near the ridges and downsinking currents near the convergent plate boundaries if they are to explain observed plate motions. This causes difficulty because of the irregular pattern of plate boundaries which migrate relative to each other. For instance, ridges and subduction zones can collide as they did in the north-eastern Pacific during the Tertiary. Thus the geometry of past and present plate motions is somewhat difficult to reconcile with a simple cellular pattern of mantle convection. Further difficulty is caused by the rapid motion of certain small plates in the Pacific region, whose length scale is too small to allow effective stressing by underlying convection. Yet another problem is raised by the lack of much relative motion between the hot spots around the world (p. 138), suggesting that large vertical velocity gradients do not occur in the upper mantle.

The most decisive argument against the viability of the classical convection mechanism comes from calculation of the strain energy dissipation within the low viscosity channel as modelled (Fig. 9.10). The viscous dissipation per unit volume is $\frac{1}{2}\eta\dot{\varepsilon}^2$. This works out to be $2\cdot7 \times 10^{12}$ W for a worldwide layer, which is about 10 % of the heat loss from the mantle. The total viscous loss in the whole convecting system must be several times larger. This seems to be quite unacceptable on thermodynamic grounds, even if the viscosity distribution differs from our assumptions. Thus the answer to the question – are the plates driven by mantle drag? – is, probably not.

Are plates driven by forces on their edges?

The idea that the oceanic parts of plates form the fast-moving cooling parts of the main mantle convecting system, and consequently that plates move in response to forces applied at their edges rather than drag on their bottom, has developed from suggestions by OROWAN (1965), ELSASSER (1969), TURCOTTE and OXBURGH (1967) and others (Fig. 9.9b). The Orowan-Elsasser concept of convection has often been associated with return flow occurring in the upper mantle. This restriction is not necessary, as part or even most of the return flow may occur below 400 km depth without invalidating the idea. The mechanism has two great advantages over the classical convection hypothesis. Firstly, it is readily reconcilable with the observed plate motions as these are the controlling influence on the pattern of circulation. Secondly, it is much more acceptable thermodynamically because it involves much slower average strain rates throughout the mantle, so that the viscous dissipation is not excessive in relation to the heat transported. In the presence of a strong lithosphere, it appears to be a much more efficient mechanism than classical mantle convection for removing heat from the Earth's deep interior.

If plates are driven by forces on their edges, then the power available at ocean ridges and subduction zones must be sufficient to overcome all resistance to plate motions. It was shown earlier in the chapter (p. 356) that the gravitational power released by upwelling at ridges and sinking at subduction zones amounts to about 4×10^{12} W, assuming the average circulation extends down to 1500 km depth. Only a fraction of this power, however, is available to drive the plates as much of it is lost by viscous dissipation in the mantle. A more realistic estimate of the power available at ocean ridges and subduction zones is now attempted, although it should be recognized that isolation of two such parts of the complete system is somewhat artificial.

The ridge push force originates from upwelling of hot asthenospheric material of significantly lower density than normal oceanic lithosphere. The ridge exists as a topographic elevation as the isostatic response to the low density material below it. To support a ridge standing 3 km above the adjacent ocean basin, the upwelling asthenosphere needs to be $\Delta\rho = 80$ kg m^{-3} lower in density than the mantle part of the lithosphere down to a depth of $D = 80$ km. The rate at which gravitational energy is released by the upwelling is given by $\frac{1}{2}LD^2\Delta\rho sg$ where L is the length of the ridge system (taken as 55 000 km) and s is the average rate of plate separation at ridges (56 mm y^{-1}). Substituting in the formula yields an estimated power release of 2.5×10^{11} W. The rise of magma to form new oceanic crust contributes a possible further 1.0×10^{11} W. The average excess pressure exerted on the plate edges is about 30 MPa without the magma effect or 40 MPa with it. Even allowing for 30% viscous dissipation during upwelling, the power available exceeds the global strain energy release by earthquakes by a factor of ten, demonstrating that ridge push is probably a significant contributor to the plate driving mechanism.

The total power released by the worldwide system of subducting slabs is given by $Vu\Delta\rho g$ where V is the volume of all subducting slabs, u is the average rate of vertical sinking and $\Delta\rho$ is the average excess density above the adjacent normal mantle. An approximate estimate can be obtained by substituting plausible values. Taking the average depth extent of subduction zones to be 300 km below the base of the normal lithosphere, the average excess density to be 40 kg m^{-3}, and assuming that the oceanic lithosphere is recycled at the rate it is formed at the ridges (this will be a slight overestimate because of collision mountain belts), the rate of energy release is calculated to be about 10^{12} W. This ignores the additional power available where the olivine-spinel transition is penetrated. According to our calculations, the sinking slabs contribute at least four times more power than the ridges, but some of this power may be wasted in overcoming slab resistance. The resulting slab pull and trench suction forces may thus be comparable in magnitude, although opposite in sign, to the ridge push force. Taken together, it seems clear that adequate power is available at the ridges and subduction zones to drive the present plate motions.

Having established that adequate power is available, the edge force hypothesis next needs to be tested for consistency with the observed motions of the plates. Some simple correlations made by

FORSYTH and UYEDA (1975) are revealing. Firstly, plate velocity appears to be almost independent of plate area; it has been suggested that this observation argues against the edge force driving mechanism, but the criticism is only valid if the sub-oceanic mantle drag offers substantial resistance to plate motions. Secondly, plates which are attached to downsinking slabs generally move faster than those which are not. Thirdly, plates with large areas of continental crust move slower than those without. The latter two observations are difficult to untangle as plates with downgoing slabs in general do not have large continental areas. Either subducting slabs exert a large force on the attached plate or mantle drag is significantly higher beneath continents than oceans or both.

Quantitative studies of the feasibility of driving plates by forces on their edges have been carried out by FORSYTH and UYEDA (1975) and by CHAPPLE and TULLIS (1977). The basic idea of these analyses is that the driving and resistive forces acting on each plate must be in equilibrium. The equilibrium conditions are satisfied if the torques acting on each plate taken about each of three mutually perpendicular axes passing through the Earth's centre are zero. In the analysis of Forsyth and Uyeda there are eight forces of unknown magnitude: ridge push, slab pull, slab resistance, collision resistance (applying on both sides of trenches as well as at collision mountain ranges), mantle drag, excess continental mantle drag, and transform fault resistance. The three equilibrium equations for each of twelve plates gave thirty six equations in eight unknowns, which were solved by least squares. Despite some differences in formulating the problem, the analysis of Forsyth and Uyeda gave very similar overall results to that of Chapple and Tullis.

Both these quantitative analyses demonstrate that a system of forces acting on plate boundaries at ridges and subduction zones can satisfactorily account for present-day plate motions. Some other important inferences can be drawn from the results. By far the two largest forces are found to be the downward pull of the sinking slabs and the corresponding slab resistance; these two forces oppose each other, so that the net slab pull force acting on the attached plate is comparable in magnitude to the other forces. Both ridge push and trench suction appear to be significant driving forces, although their contributions are difficult to assess with accuracy because of the dominance of the forces associated with the sinking slabs. Mantle drag is found to be relatively insignificant beneath the oceans but is somewhat larger beneath continents, as might be expected. The resistance offered by transform faults is negligible, but that encountered on both sides of trenches and in collision mountain belts is substantial.

There should be reasonable agreement between the stresses observed within plates (p. 334) and those calculated for an acceptable driving mechanism, bearing in mind that other factors such as crustal thickness variation may contribute to the overall state of stress. Plates driven by edge forces should be in compression near the ridges, with the compression decreasing towards the trenches and possibly becoming a tension. Plates driven by mantle drag should be in tension near the ridges and in compression near convergent plate margins, which is the opposite situation. Intra-plate earthquakes indicate that oceanic regions are in compression, supporting the edge force hypothesis. A more rigorous analysis of the theoretical stress within plates produced by various systems of boundary forces has been carried out by RICHARDSON and others (1979) using finite element analysis (Fig. 8.22). This showed reasonably good agreement with the observations if plates are driven by edge forces but poor agreement if they are driven by mantle drag. As inferred by Forsyth and Uyeda, they found that the pull exerted by the downsinking lithosphere is mostly balanced out by resistance near the trench so that stress within plate interiors is dominated by ridge push except in eastern Asia and the western Pacific where subduction is most vigorous.

In summary, the hypothesis that plates are driven by edge forces acting at trenches and ridges is thermodynamically acceptable and adequate power appears to be available. Present day plate motions and observed stresses within plates are both consistent with a realistic pattern of edge forces. The alternative hypothesis that plates are driven by mantle drag meets serious thermo-dynamic difficulties and is difficult to reconcile with past and present plate motions and the present stress field within plates. The evidence, though possibly not yet conclusive, strongly favours the edge force hypothesis.

9.6 Mechanism of continental splitting

What mechanism causes the persistent splitting of continents which has probably occurred episodically over much of geological time? At each stage of continental splitting, a new divergent plate boundary forms within a continental interior region and develops into an active spreading centre as the new ocean along the split starts to widen. Thus the early Mesozoic supercontinent Pangaea has broken up by stages over the last 280 My as the Atlantic and Indian Oceans have opened up between the separating fragments.

It has been shown that the normal steady plate motions are driven by some form of convection in the mantle involving the oceanic lithosphere which undergoes steady change in pattern as the various plate boundaries migrate relative to each other. A much more radical change in the pattern of mantle convection must occur at the time of continental splitting when new plate boundaries are formed. One of the controversial questions is whether a change in the pattern of mantle convection initiates a new split or vice-versa.

In this section the evidence on continental break-up is briefly reviewed and some of the mechanisms which have been proposed are described. It is lastly suggested that major continental splitting may be caused by local weakening of the continental lithosphere by hot spot activity in the underlying mantle at periods in Earth history when a continental region is subjected to widespread tensile stress. The tension may be predominantly caused by the suction force acting on the overriding continental lithosphere at subduction zones during periods when a large continental mass is bordered by subduction zones on opposite sides.

Evidence on how continents break-up

Evidence on how the Mesozoic continental break-up of Pangaea occurred lies deeply buried beneath later sediments at the passive continental margins around the Atlantic and Indian Oceans. It is often suggested, however, that the much more accessible rift systems of the world, such as that of East Africa, may represent continental splitting in the process of occurring at the present time.

Present-day rift systems (p. 63) are typically associated with crustal doming and volcanism. Graben formation is the response of the brittle upper crust to horizontal deviatoric tension. The tension may also cause crustal thinning beneath the rift fault zone. The uplift is the isostatic response of the region to the occurrence of anomalously low densities in the underlying upper mantle, attributable to raised temperatures. The volcanism comes from a hot, partially fused region in the upper mantle. The necessary ingredients for continental splitting – an abundant magma source and upper crustal tension – appear to be present. It is easy to see that such a structure might readily develop into a continental split.

Are the passive continental margins underlain by ancient rift systems of East African type? Graben-like structures associated with the initial rupture do occur extensively beneath the Atlantic margins but are not universally present. Furthermore, lengthy sections of the passive margins lack evidence of uplift or widespread volcanism occurring at the time of the split. Evidently the continental splitting process occurs in some regions under less active tectonic and volcanic conditions than those of modern rift systems. On the other hand, other parts of the passive margin system underwent major continental volcanism just prior to and during the continental splitting event. One of the best studied examples is the Palaeocene igneous activity occurring from 60 to 50 My ago in the Greenland-Scotland region, spanning the time when Greenland and North Europe split apart 54 My ago. Intense igneous activity later continued in the newly-forming oceanic region to produce the anomalously thick oceanic crust along the Icelandic transverse ridge (p. 124). It is clear that a hot region in the underlying upper mantle developed here prior to the continental splitting and it is possible that this was the factor which caused the new split between Greenland and North Europe to take place.

Two contrasting types of passive margin can thus be recognized, one type associated with extensive continental volcanism and the other type where development of the split occurred in a more subdued way. It is only the volcanically active type which bears some resemblance to East Africa. Continental splitting is probably initiated in the hot spot regions displaying abundant volcanism as a result of the tension and dyke intrusion and the split subsequently spreads laterally beyond the hot region to join onto pre-existing plate boundaries. This may occur by progressive lateral extension of the dykes and by associated upwelling of asthenospheric material into the propagating crack. The exact line of splitting is likely to be influenced by pre-existing lines of weakness in the continental crust as well as by the orientation of the regional tension.

According to the above evidence, a new continental split appears to require a substantial tensile stress system and a sufficient source of basaltic magma in the underlying upper mantle to initiate the sea-floor spreading process. The following four sources of tension have been suggested: (*i*) drag on the base of the lithosphere exerted by upwelling and diverging convection currents; (*ii*) membrane stresses developing in the interior of a plate migrating towards the equator; (*iii*) stresses associated with surface loading and associated isostatic upthrust in uplifted continental regions such as East Africa; and (*iv*) the trench suction force acting at convergent plate boundaries. It is suggested later that the tension which predominantly causes continental splitting is caused by the trench suction force.

The best explanation of the occasional development of a hot and magma saturated upper mantle beneath a region of continental lithosphere is that an episode of convective upwelling brings hot material from much deeper in the mantle into the asthenosphere. Partial fusion takes place by adiabatic decompression in the upwelling material (p. 188). The spasmodic occurrence of such hot regions below the continents is well documented from geological evidence (WILSON, 1963; MORGAN, 1971; CROUGH, 1979). Such hot spot activity appears to be a significant mechanism of escape of heat from the deep mantle through the continental lithosphere (p. 293), contrasting with the more steady loss by upwelling at ocean ridges.

What happened before the Mesozoic to Tertiary episode of continental splitting started? Most of the oceanic crust produced in earlier periods of spreading has now been recycled into the mantle. The tectonic history of the continents, however, suggests that major earth movements go back into the early Precambrian and palaeomagnetic results show that the continents were moving relative to each other in the Precambrian. The simplest interpretation is that the present episode of continental break-up is one of a series of episodes occurring over the Earth's geological history. Each episode involved the break-up of a large continental mass with the formation of new areally increasing oceans such as the Atlantic, with the eventual convergence of some or all of the continental fragments onto a contracting ocean such as the Pacific to form a new major continental mass. Support for the occurrence of four major tectonic events during the Earth's geological history comes from a histogram of radiometric age dates shown in Fig. 9.11. This shows four main peaks at 350, 1000, 1800 and 2600 My ago. These peaks probably mark periods of maximum tectonic activity. It is possible that they represent four past episodes of major continental break-up.

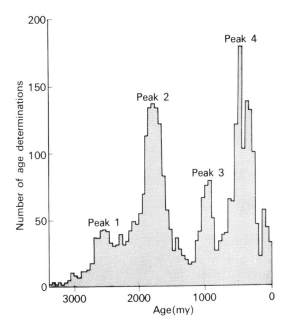

Fig. 9.11 Frequency histogram of igneous and metamorphic age determinations plotted against geological time. Redrawn from DEARNLEY (1966), *Physics Chem. Earth*, 7, 6, Pergamon Press.

Towards a hypothesis of continental splitting

One of the earliest hypotheses as to how continental splitting has occurred was that of RUNCORN (1962), who suggested that the convection pattern in the mantle might have changed as the core grew in size over the Earth's lifespan (Fig. 9.12). Following Urey, he supposed that the core formed slowly over geological time by continuing separation of the iron phase from the mantle. This would cause the core-mantle boundary to increase in radius from zero to its present value. According to theoretical calculations of Newtonian convection in a spherical shell made by Chandrasekhar, the convection pattern would progressively change from an $n = 1$ pattern at the start to an $n = 5$ pattern at present, as the mantle became progressively thinner. This suggested that there would be four major re-adjustments of the convection pattern during the life of the Earth, at each of which the continents would newly settle over regions of downsinking currents. Runcorn identified these periods of changing pattern with the four radioactive dating peaks shown in Fig. 9.11. Although this hypothesis did not examine the process of lithospheric break-up, it appears to imply that the tensional stresses occurring above upwelling currents cause the split. This hypothesis is no longer widely held as it is now recognized that the core probably formed rapidly at the beginning of the Earth's history.

An alternative hypothesis which is also based on a changing pattern of convection in the mantle (Fig. 9.13) was suggested by BOTT (1964), who called attention to the greater loss of heat from the deep Earth's interior through the oceans than through the continents, resulting from the blanketing effect of the radioactive heat production in the continental crust. It was suggested that more vigorous thermal convection must therefore be occurring in the mantle beneath the oceans than beneath the continents. Consequently radioactive heat produced in the mantle below the continents would accumulate, eventually being released by the initiation of vigorous convective upwelling beneath the continental region. It was suggested that such upwelling would start a new pattern of mantle convection which would cause the continental break-up. This hypothesis was suggested prior to our modern understanding of plate tectonics. Nevertheless it fits remarkably well with our much improved modern ideas of escape of heat through the continental and oceanic

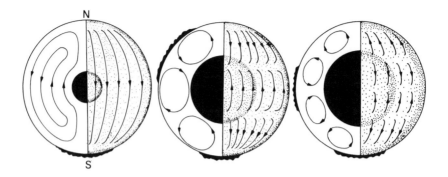

Fig. 9.12   Mantle-wide convection with different sizes of the core. *Left, n* = 1; *middle, n* = 3; *right, n* = 4. After RUNCORN (1962), *Continental drift*, p. 35, Academic Press.

lithosphere through steady upwelling at ocean ridges and by spasmodic hot spot activity through the continental lithosphere.

More recent hypotheses of continental splitting have tended to assume that a continental rift system such as that of East Africa may subsequently develop into a continental split. OXBURGH and TURCOTTE (1974) suggested that membrane stresses in the continental lithosphere may be the primary cause of continental cracking. As a plate moves towards the equator from higher latitudes, tensile membrane stresses develop in the interior and compressive stresses around the periphery. Palaeomagnetic evidence has shown that the African plate has moved rapidly northwards during the last 100 My after a relatively stationary period. Oxburgh and Turcotte therefore suggested that the East African rift system was initiated in response to tensile stresses of around 13·5 MPa produced in the interior of the African plate as it moved northwards towards the equator. Ethiopia would be the first region to be affected as it would be the first part to

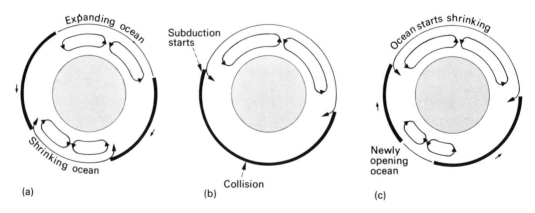

Fig. 9.13   Stages in the initiation of a new continental split, developed from BOTT (1964).
  (a) Previous episode of sea-floor spreading, showing expanding and shrinking oceans with dominantly sub-oceanic convection.
  (b) Continental collision producing large continental mass.
  (c) Initiation of new split by upwelling in sub-continental mantle.
The process is here shown for cellular mantle convection but it would equally apply to Orowan-Elsasser type of convection.

approach the equator from the south, and the crack would subsequently propagate southwards at about the rate of the northward motion of East Africa. According to this hypothesis, the lithospheric crack is the primary cause of splitting and the upwelling in the mantle causing partial fusion is a secondary consequence in response to dilatation of the lithosphere. The radiometric dating of the onset of volcanism along the rift system supports the idea of a southward propagating crack. The principal difficulty facing this hypothesis is the geological evidence that volcanism appears to predate rifting in East Africa (p. 68). Also, the tensile stress appears to be too small to cause graben subsidence of several kilometres (p. 66).

BOTT and KUSZNIR (1979) alternatively suggested that the development of a hot and magma saturated upper mantle is the primary cause of continental doming, rifting and possible later continental break-up. The hot upper mantle first produces early volcanism and then causes the overlying lithosphere to be heated and thinned, producing isostatic uplift. Tension develops in the crust in response to the loading of the uplifted topography and the associated isostatic upthrust of the underlying low density region. This tension causes rift faulting and dyke intrusion. Within this setting, it is possible that extensive dyke intrusion along the rift zone may initiate the break-up of the continent. The main difficulty of extending this hypothesis to continental splitting is that the crustal tension produced by the elevated topography and its isostatic compensation is restricted to the uplifted region, so that the initial dyke-filled crack might not be able to propagate into the adjacent regions.

The above difficulties are avoided by a new hypothesis (BOTT, 1982) that continental splitting results from interaction between regional tension produced by plate boundary forces and independent localized weakening of the continental lithosphere by local hot spot activity in the underlying mantle (Fig. 9.14). Such regional tension may possibly be produced by the trench suction (p. 358) acting on overriding continental plates at active continental margins at certain periods in earth history when a large continental region is bordered by subduction zones on opposite sides. With tensile forces acting on both edges of a plate of continental lithosphere, the whole plate will be stretched. This situation probably applied to the supercontinent Pangaea during early Mesozoic time. The thinning and weakening of the lithosphere above a local hot spot would give rise to considerable local enhancement of the tensile stress in the brittle upper part of the crust in the affected region. This may cause cracking of the lithosphere by normal faulting and dyke emplacement comparable to that of present-day rift systems. However, the crack would be able to propagate laterally into the cooler adjacent regions by the intense concentration of tensile stress at the tips of the spreading crack as a result of the general tension affecting the continental plate. Eventually the crack would connect up with existing plate boundaries and a new ocean would start to form along the line of the crack. A ridge push force (p. 357) would develop at the new plate boundary, changing the stress regime at the newly formed passive margins from one of intense tension (rifting stage) to one of slight compression (drifting stage).

9.7 General conclusions

Although knowledge of the structure of the Earth's interior has been growing at an increasing rate throughout the twentieth century, it had only been during the last twenty years that a satisfying theory of the origin of the Earth's surface features has started to emerge. It has been a long-standing puzzle to understand the fundamental process of energy release within the Earth which has made the surface of our planet so strikingly different from those of the other inner planets and the Moon. After many false starts, it has been the discovery of sea-floor spreading and plate tectonics which has led us to a new and unified understanding of this process.

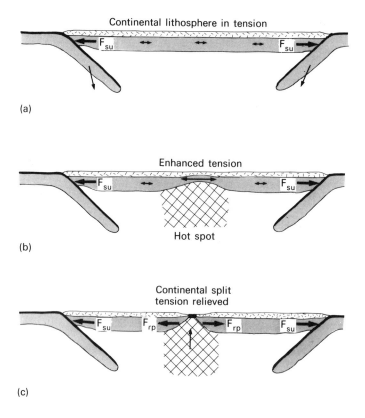

Fig. 9.14 A hypothesis of continental splitting (BOTT, 1982).
  (**a**) Continental lithosphere subjected to regional tension, possibly as a result of trench suction force acting at opposite edges of a plate formed entirely of continental lithosphere.
  (**b**) Development of a hot spot in the upper mantle beneath the stretched continental lithosphere as a result of local upwelling of material from deeper in the mantle. The resulting thinning of the lithosphere causes increased tension above the hot spot.
  (**c**) Continental splitting starts above the hot spot by dyke intrusion and propagates laterally to connect to existing plate boundaries. The tensile stress system immediately gives way to compression near the new split as a result of the ridge push force introduced at the new divergent plate boundary.

The mechanism causing plate motions has been attributed to the escape of heat from the Earth's interior. The Earth is acting as an inefficient heat-engine and a fraction of around 3% of the escaping heat is converted to strain energy in the lithosphere which is released mainly by tectonic activity concentrated along the mobile belts. The only known process by which the lithosphere can be strained continually in this way is by some sort of convection in the mantle. It has been suggested that the oceanic lithosphere itself forms the upper, cold boundary layer of the convecting system, so that the forces which drive the plates are applied at their edges rather than by drag on their underside. We have seen that the continental lithosphere is not recycled into the mantle like the oceanic lithosphere, but is moved around the Earth's surface as it has been repeatedly broken up and then re-formed into large masses by collision over geological time. The mechanism of continental break-up has also been attributed to the escape of heat from the Earth's interior, which appears to occur through the continental lithosphere by spasmodic rather than steady upwelling.

What is the next stage? Undoubtedly our knowledge of the structure of the Earth's interior will increase and we shall have further insight into the mechanisms of global tectonics as knowledge of the state of stress in the lithosphere improves. A new lease of life has already been given to our understanding of geological processes but a fundamental understanding of many features still eludes us. For instance, we still do not know how sedimentary basins form or how the continental splitting mechanism actually takes place.

References

Chapter one

ANDERS, E. (1977). Chemical compositions of the Moon, Earth, and eucrite parent body. *Phil. Trans. R. Soc.*, **285A**, 23–40.

BULLEN, K. E. (1963). *An introduction to the theory of seismology*, Third edition, Cambridge University Press, London and New York, 381 pp.

CAMERON, A. G. W. (1973a). Abundances of the elements in the solar system. *Space Sci. Rev.*, **15**, 121–146.

CAMERON, A. G. W. (1973b). Accumulation processes in the primitive solar nebula. *Icarus*, **18**, 407–450.

CAMERON, A. G. W. (1975). The origin and evolution of the solar system. *Scient. Am.*, **233**, No. 3, 33–41.

CAMERON, A. G. W. and PINE, M. R. (1973). Numerical models of the primitive solar nebula. *Icarus*, **18**, 377–406.

CLAYTON, D. D. (1964). Cosmoradiogenic chronologies of nucleosynthesis. *Astrophys. J.*, **139**, 637–663.

COOK, A. H. (1967). Gravitational considerations. In *The Earth's mantle*, pp. 63–87, edited by Gaskell, T. F., Academic Press, London and New York.

DAVIES, D. (1968). A comprehensive test ban. *Sci. J., Lond.*, November 1968, 78–84.

GERSTENKORN, H. (1955). Über Gezeitenreibung beim Zweikörperproblem. *Z. Astrophys.*, **36**, 245–274.

GROSSMAN, L. and LARIMER, J. W. (1974). Early chemical history of the solar system. *Rev. Geophys. space Phys.*, **12**, 71–101.

GUTENBERG, B. (1953). Wave velocities at depths between 50 and 600 kilometers. *Bull. seism. Soc. Am.*, **43**, 223–232.

GUTENBERG, B. (1959). *Physics of the Earth's interior*, Academic Press, New York and London, 240 pp.

HANES, D. A. (1979). A new determination of the Hubble constant. *Mon. Not. R. astr. Soc.*, **188**, 901–909.

HARLAND, W. B., SMITH, A. G. and WILCOCK, B. (editors) (1964). *The Phanerozoic time-scale*, *Q. Jl geol. Soc. Lond.*, **120s**, 458 pp.

HART, R. S., ANDERSON, D. L. and KANAMORI, H. (1977). The effect of attenuation on gross earth models. *J. geophys. Res.*, **82**, 1647–1654.

JEFFREYS, H. (1939a). The times of *P*, *S* and *SKS*, and the velocities of *P* and *S*. *Mon. Not. R. astr. Soc. geophys. Suppl.*, **4**, 498–533.

JEFFREYS, H. (1959). *The Earth: its origin, history and physical constitution*, Fourth edition, Cambridge University Press, London and New York, 420 pp.

JEFFREYS, H. (1963). On the hydrostatic theory of the figure of the Earth. *Geophys. J. R. astr. Soc.*, **8**, 196–202.

JEFFREYS, H. (1976). *The Earth: its origin, history and physical constitution*, Sixth edition, Cambridge University Press, London and New York, 574 pp.

KAULA, W. M. and HARRIS, A. W. (1975). Dynamics of lunar origin and orbital evolution. *Rev. Geophys. space Phys.*, **13**, 363–371.

KHAN, M. A. (1969). General solution of the problem of hydrostatic equilibrium of the Earth. *Geophys. J. R. astr. Soc.*, **18**, 177–188.

KING–HELE, D. (1967). The shape of the Earth. *Scient. Am.*, **217**, No. 4, 67–76.

KING–HELE, D. G. (1969). The shape of the Earth. *Royal Aircraft Establishment Technical memorandum Space* 130, 1–10.

LAMMLEIN, D. R. (1977). Lunar seismicity, structure, and tectonics. *Phil. Trans. R. Soc.*, **285A**, 451–461.

LAMMLEIN, D. R., LATHAM, G. V., DORMAN, J., NAKAMURA, Y. and EWING, M. (1974). Lunar seismicity, structure, and tectonics. *Rev. Geophys. space Phys.*, **12**, 1–21.

LATHAM, G. V. and others (1970). Passive seismic experiment. *Science, N.Y.*, **167**, 455–457.

LATTIMER, J. M., SCHRAMM, D. N. and GROSSMAN, L. (1977). Supernovae, grains and the formation of the solar system. *Nature, Lond.*, **269**, 116–118.

LEE, T., PAPANASTASSIOU, D. A. and WASSERBURG, G. J. (1976). Demonstration of ^{26}Mg excess in Allende and evidence for ^{26}Al. *Geophys. Res. Lett.*, **3**, 109–112.

LEHMANN, I. (1936). *P'. Bur. Centr. seism. Internat. A*, **14**, 3–31.

LUCK, J., BIRCK, J. and ALLEGRE, C. (1980). ^{187}Re-^{187}Os systematics in meteorites: early chronology of the solar system and age of the galaxy. *Nature, Lond.*, **283**, 256–259.

MANSINHA, L. and SMYLIE, D. E. (1968). Earthquakes and the Earth's wobble. *Science, N.Y.*, **161**, 1127–1129.

MARSH, J. G., DOUGLAS, B. C., VINCENT, S. and WALLS, D. M. (1976). Tests and comparisons of satellite-derived geoids with skylab altimeter data. *J. geophys. Res.*, **81**, 3594–3598.

MASSEY, H., BROWN, G. M., EGLINTON, G., RUNCORN, S. K. and UREY, H. C. (editors) (1977). The Moon – a new appraisal from space missions and laboratory analyses. *Phil. Trans. R. Soc.*, **285A**, 606 pp.

MELCHIOR, P. J. (1957). Latitude variation. *Physics Chem. Earth*, **2**, 212–243.

MOHOROVIČIĆ, A. (1909). Das Beben vom 8.x.1909. *Jb. met. Obs. Zagreb (Agram.)*, **9**, 1–63.

MORITZ, H. (1976). Fundamental geodetic constants. Report of the special study groups No 5–39 of the International Association of Geodesy. *Chron. U.G.G.I.*, No. 108, 72–78.

MULLER, P. M. and STEPHENSON, F. R. (1975). The accelerations of the Earth and Moon from early astronomical observations. In *Growth rhythms and the history of the Earth's rotation*, pp. 459–534, edited by Rosenberg, G. D. and Runcorn, S. K., John Wiley & Sons, London, New York, Sydney and Toronto.

MUNK, W. (1968). Once again – tidal friction. *Q. Jl R. astr. Soc.*, **9**, 352–375.

MUNK, W. H. and MACDONALD, G. J. F. (1960). *The rotation of the Earth*, Cambridge University Press, London and New York, 323 pp.

OLDHAM, R. D. (1906). The constitution of the interior of the Earth, as revealed by earthquakes. *Q. Jl geol. Soc. Lond.*, **62**, 456–475.

PANNELLA, G. (1975). Palaeontological clocks and the history of the Earth's rotation. In *Growth rhythms and the history of the Earth's rotation*, pp. 253–284, edited by Rosenberg, G. D. and Runcorn, S.K., John Wiley & Sons, London, New York, Sydney and Toronto.

PATTERSON, C. (1956). Age of meteorites and the Earth. *Geochim. cosmochim. Acta*, **10**, 230–237.

RAMSEY, A. S. (1940). *An introduction to the theory of Newtonian attraction*, Cambridge University Press, London and New York, 184 pp.

RINGWOOD, A. E. (1970). Origin of the Moon: the precipitation hypothesis. *Earth & planet. Sci. Lett. (Neth.)*, **8**, 131–140.

ROCHESTER, M. G. (1973). The Earth's rotation. *Trans. Am. Geophys. Un.*, **54**, 769–780.

RUNCORN, S. K. (1964). Changes in the Earth's moment of inertia. *Nature, Lond.*, **204**, 823–825.

SCHRAMM, D. N. and WASSERBURG, G. J. (1970). Nucleochronologies and the mean age of the elements. *Astrophys. J.*, **162**, 57–69.

SCRUTTON, C. T. (1964). Periodicity in Devonian coral growth. *Palaeontology*, **7**, 552–558.

SCRUTTON, C. T. (1967). Absolute time data from palaeontology. In *International dictionary of geophysics*, volume 1, p. 1, edited by Runcorn, S. K. and others, Pergamon Press, Oxford, London, Edinburgh, New York, Toronto, Sydney, Paris and Braunschweig.

SPENCER JONES, H. (1939). The rotation of the Earth, and the secular accelerations of the Sun, Moon and planets. *Mon. Not. R. astr. Soc.*, **99**, 541–558.

SPITZER, L., Jr. (1939). The dissipation of planetary filaments. *Astrophys. J.*, **90**, 675–688.

SPITZER, L., Jr. (1963). Star formation. In *Origin of the solar system*, pp. 39–53, edited by Jastrow, R. and Cameron, A. G. W., Academic Press, New York and London.

STEIGER, R. H. and JÄGER, E. (1977). Subcommission on geochronology: convention on the use of decay constants in geo- and cosmochronology. *Earth & planet. Sci. Lett. (Neth.)*, **36**, 359–362.

TAYLOR, S. R. (1975). *Lunar science: a post-Apollo view*, Pergamon Press, New York, Toronto, Oxford, Sydney and Braunschweig, 372 pp.

TRURAN, J. W. (1973). Theories of nucleosynthesis. *Space Sci. Rev.*, **15**, 23–49.

TRURAN, J. W. and CAMERON, A. G. W. (1978). ^{26}Al production in explosive carbon burning. *Astrophys. J.*, **219**, 226–229.

UREY, H. C. (1952). *The planets, their origin and development*, Yale University Press, 245 pp.

WAGNER, C. A., LERCH, F. J., BROWND, J. E. and RICHARDSON, J. A. (1977). Improvement in the geopotential derived from satellite and surface data (GEM 7 and 8). *J. geophys. Res.*, **82**, 901–914.

WASSERBURG, G. J., PAPANASTASSIOU, D. A., TERA, F. and HUNEKE, J. C. (1977). Outline of a lunar chronology. *Phil. Trans. R. Soc.*, **285A**, 7–22.

WASSON, J. T. (1972). Formation of ordinary chondrites. *Rev. Geophys. space Phys.*, **10**, 711–759.

WELLS, J. W. (1963). Coral growth and geochronometry. *Nature, Lond.*, **197**, 948–950.

YUKUTAKE, T. (1972). The effect of change in the geomagnetic dipole moment on the rate of the Earth's rotation. *J. Geomagn. Geoelect., Kyoto*, **24**, 19–47.

Chapter two

AIRY, G. B. (1855). On the computation of the effect of the attraction of mountain-masses, as disturbing the apparent astronomical latitude of stations in geodetic surveys. *Phil. Trans. R. Soc.*, **145**, 101–104.

ANDERSON, D. L. (1967a). A seismic equation of state. *Geophys. J. R. astr. Soc.*, **13**, 9–30.

ARTEMJEV, M. E. and ARTYUSHKOV, E. V. (1971). Structure and isostasy of the Baikal rift and the mechanism of rifting. *J. geophys. Res.*, **76**, 1197–1211.

BAKER, H. B. (1911). The origin of the Moon. *Detroit Free Press*, 23 April, 1911.

BAMFORD, D. (1977). P_n velocity anisotropy in a continental upper mantle. *Geophys. J. R. astr. Soc.*, **49**, 29–48.

BAMFORD, D., NUNN, K., PRODEHL, C. and JACOB, B. (1978). LISPB-IV. Crustal structure of northern Britain. *Geophys. J. R. astr. Soc.*, **54**, 43–60.

BERRY, M. J. and WEST, G. F. (1966). Reflected and head wave amplitudes in a medium of several layers. In *The Earth beneath the continents*, pp. 464–481, edited by Steinhart, J. S. and Smith, T. J., *Geophys. Monogr.* No. 10, American Geophysical Union, Washington, D. C.

BIRCH, F. (1958). Interpretation of the seismic structure of the crust in the light of experimental studies of wave velocities in rocks. In *Contributions in geophysics in honor of Beno Gutenberg*, pp. 158–170, edited by Benioff, H., Ewing, M., Howell, B. F., Jr. and Press, F., Pergamon Press, London, New York, Paris and Los Angeles.

BIRCH, F. (1960). The velocity of compressional waves in rocks to 10 kilobars, part 1. *J. geophys. Res.*, **65**, 1083–1102.

BIRCH, F. (1961a). The velocity of compressional waves in rocks to 10 kilobars, part 2. *J. geophys. Res.*, **66**, 2199–2224.

BLUNDELL, D. J. and PARKS, R. (1969). A study of the crustal structure beneath the Irish Sea. *Geophys. J. R. astr. Soc.*, **17**, 45–62.

BOTT, M. H. P. (1954). Interpretation of the gravity field of the eastern Alps. *Geol. Mag.*, **91**, 377–383.

BOTT, M. H. P. (1961). The granitic layer. *Geophys. J. R. astr. Soc.*, **5**, 207–216.

BOTT, M. H. P. (1967a). Geophysical investigations of the northern Pennine basement rocks. *Proc. Yorks. geol. Soc.*, **36**, 139–168.

BOTT, M. H. P. (1976). Formation of sedimentary basins of graben type by extension of the continental crust. *Tectonophysics*, **36**, 77–86.

BOTT, M. H. P. (1981). Crustal doming and the mechanism of continental rifting. *Tectonophysics*, **73**, 1–8.

BOTT, M. H. P., HOLDER, A. P., LONG, R. E. and LUCAS, A. L. (1970). Crustal structure beneath the granites of south-west England. In *Mechanism of igneous intrusion*, pp. 93–102, edited by Newall, G. and Rast. N., *Geol. J. Spec. Issue* No. 2.

BOTT, M. H. P. and KUSZNIR, N. J. (1979). Stress distributions associated with compensated plateau uplift structures with application to the continental splitting mechanism. *Geophys. J. R. astr. Soc.*, **56**, 451–459.

BOTT, M. H. P. and SCOTT, P. (1964). Recent geophysical studies in south-west England. In *Present views of some aspects of the geology of Cornwall and Devon*, pp. 25–44, edited by Hosking, K. F. G. and Shrimpton, G. J., Royal Geological Society of Cornwall.

BRAILE, L. W. and SMITH, R. B. (1975). Guide to the interpretation of crustal refraction profiles. *Geophys. J. R. astr. Soc.*, **40**, 145–176.

BROWN, L. D., CHAPIN, C. E., SANFORD, A. R., KAUFMAN, S. and OLIVER, J. (1980). Deep structure of the Rio Grande rift from seismic reflection profiling. *J. geophys. Res.*, **85**, 4773–4800.

BULLARD, E. C. (1936). Gravity measurements in East Africa. *Phil. Trans. R. Soc.*, **235A**, 445–531.

BULLARD, E. C., EVERETT, J. E. and SMITH, A. G. (1965). The fit of the continents around the Atlantic. *Phil. Trans. R. Soc.*, **258A**, 41–51.

BULLARD, E. C. and GRIGGS, D. T. (1961). The nature of the Mohorovičić discontinuity. *Geophys. J. R. astr. Soc.*, **6**, 118–123.

BYERLY, P. (1956). Subcontinental structure in the light of seismological evidence. *Adv. Geophys.*, **3**, 105–152.

CAREY, S. W. (1958). A tectonic approach to continental drift. In *Continental drift, a symposium*, pp. 177–355, edited by Carey, S. W., University of Tasmania, Hobart.

ČERVENÝ, V. (1966). On dynamic properties of reflected and head waves in the n-layered Earth's crust. *Geophys. J. R. astr. Soc.*, **11**, 139–147.

CLARK, S. P., Jr. (editor) (1966). *Handbook of physical constants*, Revised edition, *Mem. geol. Soc. Am.*, **97**, 587 pp.

CONRAD, V. (1925). Laufzeitkurven des Tauernbebens vom 28. November 1923. *Mitt. Erdb–Kommn. Wien*, No. 59, 1–23.

COOK, F. A., ALBAUGH, D. S., BROWN, L. D., KAUFMAN, S. and OLIVER, J. E. (1979). Thin-skinned tectonics in the crystalline southern Appalachians; COCORP seismic-reflection profiling of the Blue Ridge and Piedmont. *Geology*, **7**, 563–567.

CORON, S. (1963). Aperçu gravimétrique sur les Alpes occidentales. In *Recherches séismologique dans les Alpes occidentales au moyen des grandes explosions en 1956, 1958 et 1960*, pp. 31–37, edited by Closs, H. and Labrouste, Y., *Séismologie*, sér. XXII, **2**, Centre National de la Recherche Scientifique.

CREER, K. M. (1957). Palaeomagnetic investigations in Great Britain V. The remanent magnetization of unstable Keuper marls. *Phil. Trans. R. Soc.*, **250A**, 130–143.

CREER, K. M. (1965). Palaeomagnetic data from the Gondwanic continents. *Phil. Trans. R. Soc.*, **258A**, 27–40.

CREER, K. M. (1970). A review of palaeomagnetism. *Earth Sci. Rev.*, **6**, 369–466.

DAVIDSON, A. and REX, D. C. (1980). Age of volcanism and rifting in southwestern Ethiopia. *Nature, Lond.*, **283**, 657–658.

DEN TEX, E. (1965). Metamorphic lineages of orogenic plutonism. *Geologie Mijnb.*, **44e**, 105–132.

DEWEY, J. F. (1969). Evolution of the Appalachian/Caledonian orogen. *Nature, Lond.*, **222**, 124–129.

DIX, C. H. (1965). Reflection seismic crustal studies. *Geophysics*, **30**, 1068–1084.

DOBRIN, M. B. (1976). *Introduction to geophysical prospecting*, Third edition, McGraw-Hill Book Company, New York, 630 pp.

DOHR, G. and FUCHS, K. (1967). Statistical evaluation of deep crustal reflections in Germany. *Geophysics*, **32**, 951–967.

DU TOIT, A. L. (1937). *Our wandering continents, an hypothesis of continental drifting*, Oliver and Boyd, Edinburgh and London, 366 pp.

EADE, K. E., FAHRIG, W. F. and MAXWELL, J. A. (1966). Composition of crystalline shield rocks and fractionating effects of regional metamorphism. *Nature, Lond.*, **211**, 1245–1249.

EATON, J. P. (1963). Crustal structure from San Francisco, California, to Eureka, Nevada, from seismic-refraction measurements. *J. geophys. Res.*, **68**, 5789–5806.

EWING, M. and PRESS, F. (1959). Determination of crustal structure from phase velocity of Rayleigh waves Part III: the United States. *Bull. geol. Soc. Am.*, **70**, 229–244.

FUCHS, K. and MÜLLER, G. (1971). Computation of synthetic seismograms with the reflectivity method and comparison with observations. *Geophys. J. R. astr. Soc.*, **23**, 417–433.

GIESE, P. (1976a). Problems and tasks of data generalization. In *Explosion seismology in central Europe*, pp. 137–145, edited by Giese, P., Prodehl, C. and Stein, A., Springer–Verlag, Berlin, Heidelberg and New York.

GIESE, P. (1976b). General remarks on travel time data and principles of correlation. In *Explosion seismology in central Europe*, pp. 130–136, edited by Giese, P., Prodehl, C. and Stein, A., Springer–Verlag, Berlin, Heidelberg and New York.

GIESE, P. (1976c). Depth calculation. In *Explosion seismology in central Europe*, pp. 146–161, edited by Giese, P., Prodehl, C. and Stein, A., Springer–Verlag, Berlin, Heidelberg and New York.

GIESE, P. (1976d). Models of crustal structure and main wave groups. In *Explosion seismology in central Europe*, pp. 196–200, edited by Giese, P., Prodehl, C. and Stein, A., Springer–Verlag, Berlin, Heidelberg and New York.

GIESE, P. (1976e). Results of the generalized interpretation of the deep-seismic sounding data. In *Explosion seismology in central Europe*, pp. 201–214, edited by Giese, P., Prodehl, C. and Stein, A., Springer–Verlag, Berlin, Heidelberg and New York.

GIESE, P. and PRODEHL, C. (1976). Main features of crustal structure in the Alps. In *Explosion seismology in central Europe*, pp. 347–375, edited by Giese, P., Prodehl, C. and Stein, A., Springer–Verlag, Berlin, Heidelberg and New York.

GIESE, P., PRODEHL, C. and STEIN, A. (editors) (1976). *Explosion seismology in central Europe*, Springer–Verlag, Berlin, Heidelberg and New York, 429 pp.

GIRDLER, R. W. (1964). Geophysical studies of rift valleys. *Physics Chem. Earth*, **5**, 121–156.

GIRDLER, R. W., FAIRHEAD, J. D., SEARLE, R. C. and SOWERBUTTS, W. T. C. (1969). Evolution of rifting in Africa. *Nature, Lond.*, **224**, 1178–1182.

GLOCKE, A. and MEISSNER, R. (1976). Near-vertical reflections recorded at the wide-angle profile in the Rhenish massif. In *Explosion seismology in central Europe*, pp. 252–256, edited by Giese, P., Prodehl, C. and Stein, A., Springer–Verlag, Berlin, Heidelberg and New York.

GREEN, D. H. and RINGWOOD, A. E. (1967). An experimental investigation of the gabbro to eclogite transformation and its petrological applications. *Geochim. cosmochim. Acta*, **31**, 767–833.

GREEN, T. H. (1970). High pressure experimental studies on the mineralogical constitution of the lower crust. *Phys. Earth planet. Interiors*, **3**, 441–450.

GRIFFITHS, D. H., KING, R. F., KHAN, M. A. and BLUNDELL, D. J. (1971). Seismic refraction line in the Gregory rift. *Nature Phys. Sci., Lond.*, **229**, 69–71.

GUTENBERG, B. (1954). Effects of low velocity layers. *Geofis. pura appl.*, **29**, 1–10.

GUTENBERG, B. (1959). *Physics of the Earth's interior*, Academic Press, New York and London, 240 pp.

HAGEDOORN, J. G. (1959). The plus-minus method of interpreting seismic refraction sections. *Geophys. Prospect.*, **7**, 158–182.

HALL, D. H. and HAJNAL, Z. (1973). Deep seismic crustal studies in Manitoba. *Bull. seism. Soc. Am.*, **63**, 885–910.

HALLAM, A. (1973). *A revolution in the Earth sciences*, Oxford University Press, London, 127 pp.

HART, P. J. (editor) (1969). *The Earth's crust and upper mantle, Geophys. Monogr.* No. 13, American Geophysical Union, Washington, D. C., 735 pp.

HEACOCK, J. G. (editor) (1977). *The Earth's crust, Geophys. Monogr.* No. 20, American Geophysical Union, Washington, D. C., 754 pp.

HEISKANEN, W. A. and VENING MEINESZ, F. A. (1958). *The Earth and its gravity field*, McGraw-Hill Book Company, New York, Toronto and London, 470 pp.

HILL, D. P. and PAKISER, L. C. (1966). Crustal structure between the Nevada test site and Boise, Idaho, from seismic refraction measurements. In *The Earth beneath the continents*, pp. 391–419, edited by Steinhart, J. S. and Smith, T. J., *Geophys. Monogr.* No. 10, American Geophysical Union, Washington, D. C.

HOLOPAINEN, P. E. (1947). On the gravity field and the isostatic structure of the Earth's crust in the East Alps. *Publs isostatic Inst. int. Ass. Geod.*, No. 16.

ILLIES, J. H. (1977). Ancient and recent rifting in the Rhinegraben. *Geologie Mijnb.*, **56**, 329–350.

IRVING, E. (1964). *Palaeomagnetism and its application to geological and geophysical problems*, John Wiley & Sons, New York, London and Sydney, 399 pp.

IRVING, E. (1979). Pole positions and continental drift since the Devonian. In *The Earth: its origin, structure and evolution*, pp. 567–593, edited by McElhinny, M. W., Academic Press, London, New York and San Francisco.

IRVING, E. and DUNLOP, D. J. (1982). *Palaeomagnetism*, John Wiley & Sons, London, New York, Sydney and Toronto, (in the press).

IRVING, E., ROBERTSON, W. A. and STOTT, P. M. (1963). The significance of the palaeomagnetic results from Mesozoic rocks of eastern Australia. *J. geophys. Res.*, **68**, 2313–2317.

ITO, K. and KENNEDY, G. C. (1971). An experimental study of the basalt-garnet granulite-eclogite transition. In *The structure and physical properties of the Earth's crust*, pp. 303–314, edited by Heacock, J. G., *Geophys. Monogr.* No. 14, American Geophysical Union, Washington, D. C.

JAMES, D. E. (1971). Andean crustal and upper mantle structure. *J. geophys. Res.*, **76**, 3246–3271.

JAMES, D. E. and STEINHART, J. S. (1966). Structure beneath continents: a critical review of explosion studies 1960–1965. In *The Earth beneath the continents*, pp. 293–333, edited by Steinhart, J. S. and Smith, T. J., *Geophys. Monogr.* No. 10, American Geophysical Union, Washington, D. C.

JEFFREYS, H. (1959). *The Earth: its origin, history and physical constitution*, Fourth edition, Cambridge University Press, London and New York, 420 pp.

KISELEV, A. I., GOLOVKO, H. A. and MEDVEDEV, M. E. (1978). Petrochemistry of Cenozoic basalts and associated rocks in the Baikal rift zone. *Tectonophysics*, **45**, 49–59.

KOSMINSKAYA, I. P., BELYAEVSKY, N. A. and VOLVOVSKY, I. S. (1969). Explosion seismology in the USSR. In *The Earth's crust and upper mantle*, pp. 195–208, edited by Hart, P. J., *Geophys. Monogr.* No. 13, American Geophysical Union, Washington, D. C.

LAPADU–HARGUES, P. (1953). Sur la composition chimique moyenne des amphibolites. *Bull. Soc. géol. Fr.*, sér 6, **3**, 153–173.

LIEBERMANN, R. C. and RINGWOOD, A. E. (1973). Birch's law and polymorphic phase transformations. *J. geophys. Res.*, **78**, 6926–6932.

LIEBSCHER, H. J. (1964). Deutungsversuche für die Struktur der tieferen Erdkruste nach reflexionsseismischen und gravimetrischen Messungen im deutschen Alpenvorland. *Z. Geophys.*, **30**, 51–96.

LONG, R. E. and BACKHOUSE, R. W. (1976). The structure of the western flank of the Gregory rift. Part II. The mantle. *Geophys. J. R. astr. Soc.*, **44**, 677–688.

McELHINNY, M. W. (1973). *Palaeomagnetism and plate tectonics*, Cambridge University Press, London and New York, 358 pp.

MAGUIRE, P. K. H. and LONG, R. E. (1976). The structure on the western flank of the Gregory rift (Kenya). Part I. The crust. *Geophys. J. R. astr. Soc.*, **44**, 661–675.

MAIR, J. A. and LYONS, J. A. (1976). Seismic reflection techniques for crustal structure studies. *Geophysics*, **41**, 1272–1290.

MOHOROVIČIĆ, A. (1909). Das Beben vom 8.x.1909. *Jb. met. Obs. Zagreb (Agram.)*, **9**, 1–63.

MOORBATH, S. (1976). Age and isotope constraints for the evolution of Archean crust. In *The early history of the Earth*, pp. 351–373, edited by Windley, B. F., John Wiley & Sons, London, New York, Sydney and Toronto.

MUELLER, S. (editor) (1973). *The structure of the Earth's crust based on seismic data, Tectonophysics*, **20**, 391 pp.

MUELLER, S. (1977). A new model of the continental crust. In *The Earth's crust*, pp. 289–317, edited by Heacock, J. G., *Geophys. Monogr.* No. 20, American Geophysical Union, Washington, D. C.

MUELLER, S. and LANDISMAN, M. (1966). Seismic studies of the Earth's crust in continents I: Evidence for a low-velocity zone in the upper part of the lithosphere. *Geophys. J. R. astr. Soc.*, **10**, 525–538. *J. R. astr. Soc.*, **10**, 525–538.

NAFE, J. E. and DRAKE, C. L. (1963). Physical properties of marine sediments. In *The sea*, volume 3, pp. 794–815, edited by Hill, M. N., Interscience Publishers, New York and London.

NOCKOLDS, S. R. (1954). Average chemical compositions of some igneous rocks. *Bull. geol. Soc. Am.*, **65**, 1007–1032.

NORTON, I. O. and SCLATER, J. G. (1979). A model for the evolution of the Indian Ocean and the breakup of Gondwanaland. *J. geophys. Res.*, **84**, 6803–6830.

OCOLA, L. C. and MEYER, R. P. (1972). Crustal low-velocity zones under the Peru-Bolivia altiplano. *Geophys. J. R. astr. Soc.*, **30**, 199–209.

O'NIONS, R. K., EVENSEN, N. M. and HAMILTON, P. J. (1979). Geochemical modeling of mantle differentiation and crustal growth. *J. geophys. Res.*, **84**, 6091–6101.

OXBURGH, E. R. and TURCOTTE, D. L. (1974). Membrane tectonics and the East African rift. *Earth & planet. Sci. Lett. (Neth.)*, **22**, 133–140.

PAKISER, L. C. (1963). Structure of the crust and upper mantle in the western United States. *J. geophys. Res.*, **68**, 5747–5756.

POLDERVAART, A. (1955). Chemistry of the Earth's crust. *Spec. Pap. geol. Soc. Am.*, **62**, 119–144.

PRATT, J. H. (1855). On the attraction of the Himalaya mountains, and of the elevated regions beyond them, upon the plumb-line in India. *Phil. Trans. R. Soc.*, **145**, 53–100.

PRESS, F. and EWING, M. (1955). Earthquake surface waves and crustal structure. *Spec. Pap. geol. Soc. Am.*, **62**, 51–60.

PRODEHL, C. (1970). Seismic refraction study of crustal structure in the western United States. *Bull. geol. Soc. Am.*, **81**, 2629–2645.

PRODEHL, C. (1976). Comparison of seismic-refraction studies in central Europe and the western United States. In *Explosion seismology in central Europe*, pp. 385–395, edited by Giese, P., Prodehl, C. and Stein, A., Springer–Verlag, Berlin, Heidelberg and New York.

PRODEHL, C., ANSORGE, J., EDEL, J. B., EMTER, D., FUCHS, K., MUELLER, S. and PETERSCHMITT, E. (1976). Explosion-seismology research in the central and southern Rhine graben – a case history. In *Explosion seismology in central Europe*, pp. 313–328, edited by Giese, P., Prodehl, C. and Stein, A., Springer–Verlag, Berlin, Heidelberg and New York.

PUZYREV, N. N., MANDELBAUM, M. M., KRYLOV, S. V., MISHENKIN, B. P., PETRIK, G. V. and KRUPSKAYA, G. V. (1978). Deep structure of the Baikal and other continental rift zones from seismic data. *Tectonophysics*, **45**, 15–22.

RAMBERG, I. B. (1976). Gravity interpretation of the Oslo graben and associated igneous rocks. *Norg. geol. Unders.*, **325**, 1–194.

RINGWOOD, A. E. (1975). *Composition and petrology of the Earth's mantle*, McGraw-Hill Book Company, New York, 618 pp.

RINGWOOD, A. E. and GREEN, D. H. (1966). Petrological nature of the stable continental crust. In *The Earth beneath the continents*, pp. 611–619, edited by Steinhart, J. S. and Smith, T. J., *Geophys. Monogr.* No. 10, American Geophysical Union, Washington, D. C.

ROBERTS, D. and GALE, G. H. (1978). The Caledonian-Appalachian Iapetus ocean. In *Evolution of the Earth's crust*, pp. 255–342, edited by Tarling, D. H., Academic Press, London, New York and San Francisco.

RUNCORN, S. K. (1956). Palaeomagnetic comparisons between Europe and North America. *Proc. geol. Assoc. Canada*, **8**, 77–85.

SCHILT, S., OLIVER, J., BROWN, L., KAUFMAN, S., ALBAUGH, D., BREWER, J., COOK, F., JENSEN, L., KRUMHAUSL, P., LONG, G. and STEINER, D. (1979). The heterogeneity of the continental crust: results from deep crustal seismic reflection profiling using the vibroseis technique. *Rev. Geophys. space Phys.*, **17**, 354–368.

SEARLE, R. C. (1970). Evidence from gravity anomalies for thinning of the lithosphere beneath the rift valley in Kenya. *Geophys. J. R. astr. Soc.*, **21**, 13–31.

SHAW, D. M. (1976). Development of the early continental crust Part 2: Prearchean, protoarchean and later eras. In *The early history of the Earth*, pp. 33–53, edited by Windley, B. F., John Wiley & Sons, London, New York, Sydney and Toronto.

SMITH, A. G. and HALLAM, A. (1970). The fit of the southern continents. *Nature, Lond.*, **225**, 139–144.

SMITH, P. J. and BOTT, M. H. P. (1975). Structure of the crust beneath the Caledonian foreland and Caledonian belt of the north Scottish shelf region. *Geophys. J. R. astr. Soc.*, **40**, 187–205.

SMITHSON, S. B., BREWER, J. A., KAUFMAN, S., OLIVER, J. E. and HURICH, C. A. (1979). Structure of the Laramide Wind River uplift, Wyoming, from COCORP deep reflection data and from gravity data. *J. geophys. Res.*, **84**, 5955–5972.

SMITHSON, S. B. and DECKER, E. R. (1974). A continental crustal model and its geothermal implications. *Earth & planet. Sci. Lett. (Neth.)*, **22**, 215–225.

STEINHART, J. S. (1967). Mohorovičić discontinuity. In *International dictionary of geophysics*, volume 2, pp. 991–994, edited by Runcorn, S. K. and others, Pergamon Press, Oxford, London, Edinburgh, New York, Toronto, Sydney, Paris and Braunschweig.

STEINHART, J. S. and MEYER, R. P. (1961). *Explosion studies of continental structure*, Carnegie Institution of Washington Publication 622, 409 pp.

STEINHART, J. S. and SMITH, T. J. (editors) (1966). *The Earth beneath the continents*, a volume of geophysical studies in honor of Merle A. Tuve, *Geophys. Monogr.* No. 10, American Geophysical Union, Washington, D. C., 663 pp.

TARLING, D. H. (1971). *Principles and applications of palaeomagnetism*, Chapman and Hall, London, 164 pp.

TARLING, D. H. and TARLING, M. P. (1971). *Continental drift*, G. Bell and Sons, London, 112 pp.

TATEL, H. E. and TUVE, M. A. (1955). Seismic exploration of a continental crust. *Spec. Pap. geol. Soc. Am.*, **62**, 35–50.

TAYLOR, F. B. (1910). Bearing of the Tertiary mountain belt on the origin of the Earth's plan. *Bull. geol. Soc. Am.*, **21**, 179–226.

VAN DER VOO, R. (1979). Palaeomagnetism related to continental drift and plate tectonics. *Rev. Geophys. space Phys.*, **17**, 227–235.

WANG, C. (1968). Equation of state of periclase and Birch's relationship between velocity and density. *Nature, Lond.*, **218**, 74–76.

WEGENER, A. (1912). Die Entstehung der Kontinente. *Peterm. Mitt.*, 1912, 185–195, 253–256, 305–309.

WILLMORE, P. L. and BANCROFT, A. M. (1960). The time term approach to refraction seismology. *Geophys. J. R. astr. Soc.*, **3**, 419–432.

WINDLEY, B. F. (editor) (1976). *The early history of the Earth*, John Wiley & Sons, London, New York, Sydney and Toronto, 619 pp.

WOOD, H. O. and RICHTER, C. F. (1931). A study of blasting recorded in southern California. *Bull. seism. Soc. Am.*, **21**, 28–46.

WOOD, H. O. and RICHTER, C. F. (1933). A second study of blasting recorded in southern California. *Bull. seism. Soc. Am.*, **23**, 95–110.

WOOLLARD, G. P. (1966). Regional isostatic relations in the United States. In *The Earth beneath the continents*, pp. 557–594, edited by Steinhart, J. S. and Smith, T. J., *Geophys. Monogr.* No 10, American Geophysical Union, Washington, D. C.

ZIJDERVELD, J. D. A. and VAN DER VOO, R. (1973). Palaeomagnetism in the Mediterranean area. In *Implications of continental drift to the earth sciences*, volume 1, pp. 133–161, edited by Tarling, D. H. and Runcorn, S. K., Academic Press, London and New York.

Chapter three

ALVAREZ, W., ARTHUR, M. A., FISCHER, A. G., LOWRIE, W., NAPOLEONE, G., PREMOLI SILVA, I. and ROGGENTHEN, W. M. (1977). Upper Cretaceous–Palaeocene magnetic stratigraphy at Gubbio, Italy V. Type section for the late Cretaceous–Palaeocene geomagnetic reversal time scale. *Bull. geol. Soc. Am.*, **88**, 383–389.

ATWATER, T. and MUDIE, J. D. (1973). Detailed near-bottom geophysical study of the Gorda rise. *J. geophys. Res.*, **78**, 8665–8686.

BALLARD, R. D. and VAN ANDEL, T. H. (1977). Morphology and tectionics of the inner rift valley at lat 36° 50′ N on the mid-Atlantic ridge. *Bull. geol. Soc. Am.*, **88**, 507–530.

BARAZANGI, M. and DORMAN, J. (1969). World seismicity maps compiled from ESSA, Coast and Geodetic Survey, epicenter data, 1961–1967. *Bull. seism. Soc. Am.*, **59**, 369–380.

BELOUSSOV, V. V. and MILANOVSKY, Y. Y. (1977). On tectonics and tectonic position of Iceland. *Tectonophysics*, **37**, 25–40.

BOTT, M. H. P. (1967b). Solution of the linear inverse problem in magnetic interpretation with application to oceanic magnetic anomalies. *Geophys. J. R. astr. Soc.*, **13**, 313–323.

BOTT, M. H. P. (1974). Deep structure, evolution and origin of the Icelandic transverse ridge. In *Geodynamics of Iceland and the North Atlantic area*, pp. 33–47, edited by Kristjansson, L., Reidel Publishing Company, Dordrecht and Boston.

BOTT, M. H. P. and GUNNARSSON, K. (1980). Crustal structure of the Iceland-Faeroe ridge. *J. Geophys.*, **47**, 221–227.

BOTTINGA, Y. and ALLEGRE, C. J. (1973). Thermal aspects of sea-floor spreading and the nature of the oceanic crust. *Tectonophysics*, **18**, 1–17.

BURKE, K. and WILSON, J. T. (1972). Is the African plate stationary? *Nature, Lond.*, **239**, 387–390.

CANDE, S. C. and KENT, D. V. (1976). Constraints imposed by the shape of marine magnetic anomalies on the magnetic source. *J. geophys. Res.*, **81**, 4157–4162.

CANN, J. R. (1968). Geological processes at mid-ocean ridge crests. *Geophys. J. R. astr. Soc.*, **15**, 331–341.

CANN, J. R. (1970). New model for the structure of the ocean crust. *Nature, Lond.*, **226**, 928–930.

CANN, J. R. (1974). A model for oceanic crustal structure developed. *Geophys. J. R. astr. Soc.*, **39**, 169–187.

CHASE, C. G. (1978). Plate kinematics: the Americas, East Africa, and the rest of the world. *Earth & planet. Sci. Lett. (Neth.)*, **37**, 355–368.

CHRISTENSEN, N. I. (1972). Seismic anisotropy in the lower oceanic crust. *Nature, Lond.*, **237**, 450–451.

CHRISTENSEN, N. I. and SALISBURY, M. H. (1975). Structure and constitution of the lower oceanic crust. *Rev. Geophys. space Phys.*, **13**, 57–86.

COX, A., DOELL, R. R. and DALRYMPLE, G. B. (1964). Reversals of the Earth's magnetic field. *Science, N.Y.*, **144**, 1537–1543.

DAVIS, E. E. and LISTER, C. R. B. (1974). Fundamentals of ridge crest topography. *Earth & planet. Sci. Lett. (Neth.)*, **21**, 405–413.

DIETZ, R. S. (1961). Continent and ocean basin evolution by spreading of the sea floor. *Nature, Lond.*, **190**, 854–857.

EDGAR, N. T. (1974). Acoustic stratigraphy in the deep oceans. In *The geology of continental margins*, pp. 243–246, edited by Burk, C. A. and Drake, C. L., Springer-Verlag, Berlin, Heidelberg and New York.

EWING, J. and HOUTZ, R. (1969). Mantle reflections in airgun-sonobuoy profiles. *J. geophys. Res.*, **74**, 6706–6709.

EWING, J., WORZEL, J. L., EWING, M. and WINDISCH, C. (1966). Ages of horizon A and the oldest Atlantic sediments. *Science, N.Y.*, **154**, 1125–1132.

EWING, M. (1965). The sediments of the Argentine basin. *Q. Jl R. astr. Soc.*, **6**, 10–27.

FOWLER, C. M. R. (1976). Crustal structure of the mid-Atlantic ridge crest at 37° N. *Geophys. J. R. astr. Soc.*, **47**, 459–491.

FRANCIS, T. J. G. (1968). The detailed seismicity of mid-ocean ridges. *Earth & planet. Sci. Lett. (Neth.)*, **4**, 39–46.

FRANCIS, T. J. G. and PORTER, I. T. (1973). Median valley seismology: the mid-Atlantic ridge near 45° N. *Geophys. J. R. astr. Soc.*, **34**, 279–311.

GASS, I. G. (1968). Is the Troodos massif of Cyprus a fragment of Mesozoic ocean floor? *Nature, Lond.*, **220**, 39–42.

GIRDLER, R. W. and STYLES, P. (1974). Two stage Red Sea floor spreading. *Nature, Lond.*, **247**, 7–11.

GUTENBERG, B. (1924). Der Aufbau der Erdkruste auf Grund geophysikalischer Betrachtungen. *Z. Geophys.*, **1**, 94–108.

HALL, J. M. (1976). Major problems regarding the magnetization of oceanic crustal layer 2. *J. geophys. Res.*, **81**, 4223–4230.

HAYS, J. D. and OPDYKE, N. D. (1967). Antarctic radiolaria, magnetic reversals, and climatic change. *Science, N.Y.*, **158**, 1001–1011.

HEEZEN, B. C. (1962). The deep-sea floor. In *Continental drift*, pp. 235–288, edited by Runcorn, S. K., Academic Press, New York and London.

HEIRTZLER, J. R., DICKSON, G. O., HERRON, E. M., PITMAN, W. C., III and LE PICHON, X. (1968). Marine magnetic anomalies, geomagnetic field reversals, and motions of the ocean floor and continents. *J. geophys. Res.*, **73**, 2119–2136.

HEIRTZLER, J. R. and VAN ANDEL, T. H. (1977). Project FAMOUS: its origin, programs, and setting. *Bull. geol. Soc. Am.*, **88**, 481–487.

HERRON, T. J., STOFFA, P. L. and BUHL, P. (1980). Magma chamber and mantle reflections – East Pacific rise. *Geophys. Res. Lett.*, **7**, 989–992.

HESS, H. H. (1962). History of ocean basins. In *Petrologic studies: a volume in honor of A. F. Buddington*, pp. 599–620, Geological Society of America.

HESS, H. H. (1964). Seismic anisotropy of the uppermost mantle under oceans. *Nature, Lond.*, **203**, 629–631.

HESS, H. H. (1965). Mid-oceanic ridges and tectonics of the sea-floor. In *Submarine geology and geophysics*, pp. 317–333, edited by Whittard, W. F. and Bradshaw, R., Colston Papers No. 17, Butterworths, London.

HILL, M. N. (1963). Single-ship seismic refraction shooting. In *The sea*, volume 3, pp. 39–46, edited by Hill, M. N., Interscience Publishers, New York and London.

HODGES, F. N. and PAPIKE, J. J. (1976). DSDP site 334: magmatic cumulates from oceanic layer 3. *J. geophys. Res.*, **81**, 4135–4151.

HOUTZ, R. and EWING, J. (1976). Upper crustal structure as a function of plate age. *J. geophys. Res.*, **81**, 2490–2498.

ISACKS, B. OLIVER, J. and SYKES, L. R. (1968). Seismology and the new global tectonics. *J. geophys. Res.*, **73**, 5855–5899.

JONES, E. J. W., EWING, M., EWING, J. I. and EITTREIM, S. L. (1970). Influences of Norwegian Sea overflow water on sedimentation in the northern North Atlantic and Labrador Sea. *J. geophys. Res.*, **75**, 1655–1680.

JORDAN, T. H. (1975). The present-day motions of the Caribbean plate. *J. geophys. Res.*, **80**, 4433–4439.

KEEN, C. and TRAMONTINI, C. (1970). A seismic refraction survey on the mid-Atlantic ridge. *Jl R. astr. Soc. Can.*, **20**, 473–491.

KENNETT, B. L. N. (1977). Towards a more detailed seismic picture of the oceanic crust and mantle. *Marine Geophys. Res.*, **3**, 7–42.

KIDD, R. G. W. (1977). A model for the process of formation of the upper oceanic crust. *Geophys. J. R. astr. Soc.*, **50**, 149–183.

KUSZNIR, N. J. and BOTT, M. H. P. (1976). A thermal study of the formation of oceanic crust. *Geophys. J. R. astr. Soc.*, **47**, 83–95.

LABRECQUE, J. L., KENT, D. V. and CANDE, S. C. (1977). Revised magnetic polarity time scale for late Cretaceous and Cenozoic time. *Geology*, **5**, 330–335.

LANCELOT, Y., HATHAWAY, J. C. and HOLLISTER, C. D. (1972). Lithology of sediments from the western North Atlantic leg 11 deep sea drilling project. In Hollister, C. B., Ewing, J. I. and others, *Initial reports of the Deep Sea Drilling Project*, volume 11, pp. 901–949, Washington (U.S. Government Printing Office).

LARSON, R. L. and HILDE, T. W. C. (1975). A revised time scale of magnetic reversals for the early Cretaceous and late Jurassic. *J. geophys. Res.*, **80**, 2586–2594.

LARSON, R. L. and PITMAN, W. C., III (1972). World-wide correlation of Mesozoic magnetic anomalies, and its implications. *Bull. geol. Soc. Am.*, **83**, 3645–3662.

LE PICHON, X. (1968). Sea-floor spreading and continental drift. *J. geophys. Res.*, **73**, 3661–3697.

LE PICHON, X., EWING, J. and HOUTZ, R. E. (1968). Deep-sea sediment velocity determination made while reflection profiling. *J. geophys. Res.*, **73**, 2597–2614.

LE PICHON, X., FRANCHETEAU, J. and BONNIN, J. (1973). *Plate tectonics*, Elsevier Scientific Publishing Company, Amsterdam, London and New York, 300 pp.

LEWIS, B. T. R. (1978). Evolution of ocean crust seismic velocities. *Ann. Rev. Earth & planet. Sci.*, **6**, 377–404.

LEWIS, B. T. R. and SNYDSMAN, W. E. (1979). Fine structure of the lower oceanic crust on the Cocos plate. *Tectonophysics*, **55**, 87–105.

LLIBOUTRY, L. (1974). Plate movement relative to rigid lower mantle. *Nature, Lond.*, **250**, 298–300.

MANKINEN, E. A. and DALRYMPLE, G. B. (1979). Revised geomagnetic polarity time scale for the interval 0-5 m.y. B.P. *J. geophys. Res.*, **84**, 615–626.

MASON, R. G. and RAFF, A. D. (1961). Magnetic survey off the west coast of North America, 32° N. latitude to 42° N. latitude. *Bull. geol. Soc. Am.*, **72**, 1259–1266.

MAXWELL, A. E. and others (1970). Deep sea drilling in the South Atlantic. *Science, N. Y.*, **168**, 1047–1059.

McKENZIE, D. P. and PARKER, R. L. (1967). The north Pacific: an example of tectonics on a sphere. *Nature, Lond.*, **216**, 1276–1280.

MENARD, H. W. (1964). *Marine geology of the Pacific*, McGraw-Hill Book Company, New York, San Francisco, Toronto and London, 271 pp.

MINSTER, J. B. and JORDAN, T. H. (1978). Present-day plate motions. *J. geophys. Res.*, **83**, 5331–5354.

MORGAN, W. J. (1968). Rises, trenches, great faults, and crustal blocks. *J. geophys. Res.*, **73**, 1959–1982.

MORGAN, W. J. (1971). Convection plumes in the lower mantle. *Nature, Lond.*, **230**, 42–43.

NESS, G., LEVI, S. and COUCH, R. (1980). Marine magnetic anomaly timescales for the Cenozoic and late Cretaceous: a précis, critique, and synthesis. *Rev. Geophys. space Phys.*, **18**, 753–770.

NIELL, A. E., ONG, K. M., MACDORAN, P. F., RESCH, G. M., MORABITO, D. D., CLAFLIN, E. S. and DRACUP, J. F. (1979). Comparison of a radiointerferometric differential baseline measurement with conventional geodesy. *Tectonophysics*, **52**, 49–58.

OPDYKE, N. D., BURCKLE, L. H. and TODD, A. (1974). The extension of the magnetic time scale in sediments of the central Pacific Ocean. *Earth & planet. Sci. Lett. (Neth.)*, **22**, 300–306.

OPDYKE, N. D., GLASS, B., HAYS, J. D. and FOSTER, J. (1966). Palaeomagnetic study of Antarctic deep-sea cores. *Science, N.Y.*, **154**, 349–357.

ORCUTT, J. A., KENNETT, B. L. N. and DORMAN, L. M. (1976). Structure of the East Pacific rise from an ocean bottom seismometer survey. *Geophys. J. R. astr. Soc.*, **45**, 305–320.

OSMASTON, M. F. (1971). Genesis of ocean ridge median valleys and continental rift valleys. *Tectonophysics*, **11**, 387–405.

PALMASON, G. (1971). Crustal structure of Iceland from explosion seismolog. *Rit Visindafj. isl.*, **40**, 1–187.

PARSONS, B. and SCLATER, J. G. (1977). An analysis of the variation of ocean floor bathymetry and heat flow with age. *J. geophys. Res.*, **82**, 803–827.

PITMAN, W. C., III and HEIRTZLER, J. R. (1966). Magnetic anomalies over the Pacific-Antarctic ridge. *Science, N.Y.*, **154**, 1164–1171.

RAFF, A. D. and MASON, R. G. (1961). Magnetic survey off the west coast of North America, 40° N. latitude to 52° N. latitude. *Bull. geol. Soc. Am.*, **72**, 1267–1270.

RAIT, R. W., SHOR, G. G., Jr., MORRIS, G. B. and KIRK, H. K. (1971). Mantle anisotropy in the Pacific Ocean. *Tectonophysics*, **12**, 173–186.

ROSENDAHL, B. R., RAITT, R. W., DORMAN, L. M., BIBEE, L. D., HUSSONG, D. M. and SUTTON, G. H. (1976). Evolution of oceanic crust 1. A physical model of the East Pacific rise crest derived from seismic refraction data. *J. geophys. Res.*, **81**, 5294–5304.

SALISBURY, M. H. and CHRISTENSEN, N. I. (1978). The seismic velocity structure of a traverse through the Bay of Islands ophiolite complex, Newfoundland, an exposure of oceanic crust and upper mantle. *J. geophys. Res.*, **83**, 805–817.

SCLATER, J. G. and FRANCHETEAU, J. (1970). The implications of terrestrial heat flow observations on current tectonic and geochemical models of the crust and upper mantle of the Earth. *Geophys. J. R. astr. Soc.*, **20**, 509–542.

SHOR, G. G., Jr. (1963). Refraction and reflection techniques and procedure. In *The sea*. volume 3, pp. 20–38, edited by Hill, M. N., Interscience Publishers, New York and London.

SLEEP, N. H. (1969). Sensitivity of heat flow and gravity to the mechanism of sea-floor spreading. *J. geophys. Res.*, **74**, 542–549.

SLEEP, N. H. (1975). Formation of oceanic crust: some thermal constraints. *J. geophys. Res.*, **80**, 4037–4042.

SMITH, D. E., KOLENKIEWICZ, R., DUNN, P. J. and TORRENCE, M. H. (1979). The measurement of fault motion by satellite laser ranging. *Tectonophysics*, **52**, 59–67.

SPUDICH, P. and ORCUTT, J. (1980). A new look at the seismic velocity structure of the oceanic crust. *Rev. Geophys. space Phys.*, **18**, 627–645.

STOFFA, P. L. and BUHL, P. (1979). Two-ship multichannel seismic experiments for deep crustal studies: expanded spread and constant offset profiles. *J. geophys. Res.*, **84**, 7645–7660.

SUTTON, G. H., MAYNARD, G. L. and HUSSONG, D. M. (1971). Physical properties of the oceanic crust. In *The structure and physical properties of the Earth's crust*, pp. 193–209, edited by Heacock, J. G., *Geophys. Monogr.* No. 14, American Geophysical Union, Washington, D. C.

SYKES, L. R. (1967). Mechanism of earthquakes and nature of faulting on the mid-oceanic ridges. *J. geophys. Res.*, **72**, 2131–2153.

TALWANI, M. (1964). A review of marine geophysics. *Marine Geol.*, **2**, 29–80.

TALWANI, M., LE PICHON, X. and EWING, M. (1965). Crustal structure of the mid-ocean ridges 2. Computed model from gravity and seismic refraction data. *J. geophys. Res.*, **70**, 341–352.

TALWANI, M., WINDISCH, C. C. and LANGSETH, M. G., Jr. (1971). Reykjanes ridge crest: a detailed geophysical study. *J. geophys. Res.*, **76**, 473–517.

THEYER, F. and HAMMOND, S. R. (1974). Palaeomagnetic polarity sequence and radiolarian zones, Brunhes to polarity epoch 20. *Earth & planet. Sci. Lett. (Neth.)*, **22**, 307–319.

VACQUIER, V. (1965). Transcurrent faulting in the ocean floor. *Phil. Trans. R. Soc.*, **258A**, 77–81.

VENING MEINESZ, F. A. (1948). *Gravity expeditions at sea, 1923–1938*, volume IV, *Publ. Netherlands Geod. Comm., Delft.*

VINE, F. J. (1966). Spreading of the ocean floor: new evidence. *Science, N.Y.*, **154**, 1405–1415.

VINE, F. J. and MATTHEWS, D. H. (1963). Magnetic anomalies over oceanic ridges. *Nature, Lond.*, **199**, 947–949.

VOGT, P. R. and AVERY, O. E. (1974). Detailed magnetic surveys in the northeast Atlantic and Labrador Sea. *J. geophys. Res.*, **79**, 363–389.

WHITMARSH, R. B. (1975). Axial intrusion zone beneath the median valley of the mid-Atlantic ridge at 37° ·N detected by explosion seismology. *Geophys. J. R. astr. Soc.*, **42**, 189–215.

WILLIAMS, H. (1973). The Bay of Islands map area, Newfoundland. *Geol. Surv. Can. Pap.*, **72–34**, 1–7.

WILSON, J. T. (1963). Evidence from islands on the spreading of ocean floors. *Nature, Lond.*, **197**, 536–538.

WILSON, J. T. (1965). A new class of faults and their bearing on continental drift. *Nature, Lond.*, **207**, 343–347.

WORZEL, J. L. (1965). *Pendulum gravity measurements at sea 1936–1939*, John Wiley & Sons, New York, London and Sydney, 422 pp.

Chapter four

AKI, K. and PRESS, F. (1961). Upper mantle structure under oceans and continents from Rayleigh waves. *Geophys. J. R. astr. Soc.*, **5**, 292–305.

AKIMOTO, S. and FUJISAWA, H. (1965). Demonstration of the electrical conductivity jump produced by the olivine-spinel transition. *J. geophys. Res.*, **70**, 443–449.

AKIMOTO, S. and FUJISAWA, H. (1968). Olivine-spinel solid solution equilibria in the system Mg_2SiO_4-Fe_2SiO_4. *J. geophys. Res.*, **73**, 1467–1479.

ANDERSON, D. L. (1967b). The anelasticity of the mantle. *Geophys. J. R. astr. Soc.*, **14**, 135–164.

ANDERSON, D. L. (1967c). Phase changes in the upper mantle. *Science, N.Y.*, **157**, 1165–1173.

ANDERSON, D. L. and HART, R. S. (1976). An earth model based on free oscillations and body waves. *J. geophys. Res.*, **81**, 1461–1475.

ANDERSON, D. L. and HART, R. S. (1978). *Q* of the Earth. *J. geophys. Res.*, **83**, 5869–5882.

ANDERSON, D. L. and SAMMIS, C. (1970). Partial melting in the upper mantle. *Phys. Earth planet. Interiors*, **3**, 41–50.

BACKUS, G. E. and GILBERT, J. F. (1967). Numerical applications of a formalism for geophysical inverse problems. *Geophys. J. R. astr. Soc.*, **13**, 247–276.

BACKUS, G. and GILBERT, F. (1968). The resolving power of gross earth data. *Geophys. J. R. astr. Soc.*, **16**, 169–205.

BACKUS, G. and GILBERT, F. (1970). Uniqueness in the inversion of inaccurate gross earth data. *Phil. Trans. R. Soc.*, **266A**, 123–192.

BAILEY, D. K., TARNEY, J. and DUNHAM, K. (editors) (1980). The evidence for chemical heterogeneity in the Earth's mantle. *Phil. Trans. R. Soc.*, **297A**, 135–493.

BANKS, R. J. (1969). Geomagnetic variations and the electrical conductivity of the upper mantle. *Geophys. J. R. astr. Soc.*, **17**, 457–487.

BANKS, R. J. (1972). The overall conductivity distribution of the Earth. *J. Geomagn. Geoelect., Kyoto*, **24**, 337–351.

BANKS, R. J. and BULLARD, E. C. (1966). The annual and 27 day magnetic variations. *Earth & planet. Sci. Lett. (Neth.)*, **1**, 118–120.

BANKS, R. J. and OTTEY, P. (1974). Geomagnetic deep sounding in and around the Kenya rift valley. *Geophys. J. R. astr. Soc.*, **36**, 321–335.

BERNAL, J. D. (1936). Geophysical discussion. *Observatory*, **59**, 268.

BIRCH, F. (1952). Elasticity and constitution of the Earth's interior. *J. geophys. Res.*, **57**, 227–286.

BIRCH, F. (1964). Density and composition of mantle and core. *J. geophys. Res.*, **69**, 4377–4388.

BOTT, M. H. P. (1971b). The mantle transition zone as possible source of global gravity anomalies. *Earth & planet. Sci. Lett. (Neth.)*, **11**, 28–34.

BOYD, F. R. (1973). A pyroxene geotherm. *Geochim. cosmochim. Acta*, **37**, 2533–2546.

BRADLEY, J. J. and FORT, A. N., Jr. (1966). Internal friction in rocks. In *Handbook of physical constants*, Revised edition, pp. 175–193, edited by Clark, S. P., Jr., *Mem. geol. Soc. Am.*, **97**.

BRUNE, J. and DORMAN, J. (1963). Seismic waves and earth structure in the Canadian Shield. *Bull. seism. Soc. Am.*, **53**, 167–209.

BUCHBINDER, G. G. R. (1971). A velocity structure of the Earth's core. *Bull. seism. Soc. Am.*, **61**, 429–456.

BULLARD, E. C. (1967). Electromagnetic induction in the Earth. *Q. Jl R. astr. Soc.*, **8**, 143–160.

BULLEN, K. E. (1963). *An introduction to the theory of seismology*, Third edition, Cambridge University Press, London and New York, 381 pp.

BULLEN, K. E. (1975). *The Earth's density*, Chapman and Hall, London, 420 pp.

BURDICK, L. J. and HELMBERGER, D. V. (1978). The upper mantle *P* velocity structure of the western United States. *J. geophys. Res.*, **83**, 1699–1712.

CAGNIARD, L. (1953). Basic theory of the magneto-telluric method of geophysical prospecting. *Geophysics*, **18**, 605–635.

CAMFIELD, P. A. and GOUGH, D. I. (1977). A possible Proterozoic plate boundary in North America. *Can. J. Earth Sci.*, **14**, 1229–1238.

CAPON, J., GREENFIELD, R. J. and LACOSS, R. T. (1969). Long-period signal processing results for the large aperture seismic array. *Geophysics*, **34**, 305–329.

CARSWELL, D. A. and DAWSON, J. B. (1970). Garnet peridotite xenoliths in South African kimberlite pipes and their petrogenesis. *Contr. Miner. Petrogr.*, **25**, 163–184.

CHAPMAN, S. (1919). The solar and lunar diurnal variations of terrestrial magnetism. *Phil. Trans. R. Soc.*, **218A**, 1–118.

CLARK, S. P., Jr. and RINGWOOD, A. E. (1964). Density distribution and constitution of the mantle. *Rev. Geophys.*, **2**, 35–88.

CLEARY, J. and HALES, A. L. (1966). An analysis of the travel times of *P* waves to North American stations, in the distance range 32° to 100°. *Bull. seism. Soc. Am.*, **56**, 467–489.

CRAMPIN, S. (1977). A review of the effects of anisotropic layering on the propagation of seismic waves. *Geophys. J. R. astr. Soc.*, **49**, 9–27.

DICKINSON, W. R. and LUTH, W. C. (1971). A model for plate tectonic evolution of mantle layers. *Science, N.Y.*, **174**, 400–404.

DORMAN, J., EWING, M. and OLIVER, J. (1960). Study of shear-velocity distribution in the upper mantle by mantle Rayleigh waves. *Bull. seism. Soc. Am.*, **50**, 87–115.

DUCRUIX, J., COURTILLOT, V. and LE MOUËL, J. (1980). The late 1960s secular variation impulse, the eleven year magnetic variation and the electrical conductivity of the deep mantle. *Geophys. J. R. astr. Soc.*, **61**, 73–94.

DUSCHENES, J. D. and SOLOMON, S. C. (1977). Shear wave travel time residuals from oceanic earthquakes and the evolution of oceanic lithosphere. *J. geophys. Res.*, **82**, 1985–2000.

DZIEWONSKI, A. M. (1971). Upper mantle models from 'pure-path' dispersion data. *J. geophys. Res.*, **76**, 2587–2601.

DZIEWONSKI, A. M. (1979). Elastic and anelastic structure of the Earth. *Rev. Geophys. space Phys.*, **17**, 303–312.

DZIEWONSKI, A. M. and ANDERSON, D. L. (1981). Preliminary reference earth model. *Phys. Earth planet. Interiors*, **25**, 297–356.

DZIEWONSKI, A. M., HAGER, B. H. and O'CONNELL, R. J. (1977). Large-scale heterogeneities in the lower mantle. *J. geophys. Res.*, **82**, 239–255.

DZIEWONSKI, A. M., HALES, A. L. and LAPWOOD, E. R. (1975). Parametrically simple earth models consistent with geophysical data. *Phys. Earth planet. Interiors*, **10**, 12–48.

EATON, J. P. and MURATA, K. J. (1960). How volcanoes grow. *Science, N.Y.*, **132**, 925–938.

FILLOUX, J. H. (1977). Ocean-floor magnetotelluric sounding over north central Pacific. *Nature, Lond.*, **269**, 297–301.

FORSYTH, D. W. (1977). The evolution of the upper mantle beneath mid-ocean ridges. *Tectonophysics*, **38**, 89–118.

FRAZER, M. C. (1974). Geomagnetic deep sounding with arrays of magnetometers. *Rev. Geophys. space Phys.*, **12**, 401–420.

FUCHS, K. (1979). Structure, physical properties and lateral heterogeneities of the subcrustal lithosphere from long-range deep seismic sounding observations on continents. *Tectonophysics*, **56**, 1–15.

GIVEN, J. W. and HELMBERGER, D. V. (1980). Upper mantle structure of northwestern Eurasia. *J. geophys. Res.*, **85**, 7183–7194.

GOUGH, D. I. (1974). Electrical conductivity under western North America in relation to heat flow, seismology, and structure. *J. Geomagn. Geoelect., Kyoto*, **26**, 105–123.

GREEN, D. H. and RINGWOOD, A. E. (1963). Mineral assemblages in a model mantle composition. *J. geophys. Res.*, **68**, 937–945.

GUTENBERG, B. (1926). Untersuchungen zur Frage, bis zu welcher Tiefe die Erde kristallin ist. *Z. Geophys.*, **2**, 24–29.

GUTENBERG, B. (1948). On the layer of relatively low wave velocity at a depth of about 80 kilometers. *Bull. seism. Soc. Am.*, **38**, 121–148.

GUTENBERG, B. (1953). Wave velocities at depths between 50 and 600 kilometers. *Bull. seism. Soc. Am.*, **43**, 223–232.

GUTENBERG (1959). *Physics of the Earth's interior*, Academic Press, New York and London, 240 pp.

HADDON, R. A. W. and BULLEN, K. E. (1969). An earth model incorporating free earth oscillation data. *Phys. Earth planet. Interiors*, **2**, 35–49.

HALES, A. L. and DOYLE, H. A. (1967). *P* and *S* travel time anomalies and their interpretation. *Geophys. J. R. astr. Soc.*, **13**, 403–415.

HART, R. S., ANDERSON, D. L. and KANAMORI, H. (1977). The effect of attenuation on gross earth models. *J. geophys. Res.*, **82**, 1647–1654.

HARTE, B. (1978). Kimberlite nodules, upper mantle petrology, and geotherms. *Phil. Trans. R. Soc.*, **288A**, 487–500.

HASKELL, N. A. (1953). The dispersion of surface waves on multilayered media. *Bull. seism. Soc. Am.*, **43**, 17–34.

HELMBERGER, D. V. and ENGEN, G. R. (1974). Upper mantle shear structure. *J. geophys. Res.*, **79**, 4017–4028.

HELMBERGER, D. and WIGGINS, R. A. (1971). Upper mantle structure of midwestern United States. *J. geophys. Res.*, **76**, 3229–3245.

HERRIN, E. (1968). Introduction to "1968 seismological tables for *P* phases". *Bull. seism. Soc. Am.*, **58**, 1193–1241.

HERRIN, E. and TAGGART, J. (1962). Regional variations in P_n velocity and their effect on the location of epicenters. *Bull. seism. Soc. Am.*, **52**, 1037–1046.

HIRN, A., STEINMETZ, L., KIND, R. and FUCHS, K. (1973). Long range profiles in western Europe: II. Fine structure of the lower lithosphere in France (southern Bretagne). *Z. Geophys.*, **39**, 363–384.

HUTTON, V. R. S. (1976). The electrical conductivity of the Earth and planets. *Rep. Prog. Phys.*, **39**, 487–572.

HUTTON, V. R. S., DAWES, G., INGHAM, M., KIRKWOOD, S., MBIPOM, E. W. and SIK, J. (1981). Recent studies of time variations of natural electromagnetic fields in Scotland. *Phys. Earth planet. Interiors*, **24**, 66–87.

JEFFREYS, H. (1939a). The times of *P*, *S* and *SKS*, and the velocities of *P* and *S*. *Mon. Not. R. astr. Soc. geophys. Suppl.*, **4**, 498–533.

JEFFREYS, H. (1958). A modification of Lomnitz's law of creep in rocks. *Geophys. J. R. astr. Soc.*, **1**, 92–95.

JEFFREYS, H. and BULLEN, K. E. (1940). *Seismological tables*, British Association Gray-Milne Trust.

JULIAN, B. R. and ANDERSON, D. L. (1968). Travel times, apparent velocities and amplitudes of body waves. *Bull. seism. Soc. Am.*, **58**, 339–366.

KENNEDY, G. C. and HIGGINS, G. H. (1972). Melting temperatures in the Earth's mantle. *Tectonophysics*, **13**, 221–232.

KOVACH, R. L. (1978). Seismic surface waves and crustal and upper mantle structure. *Rev. Geophys. space Phys.*, **16**, 1–13.

KRAUT, E. A. and KENNEDY, G. C. (1966). New melting law at high pressures. *Phys. Rev.*, **151**, 668–675.

KUNO, H. (1969). Mafic and ultramafic nodules in basaltic rocks of Hawaii. *Mem. geol. Soc. Am.*, **115**, 189–234.

LAHIRI, B. N. and PRICE, A. T. (1939). Electromagnetic induction in non-uniform conductors, and the determination of the conductivity of the Earth from terrestrial magnetic variations. *Phil. Trans. R. Soc.*, **237A**, 509–540.

LARSEN, J. C. (1975). Low frequency (0.1–6.0. cpd) electromagnetic study of deep mantle electrical conductivity beneath the Hawaiian Islands. *Geophys. J. R. astr. Soc.*, **43**, 17–46.

LEE, W. B. and SOLOMON, S. C. (1978). Simultaneous inversion of surface wave phase velocity and attenuation: Love waves in western North America. *J. geophys. Res.*, **83**, 3389–3400.

LEHMANN, I. (1962). The travel times of the longitudinal waves of the Logan and Blanca atomic explosions and their velocities in the upper mantle. *Bull. seism. Soc. Am.*, **52**, 519–526.

LEHMANN, I. (1964). On the travel times of P as determined from nuclear explosions. *Bull. seism. Soc. Am.*, **54**, 123–139.

LÉVÊQUE, J. J. (1980). Regional upper mantle S-velocity models from phase velocities of great-circle Rayleigh waves. *Geophys. J. R. astr. Soc.*, **63**, 23–43.

LIU, L. (1976a). The post-spinel phase of forsterite. *Nature, Lond.*, **262**, 770–772.

LIU, L. (1976b). The high-pressure phases of $MgSiO_3$. *Earth & planet. Sci. Lett. (Neth.)*, **31**, 200–208.

LIU, L. (1978). A new high-pressure phase of spinel. *Earth & planet. Sci. Lett. (Neth.)*, **41**, 398–404.

LIU, L. (1980). The mineralogy of an eclogitic earth mantle. *Phys. Earth planet. Interiors*, **23**, 262–267.

MACGREGOR, I. D. (1974). The system $MgO-Al_2O_3-SiO_2$: solubility of Al_2O_3 in enstatite for spinel and garnet peridotite compositions. *Am. Miner.*, **59**, 110–119.

MITCHELL, B. J. (1976). Anelasticity of the crust and upper mantle beneath the Pacific Ocean from the inversion of observed surface wave attenuation. *Geophys. J. R. astr. Soc.*, **46**, 521–533.

NIAZI, M. and ANDERSON, D. L. (1965). Upper mantle structure of western North America from apparent velocities of P waves. *J. geophys. Res.*, **70**, 4633–4640.

O'CONNELL, R. J. and BUDIANSKY, B. (1977). Viscoelastic properties of fluid-saturated cracked solids. *J. geophys. Res.*, **82**, 5719–5735.

O'HARA, M. J., SAUNDERS, M. J. and MERCY, E. L. P. (1975). Garnet-peridotite, primary ultrabasic magma and eclogite; interpretation of upper mantle processes in kimberlite. *Physics Chem. Earth*, **9**, 571–604.

OKAL, E. A. (1978). Observed very long period Rayleigh-wave phase velocities across the Canadian shield. *Geophys. J. R. astr. Soc.*, **53**, 663–668.

OKAL, E. A. and ANDERSON, D. L. (1975). A study of lateral inhomogeneities in the upper mantle by multiple ScS travel-time residuals. *Geophys. Res. Lett.*, **2**, 313–316.

OLDENBURG, D. W. (1981). Conductivity structure of oceanic upper mantle beneath the Pacific plate. *Geophys. J. R. astr. Soc.*, **65**, 359–394.

O'NIONS, R. K., EVENSEN, N. M. and HAMILTON, P. J. (1979). Geochemical modeling of mantle differentiation and crustal growth. *J. geophys. Res.*, **84**, 6091–6101.

O'NIONS, R. K., EVENSEN, N. M. and HAMILTON, P. J. (1980). Differentiation and evolution of the mantle. *Phil. Trans. R. Soc.*, **297A**, 479–493.

OXBURGH, E. R. and PARMENTIER, E. M. (1977). Compositional and density stratification in oceanic lithosphere – causes and consequences. *Jl geol. Soc. Lond.*, **133**, 343–355.

OXBURGH, E. R. and PARMENTIER, E. M. (1978). Thermal processes in the formation of continental lithosphere. *Phil. Trans. R. Soc.*, **288A**, 415–429.

PAKISER, L. C. (1963). Structure of the crust and upper mantle in the western United States. *J. geophys. Res.*, **68**, 5747–5756.

PARKER, R. L. (1971). The inverse problem of electrical conductivity in the mantle. *Geophys. J. R. astr. Soc.*, **22**, 121–138.

PARKER, R. L. (1977). Understanding inverse theory. *Ann. Rev. Earth & planet. Sci.*, **5**, 35–64.

PARKINSON, W. D. (1962). The influence of continents and oceans on geomagnetic variations. *Geophys. J. R. astr. Soc.*, **6**, 441–449.

PATTON, H. (1980). Crust and upper mantle structure of the Eurasian continent from the phase velocity and Q of surface waves. *Rev. Geophys. space Phys.*, **18**, 605–625.

PHILLIPS, R. J. and LAMBECK, K. (1980). Gravity fields of the terrestrial planets: long-wavelength anomalies and tectonics. *Rev. Geophys. space Phys.*, **18**, 27–76.

POWELL, R. (1978). The thermodynamics of pyroxene geotherms. *Phil. Trans. R. Soc.*, **288A**, 457–469.

PRESS, F. (1970). Earth models consistent with geophysical data. *Phys. Earth planet. Interiors*, **3**, 3–22.

RINGWOOD, A. E. (1962). A model for the upper mantle, 2. *J. geophys. Res.*, **67**, 4473–4477.

RINGWOOD, A. E. (1975). *Composition and petrology of the Earth's mantle*, McGraw-Hill Book Company, New York, 618 pp.

RINGWOOD, A. E. and MAJOR, A. (1970). The system $Mg_2SiO_4 - Fe_2SiO_4$ at high pressures and temperatures. *Phys. Earth planet. Interiors*, **3**, 89–108.

ROSS, M. (1969). Generalized Lindemann melting law. *Phys. Rev.*, **184**, 233–242.

SAWAMOTO, H. (1977). Orthorhombic perovskite (Mg, Fe)SiO$_3$ and constitution of the lower mantle. In *High-pressure research, application in geophysics*, pp. 219–244, edited by Manghnani, M. H. and Akimoto, S., Academic Press, New York, San Francisco and London.

SENGUPTA, M. K. and JULIAN, B. R. (1978). Radial variation of compressional and shear velocities in the Earth's lower mantle. *Geophys. J. R. astr. Soc.*, **54**, 185–219.

SERSON, P. H. (1973). Instrumentation for induction studies on land. *Phys. Earth planet. Interiors*, **7**, 313–322.

SHANKLAND, T. J. and WAFF, H. S. (1977). Partial melting and electrical conductivity anomalies in the upper mantle. *J. geophys. Res.*, **82**, 5409–5417.

SHAW, G. H. (1978). Interpretation of the low velocity zone in terms of the presence of thermally activated point defects. *Geophys. Res. Lett.*, **5**, 629–632.

SIPKIN, S. A. and JORDAN, T. H. (1980). Multiple ScS travel times in the western Pacific: implications for mantle heterogeneity. *J. geophys. Res.*, **85**, 853–861.

SMITH, M. L. and DAHLEN, F. A. (1981). The period and Q of the Chandler wobble. *Geophys. J. R. astr. Soc.*, **64**, 223–281.

SMITH, S. W. (1966). Free oscillations excited by the Alaskan earthquake. *J. geophys. Res.*, **71**, 1183–1193.

STACEY, F. D. (1977). A thermal model of the Earth. *Phys. Earth planet. Interiors*, **15**, 341–348.

THAYER, R. E., BJORNSSON, A., ALVAREZ, L. and HERMANCE, J. F. (1981). Magma genesis and crustal spreading in the northern neovolcanic zone of Iceland: telluric-magnetotelluric constraints. *Geophys. J. R. astr. Soc.*, **65**, 423–442.

TOKSÖZ, M. N., CHINNERY, M. A. and ANDERSON, D. L. (1967). Inhomogeneities in the Earth's mantle. *Geophys. J. R. astr. Soc.*, **13**, 31–59.

TRUSCOTT, J. R. (1964). The Eskdalemuir seismological station. *Geophys. J. R. astr. Soc.*, **9**, 59–68.

TUCKER, W., HERRIN, E. and FREEDMAN, H. W. (1968). Some statistical aspects of the estimation of seismic travel times. *Bull. seism. Soc. Am.*, **58**, 1243–1260.

USSELMAN, T. M. (1975a). Experimental approach to the state of the core: Part I. The liquidus relations of the Fe-rich portion of the Fe-Ni-S system from 30 to 100 kb. *Am. J. Sci.*, **275**, 278–290.

VERMA, R. K. (1960). Elasticity of some high-density crystals. *J. geophys. Res.*, **65**, 757–766.

WANG, C. (1968). Constitution of the lower mantle as evidenced from shock wave data for some rocks. *J. geophys. Res.*, **73**, 6459–6476.

WIGGINS, R. A. and HELMBERGER, D. V. (1973). Upper mantle structure of the western United States. *J. geophys. Res.*, **78**, 1870–1880.

WOOLLARD, G. P. (1970). Evaluation of the isostatic mechanism and role of mineralogic transformations from seismic and gravity data. *Phys. Earth planet. Interiors*, **3**, 484–498.

WORTHINGTON, M. H., CLEARY, J. R. and ANDERSSEN, R. S. (1972). Density modelling by Monte Carlo inversion – II Comparison of recent earth models. *Geophys. J. R. astr. Soc.*, **29**, 445–457.

WRIGHT, C. and CLEARY, J. R. (1972). *P* wave travel-time gradient measurements for the Warramunga seismic array and lower mantle structure. *Phys. Earth planet. Interiors*, **5**, 213–230.

Chapter five

BARAZANGI, M. and ISACKS, B. (1971). Lateral variations of seismic-wave attenuation in the upper mantle above the inclined earthquake zone of the Tonga island arc: deep anomaly in the upper mantle. *J. geophys. Res.*, **76**, 8493–8516.

BARAZANGI, M. PENNINGTON, W. and ISACKS, B. (1975). Global study of seismic wave attenuation in the upper mantle behind island arcs using *pP* waves. *J. geophys. Res.*, **80**, 1079–1092.

BENIOFF, H. (1955). Seismic evidence for crustal structure and tectonic activity. *Spec. Pap. geol. Soc. Am.*, **62**, 61–74.

BOTT, M. H. P. (1971c). Evolution of young continental margins and formation of shelf basins. *Tectonophysics*, **11**, 319–327.

BOTT, M. H. P. (1979). Subsidence mechanisms at passive continental margins. In *Geological and geophysical investigations of continental margins*, pp. 3–9, edited by Watkins, J. S., Montadert, L. and Dickerson, P. W., AAPG Memoir 29, American Association of Petroleum Geologists, Tulsa, Oklahoma.

BOYNTON, C. H., WESTBROOK, G. K., BOTT, M. H. P. and LONG, R. E. (1979). A seismic refraction investigation of crustal structure beneath the Lesser Antilles island arc. *Geophys. J. R. astr. Soc.*, **58**, 371–393.

BULLARD, E. C., EVERETT, J. E. and SMITH, A. G. (1965). The fit of the continents around the Atlantic. *Phil. Trans. R. Soc.*, **258A**, 41–51.

BURK, C. A. and DRAKE, C. L. (editors) (1974). *The geology of continental margins*, Springer–Verlag, Berlin, Heidelberg and New York, 1009 pp.

BURKE, K. (1976). Development of graben associated with the initial ruptures of the Atlantic Ocean. *Tectonophysics*, **36**, 93–112.

CHAPPLE, W. M. and FORSYTH, D. W. (1979). Earthquakes and bending of plates at trenches. *J. geophys. Res.*, **84**, 6729–6749.

DE CHARPAL, O., GUENNOC, P., MONTADERT, L. and ROBERTS, D. G. (1978). Rifting, crustal attenuation and subsidence in the Bay of Biscay. *Nature, Lond.*, **275**, 706–711.

FALVEY, D. A. (1974). The development of continental margins in plate tectonic theory. *J. Aust. pet. Explor. Assoc.*, **14**, 95–106.

FEATHERSTONE, P. S., BOTT, M. H. P. and PEACOCK, J. H. (1977). Structure of the continental margin of southeastern Greenland. *Geophys. J. R. astr. Soc.*, **48**, 15–27.

FRANK, F. C. (1968). Curvature of island arcs. *Nature, Lond.*, **220**, 363.

GROW, J. A. and BOWIN, C. O. (1975). Evidence for high-density crust and mantle beneath the Chile trench due to the descending lithosphere. *J. geophys. Res.*, **80**, 1449–1458.

GROW, J. A., MATTICK, R. E. and SCHLEE, J. S. (1979). Multichannel seismic depth sections and interval velocities over outer continental shelf and upper continental slope between Cape Hatteras and Cape Cod. In *Geological and geophysical investigations of continental margins*, pp. 65–83, edited by Watkins, J. S., Montadert, L. and Dickerson, P. W., AAPG Memoir 29, American Association of Petroleum Geologists, Tulsa, Oklahoma.

GUTENBERG, B. and RICHTER, C. F. (1954). *Seismicity of the Earth and associated phenomena*, Second edition, Princeton University Press, Princeton, New Jersey, 310 pp.

HASEGAWA, A., UMINO, N. and TAKAGI, A. (1978). Double-planed deep seismic zone and upper-mantle structure in the northeastern Japan arc. *Geophys. J. R. astr. Soc.*, **54**, 281–296.

HAXBY, W. F., TURCOTTE, D.L. and BIRD, J. M. (1976). Thermal and mechanical evolution of the Michigan basin. *Tectonophysics*, **36**, 57–75.

HAYES, D. E. (1966). A geophysical investigation of the Peru-Chile trench. *Marine Geol.*, **4**, 309–351.

ISACKS, B. L. and BARAZANGI, M. (1977). Geometry of Benioff zones: lateral segmentation and downward bending of the subducted lithosphere. In *Island arcs deep sea trenches and back-arc basins*, pp. 99–114, edited by Talwani, M. and Pitman, W. C., III, Maurice Ewing Series 1, American Geophysical Union, Washington, D. C.

ISACKS, B. and MOLNAR, P. (1969). Mantle earthquake mechanisms and the sinking of the lithosphere. *Nature, Lond.*, **223**, 1121–1124.

ISACKS, B., SYKES, L. R. and OLIVER, J. (1969). Focal mechanisms of deep and shallow earthquakes in the Tonga-Kermadec region and the tectonics of island arcs. *Bull. geol. Soc. Am.*, **80**, 1443–1469.

KARIG, D. E. (1971). Origin and development of marginal basins in the western Pacific. *J. geophys. Res.*, **76**, 2542–2561.

KEAREY, P. (1974). Gravity and seismic reflection investigations into the crustal structure of the Aves ridge, eastern Caribbean. *Geophys. J. R. astr. Soc.*, **38**, 435–448.

KENT, P., LAUGHTON, A. S., ROBERTS, D. G. and JONES, E. W. J. (editors) (1980). The evolution of passive continental margins in the light of recent deep drilling results. *Phil. Trans. R. Soc.*, **294A**, 1–208.

KINSMAN, D. J. J. (1975). Rift valley basins and sedimentary history of trailing continental margins. In *Petroleum and global tectonics*, pp. 83–126, edited by Fischer, A. G. and Judson, S., Princeton University Press, Princeton, New Jersey.

McKENZIE, D. (1978). Some remarks on the development of sedimentary basins. *Earth & planet. Sci. Lett. (Neth.)*, **40**, 25–32.

MONTADERT, L., ROBERTS, D. G. and others (1979). *Initial reports of the Deep Sea Drilling Project*, volume 48, Washington (U.S. Government Printing Office), 1183 pp.

NICHOLS, I. A. and RINGWOOD, A. E. (1973). Effect of water on olivine stability in tholeiites and the production of silica-saturated magmas in the island-arc environment. *J. Geol.*, **81**, 285–300.

OLIVER, J. and ISACKS, B. (1967). Deep earthquake zones, anomalous structures in the upper mantle, and the lithosphere. *J. geophys. Res.*, **72**, 4259–4275.

RINGWOOD, A. E. (1977). Petrogenesis in island arc systems. In *Island arcs deep sea trenches and back-arc basins*, pp. 311–324, edited by Talwani, M. and Pitman, W. C., III, Maurice Ewing Series 1, American Geophysical Union, Washington, D. C.

ROBERTS, D. G. (1975). Marine geology of the Rockall plateau and trough. *Phil. Trans. R. Soc.*, **278A**, 447–509.

ROBERTS, D. G., MONTADERT, L. and SEARLE, R. C. (1979). The western Rockall plateau: stratigraphy and structural evolution. In Montadert, L, Roberts, D. G. and others, *Initial reports of the Deep Sea Drilling Project*, volume 48, pp. 1061–1088, Washington (U.S. Government Printing Office).

SCHUBERT, G., YUEN, D. A. and TURCOTTE, D. L. (1975). Role of phase transitions in a dynamic mantle. *Geophys. J. R. astr. Soc.*, **42**, 705–735.

SCRUTTON, R. A. (1972). The crustal structure of Rockall plateau microcontinent. *Geophys. J. R. astr. Soc.*, **27**, 259–275.

SCRUTTON, R. A. (1976). Continental breakup and deep crustal structure at the margins of southern Africa. In *Continental margins of Atlantic type*, pp. 275–286, edited by de Almeida, F. F. M., *An. Acad. bras. Ciênc.*, **48** (Suplemento).

SEELY, D. R. (1979). The evolution of structural highs bordering major forearc basins. In *Geological and geophysical investigations of continental margins*, pp. 245–260, edited by Watkins, J. S., Montadert, L. and Dickerson, P. W., AAPG Memoir 29, American Association of Petroleum Geologists, Tulsa, Oklahoma.

SLEEP, N. H. (1971). Thermal effects of the formation of Atlantic continental margins by continental break up. *Geophys. J. R. astr. Soc.*, **24**, 325–350.

SLEEP, N. H. (1973). Crustal thinning on Atlantic continental margins: evidence from older margins. In *Implications of continental drift to the earth sciences*, volume 2, pp. 685–692, edited by Tarling, D. H. and Runcorn, S. K., Academic Press, London and New York.

SLEEP, N. H. and SNELL, N. S. (1976). Thermal contraction and flexure of mid-continent and Atlantic marginal basins. *Geophys. J. R. astr. Soc.*, **45**, 125–154.

STAUDER, W. (1968). Tensional character of earthquake foci beneath the Aleutian trench with relation to sea-floor spreading. *J. geophys. Res.*, **73**, 7693–7701.

SYKES, L. R. (1966). The seismicity and deep structure of island arcs. *J. geophys. Res.*, **71**, 2981–3006.

TALWANI, M. and PITMAN, W. C., III (editors) (1977). *Island arcs deep sea trenches and back-arc basins*, Maurice Ewing Series 1, American Geophysical Union, Washington, D. C., 470 pp.

TÖKSOZ, M. N., MINEAR, J. W. and JULIAN, B. R. (1971). Temperature field and geophysical effects of a downgoing slab. *J. geophys. Res.*, **76**, 1113–1138.

TURCOTTE, D. L., McADOO, D. C. and CALDWELL, J. G. (1978). An elastic-perfectly plastic analysis of the bending of the lithosphere at a trench. *Tectonophysics*, **47**, 193–205.

TURCOTTE, D. L. and SCHUBERT, G. (1973). Frictional heating of the descending lithosphere. *J. geophys. Res.*, **78**, 5876–5886.

UYEDA, S. and KANAMORI, H. (1979). Back-arc opening and the mode of subduction. *J. geophys. Res.*, **84**, 1049–1061.

VOGT, P. R. and AVERY, O. E. (1974). Detailed magnetic surveys in the northeast Atlantic and Labrador Sea. *J. geophys. Res.*, **79**, 363–389.

WALCOTT, R. I. (1972). Gravity, flexure, and the growth of sedimentary basins at a continental edge. *Bull. geol. Soc. Am.*, **83**, 1845–1848.

WATKINS, J. S., MONTADERT, L. and DICKERSON, P. W. (editors) (1979). *Geological and geophysical investigations of continental margins*, AAPG Memoir 29, American Association of Petroleum Geologists, Tulsa, Oklahoma, 472 pp.

WATTS, A. B. and RYAN, W. B. F. (1976). Flexure of the lithosphere and continental margin basins. *Tectonophysics*, **36**, 25–44.

WESTBROOK, G. K. (1975). The structure of the crust and upper mantle in the region of Barbados and the Lesser Antilles. *Geophys. J. R. astr. Soc.*, **43**, 201–242.

WESTBROOK, G. K. (1982). The Barbados Ridge complex: tectonics of a mature fore-arc system, In *Trench and fore-arc sedimentation and tectonics in modern and ancient subduction zones*, edited by Leggett, J. K., *Spec. Pub. geol. Soc. Lond.* (in the press).

WILSON, J. T. (1965). A new class of faults and their bearing on continental drift. *Nature, Lond.*, **207**, 343–347.

WOODWARD, D. J. (1977). Stresses due to phase changes in subduction zones and an empirical equation of state for the mantle. *Geophys. J. R. astr. Soc.*, **50**, 459–472.

YUEN, D. A., FLEITOUT, L., SCHUBERT, G. and FROIDEVAUX, C. (1978). Shear deformation zones along major transform faults and subducting slabs. *Geophys. J. R. astr. Soc.*, **54**, 93–119.

Chapter six

ALSOP, L. E. and KUO, J. T. (1964). The characteristic numbers of the semi-diurnal earth tidal components for various earth models. *Annls Géophys.*, **20**, 286–300.

ANDERSON, D. L. and HART, R. S. (1978). *Q* of the Earth. *J. geophys. Res.*, **83**, 5869–5882.

BACKUS, G. (1958). A class of self-sustaining dissipative spherical dynamos *Annls Phys.*, **4**, 372–447.

BIRCH, F. (1961*b*). Composition of the Earth's mantle. *Geophys. J. R. astr. Soc.*, **4**, 295–311.

BIRCH, F. (1972). The melting relations of iron, and temperatures in the Earth's core. *Geophys. J. R. astr. Soc.*, **29**, 373–387.

BLACKETT, P. M. S. (1947). The magnetic field of massive rotating bodies. *Nature, Lond.*, **159**, 658–666.

BOLT, B. A. (1962). Gutenberg's early *PKP* observations. *Nature, Lond.*, **196**, 122–124.

BOLT, B. A. (1964). The velocity of seismic waves near the Earth's center. *Bull. seism. Soc. Am.*, **54**, 191–208.

BOSCHI, E., MULARGIA, F. and BONAFEDE, M. (1979). The dependence of the melting temperatures of iron upon the choice of the interatomic potential. *Geophys. J. R. astr. Soc.*, **58**, 201–208.

BRETT, R. (1976). The current status of speculations on the composition of the core of the Earth. *Rev. Geophys. space Phys.*, **14**, 375–383.

BUCHBINDER, G. G. R. (1971). A velocity structure of the Earth's core. *Bull. seism. Soc. Am.*, **61**, 429–456.

BUCHBINDER, G. G. R. (1972). Travel times and velocities in the outer core from P_mKP. *Earth & planet. Sci. Lett. (Neth.)*, **14**, 161–168.

BULLARD, E. C., FREEDMAN, C., GELLMAN, H. and NIXON, J. (1950). The westward drift of the Earth's magnetic field. *Phil. Trans. R. Soc.*, **243A**, 67–92.

BULLARD, E. C. and GELLMAN, H. (1954). Homogeneous dynamos and terrestrial magnetism. *Phil. Trans. R. Soc.*, **247A**, 213–278.

BULLEN, K. E. (1963). *An introduction to the theory of seismology*, Third edition, Cambridge University Press, London and New York, 381 pp.

BUSSE, F. H. (1975). A model of the geodynamo. *Geophys. J. R. astr. Soc.*, **42**, 437–459.

CHANG, A. C. and CLEARY, J. R. (1978). Precursors to *PKKP*. *Bull. seism. Soc. Am.*, **68**, 1059–1079.

CHAPMAN, S. and BARTELS, J. (1940). *Geomagnetism*, volumes I and II, Oxford University Press, London and New York, 1049 pp.

CLARK, S. P., Jr. (editor) (1966). *Handbook of physical constants*, Revised edition, *Mem. geol. Soc. Am.*, **97**, 587 pp.

CLEARY, J. R. and HADDON, R. A. W. (1972). Seismic wave scattering near the core-mantle boundary: a new interpretation of precursors to PKP. *Nature, Lond.*, **240**, 549–551.

COWLING, T. G. (1934). The magnetic field of sunspots. *Mon. Not. R. astr. Soc.*, **94**, 39–48.

COX, A. (1975). The frequency of geomagnetic reversals and the symmetry of the nondipole field. *Rev. Geophys. space Phys.*, **13**, No. 3 (special issue), 35–51.

COX, A. and DOELL, R. R. (1960). Review of palaeomagnetism. *Bull. geol. Soc. Am.*, **71**, 645–768.

DOORNBOS, D. J. (1978). On seismic-wave scattering by a rough core-mantle boundary. *Geophys. J. R. astr. Soc.*, **53**, 643–662.

DOORNBOS, D. J. and MONDT, J. C. (1979*a*). Attenuation of *P* and *S* waves diffracted around the core. *Geophys. J. R. astr. Soc.*, **57**, 353–379.

DOORNBOS, D. J. and MONDT, J. C. (1979*b*). *P* and *S* waves diffracted around the core and the velocity structure at the base of the mantle. *Geophys. J. R. astr. Soc.*, **57**, 381–395.

DUBROVSKIY, V. A. and PAN'KOV V. L. (1972). On the composition of the Earth's core. *Izv. Phys. solid Earth*, Engl. Trans., **1972**, 452–455.

DZIEWONSKI, A. M. and GILBERT, F. (1971). Solidity of the inner core of the Earth inferred from normal mode observations. *Nature, Lond.*, **234**, 465–466.

DZIEWONSKI, A. M. and HADDON, R. A. W. (1974). The radius of the core-mantle boundary inferred from travel time and free oscillation data; a critical review. *Phys. Earth planet. Interiors*, **9**, 28–35.

ENGDAHL, E. R. FLINN, E. A. and MASSÉ, R. P. (1974). Differential *PKiKP* travel times and the radius of the inner core. *Geophys. J. R. astr. Soc.*, **39**, 457–463.

ENGDAHL, E. R., FLINN, E. A. and ROMNEY, C. F. (1970). Seismic waves reflected from the Earth's inner core. *Nature, Lond.*, **228**, 852–853.

FULLER, M., WILLIAMS, I. and HOFFMAN, K. A. (1979). Palaeomagnetic records of geomagnetic field reversals and the morphology of the transitional fields. *Rev. Geophys. space Phys.*, **17**, 179–203.

GOETTEL, K. A. (1976). Models for the origin and composition of the Earth, and the hypothesis of potassium in the Earth's core. *Geophys. Surv.*, **2**, 369–397.

GROSSMAN, L. (1972). Condensation in the primitive solar nebula. *Geochim. cosmochim. Acta*, **36**, 597–619.

GUBBINS, D. (1974). Theories of the geomagnetic and solar dynamos. *Rev. Geophys. space Phys.*, **12**, 137–154.

GUBBINS, D. (1976). Observational constraints on the generation process of the Earth's magnetic field. *Geophys. J. R. astr. Soc.*, **47**, 19–39.

GUTENBERG, B. (1959). *Physics of the Earth's interior*, Academic Press, New York and London, 240 pp.

HALES, A. L. and ROBERTS, J. L. (1971). The velocities in the outer core. *Bull. seism. Soc. Am.*, **61**, 1051–1059.

HANKS, T. C. and ANDERSON, D. L. (1969). The early thermal history of the Earth. *Phys. Earth planet. Interiors*, **2**, 19–29.

HART, R. S., ANDERSON, D. L. and KANAMORI, H. (1977). The effect of attenuation on gross earth models. *J. geophys. Res.*, **82**, 1647–1654.

HERZENBERG, A. (1958). Geomagnetic dynamos. *Phil. Trans. R. Soc.*, **250A**, 543–583.

HIDE, R. (1966). Free hydromagnetic oscillations of the Earth's core and the theory of the geomagnetic secular variation. *Phil. Trans. R. Soc.*, **259A**, 615–647.

HIDE, R. (1970). On the Earth's core-mantle interface. *Q. Jl R. met. Soc.*, **96**, 579–590.

HIDE, R. and MALIN, S. R. C. (1970). Novel correlations between global features of the Earth's gravitational and magnetic fields. *Nature, Lond.*, **225**, 605–609.

HOSPERS, J. (1951). Remanent magnetism of rocks and the history of the geomagnetic field. *Nature, Lond.*, **168**, 1111–1112.

JACOBS, J. A. (1953). The Earth's inner core. *Nature, Lond.*, **172**, 297–298.

JACOBS, J. A. (1975). *The Earth's core*, Academic Press, London, New York and San Francisco, 253 pp.

JEFFREYS, H. (1939a). The times of *P*, *S* and *SKS*, and the velocities of *P* and *S*. *Mon. Not. R. astr. Soc. geophys. Suppl.*, **4**, 498–533.

JEFFREYS, H. (1939b). The times of *PcP* and *ScS*. *Mon. Not. R. astr. Soc. geophys. Suppl.*, **4**, 537–547.

JULIAN, B. R., DAVIES, D. and SHEPPARD, R. M. (1972). PKJKP. *Nature, Lond.*, **235**, 317–318.

KENNEDY, G. C. and HIGGINS, G. H. (1973). The core paradox. *J. geophys. Res.*, **78**, 900–904.

KIND, R. and MÜLLER, G. (1977). The structure of the outer core from SKS amplitudes and travel times. *Bull. seism. Soc. Am.*, **67**, 1541–1554.

KUHN, W. and RITTMANN, A. (1941). Über den Zustand des Erdinnern und seine Entstehung aus einem homogenen Urzustand. *Geol. Rdsch.*, **32**, 215–256.

LARMOR, J. (1920). How could a rotating body such as the Sun become a magnet? *Rep. Br. Ass. Advmt Sci.* (for 1919), 159–160.

LEHMANN, I. (1936). *P′*. *Bur. Centr. seism. Internat. A*, **14**, 3–31.

LEVY, E. H. (1972a). Effectiveness of cyclonic convection for producing the geomagnetic field. *Astrophys. J.*, **171**, 621–633.

LEVY, E. H. (1972b). Kinematic reversal schemes for the geomagnetic dipole. *Astrophys. J.*, **171**, 635–642.

LEVY, E. H. (1976). Generation of planetary magnetic fields. *Ann. Rev. Earth & planet. Sci.*, **4**, 159–185.

LEVY, E. H. (1979). Dynamo magnetic field generation. *Rev. Geophys. space Phys.*, **17**, 277–281.

LEWIS, J. S. (1971). Consequences of the presence of sulfur in the core of the Earth. *Earth & planet. Sci. Lett. (Neth.)*, **11**, 130–134.

LIU, L. (1975). On the (γ, ε, l) triple point of iron and the Earth's core. *Geophys. J. R. astr. Soc.*, **43**, 697–705.

LOPER, D. E. (1978). The gravitationally powered dynamo. *Geophys. J. R. astr. Soc.*, **54**, 389–404.

LOWES, F. J. and RUNCORN, S. K. (1951). The analysis of the geomagnetic secular variation. *Phil. Trans. R. Soc.*, **243A**, 525–546.

LOWES, F. J. and WILKINSON, I. (1963). Geomagnetic dynamo: a laboratory model. *Nature, Lond.*, **198**, 1158–1160.

MALIN, S. R. C. and BULLARD, E. (1981). The direction of the Earth's magnetic field at London, 1570–1975. *Phil. Trans. R. Soc.*, **299A**, 357–423.

MALKUS, W. V. R. (1963). Precessional torques as the cause of geomagnetism. *J. geophys. Res.*, **68**, 2871–2886.

McELHINNY, M. W. and MERRILL, R. T. (1975). Geomagnetic secular variation over the past 5 m.y. *Rev. Geophys. space Phys.*, **13**, 687–708.

McQUEEN, R. G., FRITZ, J. N. and MARSH, S. P. (1964). On the composition of the Earth's interior *J. geophys. Res.*, **69**, 2947–2965.

MELCHIOR, P. (1966). *The Earth tides*, Pergamon Press, Oxford, London, Edinburgh, New York, Paris and Frankfurt, 458 pp.

MELCHIOR, P. (1978). *The tides of the planet Earth*, Pergamon Press, Oxford, New York, Toronto, Sydney, Paris and Frankfurt, 609 pp.

MORRISON, L. V. (1979). Re-determination of the decade fluctuations in the rotation of the Earth in the period 1861–1978. *Geophys. J. R. astr. Soc.*, **58**, 349–360.

MÜLLER, G. (1973). Amplitude studies of core phases. *J. geophys. Res.*, **78**, 3469–3490.

MÜLLER, G., MULA, A. H. and GREGERSEN, S. (1977). Amplitudes of long-period PcP and the core-mantle boundary. *Phys. Earth planet. Interiors*, **14**, 30–40.

MURTHY, V. R. (1976). Composition of the core and the early chemical history of the Earth. In *The early history of the Earth*, pp. 21–31, edited by Windley, B. F., John Wiley & Sons, London, New York, Sydney and Toronto.

MURTHY, V. R. and HALL, H. T. (1970). The chemical composition of the Earth's core: possibility of sulphur in the core. *Phys. Earth planet. Interiors*, **2**, 276–282.

MURTHY, V. R. and HALL, H. T. (1972). The origin and chemical composition of the Earth's core. *Phys. Earth planet. Interiors*, **6**, 123–130.

NAGATA, T. (1953). Self-reversal of thermo-remanent magnetization of igneous rocks. *Nature, Lond.*, **172**, 850–852.

NÉEL, L. (1951). L'inversion de l'aimantation permanente des roches. *Annls Géophys.*, **7**, 90–102.

OBERG, C. J. and EVANS, M. E. (1977). Spectral analysis of Quaternary palaeomagnetic data from British Columbia and its bearing on geomagnetic secular variation. *Geophys. J. R. astr. Soc.*, **51**, 691–699.

OLDHAM, R. D. (1906). The constitution of the interior of the Earth, as revealed by earthquakes. *Q. Jl geol. Soc. Lond.*, **62**, 456–475.

OROWAN, E. (1969). Density of the Moon and nucleation of planets. *Nature, Lond.*, **222**, 867.

OVERSBY, V. M. and RINGWOOD, A. E. (1971). Time of formation of the Earth's core. *Nature, Lond.*, **234**, 463–465.

OVERSBY, V. M. and RINGWOOD, A. E. (1972). Potassium distribution between metal and silicate and its bearing on the occurrence of potassium in the Earth's core. *Earth & planet. Sci. Lett. (Neth.)*, **14**, 345–347.

PARKER, E. N. (1955). Hydromagnetic dynamo models. *Astrophys. J.*, **122**, 293–314.

PARKER, E. N. (1969). The occasional reversal of the geomagnetic field. *Astrophys. J.*, **158**, 815–827.

QAMAR, A. (1973). Revised velocities in the Earth's core. *Bull. seism. Soc. Am.*, **63**, 1073–1105.

QAMAR, A. and EISENBERG, A. (1974). The damping of core waves. *J. geophys. Res.*, **79**, 758–765.

RAMSEY, W. H. (1949). On the nature of the Earth's core. *Mon. Not. R. astr. Soc. geophys. Suppl.*, **5**, 409–426.

RINGWOOD, A. E. (1966). Chemical evolution of the terrestrial planets. *Geochim. cosmochim. Acta*, **30**, 41–104.

RINGWOOD, A. E. (1975). *Composition and petrology of the Earth's mantle*, McGraw-Hill Book Company, New York, 618 pp.

RINGWOOD, A. E. (1977). Composition of the core and implications for origin of the Earth. *Geochem. J.*, **11**, 111–135.

RINGWOOD, A. E. (1979). Composition and origin of the Earth. In *The Earth: its origin, structure and evolution*, pp. 1–58, edited by McElhinny, M. W., Academic Press, London, New York and San Francisco.

ROBERTS, G. O. (1972). Dynamo action of fluid motions with two-dimensional periodicity. *Phil. Trans. R. Soc.*, **271A**, 411–454.

ROCHESTER, M. G., JACOBS, J. A., SMYLIE, D. E. and CHONG, K. F. (1975). Can precession power the geomagnetic dynamo? *Geophys. J. R. astr. Soc.*, **43**, 661–678.

RUNCORN, S. K., BENSON, A. C., MOORE, A. F. and GRIFFITHS, D. H. (1951). Measurements of the variation with depth of the main geomagnetic field. *Phil. Trans. R. Soc.*, **244A**, 113–151.

STACEY, F. D. (1977). A thermal model of the Earth. *Phys. Earth planet. Interiors*, **15**, 341–348.

TAKEUCHI, H. (1950). On the Earth tide of the compressible Earth of variable density and elasticity. *Trans. Am. geophys. Un.*, **31**, 651–689.

USSELMAN, T. M. (1975a). Experimental approach to the state of the core: Part I. The liquidus relations of the Fe-rich portion of the Fe-Ni-S system for 30 to 100 kb. *Am. J. Sci.*, **275**, 278–290.

USSELMAN, T. M. (1975b). Experimental approach to the state of the core: Part II. Composition and thermal regime. *Am. J. Sci.*, **275**, 291–303.

VESTINE, E. H. (1962). Influence of the Earth's core upon the rate of the Earth's rotation. In *Benedum Earth magnetism symposium*, pp. 57–67, edited by Nagata, T., University of Pittsburgh Press.

VESTINE, E. H., LAPORTE, L., COOPER, C., LANGE, I. and HENDRIX, W. C. (1947). *Description of the Earth's main magnetic field and its secular change, 1905–1945*, Carnegie Institution of Washington Publication 578, 532 pp.

Chapter seven

ANDERS, E. (1977). Chemical compositions of the Moon, Earth and eucrite parent body. *Phil. Trans. R. Soc.*, **285A**, 23–40.

ANDERSON, O. L. (1965). Lattice dynamics in geophysics. *Trans. N.Y. Acad. Sci.*, ser. II, **27**, 298–308.

ANDERSON, R. N., LANGSETH, M.G. and SCLATER, J. G. (1977). The mechanisms of heat transfer through the floor of the Indian Ocean. *J. geophys. Res.*, **82**, 3391–3409.

BECK, A. E. (1965). Techniques of measuring heat flow on land. In *Terrestrial heat flow*, pp. 24–57, edited by Lee, W. H. K., *Geophys. Monogr.* No. 8, American Geophysical Union, Washington, D. C.

BIRCH, F. and CLARK, H. (1940). The thermal conductivity of rocks and its dependence upon temperature and composition. *Am. J. Sci.*, **238**, 529–558 and 613–635.

BIRCH, F., ROY, R. F. and DECKER, E. R. (1968). Heat flow and thermal history in New England and New York. In *Studies of Appalachian geology; northern and maritime*, pp. 437–451, edited by Zen, E., White, W. S., Hadley, J. B. and Thompson, J. B., Jr., Interscience, New York.

BLACKWELL, D. D. (1971). The thermal structure of the continental crust. In *The structure and physical properties of the Earth's crust*, pp. 169–184, edited by Heacock, J. G., *Geophys. Monogr.* No. 14, American Geophysical Union, Washington, D. C.

BOTT, M. H. P., JOHNSON, G. A. L., MANSFIELD, J. and WHEILDON, J. (1972). Terrestrial heat flow in north-east England. *Geophys. J. R. astr. Soc.*, **27**, 277–288.

BURKE, K. and KIDD, W. S. F. (1978). Were Archean continental geothermal gradients much steeper than those of today? *Nature, Lond.*, **272**, 240–241.

CHAPMAN, D. S. and POLLACK, H. N. (1975). Global heat flow: a new look. *Earth & planet. Sci. Lett. (Neth.)*, **28**, 23–32.

CLARK, S. P., Jr. (1957). Absorption spectra of some silicates in the visible and near infrared. *Am. Miner.*, **42**, 732–742.

CROUGH, S. T. and THOMPSON, G. A. (1976). Thermal model of continental lithosphere. *J. geophys. Res.*, **81**, 4857–4862.

DAVIES, G. F. (1980a). Review of oceanic and global heat flow estimates. *Rev. Geophys. space Phys.*, **18**, 718–722.

DAVIES, G. F. (1980b). Thermal histories of convective earth models and constraints on radiogenic heat production in the Earth. *J. geophys. Res.*, **85**, 2517–2530.

DAVIS, E. E. and LISTER, C. R. B. (1974). Fundamentals of ridge crest topography. *Earth & planet. Sci. Lett. (Neth.)*, **21**, 405–413.

DAVIS, E. E. and LISTER, C. R. B. (1977). Heat flow measured over the Juan de Fuca ridge: evidence for wide-spread hydrothermal circulation in a highly heat transportive crust. *J. geophys. Res.*, **82**, 4845–4860.

ELDER, J. W. (1965). Physical processes in geothermal areas. In *Terrestrial heat flow*, pp. 211–239, edited by Lee, W. H. K., *Geophys. Monogr.* No. 8, American Geophysical Union, Washington, D. C.

ENGLAND, P. C., OXBURGH, E. R. and RICHARDSON, S. W. (1980). Heat refraction and heat production in and around granite plutons in north-east England. *Geophys. J. R. astr. Soc.*, **62**, 439–455.

ENGLAND, P. C. and RICHARDSON, S. W. (1980). Erosion and the age dependence of continental heat flow. *Geophys. J. R. astr. Soc.*, **62**, 421–437.

FLASAR, F. M. and BIRCH, F. (1973). Energetics of core formation: a correction. *J. geophys. Res.*, **78**, 6101–6103.

GOETTEL, K. A. (1976). Models for the origin and composition of the Earth, and the hypothesis of potassium in the Earth's core. *Geophys. Surv.*, **2**, 369–397.

HANKS, T. C. and ANDERSON, D. L. (1969). The early thermal history of the Earth. *Phys. Earth planet. Interiors*, **2**, 19–29.

HEIER, K. S. and ROGERS, J. J. W. (1963). Radiometric determination of thorium, uranium and potassium in basalts and in two magmatic differentiation series. *Geochim. cosmochim. Acta*, **27**, 137–154.

JACOBS, J. A. (1975). *The Earth's core*, Academic Press, London, New York and San Francisco, 253 pp.

JESSOP, A. M., HOBART, M. A. and SCLATER, J. G. (1976). The world heat flow data collection – 1975. *Geothermal Series Number 5*, Earth Physics Branch, Ottawa, 10 pp.

KANAMORI, H., FUJII, N. and MIZUTANI, H. (1968). Thermal diffusivity measurement of rock-forming minerals from 300° to 1100° K. *J. geophys. Res.*, **73**, 595–605.

KIEFFER, S. W. (1976). Lattice thermal conductivity within the Earth and considerations of a relationship between the pressure dependence of the thermal diffusivity and the volume dependence of the Grüneisen parameter. *J. geophys. Res.*, **81**, 3025–3030.

LACHENBRUCH, A. H. (1970). Crustal temperature and heat production: implications of the linear heat-flow relation. *J. geophys. Res.*, **75**, 3291–3300.

LANGSETH, M. G. (1965). Techniques of measuring heat flow through the ocean floor. In *Terrestrial heat flow*, pp. 58–77, edited by Lee, W. H. K., *Geophys. Monogr.* No. 8, American Geophysical Union, Washington, D. C.

LARIMER, J. W. (1971). Composition of the Earth: chondritic or achondritic? *Geochim. cosmochim. Acta*, **35**, 769–786.

LEE, T., PAPANASTASSIOU, D. A. and WASSERBURG, G. J. (1977). Aluminium-26 in the early solar system: fossil or fuel? *Astrophys. J.*, **211**, L107–L110.

LEE, W. H. K. (1970). On the global variations of terrestrial heat-flow. *Phys. Earth planet. Interiors*, **2**, 332–341.

LEE, W. H. K. and UYEDA, S. (1965). Review of heat flow data. In *Terrestrial heat flow*, pp. 87–190, edited by Lee, W. H. K., *Geophys. Monogr.* No. 8, American Geophysical Union, Washington, D. C.

LEEDS, A. R. (1975). Lithospheric thickness in the western Pacific. *Phys. Earth planet. Interiors*, **11**, 61–64.

LEEDS, A. R., KNOPOFF, L. and KAUSEL, E. G. (1974). Variations of upper mantle structure under the Pacific Ocean. *Science, N. Y.*, **186**, 141–143.

LYSAK, S. V. (1978). The Baikal rift heat flow. *Tectonophysics*, **45**, 87–93.

MACDONALD, G. J. F. (1964). Tidal friction. *Rev. Geophys.*, **2**, 467–541.

MACDONALD, G. J. F. (1965). Geophysical deductions from observations of heat flow. In *Terrestrial heat flow*, pp. 191–210, edited by Lee, W. H. K., *Geophys. Monogr.* No. 8, American Geophysical Union, Washington, D. C.

McKENZIE, D. P. (1967). Some remarks on heat flow and gravity anomalies. *J. geophys. Res.*, **72**, 6261–6271.

McKENZIE, D. and WEISS, N. (1975). Speculations on the thermal and tectonic history of the Earth. *Geophys. J. R. astr. Soc.*, **42**, 131–174.

OLDENBURG, D. W. (1975). A physical model for the creation of the lithosphere. *Geophys. J. R. astr. Soc.*, **43**, 425–451.

PARKER, R. L. and OLDENBURG, D. W. (1973). Thermal model of ocean ridges. *Nature Phys. Sci., Lond.*, **242**, 137–139.

PARSONS, B. and SCLATER, J. G. (1977). An analysis of the variation of ocean floor bathymetry and heat flow with age. *J. geophys. Res.*, **82**, 803–827.

POLLACK, H. N. and CHAPMAN, D. S. (1977a). On the regional variation of heat flow, geotherms, and lithospheric thickness. *Tectonophysics*, **38**, 279–296.

POLLACK, H. N. and CHAPMAN, D. S. (1977b). The flow of heat from the Earth's interior. *Scient. Am.*, **237**, No. 2, 60–76.

POLYAK, B. G. and SMIRNOV, Y. B. (1968). Relationship between terrestrial heat flow and the tectonics of continents. *Geotectonics*, Engl. Trans., **4**, 205–213.

RICHARDSON, S. W. and OXBURGH, E. R. (1978). Heat flow, radiogenic heat production and crustal temperatures in England and Wales. *Jl geol. Soc. Lond.*, **135**, 323–337.

RICHTER, F. M. and PARSONS, B. (1975). On the interaction of two scales of convection in the mantle. *J. geophys. Res.*, **80**, 2529–2541.

ROY, R. F., BLACKWELL, D. D. and BIRCH, F. (1968). Heat generation of plutonic rocks and continental heat flow provinces. *Earth & planet. Sci. Lett. (Neth.)*, **5**, 1–12.

RUNCORN, S. K., LIBBY, L. M. and LIBBY, W. F. (1977). Primaeval melting of the Moon. *Nature, Lond.*, **270**, 676–681.

SCHATZ, J. F. and SIMMONS, G. (1972). Thermal conductivity of earth materials at high temperatures. *J. geophys. Res.*, **77**, 6966–6983.

SCHUBERT, G., FROIDEVAUX, C. and YUEN, D. A. (1976). Oceanic lithosphere and asthenosphere: thermal and mechanical structure. *J. geophys. Res.*, **81**, 3525–3540.

SCHUBERT, G., STEVENSON, D. and CASSEN, P. (1980). Whole planet cooling and the radiogenic heat source contents of the Earth and Moon. *J. geophys. Res.*, **85**, 2531–2538.

SCLATER, J. G. (1972). Heat flow and elevation of the marginal basins of the western Pacific. *J. geophys. Res.*, **77**, 5705–5719.

SCLATER, J. G., CROWE, J. and ANDERSON, R. N. (1976). On the reliability of oceanic heat flow averages. *J. geophys. Res.*, **81**, 2997–3006.

SCLATER, J. G. and FRANCHETEAU, J. (1970). The implications of terrestrial heat flow observations on current tectonic and geochemical models of the crust and upper mantle of the Earth. *Geophys. J. R. astr. Soc.*, **20**, 509–542.

SCLATER, J. G., JAUPART, C. and GALSON, D. (1980). The heat flow through oceanic and continental crust and the heat loss of the Earth. *Rev. Geophys. space Phys.*, **18**, 269–311.

SINGER, C. E. (1978). Supernovae and lunar melting. *Nature, Lond.*, **272**, 239.

SMITHSON, S. B. and DECKER, E. R. (1974). A continental crustal model and its geothermal implications. *Earth & planet. Sci. Lett. (Neth.)*, **22**, 215–225.

STACEY, F. D. (1977). A thermal model of the Earth. *Phys. Earth planet. Interiors*, **15**, 341–348.

TOZER, D. C. (1965). Thermal history of the Earth I. The formation of the core. *Geophys. J. R. astr. Soc.*, **9**, 95–112.

TOZER, D. C. (1972). The present thermal state of the terrestrial planets. *Phys. Earth planet. Interiors*, **6**, 182–197.

UREY, H. C. (1955). The cosmic abundances of potassium, uranium, and the heat balance of the Earth, the Moon and Mars. *Proc. natn. Acad. Sci. U.S.A.*, **41**, 127–144.

VITORELLO, I. and POLLACK, H. N. (1980). On the variation of continental heat flow with age and the thermal evolution of continents. *J. geophys. Res.*, **85**, 983–995.

WILLIAMS, D. L., VON HERZEN, R. P., SCLATER, J. G. and ANDERSON, R. N. (1974). The Galapagos spreading centre: lithospheric cooling and hydrothermal circulation. *Geophys. J. R. astr. Soc.*, **38**, 587–608.

Chapter eight

ANDERSON, E. M. (1951). *The dynamics of faulting and dyke formation with applications to Britain*, Second edition, Oliver and Boyd, Edinburgh and London, 206 pp.

ARTYUSHKOV, E. V. (1973). Stresses in the lithosphere caused by crustal thickness inhomogeneities. *J. geophys. Res.*, **78**, 7675–7708.

ASHBY, M. F. and VERRALL, R. A. (1978). Micromechanisms of flow and fracture, and their relevance to the rheology of the upper mantle. *Phil. Trans. R. Soc.*, **200A**, 59–95.

BÅTH, M. (1966). Earthquake energy and magnitude. *Physics Chem. Earth*, **7**, 115–165.

BENIOFF, H. (1964) Earthquake source mechanisms. *Science, N. Y.*, **143**, 1399–1406.

BOTT, M. H. P. (1959). The mechanics of oblique slip faulting. *Geol. Mag.*, **96**, 109–117.

BOTT, M. H. P. (1965). The deep structure of the northern Irish Sea – a problem of crustal dynamics. In *Submarine geology and geophysics*, pp. 179–204, edited by Whittard, W. F. and Bradshaw, R., Colston Papers No. 17, Butterworths, London.

BOTT. M. H. P. (1971a). *The interior of the Earth*, Edward Arnold, London, 316 pp.

BOTT, M. H. P. and KUSZNIR, N. J. (1979). Stress distributions associated with compensated plateau uplift structures with application to the continental splitting mechanism. *Geophys. J. R. astr. Soc.*, **56**, 451–459.

BRUNE, J. N. (1968). Seismic moment, seismicity, and rate of slip along major fault zones. *J. geophys. Res.*, **73**, 777–784.

BRUNE, J. N. (1970). Tectonic stress and the spectra of seismic shear waves from earthquakes. *J. geophys. Res.*, **75**, 4997–5009.

BYERLY, P. (1926). The Montana earthquake of June 28, 1925, G.M.C.T. *Bull. seism. Soc. Am.*, **16**, 209–265.

CARTER, N. L. (1976). Steady state flow of rocks. *Rev. Geophys. space Phys.*, **14**, 301–360.

CARTER, N. L. and AVE'LALLEMANT, H. G. (1970). High temperature flow of dunite and peridotite. *Bull. geol. Soc. Am.*, **81**, 2181–2202.

CATHLES, L. M., III (1975). *The viscosity of the Earth's mantle*, Princeton University Press, Princeton, New Jersey, 386 pp.

CHAPPLE, W. M. and FORSYTH, D. W. (1979). Earthquakes and bending of plates at trenches. *J. geophys. Res.*, **84**, 6729–6749.

CRITTENDEN, M. D., Jr. (1963a). New data on the isostatic deformation of Lake Bonneville. *Prof. Pap. U.S. geol. Surv.*, **454E**, 1–31.

CRITTENDEN, M. D., Jr. (1963b). Effective viscosity of the Earth derived from isostatic loading of Pleistocene Lake Bonneville. *J. geophys. Res.*, **68**, 5517–5530.

FITCH, T. J. (1979). Earthquakes and plate tectonics. In *The Earth: its origin, structure and evolution*, pp. 491–542, edited by McElhinny, M. W., Academic Press, London, New York and San Francisco.

GOETZE, C. (1978). The mechanisms of creep in olivine. *Phil. Trans. R. Soc.*, **288A**, 99–119.

GOLD, T. (1955). Instability of the Earth's axis of rotation. *Nature, Lond.*, **175**, 526–529.

GOLDREICH, P. and TOOMRE, A. (1969). Some remarks on polar wandering. *J. geophys. Res.*, **74**, 2555–2567.

GRIFFITH, A. A. (1921). The phenomena of rupture and flow in solids. *Phil. Trans. R. Soc.*, **221A**, 163–198.

GRIGGS, D. T., TURNER, F. J. and HEARD, H. C. (1960). Deformation of rocks at 500° to 800° C. In *Rock deformation (a symposium)*, pp. 39–104, edited by Griggs, D. and Handin, J., *Mem. geol. Soc. Am.*, **79**.

GUTENBERG, B. (1945). Amplitudes of *P, PP,* and *S* and magnitude of shallow earthquakes. *Bull. seism. Soc. Am.* **35**, 57–69.

GUTENBERG, B. (1959). *Physics of the Earth's interior*, Academic Press, New York and London, 240 pp.

GUTENBERG, B. and RICHTER, C. F. (1936). On seismic waves (third paper). *Beitr. Geophys.*, **47**, 73–131.

GUTENBERG, B. and RICHTER, C. F. (1954). *Seismicity of the Earth and associated phenomena*, Second edition, Princeton University Press, Princeton, New Jersey, 310 pp.

HANKS, T. C. and WYSS, M. (1972). The use of body-wave spectra in the determination of seismic-source parameters. *Bull. seism. Soc. Am.*, **62**, 561–589.

HASKELL, N. A. (1935). Motion of viscous fluid under a surface load, I. *Physics*, **6**, 265–269.

HASKELL, N. A. (1937). The viscosity of the asthenosphere. *Am. J. Sci.*, Ser. 5, **33**, 22–28.

HUBBERT, M. K. and RUBEY, W. W. (1959). Role of fluid pressure in mechanics of overthrust faulting I. Mechanics of fluid-filled porous solids and its application to overthrust faulting. *Bull. geol. Soc. Am.*, **70**, 115–166.

JAEGER, J. C. (1960). Shear failure of anisotropic rocks. *Geol. Mag.*, **97**, 65–72.

JEFFREYS, H. (1976). *The Earth, its origin history and physical constitution*, Sixth edition, Cambridge University Press, London and New York, 574 pp.

KANAMORI, H. (1977). The energy release in great earthquakes. *J. geophys. Res.*, **82**, 2981–2987.

KELLY, A., COOK, A. H. and GREENWOOD, G. W. (editors) (1978). Creep of engineering materials and of the Earth. *Phil. Trans. R. Soc.*, **288A**, 1–236.

KOHLSTEDT, D. L. and GOETZE, C. (1974). Low-stress high-temperature creep in olivine single crystals. *J. geophys. Res.*, **79**, 2045–2051.

KOHLSTEDT, D. L., NICHOLS, H. P. K. and HORNACK, P. (1980). The effect of pressure on the rate of dislocation recovery in olivine. *J. geophys. Res.*, **85**, 3122–3130.

KUSZNIR, N. J. and BOTT, M. H. P. (1977). Stress concentration in the upper lithosphere caused by underlying visco-elastic creep. *Tectonphysics*, **43**, 247–256.

LINDH, A., EVANS, P., HARSH, P. and BUHR, G. (1979). The Parkfield prediction experiment. *Earthquake Information Bulletin*, **11**, 209–213.

McCLINTOCK, F. A. and WALSH, J. B. (1962). Friction on Griffith cracks in rocks under pressure. *Proc. 4th U.S. Nat. Congr. appl. Mechanics*, **2**, pp. 1015–1021, American Society of Mechanical Engineers, New York.

McGARR, A. and GAY, N. C. (1978). State of stress in the Earth's crust. *Ann. Rev. Earth & planet. Sci.*, **6**, 405–436.

MUNK, W. H. and MACDONALD, G. J. F. (1960). *The rotation of the Earth*, Cambridge University Press, London and New York, 323 pp.

MURRELL, S. A. F. (1977). Natural faulting and the mechanics of brittle shear failure. *Jl geol. Soc. Lond.*, **133**, 175–189.

MURRELL, S. A. F. and DIGBY, P. J. (1970). The theory of brittle fracture initiation under triaxial stress conditions – I. *Geophys. J. R. astr. Soc.*, **19**, 309–334.

MURRELL, S. A. F. and ISMAIL, I. A. H. (1976). The effect of decomposition of hydrous minerals on the mechanical properties of rocks at high pressures and temperatures. *Tectonophysics*, **31**, 207–258.

NICOLAS, A. and POIRIER, J. P. (1976). *Crystalline plasticity and solid state flow in metamorphic rocks*, John Wiley & Sons, London, New York, Sydney and Toronto, 444 pp.

OROWAN, E. (1965). Convection in a non-Newtonian mantle, continental drift, and mountain building. *Phil. Trans. R. Soc.*, **258A**, 284–313.

PATERSON, M. S. (1978). *Experimental rock deformation – the brittle field*, Springer-Verlag, Berlin, 254 pp.

PELTIER, W. R. and ANDREWS, J. T. (1976). Glacial-isostatic adjustment – I. The forward problem. *Geophys. J. R. astr. Soc.*, **46**, 605–646.

RALEIGH, C. B. and PATERSON, M. S. (1965). Experimental deformation of serpentinite and its tectonic implications. *J. geophys. Res.*, **70**, 3965–3985.

RAMBERG, H. and STEPHANSSON, O. (1964). Compression of floating elastic and viscous plates affected by gravity, a basis for discussing crustal buckling. *Tectonophysics*, **1**, 101–120.

RANALLI, G. (1980). Regional models of the steady-state rheology of the upper mantle. In *Earth rheology, isostasy and eustasy*, pp. 111–123, edited by Mörner, N., John Wiley & Sons, Chichester, New York, Brisbane and Toronto.

RICHARDSON, R. M., SOLOMON, S. C. and SLEEP, N. H. (1979). Tectonic stress in the plates. *Rev. Geophys. space Phys.*, **17**, 981–1019.

RICHTER, C. F. (1935). An instrumental earthquake magnitude scale. *Bull. seism. Soc. Am.*, **25**, 1–32.

SELLARS, C. M. (1978). Recrystallization of metals during hot deformation. *Phil. Trans. R. Soc.*, **288A**, 147–158.

STOCKER, R. L. and ASHBY, M. F. (1973). On the rheology of the upper mantle. *Rev. Geophys. space Phys.*, **11**, 391–426.

SYKES, L. R. (1967). Mechanism of earthquakes and nature of faulting on the mid-oceanic ridges. *J. geophys. Res.*, **72**, 2131–2153.

TURCOTTE, D. L. (1974). Membrane tectonics. *Geophys. J. R. astr. Soc.*, **36**, 33–42.

VETTER, U. R. and MEISSNER, R. O. (1979). Rheologic properties of the lithosphere and applications to passive continental margins. *Tectonophysics*, **59**, 367–380.

WALCOTT, R. I. (1970a). Flexural rigidity, thickness, and viscosity of the lithosphere. *J. geophys. Res.*, **75**, 3941–3954.

WALCOTT, R. I. (1970b). An isostatic origin for basement uplifts. *Can. J. Earth Sci.*, **7**, 931–937.

WALCOTT, R. I. (1980), Rheological models and observational data of glacio-isostatic rebound. In *Earth rheology, isostasy and eustasy*, pp. 3–10, edited by Mörner, N., John Wiley & Sons, Chichester, New York, Brisbane and Toron4o.

WALLACE, R. E. (1978). Behavior of different segments of the San Andreas fault. *Earthquake Information Bulletin*, **10**, 126–130.

WANG, C. and MAO, N. (1979). Shearing of saturated clays in rock joints at high confining pressures. *Geophys. Res. Lett.*, **6**, 825–828.

WANG, H. F. and SIMMONS, G. (1978). Microcracks in crystalline rock from 5.3-km depth in the Michigan basin. *J. geophys. Res.*, **83**, 5849–5856.

WATTS, A. B. (1978). An analysis of isostasy in the world's oceans 1. Hawaiian-Emperor seamount chain. *J. geophys. Res.*, **83**, 5989–6004.

WATTS, A. B., COCHRAN, J. R. and SELZER, G. (1975). Gravity anomalies and flexure of the lithosphere: a three-dimensional study of the Great Meteor seamount, northeast Atlantic. *J. geophys. Res.*, **80**, 1391–1398.

WEERTMAN, J. (1970). The creep strength of the Earth's mantle. *Rev. Geophys. space Phys.*, **8**, 145–168.

WEERTMAN, J. (1978). Creep laws for the mantle of the Earth. *Phil. Trans. R. Soc.*, **288A**, 9–26.

WYSS, M. (1970). Stress estimates for South American shallow and deep earthquakes. *J. geophys. Res.*, **75**, 1529–1544.

Chapter nine

BÉNARD, H. (1900). Les tourbillons cellulaires dans une nappe liquide. *Revue gén. Sci. pur. appl.*, **11**, 1261–1271 and 1309–1328.

BOTT, M. H. P. (1964). Convection in the Earth's mantle and the mechanism of continental drift. *Nature, Lond.*, **202**, 583–584.

BOTT, M. H. P. (1982). The mechanism of continental splitting. *Tectonophysics*, **81**, 301–309.

BOTT, M. H. P. and KUSZNIR, N. J. (1979). Stress distributions associated with compensated plateau uplift structures with application to the continental splitting mechanism. *Geophys. J. R. astr. Soc.*, **56**, 451–459.

CAREY, S. W. (1958). A tectonic approach to continental drift. In *Continental drift, a symposium*, pp. 177–355, edited by Carey, S. W., University of Tasmania, Hobart.

CAREY, S. W. (1976). *The expanding Earth*, Elsevier Scientific Publishing Company, Amsterdam, Oxford and New York, 488 pp.

CATHLES, L. M., III (1975). *The viscosity of the Earth's mantle*, Princeton University Press, Princeton, New Jersey, 386 pp.

CHAPPLE, W. M. and TULLIS, T. E. (1977). Evaluation of the forces that drive the plates. *J. geophys. Res.*, **82**, 1967–1984.

CROUGH, S. T. (1979). Hotspot epeirogeny. *Tectonophysics*, **61**, 321–333.

DANA, J. D. (1847). Geological results of the Earth's contraction in consequence of cooling. *Am. J. Sci.*, Ser. 2, **3**, 176–188.

DARWIN, G. H. (1887). Note on Mr. Davison's paper on the straining of the Earth's crust in cooling. *Phil. Trans. R. Soc.*, **178A**, 242–249.

DAVISON, C. (1887). On the distribution of strain in the Earth's crust resulting from secular cooling; with special reference to the growth of continents and the formation of mountain chains. *Phil. Trans. R. Soc.*, **178A**, 231–242.

DEARNLEY, R. (1966). Orogenic fold-belts and a hypothesis of Earth evolution. *Physics Chem. Earth*, 7, 1–114.

DE BEAUMONT, E. (1852). *Notice sur les systèmes de montagnes*, 3 volumes, P. Bertrand, Paris.

DICKE, R. H. (1962). The Earth and cosmology. *Science, N. Y.*, **138**, 653–664.

DIRAC, P. A. M. (1938). A new basis for cosmology. *Proc. R. Soc.*, **165A**, 199–208.

DIRAC, P. A. M. (1974). Cosmological models and the large numbers hypothesis. *Proc. R. Soc.*, **338A**, 439–446.

EGYED, L. (1957). A new dynamic conception of the internal constitution of the Earth. *Geol. Rdsch.*, **46**, 101–121.

ELDER, J. (1976). *The bowels of the Earth*, Oxford University Press, London, 222 pp.

ELSASSER, W. M. (1969). Convection and stress propagation in the upper mantle. In *The application of modern physics to the Earth and planetary interiors*, pp. 223–246, edited by Runcorn, S. K., Wiley-Interscience, London, New York, Sydney and Toronto.

ELSASSER, W. M. (1971). Sea-floor spreading as thermal convection. *J. geophys. Res.*, **76**, 1101–1112.

FISHER, O. (1881). *Physics of the Earth's crust*, Macmillan & Co., London.

FORSYTH, D. and UYEDA, S. (1975). On the relative importance of the driving forces of plate motion. *Geophys. J. R. astr. Soc.*, **43**, 163–200.

HILGENBERG, O. C. (1933). *Vom wachsendem Erdball*, Geissmann and Bartsch, Berlin.

HILGENBERG, O. C. (1962). Rock magnetism and the Earth's palaeopoles. *Geofis. pura appl.*, **53**, 52–54.

HOLMES, A. (1928). Radioactivity and continental drift (Report of Glasgow Geological Society meeting 12 January, 1928). *Geol. Mag.*, **65**, 236–238.

HOLMES, A. (1931). Radioactivity and earth movements. *Trans. geol. Soc. Glasg.*, **18**, 559–606.

HOLMES, A. (1933). The thermal history of the Earth. *J. Wash. Acad. Sci.*, **23**, 169–195.

HOYLE, F. and NARLIKAR, J. V. (1971). On the nature of mass. *Nature, Lond.*, **233**, 41–44.

IRVING, E. (1964). *Palaeomagnetism and its application to geological and geophysical problems*, John Wiley & Sons, New York, London and Sydney, 399 pp.

JEANLOZ, R. and RICHTER, F. M. (1979). Convection, composition, and the thermal state of the lower mantle. *J. geophys. Res.*, **84**, 5497–5504.

JEFFREYS, H. (1930). The instability of a compressible fluid heated below. *Proc. Camb. phil. Soc. math. phys. Sci.*, **26**, 170–172.

JEFFREYS, H. (1959). *The Earth: its origin, history and physical constitution*, Cambridge University Press, London and New York, 420 pp.

KNOPOFF, L. (1964). The convection current hypothesis. *Rev. Geophys.*, **2**, 89–122.

LINDEMANN, B. (1927). *Kettengebirge, kontinentale Zerspaltung und Erdexpansion*, Jena, 186 pp.

LYTTLETON, R. A. (1965). On the phase-change hypothesis of the structure of the Earth. *Proc. R. Soc.*, **287A**, 471–493.

McELHINNY, M. W., TAYLOR, S. R. and STEVENSON, D. J. (1978). Limits to the expansion of Earth, Moon, Mars and Mercury and to changes in the gravitational constant. *Nature, Lond.*, **271**, 316–321.

McKENZIE, D. P., ROBERTS, J. M. and WEISS, N. O. (1974). Convection in the Earth's mantle: towards a numerical simulation. *J. Fluid Mech.*, **62**, 465–538.

MORGAN, W. J. (1971). Convection plumes in the lower mantle. *Nature, Lond.*, **230**, 42–43.

OROWAN, E. (1965). Convection in a non-Newtonian mantle, continental drift, and mountain building. *Phil. Trans. R. Soc.*, **258A**, 284–313.

OWEN, H. G. (1976). Continental displacement and expansion of the Earth during the Mesozoic and Cenozoic. *Phil. Trans. R. Soc.*, **281A**, 223–291.

OXBURGH, E. R. and TURCOTTE, D. L. (1974). Membrane tectonics and the East African rift. *Earth & planet. Sci. Lett. (Neth.)*, **22**, 133–140.

PARMENTIER, E. M., TURCOTTE, D. L. and TORRANCE, K. E. (1976). Studies of finite amplitude non-Newtonian thermal convection with application to convection in the Earth's mantle. *J. geophys. Res.*, **81**, 1839–1846.

RAYLEIGH, J. W. S. (LORD) (1916). On convection currents in a horizontal layer of fluid, when the higher temperature is on the under side. *Phil. Mag.*, Ser. 6, **32**, 529–546.

RICHARDSON, R. M., SOLOMON, S. C. and SLEEP, N. H. (1979).

RICHTER, F. M. and PARSONS, B. (1975). On the interaction of two scales of convection in the mantle. *J. geophys. Res.*, **80**, 2529–2541.

RUNCORN, S. K. (1962). Palaeomagnetic evidence for continental drift and its geophysical cause. In *Continental drift*, pp. 1–40, edited by Runcorn, S. K., Academic Press, New York and London.

STACEY, F. D. (1977). A thermal model of the Earth. *Phys. Earth planet. Interiors*, **15**, 341–348.

STEVENSON, D. J. and TURNER, J. S. (1979). Fluid models of mantle convection. In *The Earth: its origin, structure and evolution*, pp. 227–263, edited by McElhinny, M. W., Academic Press, London, New York and San Francisco.

TURCOTTE, D. L. and OXBURGH, E. R. (1967). Finite amplitude convective cells and continental drift. *J. Fluid Mech.*, **28**, 29–42.

WARD, M. A. (1963). On detecting changes in the Earth's radius. *Geophys. J. R. astr. Soc.*, **8**, 217–225.

WILSON, J. T. (1963). Evidence from islands on the spreading of ocean floors. *Nature, Lond.*, **197**, 536–538.

WOODWARD, D. J. (1976). *Visco-elastic finite element analysis of subduction zones*, University of Durham Ph. D. thesis, 215 pp.

WYSS, M. (1970). Stress estimates for South American shallow and deep earthquakes. *J. geophys. Res.*, **75**, 1529–1544.

Index

Page numbers given in italics refer to an entire section on the topic indexed.